VOLUME ONE HUNDRED AND NINETEEN

INTERNATIONAL REVIEW OF NEUROBIOLOGY

Adenosine Receptors in Neurology and Psychiatry

INTERNATIONAL REVIEW OF NEUROBIOLOGY

VOLUME 119

SERIES EDITORS

R. ADRON HARRIS

Waggoner Center for Alcohol and Drug Addiction Research
The University of Texas at Austin
Austin, Texas, USA

PETER JENNER

Division of Pharmacology and Therapeutics
GKT School of Biomedical Sciences
King's College, London, UK

EDITORIAL BOARD

VOLUME ONE HUNDRED AND NINETEEN

INTERNATIONAL REVIEW OF NEUROBIOLOGY

Adenosine Receptors in Neurology and Psychiatry

Edited by

AKIHISA MORI

Strategic Product Portfolio Department,
Kyowa Hakko Kirin Co., Ltd.,
Tokyo, Japan

ELSEVIER

AMSTERDAM • BOSTON • HEIDELBERG • LONDON
NEW YORK • OXFORD • PARIS • SAN DIEGO
SAN FRANCISCO • SINGAPORE • SYDNEY • TOKYO

Academic Press is an imprint of Elsevier

Academic Press is an imprint of Elsevier
32 Jamestown Road, London NW1 7BY, UK
525 B Street, Suite 1800, San Diego, CA 92101-4495, USA
225 Wyman Street, Waltham, MA 02451, USA
The Boulevard, Langford Lane, Kidlington, Oxford OX5 1GB, UK

First edition 2014

Notices

Knowledge and best practice in this field are constantly changing. As new research and experience broaden our understanding, changes in research methods, professional practices, or medical treatment may become necessary.

Practitioners and researchers must always rely on their own experience and knowledge in evaluating and using any information, methods, compounds, or experiments described herein. In using such information or methods they should be mindful of their own safety and the safety of others, including parties for whom they have a professional responsibility.

To the fullest extent of the law, neither the Publisher nor the authors, contributors, or editors, assume any liability for any injury and/or damage to persons or property as a matter of products liability, negligence or otherwise, or from any use or operation of any methods, products, instructions, or ideas contained in the material herein.

ISBN: 978-0-12-801022-8
ISSN: 0074-7742

For information on all Academic Press publications visit our website at store.elsevier.com

CONTENTS

CONTRIBUTORS

Jiang-Fan Chen
Department of Neurology, Boston University School of Medicine, Boston, Massachusetts, USA, and The Molecular Medicine Institute, Wenzhou Medical University, Wenzhou, Zhejiang, PR China

Yijuang Chern
Division of Neuroscience, Institute of Biomedical Sciences, Academia Sinica, Taipei, Taiwan

Rodrigo A. Cunha
CNC-Center for Neuroscience and Cell Biology, and Faculty of Medicine, University of Coimbra, Coimbra, Portugal

Thérèse Di Paolo
Neuroscience Research Unit, Centre de recherche du CHU de Québec, and Faculty of Pharmacy, Laval University, Quebec, Quebec, Canada

Zhi-Li Huang
Department of Pharmacology, State Key Laboratory of Medical Neurobiology, Institute of Brain Science, Shanghai Medical College of Fudan University, Shanghai, China

Kiich Ishiwata
Research Team for Neuroimaging, Tokyo Metropolitan Institute of Gerontology, Tokyo, Japan

Peter Jenner
Neurodegenerative Diseases Research Group, Institute of Pharmaceutical Sciences, School of Biomedical Sciences, King's College, London, United Kingdom

Takako Kadowaki-Horita
Central Nervous System Research Laboratories, Research & Development Division, Kyowa Hakko Kirin Co., Ltd., Sunto-gun, Shizuoka, Japan

Tomoyuki Kanda
Central Nervous System Research Laboratories, Research & Development Division, Kyowa Hakko Kirin Co., Ltd., Sunto-gun, Shizuoka, Japan

Masahito Kawamura
Department of Pharmacology, Jikei University School of Medicine, Minato-ku, Tokyo, Japan

Minoru Kobayashi
Central Nervous System Research Laboratories, CNS R&D Unit, R&D Division, Kyowa Hakko Kirin Co., Ltd., Shizuoka, Japan

Diogo R. Lara
Laboratório de Neuroquímica e Psicofarmacologia, Departamento de Biologia Celular e Molecular, Faculdade de Biociências, Pontifícia Universidade Católica do Rio Grande do Sul, Porto Alegre, Brazil

Chien-fei Lee
Division of Neuroscience, Institute of Biomedical Sciences, Academia Sinica, Taipei, Taiwan

Susan A. Masino
Department of Psychology and Neuroscience Program, Trinity College, Hartford, Connecticut, USA

Alessia Melani
Department of Neuroscience, Psychology, Drug Research and Child Health (NEUROFARBA), Division of Pharmacology and Toxicology, University of Florence, 50139 Florence, Italy

Masahiro Mishina
Department of Neurological Science, Graduate School of Medicine, Nippon Medical School, and Research Team for Neuroimaging, Tokyo Metropolitan Institute of Gerontology, Tokyo, Japan

Akihisa Mori
Strategic Product Portfolio Department, Kyowa Hakko Kirin Co., Ltd., Tokyo, Japan

Nicolas Morin
Neuroscience Research Unit, Centre de recherche du CHU de Québec, and Faculty of Pharmacy, Laval University, Quebec, Quebec, Canada

Masahiro Nagai
Department of Neurology and Clinical Pharmacology, Ehime University Graduate School of Medicine, Shitsukawa Toon, Ehime, Japan

Noriko Nishikawa
Department of Neurology and Clinical Pharmacology, Ehime University Graduate School of Medicine, Shitsukawa Toon, Ehime, Japan

Masahiro Nomoto
Department of Neurology and Clinical Pharmacology, Ehime University Graduate School of Medicine, Shitsukawa Toon, Ehime, Japan

Felicita Pedata
Department of Neuroscience, Psychology, Drug Research and Child Health (NEUROFARBA), Division of Pharmacology and Toxicology, University of Florence, 50139 Florence, Italy

Anna Maria Pugliese
Department of Neuroscience, Psychology, Drug Research and Child Health (NEUROFARBA), Division of Pharmacology and Toxicology, University of Florence, 50139 Florence, Italy

Wei-Min Qu
Department of Pharmacology, State Key Laboratory of Medical Neurobiology, Institute of Brain Science, Shanghai Medical College of Fudan University, Shanghai, China

Daniel Rial
CNC-Center for Neuroscience and Cell Biology, University of Coimbra, Coimbra, Portugal

David N. Ruskin
Department of Psychology and Neuroscience Program, Trinity College, Hartford, Connecticut, USA

Masahiko Tomiyama
Department of Neurology, Aomori Prefectural Central Hospital, Aomori, Japan

Shin-ichi Uchida
Central Nervous System Research Laboratories, Research & Development Division, Kyowa Hakko Kirin Co., Ltd., Sunto-gun, Shizuoka, Japan

Koji Yamada
Central Nervous System Research Laboratories, CNS R&D Unit, R&D Division, Kyowa Hakko Kirin Co., Ltd., Shizuoka, Japan

Ze Zhang
Department of Pharmacology, State Key Laboratory of Medical Neurobiology, Institute of Brain Science, Shanghai Medical College of Fudan University, Shanghai, China

PREFACE

Most diseases resulting from brain dysfunction are very tragic not only for the patients themselves but also for their families because these diseases deprive them of both the quality of human life and often compassion. The brain is one of the last frontiers in modern biology, with many unknown mechanisms. Thereby, the areas of neurology and psychiatry are the most challenging for scientists those whose goal is to open the gates to new cures for diseases involving the brain.

Adenosine is a purine ribonucleoside and is ubiquitously distributed throughout the mammalian body. Since the 1920s, adenosine and purine receptors have been studied extensively in biomedical research. Approximately an 80-year journey of research already traveled by great ancestors stands before us.

Since the beginning of research, there have been significant outcomes ensuring that the adenosine and purine families contribute to the mammalian cardiovascular system, including vasodilation and heart function. It should be noted that the progression of receptor science has always been supported by molecular biology for receptor classification and promoted via the appearance of pharmacological tools like selective agonists and antagonists to identify receptor function. These tools also sometimes become known in the development of medicines as new therapeutics themselves. Adenosine receptor science is, of course, not an exception. In the early 1970s, it was proposed that adenosine was an important neuromodulator in both the peripheral and central nervous systems. This was evidenced by studies using methylxanthines, such as caffeine and theophylline, as antagonists at adenosine receptors. Furthermore, in the beginning of the 1990s, we saw an explosion of adenosine receptor science because molecular biology and pharmacology identified the four adenosine receptor subtypes: the A_1, A_{2A}, A_{2B}, and A_3 receptors. At the same time, caffeine, a non-selective adenosine antagonist, was still widely employed in many studies, including human epidemiology, as a kind of unique contribution for the field, until selective receptor agents were invented (for example, a cup of coffee containing caffeine made science close to many people's mind). Recently, receptor selective agents were created to assist and accelerate the entire research effort, which often brought to light particular physiological functions of adenosine as a neuromodulator/neurotransmitter. Some of the adenosine agents also became new candidates as medicines for the treatment of brain dysfunction.

Due to the nature of adenosine, which is a ubiquitous modulator of cellular activity, it is not surprising that there is a growing body of interest for adenosine and its receptors and their role as new and broad targets for research of brain function and dysfunction.

I invited distinguished scientists, who are working to further adenosine receptor science by seeking new therapies for a number of CNS diseases as well as adenosine function in the brain, to contribute to this volume. These scientists participated enthusiastically by sharing their new insights, knowledge, therapeutic strategies, views, and future applications for adenosine receptor science from multiple disease and brain functional perspectives, such as Parkinson's disease, Huntington's disease, epilepsy, cognitive function, cerebral ischemia, sleep, depression, anxiety, and schizophrenia.

In the early 1990s, when I started to work in the research area of adenosine receptor function in the brain, I did not have any idea like I do *today* that there could be so many aspects and practical ideas for new therapies involving adenosine receptors, as described in this volume. Now, I am very confident knowing that *tomorrow* will be more fantastic and will produce practical cures for neurological and psychiatric diseases from the continued research and progression of adenosine receptor science.

Finally, I would like to acknowledge the tremendous efforts and huge insights made into adenosine receptor science by my colleagues and contributors for this volume.

Akihisa Mori
Cherry blossom season in Tokyo, 2014

CHAPTER ONE

Adenosine Receptor Neurobiology: Overview

Jiang-Fan Chen*,[1], **Chien-fei Lee**†, **Yijuang Chern**†
*Department of Neurology, Boston University School of Medicine, Boston, Massachusetts, USA
†Division of Neuroscience, Institute of Biomedical Sciences, Academia Sinica, Taipei, Taiwan
[1]Corresponding author: e-mail address: chenjf@bu.edu

Contents

International Review of Neurobiology, Volume 119
ISSN 0074-7742
http://dx.doi.org/10.1016/B978-0-12-801022-8.00001-5

Abstract

Adenosine is a naturally occurring nucleoside that is distributed ubiquitously throughout the body as a metabolic intermediary. In the brain, adenosine functions as an important upstream neuromodulator of a broad spectrum of neurotransmitters, receptors, and signaling pathways. By acting through four G-protein-coupled receptors, adenosine contributes critically to homeostasis and neuromodulatory control of a variety of normal and abnormal brain functions, ranging from synaptic plasticity, to cognition, to sleep, to motor activity to neuroinflammation, and cell death. This review begun with an overview of the gene and genome structure and the expression pattern of adenosine receptors (ARs). We feature several new developments over the past decade in our understanding of AR functions in the brain, with special focus on the identification and characterization of canonical and noncanonical signaling pathways of ARs. We provide an update on functional insights from complementary genetic-knockout and pharmacological studies on the AR control of various brain functions. We also highlight several novel and recent developments of AR neurobiology, including (i) recent breakthrough in high resolution of three-dimension structure of adenosine A_{2A} receptors (A_{2A}Rs) in several functional status, (ii) receptor–receptor heterodimerization, (iii) AR function in glial cells, and (iv) the druggability of AR. We concluded the review with the contention that these new developments extend and strengthen the support for A_1 and A_{2A}Rs in brain as therapeutic targets for neurologic and psychiatric diseases.

1. GENERAL INTRODUCTION

In the brain, adenosine does not act as a classical neurotransmitter, since it is not enriched and stored in vesicles, and is not released in response to action potential, and does not act only or predominantly in synapses to transfer information unidirectionally from the pre- to the postsynaptic components. Instead, adenosine is generated intracellularly as a metabolic intermediary for nucleic acids synthesis and the biological energy currency adenosine triphosphate (ATP) and extracellularly by enzymatical conversion from released ATP into adenosine through coupled ectonucleotidases (CD73 and CD39). As with all other cells, adenosine functions as a homeostatic factor by its paracrine function and cellular signaling. More importantly, adenosine in the brain also serves as a critical neuromodulator to control neuronal excitability, release of various neurotransmitters including glutamate, γ-aminobutyric acid

(GABA), acetylcholine, and dopamine, and to modulate synaptic plasticity and function (Sebastiao & Ribeiro, 1996). As such, the level of adenosine rises with increased neuronal activity in response to hypoxia or ischemia. The increased extracellular adenosine tends to reset the balance (as a "retaliatory metabolite") and functions as an endogenous neuroprotectant (de Mendonça, Sebastião, & Ribeiro, 2000). Extracellular adenosine exerts physiological and pathological effects by acting at four types of evolutionarily conserved and pharmacologically well-characterized adenosine receptor (AR) subtypes, namely A_1, A_{2A}, A_{2B}, and A_3 receptors (Cirera & Fredholm, 2001; Fredholm, IJzerman, Jacobson, Klotz, & Linden, 2001). Each AR subtype has unique tissue distribution, pharmacological properties, different G-protein coupling, and distinct signaling pathways (Fredholm, IJzerman, et al., 2001; Fredholm, IJzerman, Jacobson, Linden, & Muller, 2011). The relative potency of adenosine (EC50) for its receptors, as estimated in Chinese hamster ovary cells stably expressing the receptors, is 73, 150, 5100, and 6500 nM for the A_1R (A_1 receptor), $A_{2A}R$ (A_{2A} receptor), $A_{2B}R$ (A_{2B} adenosine receptor), and A_3R (A_3 adenosine receptor), respectively (Daly & Padgett, 1992; Peakman & Hill, 1994; Zhou et al., 1992). At physiological condition, endogenous adenosine level is estimated at 20–200 nM in the brain, which is sufficient to activate A_1, A_{2A}, and A_3 receptors with approximately equipotent, with exception of the A_{2B} receptors, which require higher agonist concentrations (Fredholm, Irenius, Kull, & Schulte, 2001). The effects of adenosine depend on the level of adenosine as well as the expression level of subtypes in a given cell type or subregion of the brain. The A_1 and A_3 receptors couple to the G_i family of G proteins, and thus stimulate K^+ channels, reduce transient voltage-dependent Ca^{2+} channels and inhibit cAMP formation; $A_{2A}Rs$ couple to the members of the G_s family (G_{olf} in striatal neurons), whereas A_{2B} receptors couple to many G proteins including G_s, G_q, and G_{12}. These receptors have been implicated in a variety of biological functions, both physiological and pathological (Fredholm, IJzerman, et al., 2001; Fredholm et al., 2011), ranging from motor activity (Schwarzschild, Agnati, Fuxe, Chen, & Morelli, 2006), sleep regulation (Huang et al., 2005; Lazarus et al., 2011), cognition (Augusto et al., 2013; Wei et al., 2014; Wei, Singer, et al., 2011), and angiogenesis (Liu et al., 2010), as well as involvement in inflammatory diseases (Eltzschig & Carmeliet, 2011; Hasko, Linden, Cronstein, & Pacher, 2008), ischemia and reperfusion (Eltzschig & Eckle, 2011; Yu, Huang, et al., 2004), and neurodegenerative disorders (Fredholm, 2007). Lastly, one of the major impetuses for studying AR function in the brain is the realization that many of the effects of caffeine, the most widely used of all psychoactive compounds, are exerted by blocking effects of brain ARs (Fredholm, Battig, Holmen, Nehlig, & Zvartau, 1999).

2. SOURCE AND REGULATION OF EXTRACELLULAR ADENOSINE LEVEL

Unlike classic neurotransmitters, intracellular adenosine is not stored and released from vesicles but rather is generated by the highly regulated intracellular metabolic pathways (including ATP production via AMP by adenosine kinase (ADK), nucleotide/DNA synthesis, and *S*-adenosylhomocysteine pathway). Most if not all cells possess highly efficient equilibrative nucleoside transporters, ENT1 and ENT2, which allow adenosine to bidirectionally and quickly cross the cell membrane (King, Ackley, Cass, Young, & Baldwin, 2006). Consequently, extracellular concentration of adenosine can never be zero and, even under the most basal conditions, there will always be a finite level of adenosine in the extracellular space, which is estimated to fall within the 25–250 nM range in the brain (Ballarin, Fredholm, Ambrosio, & Mahy, 1991; Dunwiddie & Masino, 2001; Pedata, Corsi, Melani, Bordoni, & Latini, 2001). This intracellular production of adenosine is traditionally considered as the most important source of extracellular adenosine. Over the past decade, it is increasingly recognized that the rapid (1 ms) extracellular conversion of locally released ATP to adenosine through two-step ectonucleotidases (i.e., first conversion of ATP/ADP to AMP by the ectonucleoside triphosphate diphosphohydrolase-1, CD39, followed by AMP hydrolysis to adenosine by ecto-5′-nucleotidase, CD73) represents another important source of extracellular adenosine in tissue (Delaney & Geiger, 1998; Dunwiddie & Masino, 2001; Fredholm, Chern, Franco, & Sitkovsky, 2007). ATP can be released from various cell types by multiple mechanisms: coreleased from storage vesicles together with other hormones (neurotransmitter), by a "kiss-and-run" mechanism (MacDonald, Braun, Galvanovskis, & Rorsman, 2006), from the lysosome exocytosis (Zhang et al., 2007), controlled release through pannexin hemichannels (Chekeni et al., 2010; Elliott et al., 2009), release from inflammatory cells or vascular endothelia through connexin hemichannels and channels such as P2X7 receptor (Chen et al., 2006; Faigle, Seessle, Zug, El Kasmi, & Eltzschig, 2008; Linden, 2006), and uncontrolled leakage from necrotic cells (Eltzschig, 2009). It should be noted that the extracellular adenosine might originate not only from neurons (both nerve terminals and postsynaptic component) but also from surrounding nonneuronal cells such as glial cells (Halassa, Fellin, & Haydon, 2007; Halassa et al., 2009). Antagonist studies have demonstrated that adenosine exerts a tonic inhibitory effect on synaptic transmission, implying that under basal conditions, adenosine levels are sufficient to tonically activate relevant AR subtypes in the brain (Cunha, 2001).

3. ARs SUBTYPES: CLASSIFICATION AND GENE STRUCTURES

ARs were initially classified based on their pharmacological response profiles to inhibit (i.e., A_1 and A_3 subtypes) or stimulate (i.e., A_2 subtype) adenylate cyclase (AC) by adenosine analogues and antagonism by methylxanthines (Londos, Cooper, & Wolff, 1980; van Calker, Muller, & Hamprecht, 1979). The A_2 receptor subtype was subsequently further categorized according to the presence of high-affinity (A_{2A}) or low-affinity (A_{2B}) binding sites for adenosine in brain (Daly, Butts-Lamb, & Padgett, 1983). To date, four AR subtypes have been identified, purified, and cloned from mouse, rat, human, and other mammalian as well as nonmammalian species (Fredholm, IJzerman, et al., 2001; Fredholm et al., 2000). Molecular cloning and expression of these receptors have not only definitively verified the presence of four different subtypes but also segued to genetic-knockout (KO) and transgenic overexpression studies to further elucidate the functions of each AR subtype.

A_1, A_{2A}, and A_{2B} receptors are well conserved among mammals, but A_3 receptors show considerable structural variability. Like many G-protein-coupled receptor (GPCR), the coding region of all four ARs is split up by an intron in a region corresponding to the second intracellular loop (Xu, Bastia, & Schwarzschild, 2005). Both genome structures of the A_1R and $A_{2A}R$ genes are characterized by the use of alternative promoters and the presence of multiple mRNA species with alternative splicing. The A_1R gene has at least two promoters and different exons in the 5′-untranslated region that can assemble and give rise to different mRNA species (Deckert et al., 1995; Ren & Stiles, 1995, 1999). Similarly, the 5′-untranslated region of the $A_{2A}R$ gene also displays two alternative promoters in rats (Chern, King, Lai, & Lai, 1992; Lee, Lai, Lee, Chien, & Chern, 2014) and five alternative promoters in human, giving rise to various mRNA species (Yu, Frith, et al., 2004).

3.1. A novel protein transcribed from a different frame of the $A_{2A}R$ gene

A recent study reported that $A_{2A}R$ might transmit signal by enhancing the production of a new protein (designated uORF5) from the rat $A_{2A}R$ gene (*Adora2a*) using a reading frame that is different from the one that is used to produce the rat $A_{2A}R$ protein (Chern et al., 1992; Lee et al., 2014). The enhancing effect of $A_{2A}R$ on the production of the uORF5 protein is PKA dependent and occurs at the posttranscriptional level. Coding regions for uORF5-like proteins also exist in the human and mouse *Adora2a* genes, which suggest that the uORF5 protein is evolutionarily conserved and might have important function(s). Because the uORF5 protein is translated from the same transcript as the $A_{2A}R$ protein, the tissue expression pattern of uORF5 is likely to be similar to that of $A_{2A}R$. In fact, the endogenous uORF5 protein can be detected in the rat striatum and in *Adora2a*-expressing PC12 cells. Elevated expression of uORF5 suppresses the activity of a transcription factor (AP1) and downregulates several components of the mitogen-activated protein kinase (MAPK) pathway in PC12 cells. This is of great interest because earlier studies showed that the activation of $A_{2A}R$ alters the expression levels of c-fos, which is a major component of AP1. It is possible that the activation of $A_{2A}R$ modulates AP-1-mediated gene expression via a uORF-dependent pathway. Further investigation of uORF5 *in vivo* is needed to elucidate its contribution to the pathophysiological functions of $A_{2A}R$.

Although this new level of signal transduction has not been reported for other ARs, it is worth mentioning that genomic analyses suggest that the A_3R gene might also encode alternative proteins, in addition to the A_3R protein. At least two different transcripts can be transcribed from the human A_3R gene (*ADORA3*). These different transcripts share the same protein-coding exon at the N terminus, followed by different protein-coding exons generated by alternative splicing. In addition to the human A_3R protein, a hypothetical new protein (designated as aA_3R) containing 347 amino acids might be produced. It is interesting to note that the aA_3R protein contains three transmembrane domains that are identical to those of the A_3R protein, and an entirely distinct C terminus. Specific tools (such as antibodies) are needed to confirm the existence of aA3R *in vivo* and its contribution to the function of A_3R.

4. EXPRESSION OF ARs IN THE BRAIN

4.1. A_1 receptor

The A_1R is expressed throughout the body with the highest levels observed in brain, notably in neurons of cortex, hippocampus, and cerebellum, as well as in dorsal horn of spinal cord (Dixon, Gubitz, Sirinathsinghji, Richardson, & Freeman, 1996; Fredholm et al., 2000; Mahan et al., 1991). The A_1R is also expressed at intermediate levels elsewhere in brain (Fredholm et al., 2000). In brain, A_1Rs are found at both pre- and postsynaptic sites (Rebola, Pinheiro, Oliveira, Malva, & Cunha, 2003). Multiple sclerosis (MS) patients exhibit decreased expression of A_1Rs in mononuclear cells in peripheral blood (Johnston et al., 2001; Mayne, Shepel, Jiang, Geiger, & Power, 1999) and in CD45-positive glial cells in the brain of MS patients (Johnston et al., 2001). Some types of stress might alter A_1R expression, which can be reversed by treatment with glucocorticoids (Ren & Stiles, 1999) and thyroid hormones (Fideu, Arce, Esquifino, & Miras-Portugal, 1994).

4.2. A_{2A} receptor

Brain $A_{2A}R$ expression levels are highly concentrated in dorsal and ventral striatum (on striatopallidal medium spiny neuron (MSNs)) as well as in olfactory tubercle (Fink et al., 1992; Rosin, Robeva, Woodard, Guyenet, & Linden, 1998; Schiffmann, Jacobs, & Vanderhaeghen, 1991; Schiffmann, Libert, Vassart, & Vanderhaeghen, 1991; Svenningsson, Hall, Sedvall, & Fredholm, 1997; Svenningsson, Le Moine, et al., 1997). $A_{2A}Rs$ are also expressed at substantially lower levels outside of striatum in brain regions including hippocampus and cerebral cortex (Dixon et al., 1996; Svenningsson, Hall, et al., 1997). In brain, $A_{2A}Rs$ are found predominantly at postsynaptic neurons in striatum, but they are also detected significantly at lower levels at presynaptic sites in corticostriatal terminals and in hippocampus (Rebola, Canas, Oliveira, & Cunha, 2005). Interestingly, $A_{2A}Rs$ were found to be localized mainly to asymmetric, excitatory striatal synapses where the involved presynaptic neurons expressed VGLUT (Hettinger, Lee, Linden, & Rosin, 2001). Furthermore, $A_{2A}Rs$ at presynaptic sites are colocalized with A_1Rs to fine-tune glutamate release at these glutamatergic striatal terminals (Ciruela et al., 2006). The $A_{2A}R$ expression is sensitive to hypoxia (Kobayashi & Millhorn, 1999) and other brain insults and markedly upregulated in different brain regions after various brain insults, including spinal cord compression injury (Cassada et al., 2002), amygdala stimulation (kindling), intraperitoneal injection of kainate (Rebola, Porciuncula, et al., 2005), chronic restraint stress (Cunha, Canas, Oliveira, & Cunha, 2006), streptozotocin-induced diabetic rats (Duarte, Oliveira, Ambrosio, & Cunha, 2006), and aged animals (Canas, Duarte, Rodrigues, Kofalvi, & Cunha, 2009; Cunha, Constantino, Sebastiao, & Ribeiro, 1995; Lopes, Cunha, & Ribeiro, 1999; Rebola, Sebastiao, et al., 2003). Proinflammatory cytokines, interleukin-1β (IL-1β) and tumor necrosis factor-α (TNF-α), and lipopolysaccharide increase the expression of $A_{2A}R$ gene in PC12 cells (Trincavelli, Costa, Tuscano, Lucacchini, & Martini, 2002), human monocytic TH1 cells (Khoa et al., 2001), and isolated macrophages by as much as 50 folds (Murphree, Sullivan, Marshall, & Linden, 2005). The induction of $A_{2A}Rs$ in glial and immune cells by brain insults and inflammatory signals, coupled with a local increase in adenosine and proinflammatory cytokines (such as IL-1β, which further induces $A_{2A}R$ expression), may serve as a part of an important feed-forward mechanism to locally control neuroinflammatory responses in the brain.

4.3. A_{2B} and A_3 receptors

Regionally defining $A_{2B}R$ and A_3R expressions has largely been restricted to mRNA evidence where $A_{2B}Rs$ and A_3R are thought to be broadly expressed but at low levels in the brain (Fredholm et al., 2000).

5. STRUCTURE BIOLOGY OF ADENOSINE A_{2A} RECEPTORS

One of the most exciting breakthroughs in the field over the past 5 years is the revealing of high-resolution crystal structures of $A_{2A}R$ in several functional states (Dore et al., 2011; Gutierrez-de-Teran et al., 2013; Jaakola et al., 2008; Lebon et al., 2011; Liu et al., 2012; Xu et al., 2011). The crystal structure of human $A_{2A}R$ has been elegantly solved to 2.6–2.7 Å resolution, first in a high-affinity antagonist (ZM241385)-bound state (Jaakola et al., 2008) and then in the agonist (UK-432097)-bound state (Xu et al., 2011). These studies have provided new insight into the binding pocket for $A_{2A}R$ antagonists: the $A_{2A}R$ antagonist ZM241385 binds to the engineered $A_{2A}R$ in an extended conformation perpendicular to the plane of the membrane bilayer (Jaakola et al., 2008). In contrast, $A_{2A}R$-bound agonists are parallel to the plasma membrane, in a conformation similar to those of retinal:rhodopsin and β blocker:β-adrenergic receptor (Xu et al.,

2011). Thus, activated $A_{2A}Rs$ have structures that are significantly different from the inactive state. These models not only elucidate the 3D structure of $A_{2A}R$, particularly at the ligand- and allosteric-modulator-binding pockets, but also serve as important modeling templates for detailed homology modeling of other ARs (for review, see Jacobson, 2013). This new structural information has led to reevaluation and refined analysis of the homology modeling of the $A_{2A}R$ using the previously published data (Ivanov, Barak, & Jacobson, 2009) and will have significant implication in the model-based AR drug discovery program. Based on this finding, additional mutagenesis and functional analysis were performed to validate ligand-binding pocket and subtype selectivity of ARs (Jaakola et al., 2010). These findings provide a unique opportunity for the structure-based design of adenosine drugs for future clinical application.

6. CANONICAL AND NONCANONICAL SIGNALING PATHWAYS OF ARs

It is well known that the signal transduction pathway(s) of a specific receptor are largely dependent on the physiological environment and the cellular setting of the indicated cell types. Given the importance of these receptors, a vast amount of research on ARs has been reported. In this section, we summarize the major signaling pathways that might mediate the pathophysiological functions of ARs (Figs. 1.1–1.4).

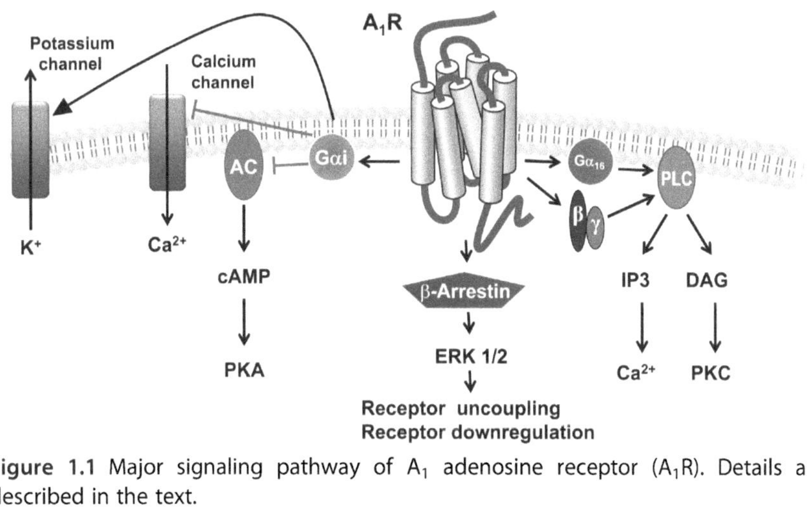

Figure 1.1 Major signaling pathway of A_1 adenosine receptor (A_1R). Details are described in the text.

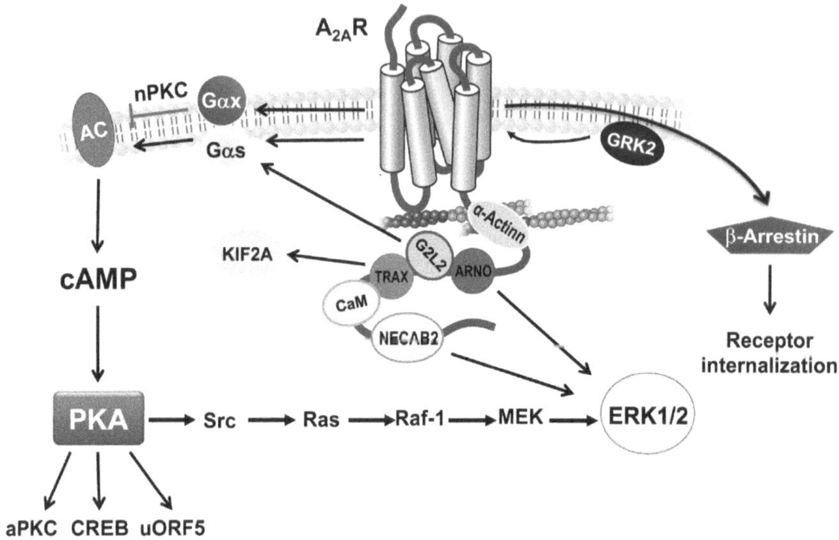

Figure 1.2 Major signaling pathway of A_{2A} adenosine receptor ($A_{2A}R$). Details are described in the text.

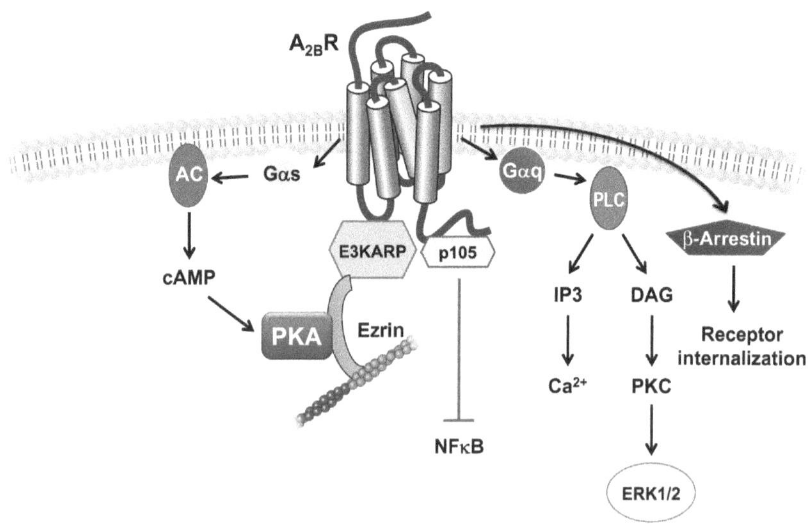

Figure 1.3 Major signaling pathway of A_{2B} adenosine receptor ($A_{2B}R$). Details are described in the text.

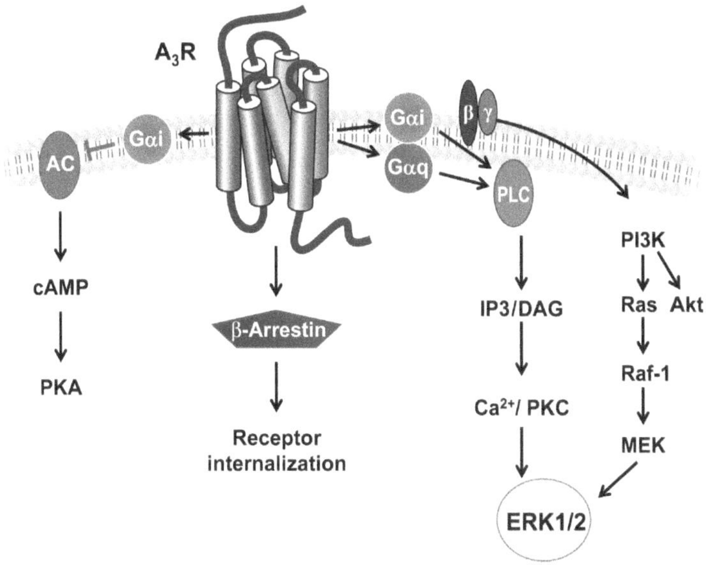

Figure 1.4 Major signaling pathway of A_3 adenosine receptor (A_3R). Details are described in the text.

6.1. A_1 adenosine receptor

6.1.1 Canonical pathways

A_1R is expressed in many different tissues (including the brain) and has a relatively high affinity for adenosine (EC_{50}, 0.31 μM; Fredholm, Irenius, et al., 2001). Ample evidence suggests that a significant portion of the functions of adenosine in the brain is mediated by A_1R. It is currently thought that G proteins mediate most of the A_1R signaling pathways. A_1R is mainly coupled to the members of the pertussis toxin (PTX)-sensitive Gαi and Gαo families (Munshi, Pang, Sternweis, & Linden, 1991). Stimulation of A_1R leads to the inhibition of AC and reduction of cAMP production, as well as the inhibition of the cAMP-dependent kinase (PKA) in neurons, to modulate the release of neurotransmitters and neuropeptides (Carruthers et al., 2001; Jeong, Jang, Nabekura, & Akaike, 2003). Other downstream targets of Gαi include K^+ channels (Kirsch, Codina, Birnbaumer, & Brown, 1990) and Ca^{2+} channels (Liu & Gao, 2007). Importantly, stimulation of A_1R also activates the phospholipase C (PLC) pathway via the release of Gβγ dimers in various cell types (Biber, Klotz, Berger, Gebicke-Harter, & van Calker, 1997; Dickenson & Hill, 1998; Fenton, Shea, Doddi, & Dobson, 2010). Moreover, the PLC/PKC signaling pathway can also be activated by

A_1R via a Gα16-mediated pathway in human lymphoblastoma cells and HEK293 cells (Liu & Wong, 2004).

The C terminus of A_1R is relatively short (36–38 amino acids; designated A_1R-C) and has been implicated in G-protein coupling and receptor folding (Pankevych, Korkhov, Freissmuth, & Nanoff, 2003). The distal portion of A_1R-C appears to function as an autoinhibitory domain that prevents the access of G proteins to A_1R. In addition, A_1R-C plays a critical role in maintaining the proper folding of A_1R, thus allowing it be exported from the endoplasmic reticulum (ER) to the plasma membrane. A more recent study suggested that DRiP78, which is an ER-resident chaperone, physically interacts with A_1R and regulates its folding and transport to the plasma membrane (Malaga-Dieguez et al., 2010).

6.1.2 Noncanonical pathways

Similar to many other GPCRs, the functional desensitization of A_1R is mediated by GPCR kinase (GRK) and arrestin (Nie, Mei, & Ramkumar, 1997). The stimulation of A_1R induces a rapid recruitment of β-arrestin 1 to the plasma membrane and the activation of ERK1/2, followed by the downregulation of A_1R (Jajoo et al., 2010). No selective biased ligands for the G-protein-independent pathway of A_1R have been developed successfully to date (Langemeijer, Verzijl, Dekker, & Ijzerman, 2013).

6.2. A_{2A} adenosine receptor

6.2.1 Canonical pathways (Fredholm et al., 2007)

$A_{2A}R$ is a class A GPCR that has a relatively high affinity for adenosine (EC_{50}, 0.7 μM; Fredholm, Irenius, et al., 2001). The best-known signaling of $A_{2A}R$ is the activation of the cAMP/PKA pathway via coupling to Gαs family members (i.e., Gαs and Gαolf; Chern, Chiou, Lai, & Tsai, 1995; Gubitz, Widdowson, Kurokawa, Kirkpatrick, & Richardson, 1996; Kull, Svenningsson, & Fredholm, 2000). The elevation of cAMP level leads to the activation of PKA and multiple downstream targets, including the transcription factor cAMP response element-binding protein (CREB). The phosphorylation of CREB by PKA at Ser^{133} promotes the transcription of several genes that are critical for neuronal plasticity and survival (Josselyn & Nguyen, 2005; Tanaka, 2001; Waltereit & Weller, 2003). Because CREB is the point of convergence of MAPKs and PKA, the stimulation of $A_{2A}R$ leads to the activation of CREB and rescues the impairment of neurite outgrowth caused by MAPK blockage in PC12 cells (Cheng, Shih, & Chern, 2002). In addition to CREB, the activation of PKA enhances the activity of several kinases. For example, the stimulation of $A_{2A}R$ triggers the Src/Ras/Raf-1/MEK/ERK pathway in a PKA-dependent manner (Che, Chan, & Cronstein, 2007; Schulte & Fredholm, 2003a). In PC12 cells, the activation of PKA by $A_{2A}R$ also leads to the stimulation of the atypical protein kinase C and protects cells from apoptosis (Huang, Lin, Huang, Messing, & Chern, 2001).

In addition to Gαs family members, the functions of $A_{2A}R$ are also mediated by other Gα proteins. It has been reported that $A_{2A}R$ activation causes the activation of a novel PKC (nPKC) via a PTX-sensitive G protein (Gαx) in PC12 cells. The activated nPKC phosphorylates and suppresses the type VI adenylyl cyclase (AC6), and in turn terminates the production of cAMP mediated by $A_{2A}R$. Such negative regulation of the Gαs/AC6 pathway by Gαx/nPKC signaling during $A_{2A}R$ stimulation is likely to play a critical role in the fine-tuning of $A_{2A}R$ signaling (Lai et al., 1997).

6.2.2 Noncanonical pathways

6.2.2.1 β-Arrestin-dependent pathway

The role of arrestin in $A_{2A}R$ signaling has not been investigated extensively. In addition to the resistance of $A_{2A}R$ to internalization after stimulation, the activated $A_{2A}R$ recruits β-arrestin to the plasma membrane via GPCR kinase 2 (GRK2) in monocyteid THP 1 cells, which results in the desensitization of $A_{2A}R$ (Khoa, Postow, Danielsson, & Cronstein, 2006). This agonist-mediated recruitment of GRK2/β-arrestin to the plasma membrane and the resultant desensitization of $A_{2A}R$ can be prevented by the TNF-α receptor, thus providing an important possibility of cross-talk between $A_{2A}R$ and TNF-α (Khoa et al., 2006).

It has long been known that $A_{2A}R$ forms a functional complex with D_2 dopamine receptor (D_2R; Fuxe et al., 2005). The stimulation of $A_{2A}R$ alters the binding affinity of D_2R to its ligands (Torvinen, Kozell, Neve, Agnati, & Fuxe, 2004). Coactivation of $A_{2A}R$ and D_2R potentiates the D_2R-mediated recruitment of β2-arrestin, which is followed by D_2R internalization and Akt activation (Borroto-Escuela et al., 2011). Conversely, blockage of $A_{2A}R$ greatly reduces D_2R translocation and ERK phosphorylation in a β2-arrestin-dependent manner (Huang, Wu, Zhang, & Feng, 2013). Collectively, these data suggest that binding to $A_{2A}R$ facilitates the β-arrestin-mediated signaling pathway of D_2R. Because the $A_{2A}R$–D_2R complex is an important target in Parkinson's disease (PD) and Huntington's disease, the identification of biased ligands for D_2R and the $D_2R/A_{2A}R$ complex is of great interest.

6.2.2.2 Accessory proteins that binds to the C terminus of $A_{2A}R$

In contrast with other AR subtypes, $A_{2A}R$ has a very long C-terminal domain (>120 amino acids) without a palmitoyl moiety, which provides multiple docking sites for signalosomes and regulatory machineries (Chern et al., 1992; Keuerleber, Gsandtner, & Freissmuth, 2011). To date, at least 10 proteins with divergent functions have been shown to bind to the C-terminal domain of $A_{2A}R$ (designated $A_{2A}R$-C). It is likely that $A_{2A}R$-C serves as a coincidence detector for signal integration (Gsandtner & Freissmuth, 2006; Keuerleber et al., 2011; Zezula & Freissmuth, 2008).

At least three $A_{2A}R$-interacting proteins (i.e., Usp4, HSP90α, and HSP70-1A) are needed for the proper folding and transport of $A_{2A}R$. Usp4 is a deubiquitinating enzyme that binds to $A_{2A}R$-C and ensures the accumulation of deubiquitinated $A_{2A}R$, which allows the proper transport of the functional $A_{2A}R$ to the cell surface (Milojevic et al., 2006). In addition, two molecular chaperones (heat-shock proteins HSP90α and HSP70-1A) associate with $A_{2A}R$-C to prevent the premature ER export of the properly folded $A_{2A}R$ (Bergmayr et al., 2013). These proteins govern the quality and the quantity of $A_{2A}R$ at plasma membranes. Once $A_{2A}R$ reaches the plasma membrane, its C terminus binds to scaffolding proteins (such as the synapse-associated protein 102, SAP102; Chen, Thomas, Sanz-Clemente, & Roche, 2011), which facilitates the retention of $A_{2A}R$ at the membrane compartment with slow mobility (Thurner et al., 2014). SAP102 also alters the signaling mode of $A_{2A}R$ and enhances the kinetics of ERK activation by $A_{2A}R$. Such interaction between $A_{2A}R$ and SAP102 was observed in primary hippocampal neurons and might contribute to the pathophysiological function of $A_{2A}R$ in the regulation of neuronal functions.

α-Actinin, a major F-actin-cross-linking protein, is another important $A_{2A}R$-C-interacting protein (Burgueno et al., 2003). Previous studies suggest that α-actinin is an important regulator of GRKs and determines the substrate specificity of GRK during GPCR desensitization (Freeman, Pitcher, Li, Bennett, & Lefkowitz, 2000). Via its association with $A_{2A}R$, α-actinin regulates the β-arrestin-dependent internalization and trafficking of $A_{2A}R$ (Burgueno et al., 2003). Moreover, $A_{2A}R$-C interacts directly with another actin-binding protein, Gas-2 like 2 (G2L2), which binds to the inactive Gαs and promotes the recruitment of the trimeric G-protein complex to the proximal position of $A_{2A}R$, to enhance the efficiency of $A_{2A}R$ activation (Wu et al., 2013). Among all the $A_{2A}R$-interacting proteins, G2L2 is the only binding partner that promotes the $A_{2A}R$-elicited cAMP signaling by forming a microdomain that is suitable for the effective coupling between $A_{2A}R$ and Gαs.

As mentioned above, the stimulation of $A_{2A}R$ activates the ERK pathway via the cAMP/PKA-dependent/Src-mediated and cAMP/PKA independent/Sos-mediated pathways (Schulte & Fredholm, 2003b). Gsandtner et al. (2005) later reported that the C terminus of $A_{2A}R$ binds directly to a nucleotide exchange factor (ARNO) for ARF6, an ADP ribosylation factor. This binding between $A_{2A}R$-C and ARNO selectively mediates the sustained phase of MAPK activation in a G-protein-independent manner without affecting cAMP induction or receptor desensitization. The underlying mechanism by which ARF6 regulates the activation of MAPK is unknown. Evidence indicates that receptor endocytosis is not involved in this process. Given that ARNO and ARF6 have been implicated in axonal elongation and branching (Hernandez-Deviez, Roth, Casanova, & Wilson, 2004), the role of $A_{2A}R$-mediated MAPK signaling in neuronal development is worthy of further investigation.

Several Ca^{2+}-binding proteins were also shown to interact with $A_{2A}R$-C. NECAB2 was the first Ca^{2+}-binding protein that was shown to coexpress with, and bind to, A_{2A}-C in striatal glutamatergic nerve terminals. The direct association with NECAB2 regulates the surface expression and internalization of $A_{2A}R$ and the MAPK pathway (Canela et al., 2007). Interestingly, another calcium-binding protein (NCS-1) also regulates the $A_{2A}R$-evoked ERK pathways without affecting the cellular distribution and internalization of $A_{2A}R$ (Navarro et al., 2012). Conversely, calmodulin (CaM) facilitates MAPK signaling in a different manner. When it interacts with $A_{2A}R$-C in the $A_{2A}R/D_2R$ complex, the calcium-binding CaM selectively alters the conformation of this complex and modulates MAPK signaling (Navarro et al., 2009; Woods et al., 2008). Although the physiological consequence of these observations needs to be further evaluated *in vivo*, the observations mentioned above collectively suggest that the elevation of calcium might potentiate the ability of $A_{2A}R$ to activate the MAPK/ERK1/2 pathway.

The translin-associated factor X (TRAX) is another $A_{2A}R$-C interacting protein that elicits novel functions of $A_{2A}R$. It was originally identified as a binding protein of translin, and it exists mainly in the nuclei of most tissues and cells examined. TRAX was identified as an $A_{2A}R$-C-binding protein via a yeast-two hybrid screening. Under conditions of impairment of the p53-dependent pathway, the stimulation of $A_{2A}R$ triggers an antiproliferative response in a TRAX-dependent manner (Sun et al., 2006). This unique property enables $A_{2A}R$ to rescue the impaired neuritogenesis caused by p53 blockage via a kinesin heavy chain member 2A-dependent pathway, and it converts the action of mitogenic factors (e.g., EGF) from proliferating to differentiating (Sun et al., 2006, 2010). Because $A_{2A}R$ expression is tightly controlled during neuronal development (Weaver, 1993), the role of TRAX in mediating the possible functions of $A_{2A}R$ during brain development is worthy of further investigation *in vivo*.

Another surprising interacting protein of $A_{2A}R$-C is the fibroblast growth factor receptor (FGFR; Flajolet et al., 2008). Multiple lines of evidence demonstrated a direct physical interaction between $A_{2A}R$-C and the FGFR. Simultaneous activation of both receptors leads to a strong activation of the MAPK/ERK pathway, alteration in spine morphogenesis, and enhanced plasticity in the corticostriatal circuit. This finding indicates the existence of a strong cross-talk between a GPCR and a tyrosine kinase trophic factor receptor and extends our current knowledge on the functions of $A_{2A}R$.

6.3. A_{2B} adenosine receptor

6.3.1 Canonical pathways

The affinity of $A_{2B}R$ for adenosine is the lowest among the four ARs (EC_{50}, 24 μM; Fredholm, Irenius, et al., 2001). Because such a high concentration of adenosine is needed to activate $A_{2B}R$, it is generally believed that $A_{2B}R$ functions mainly under pathological conditions, such as hypoxia and inflammation. Whether $A_{2B}R$ regulates the inflammatory response positively or negatively may depend on the specific cell type that is involved (Aherne, Kewley, & Eltzschig, 2011; Ham & Rees, 2008). After its induction, $A_{2B}R$ activates Gαs proteins that trigger AC-mediated cAMP production and enhance PKA activity (Feoktistov, Murray, & Biaggioni, 1994; Xaus et al., 1999). In several different cell types (e.g., human mast cells and umbilical vein endothelial cells), stimulation of $A_{2B}R$ also results in the activation of Gαq and triggers a PLC-mediated pathway to enhance the release of interleukin-8 (Feoktistov & Biaggioni, 1995; Feoktistov et al., 2002). This activation of PLCs by $A_{2B}R$ may also signal through the PKC/MAPK pathway to stimulate IL-6 secretion in some other types of cell (e.g., folliculostellate cells) and may play an important role in the regulation of the inflammatory response (Rees et al., 2003).

6.3.2 Noncanonical pathways

As demonstrated for other ARs, β-arrestin is involved in the internalization of $A_{2B}R$. Mundell, Loudon, and Benovic (1999) showed that the down-regulation of β-arrestin 2 was sufficient to cause an agonist-promoted reduction of the internalization and trafficking of $A_{2B}R$. A serine residue (Ser^{329}) located at the C-terminal end of $A_{2B}R$ plays a key role in β-arrestin binding and controls the agonist-mediated $A_{2B}R$ internalization and desensitization (Matharu, Mundell, Benovic, & Kelly, 2001). Evidently, different arrestin members dictate the kinetics of the cycling and resensitization of $A_{2B}R$ (Mundell, Matharu, Kelly, & Benovic, 2000).

Although the C terminus of $A_{2B}R$ (34–42 amino acids, designated $A_{2B}R$-C) is as short as those of A_1R and A_3R, a few $A_{2B}R$-interacting proteins have been reported. In an intestinal epithelial cell model, adenosine potentiates the recruitment of $A_{2B}R$ to the plasma membrane, where it interacts with the Na^+–H^+ exchanger regulatory cofactor 2 (NHERF2, also named E3KARP), which is a scaffold protein that is tethered to a PKA- and cytoskeleton-anchoring protein (Ezrin). This protein–protein interaction effectively positions $A_{2B}R$ at the membrane–cytoskeletal interface and facilitates the activation of PKA by $A_{2B}R$ (Sitaraman et al., 2002). Another intriguing finding is that $A_{2B}R$-C physically binds to p105, an NFκB precursor, and suppresses the degradation of p105 and subsequent NFκB activation (Sun et al., 2012). This physical association between $A_{2B}R$ and p105 provides an efficient means to control NFκB signaling and the inflammatory process.

6.4. A_3 adenosine receptor

6.4.1 Canonical pathways

The affinity of A_3R for adenosine (EC_{50}, 0.29 μM) is similar to those of A_1R and $A_{2A}R$ (Fredholm, Irenius, et al., 2001). It is expressed ubiquitously in many tissues and has a low level of expression in the brain (Zhou et al., 1992). Nevertheless, genetic removal of A_3R revealed that A_3R may modulate important neuronal activities that are involved in motor function, pain process, and neuronal survival (Fedorova, Jacobson, Basile, & Jacobson, 2003). Similar to A_1R, stimulation of A_3R triggers the Gαi-mediated suppression of AC (Fig. 1.4; Palmer, Gettys, & Stiles, 1995; Salvatore, Jacobson, Taylor, Linden, & Johnson, 1993; Zhou et al., 1992). Another important signaling pathway of A_3R is mediated by the activation of PLC, which produces inositol 1,4,5-triphosphate (IP3), elevates the cellular Ca^{2+} content, triggers PKC activation, and enhances ERK1/2 activation. These pathways were observed in a variety of cell types and tissues, including the striatum, hippocampus, microglia, and mast cells (Abbracchio et al., 1995; Hammarberg, Schulte, & Fredholm, 2003; Ramkumar, Stiles, Beaven, & Ali, 1993). In addition to the Gα proteins, Gβγ proteins can be released from the trimeric G-protein complex upon A_3R stimulation. It has been reported that Gβγ mediates the A_3R-evoked activation of phosphatidylnositol-3-kinase and the rapid activation of Akt and the Ras/MEK/ERK1/2 pathways (Schulte & Fredholm, 2002). Depending on specific cellular settings, these signaling pathways are believed to mediate many pathophysiological functions of A_3R (Cheong et al., 2013; Dumont, Pecasse, & Maenhaut, 2001; Fishman, Bar-Yehuda, Liang, & Jacobson, 2012; Rudich, Ravid, & Sagi-Eisenberg, 2012).

6.4.2 Noncanonical pathways

As has been shown for other AR subtypes, stimulation of A_3R recruits β-arrestin to plasma membranes and eventually causes desensitization. Using a panel of functional assays (including arrestin translocation, cAMP content, calcium mobilization, and ERK1/2 phosphorylation), several biased ligands that prefer A_3R-mediated β-arrestin translocation were reported. These compounds are important tools for the identification of the various signaling pathways of A_3R and are candidate therapeutic agents (Gao & Jacobson, 2008; Gao et al., 2011).

The C terminus of A_3R (designated as A_3R-C) is relatively short (30–36 amino acids). It contains a high density of serine and threonine residues, which may contribute to the desensitization and internalization of A_3R (Palmer & Stiles, 2000; Trincavelli, Tuscano, et al., 2002). To date, no A_3R-interacting protein has been identified.

7. INTERACTIONS BETWEEN ARs AND OTHER GPCRs AND NEUROTROPHIC FACTOR RECEPTORS

ARs are capable of forming oligomers among themselves or with other GPCRs. The A_1R forms an oligomer with the D_1 dopamine receptor (D_1R), which can be regulated by agonist binding (Gines et al., 2000). The $A_{2A}R$ can form higher order oligomers with itself (Vidi, Chemel, Hu, & Watts, 2008) and has multiple binding partners including the A_1R (Ciruela et al., 2006), the D_2R (Hillion et al., 2002; Kamiya, Saitoh, Yoshioka, & Nakata, 2003), the D_3 dopamine receptor (Torvinen et al., 2005), the metabotropic glutamate type 5 receptor (mGlu5R; Diaz-Cabiale et al., 2002; Kachroo et al., 2005), and the cannabinoid CB1 receptor (Carriba et al., 2007). However, the key issue whether GPCR heterodimerization actually does occur in intact animals and whether these GPCR heterodimers confer major pharmacological consequences that are not accounted for by monomeric receptors-induce signaling (Pin et al., 2007) remains to be determined. If these distinct heterodimerization partners are responsible in part for the distinct functions and differential sensitivities to pharmacological agents observed in striatal A_{2A}Rs (Ciruela et al., 2006; Quiroz et al., 2009; Schiffmann, Fisone, Moresco, Cunha, & Ferre, 2007), this offer new opportunities to distinguish A_{2A}R functions by targeting these AR-containing heteromers with novel adenosine drugs (Ciruela et al., 2011). Indeed, a few drugs targeted towards receptor dimers or oligomers were reported, including dendrimer of the A_{2A}R-selective agonist CGS 21680, which molecular modeling analysis predicts will bind to the homodimer of the A_{2A}R (Ivanov & Jacobson, 2008; Kim, Hechler, Klutz, Gachet, & Jacobson, 2008) and heterobivalent reagents composed of an A_{2A}R antagonist and a D_2R agonist bound to membranes containing both receptors with higher affinities than monovalent ligands (Soriano et al., 2009).

In addition to other GPCR, the cross-talk between AR, particularly the A_{2A}R, and the receptor for neurotrophic factors involving tyrosine receptor kinase has been demonstrated (Sebastiao & Ribeiro, 2009a). The A_{2A}R activation transactivates neurotrophin receptors and enhances their trophic functions (Lee & Chao, 2001; Sebastiao & Ribeiro, 2009b; Wiese et al., 2007) on Brain-derived neurotrophic factor (BDNF)-mediated synaptic transmission (Diogenes, Fernandes, Sebastiao, & Ribeiro, 2004; Tebano et al., 2008) and long-term potentiation (LTP; Fontinha, Diogenes, Ribeiro, & Sebastiao, 2008) and Glial cell line-derived neurotrophic factor (GDNF)-mediated action in striatal dopaminergic terminals (Gomes, Vaz, Ribeiro, & Sebastiao, 2006). Further evaluation along this line might lead to novel strategy for controlling neurotrophic receptor function by AR activity for treating neurodegenerative diseases.

8. AR FUNCTIONS: INSIGHTS FROM PHARMACOLOGICAL AND GENETIC-KO APPROACHES

Over the past decade, genetic-KO models for all four G-protein-coupled ARs have been generated by the targeted deletion of critical exons (Fredholm, Chen, Masino, & Vaugeois, 2005; Wei, Li, & Chen, 2011). These genetic-KO models have substantially advanced the field, extending findings from pharmacological studies and revealing novel and important physiological functions of the different receptor subtypes. From a drug discovery perspective, these genetic and transgenic overexpression models are instrumental in

(a) overcoming the limitations of pharmacological agents which have only partial specificity and (b) studying biological roles of a receptor for which there is no pharmacological agent available or no means for routine administration of an agonist or antagonist. AR-KO models are widely used and remain the most important experimental tools for validating ARs as drug targets for various pathological conditions. For example, use of the $A_{2A}R$-KO models overcomes concerns about the pharmacological specificity of $A_{2A}R$ antagonists and convincingly demonstrated the neuroprotective (Chen et al., 1999; Yu, Huang, et al., 2004), procognitive (Chen et al., 1999; Wei, Singer, et al., 2011; Yu, Huang, et al., 2004; Zhou et al., 2009), and sleep-promoting (Chen et al., 1999; Huang et al., 2005; Lazarus et al., 2011; Wei, Singer, et al., 2011; Yu, Huang, et al., 2004; Zhou et al., 2009) properties of $A_{2A}R$ inactivation in the brain. Interestingly, none of the AR-KO models characterized thus far produces a lethal phenotype (Fredholm et al., 2005). These findings have significant implications for drug discovery, suggesting that blockade of ARs can be achieved pharmacologically without "deadly" consequences, a conclusion consistent with the safe consumption of caffeine by 60–80% world populations (Fredholm et al., 1999).

8.1. AR modulation of neuronal excitability by presynaptic and postsynaptic mechanisms

A_1R and $A_{2A}R$ are expressed at presynaptic nerve terminals and activation of these receptors controls release of almost all neurotransmitters including glutamate, GABA, dopamine, and acetylcholine in the brain. A_1Rs are particularly enriched at presynaptic terminals and their activation efficiently inhibits the release of glutamate and aspartate release via G-protein-mediated inhibition of calcium channels in nerve endings (Dunwiddie & Fredholm, 1997; Dunwiddie & Masino, 2001). The facilitating effect of $A_{2A}R$ agonists has also been demonstrated, in the absence of powerful A_1R-mediated inhibition, and is attributed to either a direct effect on glutamatergic terminals (Nikbakht & Stone, 2001; Rosin, Hettinger, Lee, & Linden, 2003) via cAMP–PKA pathway, leading to increased Ca^{2+} influx (Dunwiddie & Fredholm, 1997; Gubitz et al., 1996) or an indirect effect via downregulation of A_1R-mediated inhibition (Lopes, Cunha, Kull, Fredholm, & Ribeiro, 2002). Thus, $A_{2A}Rs$ present at corticostriatal projection terminals (Rebola, Rodrigues, et al., 2005; Rosin et al., 2003) modulate glutamate release to control synaptic plasticity (Ciruela et al., 2006; Rebola, Rodrigues, et al., 2005). While adenosine-mediated inhibition of neural activity is largely attributed to A_1R-mediated presynaptic inhibition, activation of postsynaptic A_1Rs also results in G-protein-dependent activation of inwardly rectifying potassium channels, leading to hyperpolarization of resting membrane potential of postsynaptic neurons (Dunwiddie & Fredholm, 1997). This could potentially explain how caffeine, via blockade of A_1Rs, increases firing of different neurons, including the cholinergic neurons regulating sleep–wakefulness (Oishi, Huang, Fredholm, Urade, & Hayaishi, 2008; Rainnie, Grunze, McCarley, & Greene, 1994). Activation of postsynaptic A_{2A}Rs, however, inhibits N-Methyl-D-aspartate (NMDA) receptor-mediated synaptic currents in rat neostriatal neurons via the PKA pathway (Norenberg, Wirkner, Assmann, Richter, & Illes, 1998; Norenberg, Wirkner, & Illes, 1997; Wirkner et al., 2000). Thus, A_{2A}R antagonists potentiate the excitotoxic effect of direct NMDA receptor stimulation, while reducing neurotransmitter release (Popoli et al., 2002; Robledo, Ursu, & Mahy, 1999).

8.2. AR modulation of synaptic plasticity

LTP is thought to represent the synaptic plasticity phenomenon that sustains memory formation at the cellular level. Evaluation of synaptic plasticity both *in vitro* and *in vivo* has argued for the involvement of ARs in these processes (de Mendonça & Ribeiro, 1997). LTP in the mossy fiber pathway of hippocampal slices was shown to be impaired in A_1R-KO mice, an effect mimicked by pretreatment with an A_1R antagonist (Moore, Nicoll, & Schmitz, 2003). In the Shaffer collateral-CA1 pathway of hippocampal slices, the deletion of A_1R did not produce alterations in LTP or Long-term depression (LTD), but it did impair pair pulse facilitation, a short-term plasticity phenomena dependent on A_1Rs at these synapses (Gimenez-Llort et al., 2005). Striatal A_{2A}Rs are uniquely positioned to integrate incoming information (glutamate signals) and neuronal sensitivity to this incoming information (DA signals) to control striatal synaptic plasticity and behavior. Indeed, striatal A_{2A}Rs modulate LTP at cortico-accumbal synapses (d'Alcantara, Ledent, Swillens, & Schiffmann, 2001) and spike-timing-dependent LTP at glutamatergic synapses onto striatopallidal MSNs (Flajolet et al., 2008; Shen, Flajolet, Greengard, & Surmeier, 2008). Furthermore, concomitant stimulation of A_{2A}R and D_2R changes the striatopallidal MSN plasticity from LTD to LTP (Shen, Flajolet, et al., 2008) and modulates a form of striatal LTD that is dependent on endocannabinoid release and D_2R activation (Kreitzer & Malenka, 2007). Thus, A_{2A}Rs can selectively modulate D_2R signaling and striatal glutamatergic plasticity (LTD in particular), the putative neurophysiological correlate of striatal-based cognitive processes (Kreitzer & Malenka, 2008).

8.3. AR control of learning and memory

Since the forebrain structures play roles in learning and rewarding processes (nucleus accumbens) and in encoding and retrieval of memories (hippocampus), the regulation of synaptic plasticity by A_1R and $A_{2A}R$ could have important behavioral consequences. Indeed, $A_{2A}R$ activation in the cingulate cortex impairs memory retrieval (Pereira et al., 2005) and social recognition memory, which is prevented by $A_{2A}R$ antagonists (Prediger, Fernandes, & Takahashi, 2005; Prediger, Pamplona, Fernandes, & Takahashi, 2005; Prediger & Takahashi, 2005). Similarly, transgenic overexpression of $A_{2A}Rs$ in the cortex impairs spatial working memory without affecting spatial reference memory (Gimenez-Llort et al., 2007). Conversely, using three different $A_{2A}R$-KO models (global, forebrain-specific, and striatum-specific $A_{2A}R$ KO), we recently showed that inactivation of striatal $A_{2A}Rs$ is sufficient to enhance working memory (Wei, Singer, et al., 2011; Zhou et al., 2009), reversal learning (Wei, Singer, et al., 2011), goal-directed behavior (Yu, Gupta, Chen, & Yin, 2009), and Pavalovian fear conditioning (Wei et al., 2014), without affecting spatial reference memory, motor function, and anxiety-like behaviors. Lastly, $A_{2A}R$ blockade has been shown to attenuate memory impairments in aging (Prediger, Batista, & Takahashi, 2005) and Alzheimer's disease models (Canas, Porciuncula, et al., 2009; Cunha, Canas, et al., 2008; Dall'Igna et al., 2007), diabetic encephalopathy (Duarte, Carvalho, Cunha, & Gruetter, 2009), and attention-deficit hyperactivity disorder (Prediger, Fernandes, et al., 2005), through a control of the dysfunction of glutamatergic synapses (Canas, Duarte, et al., 2009; Cunha et al., 1995; Lopes et al., 1999; Rebola, Sebastiao, et al., 2003). Thus, $A_{2A}R$ blockade not only has procognitive effects in normal animals but also prevents or reverses memory impairments in disease models. These findings not only herald the notion that the suppression of $A_{2A}R$ activity is procognitive but also support the possibility that the $A_{2A}R$ may be a target for selectively alleviating cognitive deficits in neuropsychiatric disorders. On the other hand, the behavioral evaluation of memory performance in A_1R KO revealed only minor to no effects (Gimenez-Llort et al., 2002, 2005; Lang et al., 2003) on altered emotional status (Gimenez-Llort et al., 2002; Johansson et al., 2001).

8.4. ARs control of sleep–wake cycle

Adenosine is considered as an endogenous promoter of sleep for its homeostatic role as a consequence of neuronal activity-induced extracellular levels of adenosine (Basheer, Strecker, Thakkar, & McCarley, 2004). Accordingly, perfusion of adenosine into the basal forebrain of rats reduces wakefulness (Basheer, Porkka-Heiskanen, Stenberg, & McCarley, 1999). Both A_1Rs and A_{2A}Rs have been proposed to mediate the adenosine-induced sleep-promoting effect. A_1R agonists and antagonists induce and suppress sleep, respectively, via its action at basal forebrain (Benington, Kodali, & Heller, 1995; Ticho & Radulovacki, 1991; Virus, Ticho, Pilditch, & Radulovacki, 1990). Similarly, infusion of an antisense A_1R construct was shown to decrease the amount of rapid eye movement sleep and increase wakefulness (Thakkar, Winston, & McCarley, 2003). On the other hand, intracerebroventricular infusion of an agonist for the A_{2A}R (CGS 21680) increases sleep (Satoh, Matsumura, & Hayaishi, 1998; Satoh et al., 1999; Scammell et al., 2001), but the sleep-inducing effect of CGS 21680 is lost in the A_{2A}R KO (Urade et al., 2003). The c-fos mapping analysis pointed the rostal basal forebrain and nuclear accumbens as the possible action sites (Satoh et al., 1999). More recently, an A_{2A}R agonist has been found to modulate histaminergic transmission by increasing inhibition of these neurons through GABA release (Hong et al., 2005). Thus, both A_1Rs and A_{2A}Rs participate in the physiological regulatory mechanisms of sleep. Caffeine, the world's most widely used psychoactive drug, is shown to act mainly at the A_{2A}R but not at the A_1R (Huang et al., 2005), particularly the A_{2A}R in nucleaus accumbens, to enhance wakefulness (Lazarus et al., 2011). Using the Cre/lox technology and focal A_{2A}R KO in rats through the local infection with adeno-associated virus carrying short-hairpin RNA of A_{2A}Rs (Lazarus et al., 2011), we showed that deletion of the A_{2A}Rs selectively in the accumbal shell resulted in abrogation of caffeine-induced wakefulness. Thus, activation of A_{2A}Rs leads to the enhanced activity of GABAergic output neurons on the striatopallidal pathway to exert a tight inhibitory control of the arousal systems in the thalamus, hypothalamus, and cerebral cortex. Caffeine can override the "adenosine brake" to promote wakefulness (Lazarus, Huang, Lu, Urade, & Chen, 2012; Lazarus et al., 2011).

8.5. A_{2A}R control of motor and psychomotor activities

A_{2A}Rs are highly expressed in the dorsal and ventral striata, suggesting an involvement in the physiological control of motor activity. A large body of pharmacological evidence demonstrates that blockade of ARs (in particular the A_{2A}R) stimulates motor activity, while activation of the A_{2A}R (and to a less extend the A_1R) inhibits motor activity. Thus, the A_{2A}R agonist

CGS 21680 strongly reduced spontaneous locomotor activity in wild-type mice but had no significant effect on A_{2A}R-KO mice (Chen et al., 1999; Ledent et al., 1997). Studies with A_{2A}R-KO animals have indeed produced compelling evidence as to the requirement of A_{2A}R blockade for caffeine-stimulatory effects (El Yacoubi et al., 2000; Ledent et al., 1997). Our study with striatal A_{2A}Rs KO demonstrated that the A_{2A}R antagonists, KW6002 and caffeine, act at the striatopallidal neurons to exert its motor-stimulant effect (Yu et al., 2008). Moreover, A_{2A}R in striatopallidal MSN antagonizes D_2R function and potentiates psychomotor activity (Schiffmann et al., 2007). On the other hand, A_{2A}R is also located in the striatal glutamatergic terminals (Rosin et al., 2003), where it is involved in the modulation of glutamate release and corticostriatal synaptic transmission (Ciruela et al., 2006; Martire et al., 2011; Quiroz et al., 2009; Tozzi et al., 2007). Selective blockade of A_{2A}R in extrastriatal forebrain neurons attenuates behavioral responses to psychostimulants, such as cocaine (Shen, Coelho, et al., 2008), amphetamine (Bastia et al., 2005; Fredduzzi et al., 2002), or L-DOPA (Xiao et al., 2006). Opposite controls of psychomotor activity by A_{2A}Rs in striatopallidal neurons and A_{2A}Rs in glutamatergic terminals suggest a novel mechanism by which A_{2A}Rs fine-tune striatal activity by integrating GABAergic, dopaminergic, and glutamatergic signaling.

8.6. Adenoisne receptor control of anxiety, aggression, and depression

The anxiogenic actions of adenosine antagonists, such as caffeine, in animals and man have generally been attributed to the blockade of A_1R sites (Millan, 2003). Accordingly, pharmacological activation and antagonism of A_1Rs are associated with anxiolytic and anxiogenic actions, respectively (Baldwin & File, 1989; Florio, Prezioso, Papaioannou, & Vertua, 1998; Jain, Kemp, Adeyemo, Buchanan, & Stone, 1995; Prediger, Batista, & Takahashi, 2004), while genetic deletion of the A_1Rs enhanced anxiety (Gimenez-Llort et al., 2002; Johansson et al., 2001). A_{2A}R-KO mice developed in a CD-1 genetic background displayed an response to anxiogenic stimuli (Ledent et al., 1997). Furthermore, genetic-KO study has confirmed that A_{2A}Rs are targets for the caffeine (El Yacoubi et al., 2001; Fredholm et al., 1999; Ledent et al., 1997). Several human studies found a positive correlation between A_{2A}R gene polymorphisms and acute anxiogenic responses to caffeine (Alsene, Deckert, Sand, & de Wit, 2003) and panic attack (Hamilton et al., 2004). Lastly, pharmacological blockade and genetic KO of the A_{2A}R show an antidepressant effect in behavioral assessement of depression (such as forced swim and tail suspension tests; El Yacoubi et al., 2001; Kaster et al., 2004). A_3R-KO mice, on the contrary, show increased immobile time during the two tests (Fedorova et al., 2003).

8.7. AR control of angiogenesis

Activation of the $A_{2A}R$ in murine macrophages produces a regulation of vascular endothelial growth factor (VEGF) in addition to the antiinflammatory actions (Leibovich et al., 2002; Pinhal-Enfield et al., 2003). The ability of $A_{2A}R$ agonists to inhibit TNF-α production while increasing VEGF expression allows macrophages to switch from an inflammatory to an angiogenic phenotype. The underlying mechanism may be, at least in part, achieved by downregulating the expression of TSP1, an anti-angiogenic matrix protein (Desai et al., 2005). Recent studies in $A_{2A}R$-KO mice demonstrate that $A_{2A}R$ activation stimulates angiogenesis and increases the rate at which wounds close (Montesinos et al., 2002, 1997) and promote pathological angiogenesis in oxygen-induced retinopathay without affecting normal retinal vascularization during postnatal development (Liu et al., 2010).

8.8. AR control of cerebral blood flow and blood–brain barrier integrity

The vasodilator adenosine has been identified as an important mediator that couples cerebral blood flow to neuronal activation (Dirnagl, Niwa, Lindauer, & Villringer, 1994). Thus, adenosine mediates glutamate-induced vasodilation in the cerebral cortex (Iliff, D'Ambrosio, Ngai, & Winn, 2003; Ohata, Cao, & Koehler, 2006). Adenosine can exert important regulatory functions by the activation of vascular and astrocytic ARs (Fields & Burnstock, 2006). ATP release and degradation into adenosine via ectonucleotidases appears to mediate arteriolar dilation in response to neuronal activation (Xu & Pelligrino, 2007) by a process involving astrocytes (Xu et al., 2008). Furthermore, activation of the A_1R and $A_{2A}R$ signaling can increase blood–brain barrier (BBB) permeability via the mechanism through changing the expression of tight junction molecules (Carman, Mills, Krenz, Kim, & Bynoe, 2011). Accordingly, blockade of AR by caffeine protects against BBB dysfunction induced by 1-methyl-4-phenyl-1,2,3,6-tetrahydropyridine (MPTP) and high cholesterol in animal models of PD and Alzheimer's disease (Chen, Gawryluk, Wagener, Ghribi, & Geiger, 2008; Chen, Lan, Roche, Liu, & Geiger, 2008).

9. ARs AND GLIAL FUNCTIONS

9.1. Glial control of adenosine homeostasis

Recent findings indicate that glial release of ATP is one of the main sources for extracellular adenosine (Halassa, Fellin, & Haydon, 2007; Halassa, Fellin, Takano, Dong, & Haydon, 2007; Haydon & Carmignoto, 2006; Martin et al., 2007; Studer et al., 2006). For example, studies with transgenic mice expressing a dominant-negative Soluble NSF Attachment Protein Receptor (SNARE) domain selectively in astrocytes show that astrocytic release of ATP (followed by degradation into adenosine via ectonucleotidases) is a major source of extracellular adenosine that affects synaptic transmission and control of sleep–wake cycle (Pascual et al., 2005). Multiple lines of evidence also suggest that astrocytic ADK is of critical importance in regulating extracellular adenosine concentrations and fulfills the role of a metabolic reuptake system for adenosine (Fedele et al., 2004, 2005; Güttinger et al., 2005; Huber et al., 2001; Kowaluk & Jarvis, 2000; Pak, Haas, Decking, & Schrader, 1994; Ren et al., 2007). Since astrocytes can contact thousands of synapses and coordinate synaptic networks (Halassa, Fellin, Takano, et al., 2007; Ventura & Harris, 1999), astrocytic control of extracellular adenosine (by release of ATP and ADK activity) has a major regulatory function in setting a global adenosine-mediated inhibitory tone within a neuronal network (Haydon & Carmignoto, 2006, #6057).

9.2. Glial ARs

Many functional measurements (such as cAMP levels and cytokine release) coupled with pharmacological tools have clearly demonstrated the presence and function of all four ARs in glial cells (Hasko, Pacher, Vizi, & Illes, 2005). AR expression in glial cells is generally at the low level under physiological conditions, but is induced after brain insults and inflammatory signals, providing a mechanism to locally control neuroinflammatory responses in the brain. One of the important functions of glial ARs is the control of extracellular glutamate level and excitotoxicity. Adenosine, acting on $A_{2A}R$, can increase extracellular levels of glutamate, both by reducing glutamate uptake via glutamate transporter-1 (GLT-1) and by direct release from neuronal terminals (Li, Nomura, Aihara, & Nishizaki, 2001; Nishizaki et al., 2002). We recently showed that astrocytic $A_{2A}Rs$ control the uptake of glutamate through a dual mechanism (Matos et al., 2012): a long-term activation of

$A_{2A}R$ triggers a cAMP/PKA-dependent decrease of the expression of GLT-I and GLAST prior to the reduction of the levels and activity of both transporters, whereas the acute activation of astrocytic $A_{2A}R$ decreases the activity of GLT-1 through an mechanism involving the physical association and modulation between $A_{2A}R$ and Na^+/K^+-ATPase-α_2 and GLT-1 complex in astrocytes (Matos, Augusto, Agostinho, Cunha, & Chen, 2013). In addition, activation of A_1Rs protects astrocytes from damage and cell death (Bjorklund, Shang, Tonazzini, Dare, & Fredholm, 2008; Ciccarelli et al., 2007; D'Alimonte et al., 2007), reduces their proliferation rate in cultured astrocytes (Ciccarelli et al., 1994), and promotes myelination of oligodendroglia (Othman, Yan, & Rivkees, 2003). A_1Rs on neurons (especially at nerve terminals) are critically important in mediating the dampening effect on neuronal activity mediated by adenosine generated from ATP released from astrocytes (Halassa et al., 2009; Martin et al., 2007). Activation of A_3Rs protects astrocytes from hypoxia-induced cell death (Bjorklund et al., 2008). Adenosine is also speculated to be an important regulator of astrocytic swelling via modulation of volume-regulated anion channels.

10. PATHOPHYSIOLOGICAL FUNCTIONS OF ARs

Adenosine levels rise markedly in response to ischemia, hypoxia, excitotoxicity, inflammation, and other brain insults (Gomes, Kaster, Tome, Agostinho, & Cunha, 2011; Pedata et al., 2001; Saura et al., 2005). Brain injury triggers a huge surge in extracellular adenosine, due to increased release of ATP from neurons and exocytosis from astrocytes, leading to the degradation of adenosine. This surge in extracellular adenosine is accompanied by increases in the level of local inflammatory cytokines which lead to a delayed ($\sim$24 h), marked (20- to 50-fold), and sustained (lasting for several hours to days) increases in AR expression, primarily in inflammatory cells (Linden, 2011). The mechanism underlying the induction of the $A_{2A}R$ in inflammatory cells by hypoxia, IL-1β, and TNF-α may be activation of the transcription factor NF-kB pathway, leading to increased transcription (Khoa et al., 2001; Linden, 2011). Thus, adenosine signal is locally amplified through a feed-forward mechanism/loop with both a surge in extracellular ATP/adenosine and the coordinated induction of $A_{2A}R$ by local inflammatory cytokines. Local adenosine signaling may serve as a pathological signal (a "find-me" signal) to trigger mainly adaptive changes, leading to ultimately suppressing inflammation and resolution of pathological processes. Thus, facilitation of the adenosine signaling pathway by pharmacological activation of ARs will be therapeutic.

Increased extracellular adenosine has long been known to act predominantly as a neuroprotectant under pathological conditions (Cunha, 2001; Fredholm, Cunha, & Svenningsson, 2003; Ongini, Adami, Ferri, & Bertorelli, 1997; Ribeiro, Sebastiao, & de Mendonça, 2002; Schwarzschild, Chen, & Ascherio, 2002). However, adenosine may also contribute to neurotoxicity, neuronal damage, and cell death (de Mendonça, et al., 2000). Studies of the regulation of inflammatory responses and brain damage by $A_{2A}R$ activity have yielded some complex and seemingly paradoxical results (Chen et al., 2007; Jones, Smith, & Stone, 1998a; Li et al., 2006; Popoli et al., 2007). On one hand, $A_{2A}R$ *activation* has been shown to attenuate brain damage induced by intracerebral hemorrhage (Mayne et al., 2001), hippocampal kainate-induced excitotoxicity (Jones et al., 1998a), ischemic brain injury (Li et al., 2006; Von Lubitz, Lin, & Jacobson, 1995), striatum lesion (Chiang et al., 2009; Chou et al., 2005; Popoli, Blum, Domenici, Burnouf, & Chern, 2008), and more recently, spinal cord injury after ischemia and compression (Genovese et al., 2009; Li et al., 2006). The neuroprotective effect of $A_{2A}R$ *activation* in these injury models has been largely attributed to the inhibition of inflammatory processes and vascular effects (Chen & Pedata, 2008; Chen et al., 2007; Li et al., 2006; Phillis, 2004; Popoli et al., 2007) and to the activation of PKA-mediated cell survival signals (Blum, Hourez, Galas, Popoli, & Schiffmann, 2003; Huang et al., 2001). The adenosine-induced survival signaling is mediated by CREB phophorylation (Cheng et al., 2002) or by direct binding to intracellular G-protein-interacting protein TRAX (Sun et al., 2006) and by PKA-mediated enhancement of ubiquitin–proteasome activity (Chiang et al., 2009). On the other hand, $A_{2A}R$ *inactivation* has been consistently shown to protect the brain against various insults including ischemia, excitotoxicity, and mitochondrial toxicity (Cunha, 2005). Our genetical and pharmacological studies showed that $A_{2A}R$ *inactivation* down-regulates neuroinflammation and protects brain tissue against ischemia (Chen et al., 1999; Li et al., 2006; Melani et al., 2009; Phillis, 1995; Yu, Huang, et al., 2004), the dopaminergic neurotoxin MPTP and 6-OHDA (Chen et al., 2001; Ikeda, Kurokawa, Aoyama, & Kuwana, 2002; Yu et al., 2008), mitochondrial toxicity (Fink et al., 2004; Huang et al., 2006), excitotoxicity (Jones et al., 1998a; Jones, Smith, & Stone, 1998b), and traumatic brain injury (Li et al., 2009). The protective effect of $A_{2A}R$ *inactivation* in the brain has been attributed to the inhibition of glutamate release (Chen & Pedata, 2008; Gianfriddo, Melani, Turchi, Giovannini, & Pedata, 2004;

O'Regan, Simpson, Perkins, & Phillis, 1992; Popoli et al., 2002) and suppression of proinflammatory cytokines (Chen & Pedata, 2008; Li et al., 2009; Melani et al., 2009; Yu, Huang, et al., 2004; Yu et al., 2008). The apparent paradox of neuroprotection by both $A_{2A}R$ inactivation and activation reflects the complexity of $A_{2A}R$ actions on various neuronal, glial, and vascular components, each of which could have distinct effects in brain injury (Fredholm et al., 2003; Schwarzschild et al., 2002, 2003). Furthermore, the neuroporotective effects of $A_{2A}R$ agonists and antagonists are dependent on the different stages of pathological processes (Li et al., 2006; Popoli et al., 2008), different extents of brain injury (Blum, Galas, et al., 2003), or different routes of drug delivery (Jones et al., 1998a). Remarkably, we recently uncovered a glutamate-mediated switching mechanism whereby increasing the local level of glutamate redirected $A_{2A}R$ signaling from the PKA to the PKC pathways, switching the effect of $A_{2A}R$ activation from antiinflammatory to proinflammatory (Dai et al., 2010). Thus, extrasynaptic glutamate levels control the effect of $A_{2A}R$ activation, switching it between antiinflammatory/neuroprotective and proinflammatory/cytotoxic (Dai et al., 2010). Such findings may explain—at least in part—the opposing effects of $A_{2A}R$ ligands on tissue injury by demonstrating that $A_{2A}R$ effects on brain injury are context dependent due to their dependence on local glutamate levels.

11. ARs AS DRUG TARGETS

There have been significant progresses in structural biology, medicinal chemistry, pharmacology, and molecular biology of ARs, and important roles of adenosine in physiology and pathophysiology have been clarified. Moreover, there is evidence that ARs can be directly or indirectly targeted in human, as reviewed recently (for review see Chen, Eltzschig, & Fredholm, 2013). We highlight the following three areas of progress:

- Over the past 20 years, medicinal chemistry has generated agonists and antagonists that are high affinity (at low nM of kDa), selective (>100- to 200-fold over other AR subtypes) for the human variants of each of the four receptors (Muller & Jacobson, 2011). Moreover, radioactive and more recently fluorescent ligands of ARs have also been developed and introduced for drug screening and monitoring *in vivo* occupancy in human.
- Studies using both pharmacological as well as transgenic approaches have firmly established that ARs not only are important in physiological homeostatic regulation but also play a role in many pathophysiological

processes including inflammatory diseases (Eltzschig & Carmeliet, 2011; Hasko et al., 2008), ischemia and reperfusion (Eltzschig & Eckle, 2011), and neurodegenerative disorders (Burnstock, Fredholm, & Verkhratsky, 2011; Fredholm, 2007).

- There is a good evidence that ARs can be safely targeted for the treatment of human diseases and, consequently, that ARs can be used as drug targets: (a) adenosine itself (adenocard and adenoscan) is clinically used for the treatment of supraventricular tachycardia (Delacretaz, 2006); (b) the $A_{2A}R$ agonist regadenoson (Lexiscan) is approved by FDA for myocardiac perfusion imaging in clinic; (c) many clinically used drugs including dipyridamole and metotrexate may exert its effect by altering extracellular adenosine concentrations and signaling effects; and (d) caffeine is used for treating premature apnea in clinical setting and a majority of people in the whole world consume caffeine on a regular basis in doses that antagonize ARs but do little else.

ACKNOWLEDGMENTS

This work was supported by grants from the National Institute of Health grant NS041083-10, NS073947 (USA), the National Science Council (100-2320-B-001-0110MY3), and Institute of Biomedical Sciences, Academia Sinica (103-Academia Sinica Investigation Award-06). We are grateful to Mr. Tsung-Hung Hung for his assistance in preparing the illustrations.

REFERENCES

Abbracchio, M. P., Brambilla, R., Ceruti, S., Kim, H. O., von Lubitz, D. K., Jacobson, K. A., et al. (1995). G protein-dependent activation of phospholipase C by adenosine A3 receptors in rat brain. *Molecular Pharmacology*, *48*, 1038–1045.

Aherne, C. M., Kewley, E. M., & Eltzschig, H. K. (2011). The resurgence of A2B adenosine receptor signaling. *Biochimica et Biophysica Acta*, *1808*, 1329–1339.

Alsene, K., Deckert, J., Sand, P., & de Wit, H. (2003). Association between A2a receptor gene polymorphisms and caffeine-induced anxiety. *Neuropsychopharmacology*, *28*, 1694–1702.

Augusto, E., Matos, M., Sevigny, J., El-Tayeb, A., Bynoe, M. S., Muller, C. E., et al. (2013). Ecto-5′-nucleotidase (CD73)-mediated formation of adenosine is critical for the striatal adenosine A2A receptor functions. *The Journal of Neuroscience*, *33*, 11390–11399.

Baldwin, H. A., & File, S. E. (1989). Caffeine-induced anxiogenesis: The role of adenosine, benzodiazepine and noradrenergic receptors. *Pharmacology, Biochemistry, and Behavior*, *32*, 181–186.

Ballarin, M., Fredholm, B. B., Ambrosio, S., & Mahy, N. (1991). Extracellular levels of adenosine and its metabolites in the striatum of awake rats: Inhibition of uptake and metabolism. *Acta Physiologica Scandinavica*, *142*, 97–103.

Basheer, R., Porkka-Heiskanen, T., Stenberg, D., & McCarley, R. W. (1999). Adenosine and behavioral state control: Adenosine increases c-Fos protein and AP1 binding in basal forebrain of rats. *Brain Research. Molecular Brain Research*, *73*, 1–10.

Basheer, R., Strecker, R. E., Thakkar, M. M., & McCarley, R. W. (2004). Adenosine and sleep–wake regulation. *Progress in Neurobiology*, *73*, 379–396.

Bastia, E., Xu, Y. H., Scibelli, A. C., Day, Y. J., Linden, J., Chen, J. F., et al. (2005). A crucial role for forebrain adenosine A(2A) receptors in amphetamine sensitization. *Neuropsychopharmacology*, *30*, 891–900.

Benington, J. H., Kodali, S. K., & Heller, H. C. (1995). Stimulation of A1 adenosine receptors mimics the electroencephalographic effects of sleep deprivation. *Brain Research*, *692*, 79–85.

Bergmayr, C., Thurner, P., Keuerleber, S., Kudlacek, O., Nanoff, C., Freissmuth, M., et al. (2013). Recruitment of a cytoplasmic chaperone relay by the A2A adenosine receptor. *The Journal of Biological Chemistry*, *288*, 28831–28844.

Biber, K., Klotz, K. N., Berger, M., Gebicke-Harter, P. J., & van Calker, D. (1997). Adenosine A1 receptor-mediated activation of phospholipase C in cultured astrocytes depends on the level of receptor expression. *The Journal of Neuroscience*, *17*, 4956–4964.

Bjorklund, O., Shang, M. M., Tonazzini, I., Dare, E., & Fredholm, B. B. (2008). Adenosine A(1) and A(3) receptors protect astrocytes from hypoxic damage. *European Journal of Pharmacology*, *596*, 6–13.

Blum, D., Galas, M. C., Pintor, A., Brouillet, E., Ledent, C., Muller, C. E., et al. (2003). A dual role of adenosine A2A receptors in 3-nitropropionic acid-induced striatal lesions: Implications for the neuroprotective potential of A2A antagonists. *The Journal of Neuroscience*, *23*, 5361–5369.

Blum, D., Hourez, R., Galas, M. C., Popoli, P., & Schiffmann, S. N. (2003). Adenosine receptors and Huntington's disease: Implications for pathogenesis and therapeutics. *Lancet Neurology*, *2*, 366–374.

Borroto-Escuela, D. O., Romero-Fernandez, W., Tarakanov, A. O., Ciruela, F., Agnati, L. F., & Fuxe, K. (2011). On the existence of a possible A2A–D2–beta-Arrestin2 complex: A2A agonist modulation of D2 agonist-induced beta-arrestin2 recruitment. *Journal of Molecular Biology*, *406*, 687–699.

Burgueno, J., Blake, D. J., Benson, M. A., Tinsley, C. L., Esapa, C. T., Canela, E. I., et al. (2003). The adenosine A2A receptor interacts with the actin-binding protein alpha-actinin. *The Journal of Biological Chemistry*, *278*, 37545–37552.

Burnstock, G., Fredholm, B. B., & Verkhratsky, A. (2011). Adenosine and ATP receptors in the brain. *Current Topics in Medicinal Chemistry*, *11*, 973–1011.

Canas, P. M., Duarte, J. M., Rodrigues, R. J., Kofalvi, A., & Cunha, R. A. (2009). Modification upon aging of the density of presynaptic modulation systems in the hippocampus. *Neurobiology of Aging*, *30*, 1877–1884.

Canas, P. M., Porciuncula, L. O., Cunha, G. M., Silva, C. G., Machado, N. J., Oliveira, J. M., et al. (2009). Adenosine A2A receptor blockade prevents synaptotoxicity and memory dysfunction caused by beta-amyloid peptides via p38 mitogen-activated protein kinase pathway. *The Journal of Neuroscience*, *29*, 14741–14751.

Canela, L., Lujan, R., Lluis, C., Burgueno, J., Mallol, J., Canela, E. I., et al. (2007). The neuronal Ca(2+)-binding protein 2 (NECAB2) interacts with the adenosine A(2A) receptor and modulates the cell surface expression and function of the receptor. *Molecular and Cellular Neurosciences*, *36*, 1–12.

Carman, A. J., Mills, J. H., Krenz, A., Kim, D. G., & Bynoe, M. S. (2011). Adenosine receptor signaling modulates permeability of the blood-brain barrier. *The Journal of Neuroscience*, *31*, 13272–13280.

Carriba, P., Ortiz, O., Patkar, K., Justinova, Z., Stroik, J., Themann, A., et al. (2007). Striatal adenosine A2A and cannabinoid CB1 receptors form functional heteromeric complexes that mediate the motor effects of cannabinoids. *Neuropsychopharmacology*, *32*, 2249–2259.

Carruthers, A. M., Sellers, L. A., Jenkins, D. W., Jarvie, E. M., Feniuk, W., & Humphrey, P. P. (2001). Adenosine A(1) receptor-mediated inhibition of protein kinase A-induced calcitonin gene-related peptide release from rat trigeminal neurons. *Molecular Pharmacology*, *59*, 1533–1541.

Cassada, D. C., Tribble, C. G., Long, S. M., Kaza, A. K., Linden, J., Rieger, J. M., et al. (2002). Adenosine A2A agonist reduces paralysis after spinal cord ischemia: Correlation with A2A receptor expression on motor neurons. *The Annals of Thoracic Surgery*, *74*, 846–849, discussion 849–850.

Che, J., Chan, E. S., & Cronstein, B. N. (2007). Adenosine A2A receptor occupancy stimulates collagen expression by hepatic stellate cells via pathways involving protein kinase A, Src, and extracellular signal-regulated kinases 1/2 signaling cascade or p38 mitogen-activated protein kinase signaling pathway. *Molecular Pharmacology*, *72*, 1626–1636.

Chekeni, F. B., Elliott, M. R., Sandilos, J. K., Walk, S. F., Kinchen, J. M., Lazarowski, E. R., et al. (2010). Pannexin 1 channels mediate 'find-me' signal release and membrane permeability during apoptosis. *Nature*, *467*, 863–867.

Chen, Y., Corriden, R., Inoue, Y., Yip, L., Hashiguchi, N., Zinkernagel, A., et al. (2006). ATP release guides neutrophil chemotaxis via P2Y2 and A3 receptors. *Science*, *314*, 1792–1795.

Chen, J. F., Eltzschig, H. K., & Fredholm, B. B. (2013). Adenosine receptors as drug targets—What are the challenges? *Nature Reviews Drug Discovery*, *12*, 265–286.

Chen, X., Gawryluk, J. W., Wagener, J. F., Ghribi, O., & Geiger, J. D. (2008). Caffeine blocks disruption of blood brain barrier in a rabbit model of Alzheimer's disease. *Journal of Neuroinflammation*, *5*, 12.

Chen, J. F., Huang, Z., Ma, J., Zhu, J., Moratalla, R., Standaert, D., et al. (1999). A(2A) adenosine receptor deficiency attenuates brain injury induced by transient focal ischemia in mice. *The Journal of Neuroscience*, *19*, 9192–9200.

Chen, X., Lan, X., Roche, I., Liu, R., & Geiger, J. D. (2008). Caffeine protects against MPTP-induced blood–brain barrier dysfunction in mouse striatum. *Journal of Neurochemistry*, *107*, 1147–1157.

Chen, J. F., & Pedata, F. (2008). Modulation of ischemic brain injury and neuroinflammation by adenosine A2A receptors. *Current Pharmaceutical Design*, *14*, 1490–1499.

Chen, J. F., Sonsalla, P. K., Pedata, F., Melani, A., Domenici, M. R., Popoli, P., et al. (2007). Adenosine A2A receptors and brain injury: Broad spectrum of neuroprotection, multifaceted actions and "fine tuning" modulation. *Progress in Neurobiology*, *83*, 310–331.

Chen, B. S., Thomas, E. V., Sanz-Clemente, A., & Roche, K. W. (2011). NMDA receptor-dependent regulation of dendritic spine morphology by SAP102 splice variants. *The Journal of Neuroscience*, *31*, 89–96.

Chen, J. F., Xu, K., Petzer, J. P., Staal, R., Xu, Y. H., Beilstein, M., et al. (2001). Neuroprotection by caffeine and A(2A) adenosine receptor inactivation in a model of Parkinson's disease. *The Journal of Neuroscience*, *21*, RC143.

Cheng, H. C., Shih, H. M., & Chern, Y. (2002). Essential role of cAMP-response element-binding protein activation by A2A adenosine receptors in rescuing the nerve growth factor-induced neurite outgrowth impaired by blockage of the MAPK cascade. *The Journal of Biological Chemistry*, *277*, 33930–33942.

Cheong, S. L., Federico, S., Venkatesan, G., Mandel, A. L., Shao, Y. M., Moro, S., et al. (2013). The A3 adenosine receptor as multifaceted therapeutic target: Pharmacology, medicinal chemistry, and in silico approaches. *Medicinal Research Reviews*, *33*, 235–335.

Chern, Y., Chiou, J. Y., Lai, H. L., & Tsai, M. H. (1995). Regulation of adenylyl cyclase type VI activity during desensitization of the A2a adenosine receptor-mediated cyclic AMP response: Role for protein phosphatase 2A. *Molecular Pharmacology*, *48*, 1–8.

Chern, Y., King, K., Lai, H. L., & Lai, H. T. (1992). Molecular cloning of a novel adenosine receptor gene from rat brain. *Biochemical and Biophysical Research Communications*, *185*, 304–309.

Chiang, M. C., Chen, H. M., Lai, H. L., Chen, H. W., Chou, S. Y., Chen, C. M., et al. (2009). The A2A adenosine receptor rescues the urea cycle deficiency of Huntington's disease by enhancing the activity of the ubiquitin–proteasome system. *Human Molecular Genetics*, *18*, 2929–2942.

Chou, S. Y., Lee, Y. C., Chen, H. M., Chiang, M. C., Lai, H. L., Chang, H. H., et al. (2005). CGS21680 attenuates symptoms of Huntington's disease in a transgenic mouse model. *Journal of Neurochemistry*, *93*, 310–320.

Ciccarelli, R., D'Alimonte, I., Ballerini, P., D'Auro, M., Nargi, E., Buccella, S., et al. (2007). Molecular signalling mediating the protective effect of A1 adenosine and mGlu3 metabotropic glutamate receptor activation against apoptosis by oxygen/glucose deprivation in cultured astrocytes. *Molecular Pharmacology*, *71*, 1369–1380.

Ciccarelli, R., Di Iorio, P., Ballerini, P., Ambrosini, G., Giuliani, P., Tiboni, G. M., et al. (1994). Effects of exogenous ATP and related analogues on the proliferation rate of dissociated primary cultures of rat astrocytes. *Journal of Neuroscience Research*, *39*, 556–566.

Cirera, S., & Fredholm, M. (2001). Isolation and mapping the pig homologs survivin (BIRC5) and effector cell protease receptor 1 (EPR1) genes. *Cytogenetics and Cell Genetics*, *92*, 351–352.

Ciruela, F., Casado, V., Rodrigues, R. J., Lujan, R., Burgueno, J., Canals, M., et al. (2006). Presynaptic control of striatal glutamatergic neurotransmission by adenosine A1-A2A receptor heteromers. *The Journal of Neuroscience*, *26*, 2080–2087.

Ciruela, F., Gomez-Soler, M., Guidolin, D., Borroto-Escuela, D. O., Agnati, L. F., Fuxe, K., et al. (2011). Adenosine receptor containing oligomers: Their role in the control of dopamine and glutamate neurotransmission in the brain. *Biochimica et Biophysica Acta*, *1808*, 1245–1255.

Cunha, R. A. (2001). Adenosine as a neuromodulator and as a homeostatic regulator in the nervous system: Different roles, different sources and different receptors. *Neurochemistry International*, *38*, 107–125.

Cunha, R. A. (2005). Neuroprotection by adenosine in the brain: From A1 receptor activation to A2A receptor blockade. *Purinergic Signalling*, *1*, 111–134.

Cunha, G. M., Canas, P. M., Melo, C. S., Hockemeyer, J., Muller, C. E., Oliveira, C. R., et al. (2008). Adenosine A2A receptor blockade prevents memory dysfunction caused by beta-amyloid peptides but not by scopolamine or MK-801. *Experimental Neurology*, *210*, 776–781.

Cunha, G. M., Canas, P. M., Oliveira, C. R., & Cunha, R. A. (2006). Increased density and synapto-protective effect of adenosine A(2A) receptors upon sub-chronic restraint stress. *Neuroscience*, *141*, 1775–1781.

Cunha, R. A., Constantino, M. C., Sebastiao, A. M., & Ribeiro, J. A. (1995). Modification of A1 and A2a adenosine receptor binding in aged striatum, hippocampus and cortex of the rat. *Neuroreport*, *6*, 1583–1588.

Dai, S. S., Zhou, Y. G., Li, W., An, J. H., Li, P., Yang, N., et al. (2010). Local glutamate level dictates adenosine A2A receptor regulation of neuroinflammation and traumatic brain injury. *The Journal of Neuroscience*, *30*, 5802–5810.

d'Alcantara, P., Ledent, C., Swillens, S., & Schiffmann, S. N. (2001). Inactivation of adenosine A2A receptor impairs long term potentiation in the accumbens nucleus without altering basal synaptic transmission. *Neuroscience*, *107*, 455–464.

D'Alimonte, I., Ballerini, P., Nargi, E., Buccella, S., Giuliani, P., Di Iorio, P., et al. (2007). Staurosporine-induced apoptosis in astrocytes is prevented by A1 adenosine receptor activation. *Neuroscience Letters*, *418*, 66–71.

Dall'Igna, O. P., Fett, P., Gomes, M. W., Souza, D. O., Cunha, R. A., & Lara, D. R. (2007). Caffeine and adenosine A(2a) receptor antagonists prevent beta-amyloid (25–35)-induced cognitive deficits in mice. *Experimental Neurology*, *203*, 241–245.

Daly, J. W., Butts-Lamb, P., & Padgett, W. (1983). Subclasses of adenosine receptors in the central nervous system: Interaction with caffeine and related methylxanthines. *Cellular and Molecular Neurobiology*, *3*, 69–80.

Daly, J. W., & Padgett, W. L. (1992). Agonist activity of 2- and 5′-substituted adenosine analogs and their N6-cycloalkyl derivatives at A1- and A2-adenosine receptors coupled to adenylate cyclase. *Biochemical Pharmacology, 43*, 1089–1093.

Deckert, J., Nothen, M. M., Bryant, S. P., Ren, H., Wolf, H. K., Stiles, G. L., et al. (1995). Human adenosine A1 receptor gene: Systematic screening for DNA sequence variation and linkage mapping on chromosome 1q31-32.1 using a silent polymorphism in the coding region. *Biochemical and Biophysical Research Communications, 214*, 614–621.

Delacretaz, E. (2006). Clinical practice. Supraventricular tachycardia. *The New England Journal of Medicine, 354*, 1039–1051.

Delaney, S. M., & Geiger, J. D. (1998). Levels of endogenous adenosine in rat striatum. II. Regulation of basal and N-methyl-D-aspartate-induced levels by inhibitors of adenosine transport and metabolism. *The Journal of Pharmacology and Experimental Therapeutics, 285*, 568–572.

de Mendonça, A., & Ribeiro, J. A. (1997). Adenosine and neuronal plasticity. *Life Sciences, 60*, 245–251.

de Mendonça, A., Sebastião, A. M., & Ribeiro, J. A. (2000). Adenosine: Does it have a neuroprotective role after all? *Brain Research. Brain Research Reviews, 33*, 258–274.

Desai, A., Victor-Vega, C., Gadangi, S., Montesinos, M. C., Chu, C. C., & Cronstein, B. N. (2005). Adenosine A2A receptor stimulation increases angiogenesis by down-regulating production of the antiangiogenic matrix protein thrombospondin 1. *Molecular Pharmacology, 67*, 1406–1413.

Diaz-Cabiale, Z., Vivo, M., Del Arco, A., O'Connor, W. T., Harte, M. K., Muller, C. E., et al. (2002). Metabotropic glutamate mGlu5 receptor-mediated modulation of the ventral striopallidal GABA pathway in rats. Interactions with adenosine A(2A) and dopamine D(2) receptors. *Neuroscience Letters, 324*, 154–158.

Dickenson, J. M., & Hill, S. J. (1998). Involvement of G-protein betagamma subunits in coupling the adenosine A1 receptor to phospholipase C in transfected CHO cells. *European Journal of Pharmacology, 355*, 85–93.

Diogenes, M. J., Fernandes, C. C., Sebastiao, A. M., & Ribeiro, J. A. (2004). Activation of adenosine A2A receptor facilitates brain-derived neurotrophic factor modulation of synaptic transmission in hippocampal slices. *The Journal of Neuroscience, 24*, 2905–2913.

Dirnagl, U., Niwa, K., Lindauer, U., & Villringer, A. (1994). Coupling of cerebral blood flow to neuronal activation: Role of adenosine and nitric oxide. *The American Journal of Physiology, 267*, H296–H301.

Dixon, A. K., Gubitz, A. K., Sirinathsinghji, D. J., Richardson, P. J., & Freeman, T. C. (1996). Tissue distribution of adenosine receptor mRNAs in the rat. *British Journal of Pharmacology, 118*, 1461–1468.

Dore, A. S., Robertson, N., Errey, J. C., Ng, I., Hollenstein, K., Tehan, B., et al. (2011). Structure of the adenosine A(2A) receptor in complex with ZM241385 and the xanthines XAC and caffeine. *Structure, 19*, 1283–1293.

Duarte, J. M., Carvalho, R. A., Cunha, R. A., & Gruetter, R. (2009). Caffeine consumption attenuates neurochemical modifications in the hippocampus of streptozotocin-induced diabetic rats. *Journal of Neurochemistry, 111*, 368–379.

Duarte, J. M., Oliveira, C. R., Ambrosio, A. F., & Cunha, R. A. (2006). Modification of adenosine A1 and A2A receptor density in the hippocampus of streptozotocin-induced diabetic rats. *Neurochemistry International, 48*, 144–150.

Dumont, J. E., Pecasse, F., & Maenhaut, C. (2001). Crosstalk and specificity in signalling. Are we crosstalking ourselves into general confusion? *Cellular Signalling, 13*, 457–463.

Dunwiddie, T. V., & Fredholm, B. B. (1997). Adenosine neuromodulation. In K. A. Jacobson, & M. F. Jarvis (Eds.), *Purinergic approaches in experimental therapeutics* (pp. 359–382). New York, NY: Wiley-Liss.

Dunwiddie, T. V., & Masino, S. A. (2001). The role and regulation of adenosine in the central nervous system. *Annual Review of Neuroscience*, *24*, 31–55.
Elliott, M. R., Chekeni, F. B., Trampont, P. C., Lazarowski, E. R., Kadl, A., Walk, S. F., et al. (2009). Nucleotides released by apoptotic cells act as a find-me signal to promote phagocytic clearance. *Nature*, *461*, 282–286.
Eltzschig, H. K. (2009). Adenosine: An old drug newly discovered. *Anesthesiology*, *111*, 904–915.
Eltzschig, H. K., & Carmeliet, P. (2011). Hypoxia and inflammation. *The New England Journal of Medicine*, *364*, 656–665.
Eltzschig, H. K., & Eckle, T. (2011). Ischemia and reperfusion—From mechanism to translation. *Nature Medicine*, *17*, 1391–1401.
El Yacoubi, M., Ledent, C., Menard, J. F., Parmentier, M., Costentin, J., & Vaugeois, J. M. (2000). The stimulant effects of caffeine on locomotor behaviour in mice are mediated through its blockade of adenosine A(2A) receptors. *British Journal of Pharmacology*, *129*, 1465–1473.
El Yacoubi, M., Ledent, C., Parmentier, M., Bertorelli, R., Ongini, E., Costentin, J., et al. (2001). Adenosine A2A receptor antagonists are potential antidepressants: Evidence based on pharmacology and A2A receptor knockout mice. *British Journal of Pharmacology*, *134*, 68–77.
Faigle, M., Seessle, J., Zug, S., El Kasmi, K. C., & Eltzschig, H. K. (2008). ATP release from vascular endothelia occurs across Cx43 hemichannels and is attenuated during hypoxia. *PLoS One*, *3*, e2801.
Fedele, D. E., Gouder, N., Güttinger, M., Gabernet, L., Scheurer, L., Rulicke, T., et al. (2005). Astrogliosis in epilepsy leads to overexpression of adenosine kinase resulting in seizure aggravation. *Brain*, *128*, 2383–2395.
Fedele, D. E., Koch, P., Brüstle, O., Scheurer, L., Simpson, E. M., Mohler, H., et al. (2004). Engineering embryonic stem cell derived glia for adenosine delivery. *Neuroscience Letters*, *370*, 160–165.
Fedorova, I. M., Jacobson, M. A., Basile, A., & Jacobson, K. A. (2003). Behavioral characterization of mice lacking the A3 adenosine receptor: Sensitivity to hypoxic neurodegeneration. *Cellular and Molecular Neurobiology*, *23*, 431–447.
Fenton, R. A., Shea, L. G., Doddi, C., & Dobson, J. G., Jr. (2010). Myocardial adenosine A(1)-receptor-mediated adenoprotection involves phospholipase C, PKC-epsilon, and p38 MAPK, but not HSP27. *American Journal of Physiology. Heart and Circulatory Physiology*, *298*, H1671–H1678.
Feoktistov, I., & Biaggioni, I. (1995). Adenosine A2b receptors evoke interleukin-8 secretion in human mast cells. An enprofylline-sensitive mechanism with implications for asthma. *The Journal of Clinical Investigation*, *96*, 1979–1986.
Feoktistov, I., Goldstein, A. E., Ryzhov, S., Zeng, D., Belardinelli, L., Voyno-Yasenetskaya, T., et al. (2002). Differential expression of adenosine receptors in human endothelial cells: Role of A2B receptors in angiogenic factor regulation. *Circulation Research*, *90*, 531–538.
Feoktistov, I., Murray, J. J., & Biaggioni, I. (1994). Positive modulation of intracellular Ca2+ levels by adenosine A2b receptors, prostacyclin, and prostaglandin E1 via a cholera toxin-sensitive mechanism in human erythroleukemia cells. *Molecular Pharmacology*, *45*, 1160–1167.
Fideu, M. D., Arce, A., Esquifino, A. I., & Miras-Portugal, M. T. (1994). Thyroid hormones modulate both adenosine transport and adenosine A1 receptors in rat brain. *The American Journal of Physiology*, *267*, C1651–C1656.
Fields, R. D., & Burnstock, G. (2006). Purinergic signalling in neuron–glia interactions. *Nature Reviews Neuroscience*, *7*, 423–436.
Fink, J. S., Kalda, A., Ryu, H., Stack, E. C., Schwarzschild, M. A., Chen, J. F., et al. (2004). Genetic and pharmacological inactivation of the adenosine A2A receptor attenuates 3-nitropropionic acid-induced striatal damage. *Journal of Neurochemistry*, *88*, 538–544.

Fink, J. S., Weaver, D. R., Rivkees, S. A., Peterfreund, R. A., Pollack, A. E., Adler, E. M., et al. (1992). Molecular cloning of the rat A2 adenosine receptor: Selective co-expression with D2 dopamine receptors in rat striatum. *Brain Research. Molecular Brain Research, 14,* 186–195.

Fishman, P., Bar-Yehuda, S., Liang, B. T., & Jacobson, K. A. (2012). Pharmacological and therapeutic effects of A3 adenosine receptor agonists. *Drug Discovery Today, 17,* 359–366.

Flajolet, M., Wang, Z., Futter, M., Shen, W., Nuangchamnong, N., Bendor, J., et al. (2008). FGF acts as a co-transmitter through adenosine A(2A) receptor to regulate synaptic plasticity. *Nature Neuroscience, 11,* 1402–1409.

Florio, C., Prezioso, A., Papaioannou, A., & Vertua, R. (1998). Adenosine A1 receptors modulate anxiety in CD1 mice. *Psychopharmacology, 136,* 311–319.

Fontinha, B. M., Diogenes, M. J., Ribeiro, J. A., & Sebastiao, A. M. (2008). Enhancement of long-term potentiation by brain-derived neurotrophic factor requires adenosine A2A receptor activation by endogenous adenosine. *Neuropharmacology, 54,* 924–933.

Fredduzzi, S., Moratalla, R., Monopoli, A., Cuellar, B., Xu, K., Ongini, E., et al. (2002). Persistent behavioral sensitization to chronic L-DOPA requires A2A adenosine receptors. *The Journal of Neuroscience, 22,* 1054–1062.

Fredholm, B. B. (2007). Adenosine, an endogenous distress signal, modulates tissue damage and repair. *Cell Death and Differentiation, 14,* 1315–1323.

Fredholm, B. B., Arslan, G., Halldner, L., Kull, B., Schulte, G., & Wasserman, W. (2000). Structure and function of adenosine receptors and their genes. *Naunyn-Schmiedeberg's Archives of Pharmacology, 362,* 364–374.

Fredholm, B. B., Battig, K., Holmen, J., Nehlig, A., & Zvartau, E. E. (1999). Actions of caffeine in the brain with special reference to factors that contribute to its widespread use. *Pharmacological Reviews, 51,* 83–133.

Fredholm, B., Chen, J.-F., Masino, S. A., & Vaugeois, J.-M. (2005). Actions of adeosine at its receptors in the CNS: Insights from knckouts and drugs. *Annual Review of Pharmacology and Toxicology, 45,* 385–412.

Fredholm, B. B., Chern, Y., Franco, R., & Sitkovsky, M. (2007). Aspects of the general biology of adenosine A2A signaling. *Progress in Neurobiology, 83,* 263–276.

Fredholm, B. B., Cunha, R. A., & Svenningsson, P. (2003). Pharmacology of adenosine A(2A) receptors and therapeutic applications. *Current Topics in Medicinal Chemistry, 3,* 413–426.

Fredholm, B. B., IJzerman, A. P., Jacobson, K. A., Klotz, K. N., & Linden, J. (2001). International Union of Pharmacology. XXV. Nomenclature and classification of adenosine receptors. *Pharmacological Reviews, 53,* 527–552.

Fredholm, B. B., IJzerman, A. P., Jacobson, K. A., Linden, J., & Muller, C. E. (2011). International Union of Basic and Clinical Pharmacology. LXXXI. Nomenclature and classification of adenosine receptors—An update. *Pharmacological Reviews, 63,* 1–34.

Fredholm, B. B., Irenius, E., Kull, B., & Schulte, G. (2001). Comparison of the potency of adenosine as an agonist at human adenosine receptors expressed in Chinese hamster ovary cells. *Biochemical Pharmacology, 61,* 443–448.

Freeman, J. L., Pitcher, J. A., Li, X., Bennett, V., & Lefkowitz, R. J. (2000). Alpha-actinin is a potent regulator of G protein-coupled receptor kinase activity and substrate specificity in vitro. *FEBS Letters, 473,* 280–284.

Fuxe, K., Ferre, S., Canals, M., Torvinen, M., Terasmaa, A., Marcellino, D., et al. (2005). Adenosine A2A and dopamine D2 heteromeric receptor complexes and their function. *Journal of Molecular Neuroscience, 26,* 209–220.

Gao, Z. G., & Jacobson, K. A. (2008). Translocation of arrestin induced by human A(3) adenosine receptor ligands in an engineered cell line: Comparison with G protein-dependent pathways. *Pharmacological Research, 57,* 303–311.

Gao, Z. G., Verzijl, D., Zweemer, A., Ye, K., Goblyos, A., Ijzerman, A. P., et al. (2011). Functionally biased modulation of A(3) adenosine receptor agonist efficacy and potency by imidazoquinolinamine allosteric enhancers. *Biochemical Pharmacology*, *82*, 658–668.
Genovese, T., Melani, A., Esposito, E., Mazzon, E., Di Paola, R., Bramanti, P., et al. (2009). The selective adenosine A2A receptor agonist CGS 21680 reduces JNK MAPK activation in oligodendrocytes in injured spinal cord. *Shock*, *32*, 578–585.
Gianfriddo, M., Melani, A., Turchi, D., Giovannini, M. G., & Pedata, F. (2004). Adenosine and glutamate extracellular concentrations and mitogen-activated protein kinases in the striatum of Huntington transgenic mice. Selective antagonism of adenosine A2A receptors reduces transmitter outflow. *Neurobiology of Disease*, *17*, 77–88.
Gimenez-Llort, L., Fernandez-Teruel, A., Escorihuela, R. M., Fredholm, B. B., Tobena, A., Pekny, M., et al. (2002). Mice lacking the adenosine A1 receptor are anxious and aggressive, but are normal learners with reduced muscle strength and survival rate. *The European Journal of Neuroscience*, *16*, 547–550.
Gimenez-Llort, L., Masino, S. A., Diao, L., Fernandez-Teruel, A., Tobena, A., Halldner, L., et al. (2005). Mice lacking the adenosine A(1) receptor have normal spatial learning and plasticity in the CA1 region of the hippocampus, but they habituate more slowly. *Synapse*, *57*, 8–16.
Gimenez-Llort, L., Schiffmann, S. N., Shmidt, T., Canela, L., Camon, L., Wassholm, M., et al. (2007). Working memory deficits in transgenic rats overexpressing human adenosine A2A receptors in the brain. *Neurobiology of Learning and Memory*, *87*, 42–56.
Gines, S., Hillion, J., Torvinen, M., Le Crom, S., Casado, V., Canela, E. I., et al. (2000). Dopamine D1 and adenosine A1 receptors form functionally interacting heteromeric complexes. *Proceedings of the National Academy of Sciences of the United States of America*, *97*, 8606–8611.
Gomes, C. V., Kaster, M. P., Tome, A. R., Agostinho, P. M., & Cunha, R. A. (2011). Adenosine receptors and brain diseases: Neuroprotection and neurodegeneration. *Biochimica et Biophysica Acta*, *1808*, 1380–1399.
Gomes, C. A., Vaz, S. H., Ribeiro, J. A., & Sebastiao, A. M. (2006). Glial cell line-derived neurotrophic factor (GDNF) enhances dopamine release from striatal nerve endings in an adenosine A2A receptor-dependent manner. *Brain Research*, *1113*, 129–136.
Gsandtner, I., Charalambous, C., Stefan, E., Ogris, E., Freissmuth, M., & Zezula, J. (2005). Heterotrimeric G protein-independent signaling of a G protein-coupled receptor. Direct binding of ARNO/cytohesin-2 to the carboxyl terminus of the A2A adenosine receptor is necessary for sustained activation of the ERK/MAP kinase pathway. *The Journal of Biological Chemistry*, *280*, 31898–31905.
Gsandtner, I., & Freissmuth, M. (2006). A tail of two signals: The C terminus of the A(2A)-adenosine receptor recruits alternative signaling pathways. *Molecular Pharmacology*, *70*, 447–449.
Gubitz, A. K., Widdowson, L., Kurokawa, M., Kirkpatrick, K. A., & Richardson, P. J. (1996). Dual signalling by the adenosine A2a receptor involves activation of both N- and P-type calcium channels by different G proteins and protein kinases in the same striatal nerve terminals. *Journal of Neurochemistry*, *67*, 374–381.
Gutierrez-de-Teran, H., Massink, A., Rodriguez, D., Liu, W., Han, G. W., Joseph, J. S., et al. (2013). The role of a sodium ion binding site in the allosteric modulation of the A(2A) adenosine G protein-coupled receptor. *Structure*, *21*, 2175–2185.
Güttinger, M., Padrun, V., Pralong, W., & Boison, D. (2005). Seizure suppression and lack of adenosine A_1 receptor desensitization after focal long-term delivery of adenosine by encapsulated myoblasts. *Experimental Neurology*, *193*, 53–64.
Halassa, M. M., Fellin, T., & Haydon, P. G. (2007). The tripartite synapse: Roles for gliotransmission in health and disease. *Trends in Molecular Medicine*, *13*, 54–63.

Keuerleber, S., Gsandtner, I., & Freissmuth, M. (2011). From cradle to twilight: The carboxyl terminus directs the fate of the A(2A)-adenosine receptor. *Biochimica et Biophysica Acta, 1808*, 1350–1357.

Khoa, N. D., Montesinos, M. C., Reiss, A. B., Delano, D., Awadallah, N., & Cronstein, B. N. (2001). Inflammatory cytokines regulate function and expression of adenosine A(2A) receptors in human monocytic THP-1 cells. *Journal of Immunology, 167*, 4026–4032.

Khoa, N. D., Postow, M., Danielsson, J., & Cronstein, B. N. (2006). Tumor necrosis factor-alpha prevents desensitization of Galphas-coupled receptors by regulating GRK2 association with the plasma membrane. *Molecular Pharmacology, 69*, 1311–1319.

Kim, Y., Hechler, B., Klutz, A. M., Gachet, C., & Jacobson, K. A. (2008). Toward multivalent signaling across G protein-coupled receptors from poly(amidoamine) dendrimers. *Bioconjugate Chemistry, 19*, 406–411.

King, A. E., Ackley, M. A., Cass, C. E., Young, J. D., & Baldwin, S. A. (2006). Nucleoside transporters: From scavengers to novel therapeutic targets. *Trends in Pharmacological Sciences, 27*, 416–425.

Kirsch, G. E., Codina, J., Birnbaumer, L., & Brown, A. M. (1990). Coupling of ATP-sensitive K+ channels to A1 receptors by G proteins in rat ventricular myocytes. *The American Journal of Physiology, 259*, H820–H826.

Kobayashi, S., & Millhorn, D. E. (1999). Stimulation of expression for the adenosine A2A receptor gene by hypoxia in PC12 cells. A potential role in cell protection. *The Journal of Biological Chemistry, 274*, 20358–20365.

Kowaluk, E. A., & Jarvis, M. F. (2000). Therapeutic potential of adenosine kinase inhibitors. *Expert Opinion on Investigational Drugs, 9*, 551–564.

Kreitzer, A. C., & Malenka, R. C. (2007). Endocannabinoid-mediated rescue of striatal LTD and motor deficits in Parkinson's disease models. *Nature, 445*, 643–647.

Kreitzer, A. C., & Malenka, R. C. (2008). Striatal plasticity and basal ganglia circuit function. *Neuron, 60*, 543–554.

Kull, B., Svenningsson, P., & Fredholm, B. B. (2000). Adenosine A(2A) receptors are colocalized with and activate g(olf) in rat striatum. *Molecular Pharmacology, 58*, 771–777.

Lai, H. L., Yang, T. H., Messing, R. O., Ching, Y. H., Lin, S. C., & Chern, Y. (1997). Protein kinase C inhibits adenylyl cyclase type VI activity during desensitization of the A2a-adenosine receptor-mediated cAMP response. *The Journal of Biological Chemistry, 272*, 4970–4977.

Lang, U. E., Lang, F., Richter, K., Vallon, V., Lipp, H. P., Schnermann, J., et al. (2003). Emotional instability but intact spatial cognition in adenosine receptor 1 knock out mice. *Behavioural Brain Research, 145*, 179–188.

Langemeijer, E. V., Verzijl, D., Dekker, S. J., & Ijzerman, A. P. (2013). Functional selectivity of adenosine A1 receptor ligands? *Purinergic Signal, 9*, 91–100.

Lazarus, M., Huang, Z. L., Lu, J., Urade, Y., & Chen, J. F. (2012). How do the basal ganglia regulate sleep–wake behavior? *Trends in Neurosciences, 35*, 723–732.

Lazarus, M., Shen, H. Y., Cherasse, Y., Qu, W. M., Huang, Z. L., Bass, C. E., et al. (2011). Arousal effect of caffeine depends on adenosine A2A receptors in the shell of the nucleus accumbens. *The Journal of Neuroscience, 31*, 10067–10075.

Lebon, G., Warne, T., Edwards, P. C., Bennett, K., Langmead, C. J., Leslie, A. G., et al. (2011). Agonist-bound adenosine A2A receptor structures reveal common features of GPCR activation. *Nature, 474*, 521–525.

Ledent, C., Vaugeois, J. M., Schiffmann, S. N., Pedrazzini, T., El Yacoubi, M., Vanderhaeghen, J. J., et al. (1997). Aggressiveness, hypoalgesia and high blood pressure in mice lacking the adenosine A2a receptor. *Nature, 388*, 674–678.

Lee, F. S., & Chao, M. V. (2001). Activation of Trk neurotrophin receptors in the absence of neurotrophins. *Proceedings of the National Academy of Sciences of the United States of America, 98*, 3555–3560.

Lee, C. F., Lai, H. L., Lee, Y. C., Chien, C. L., & Chern, Y. (2014). The A2A adenosine receptor is a dual coding gene: A novel mechanism of gene usage and signal transduction. *The Journal of Biological Chemistry, 289*, 1257–1270.

Leibovich, S. J., Chen, J. F., Pinhal-Enfield, G., Belem, P. C., Elson, G., Rosania, A., et al. (2002). Synergistic up-regulation of vascular endothelial growth factor expression in murine macrophages by adenosine A(2A) receptor agonists and endotoxin. *The American Journal of Pathology, 160*, 2231–2244.

Li, W., Dai, S., An, J., Xiong, R., Li, P., Chen, X., et al. (2009). Genetic inactivation of adenosine A2A receptors attenuates acute traumatic brain injury in the mouse cortical impact model. *Experimental Neurology, 215*, 69–76.

Li, X. X., Nomura, T., Aihara, H., & Nishizaki, T. (2001). Adenosine enhances glial glutamate efflux via A2a adenosine receptors. *Life Sciences, 68*, 1343–1350.

Li, Y., Oskouian, R. J., Day, Y. J., Rieger, J. M., Liu, L., Kern, J. A., et al. (2006). Mouse spinal cord compression injury is reduced by either activation of the adenosine A2A receptor on bone marrow-derived cells or deletion of the A2A receptor on non-bone marrow-derived cells. *Neuroscience, 141*, 2029–2039.

Linden, J. (2006). Purinergic chemotaxis. *Science, 314*, 1689–1690.

Linden, J. (2011). Regulation of leukocyte function by adenosine receptors. *Advances in Pharmacology, 61*, 95–114.

Liu, W., Chun, E., Thompson, A. A., Chubukov, P., Xu, F., Katritch, V., et al. (2012). Structural basis for allosteric regulation of GPCRs by sodium ions. *Science, 337*, 232–236.

Liu, Z. W., & Gao, X. B. (2007). Adenosine inhibits activity of hypocretin/orexin neurons by the A1 receptor in the lateral hypothalamus: A possible sleep-promoting effect. *Journal of Neurophysiology, 97*, 837–848.

Liu, A. M., & Wong, Y. H. (2004). G16-mediated activation of nuclear factor kappaB by the adenosine A1 receptor involves c-Src, protein kinase C, and ERK signaling. *The Journal of Biological Chemistry, 279*, 53196–53204.

Liu, X. L., Zhou, R., Pan, Q. Q., Jia, X. L., Gao, W. N., Wu, J., et al. (2010). Genetic inactivation of the adenosine A2A receptor attenuates pathologic but not developmental angiogenesis in the mouse retina. *Investigative Ophthalmology & Visual Science, 51*, 6625–6632.

Londos, C., Cooper, D. M., & Wolff, J. (1980). Subclasses of external adenosine receptors. *Proceedings of the National Academy of Sciences of the United States of America*, 77, 2551–2554.

Lopes, L. V., Cunha, R. A., Kull, B., Fredholm, B. B., & Ribeiro, J. A. (2002). Adenosine A(2A) receptor facilitation of hippocampal synaptic transmission is dependent on tonic A(1) receptor inhibition. *Neuroscience, 112*, 319–329.

Lopes, L. V., Cunha, R. A., & Ribeiro, J. A. (1999). Increase in the number, G protein coupling, and efficiency of facilitatory adenosine A2A receptors in the limbic cortex, but not striatum, of aged rats. *Journal of Neurochemistry, 73*, 1733–1738.

MacDonald, P. E., Braun, M., Galvanovskis, J., & Rorsman, P. (2006). Release of small transmitters through kiss-and-run fusion pores in rat pancreatic beta cells. *Cell Metabolism, 4*, 283–290.

Mahan, L. C., McVittie, L. D., Smyk-Randall, E. M., Nakata, H., Monsma, F. J., Jr., Gerfen, C. R., et al. (1991). Cloning and expression of an A1 adenosine receptor from rat brain. *Molecular Pharmacology, 40*, 1–7.

Malaga-Dieguez, L., Yang, Q., Bauer, J., Pankevych, H., Freissmuth, M., & Nanoff, C. (2010). Pharmacochaperoning of the A1 adenosine receptor is contingent on the endoplasmic reticulum. *Molecular Pharmacology*, 77, 940–952.

Martin, E. D., Fernandez, M., Perea, G., Pascual, O., Haydon, P. G., Araque, A., et al. (2007). Adenosine released by astrocytes contributes to hypoxia-induced modulation of synaptic transmission. *Glia*, *55*, 36–45.

Martire, A., Tebano, M. T., Chiodi, V., Ferreira, S. G., Cunha, R. A., Kofalvi, A., et al. (2011). Pre-synaptic adenosine A2A receptors control cannabinoid CB1 receptor-mediated inhibition of striatal glutamatergic neurotransmission. *Journal of Neurochemistry*, *116*, 273–280.

Matharu, A. L., Mundell, S. J., Benovic, J. L., & Kelly, E. (2001). Rapid agonist-induced desensitization and internalization of the A(2B) adenosine receptor is mediated by a serine residue close to the COOH terminus. *The Journal of Biological Chemistry*, *276*, 30199–30207.

Matos, M., Augusto, E., Agostinho, P., Cunha, R. A., & Chen, J. F. (2013). Antagonistic interaction between adenosine A2A receptors and Na+/K+-ATPase-alpha2 controlling glutamate uptake in astrocytes. *The Journal of Neuroscience*, *33*, 18492–18502.

Matos, M., Augusto, E., Santos-Rodrigues, A. D., Schwarzschild, M. A., Chen, J. F., Cunha, R. A., et al. (2012). Adenosine A(2A) receptors modulate glutamate uptake in cultured astrocytes and gliosomes. *Glia*, *60*, 702–716.

Mayne, M., Fotheringham, J., Yan, H. J., Power, C., Del Bigio, M. R., Peeling, J., et al. (2001). Adenosine A2A receptor activation reduces proinflammatory events and decreases cell death following intracerebral hemorrhage. *Annals of Neurology*, *49*, 727–735.

Mayne, M., Shepel, P. N., Jiang, Y., Geiger, J. D., & Power, C. (1999). Dysregulation of adenosine A1 receptor-mediated cytokine expression in peripheral blood mononuclear cells from multiple sclerosis patients. *Annals of Neurology*, *45*, 633–639.

Melani, A., Cipriani, S., Vannucchi, M. G., Nosi, D., Donati, C., Bruni, P., et al. (2009). Selective adenosine A2a receptor antagonism reduces JNK activation in oligodendrocytes after cerebral ischaemia. *Brain*, *132*, 1480–1495.

Millan, M. J. (2003). The neurobiology and control of anxious states. *Progress in Neurobiology*, *70*, 83–244.

Milojevic, T., Reiterer, V., Stefan, E., Korkhov, V. M., Dorostkar, M. M., Ducza, E., et al. (2006). The ubiquitin-specific protease Usp4 regulates the cell surface level of the A2A receptor. *Molecular Pharmacology*, *69*, 1083–1094.

Montesinos, M. C., Desai, A., Chen, J. F., Yee, H., Schwarzschild, M. A., Fink, J. S., et al. (2002). Adenosine promotes wound healing and mediates angiogenesis in response to tissue injury via occupancy of A(2A) receptors. *The American Journal of Pathology*, *160*, 2009–2018.

Montesinos, M. C., Gadangi, P., Longaker, M., Sung, J., Levine, J., Nilsen, D., et al. (1997). Wound healing is accelerated by agonists of adenosine A2 (G alpha s-linked) receptors. *The Journal of Experimental Medicine*, *186*, 1615–1620.

Moore, K. A., Nicoll, R. A., & Schmitz, D. (2003). Adenosine gates synaptic plasticity at hippocampal mossy fiber synapses. *Proceedings of the National Academy of Sciences of the United States of America*, *100*, 14397–14402.

Muller, C. E., & Jacobson, K. A. (2011). Recent developments in adenosine receptor ligands and their potential as novel drugs. *Biochimica et Biophysica Acta*, *1808*, 1290–1308.

Mundell, S. J., Loudon, R. P., & Benovic, J. L. (1999). Characterization of G protein-coupled receptor regulation in antisense mRNA-expressing cells with reduced arrestin levels. *Biochemistry*, *38*, 8723–8732.

Mundell, S. J., Matharu, A. L., Kelly, E., & Benovic, J. L. (2000). Arrestin isoforms dictate differential kinetics of A2B adenosine receptor trafficking. *Biochemistry*, *39*, 12828–12836.

Munshi, R., Pang, I. H., Sternweis, P. C., & Linden, J. (1991). A1 adenosine receptors of bovine brain couple to guanine nucleotide-binding proteins Gi1, Gi2, and Go. *The Journal of Biological Chemistry*, *266*, 22285–22289.

Murphree, L. J., Sullivan, G. W., Marshall, M. A., & Linden, J. (2005). Lipopolysaccharide rapidly modifies adenosine receptor transcripts in murine and human macrophages: Role of NF-kappaB in A(2A) adenosine receptor induction. *The Biochemical Journal, 391*, 575–580.

Navarro, G., Aymerich, M. S., Marcellino, D., Cortes, A., Casado, V., Mallol, J., et al. (2009). Interactions between calmodulin, adenosine A2A, and dopamine D2 receptors. *The Journal of Biological Chemistry, 284*, 28058–28068.

Navarro, G., Hradsky, J., Lluis, C., Casado, V., McCormick, P. J., Kreutz, M. R., et al. (2012). NCS-1 associates with adenosine A(2A) receptors and modulates receptor function. *Frontiers in Molecular Neuroscience, 5*, 53.

Nie, Z., Mei, Y., & Ramkumar, V. (1997). Short term desensitization of the A1 adenosine receptors in DDT1MF-2 cells. *Molecular Pharmacology, 52*, 456–464.

Nikbakht, M. R., & Stone, T. W. (2001). Suppression of presynaptic responses to adenosine by activation of NMDA receptors. *European Journal of Pharmacology, 427*, 13–25.

Nishizaki, T., Nagai, K., Nomura, T., Tada, H., Kanno, T., Tozaki, H., et al. (2002). A new neuromodulatory pathway with a glial contribution mediated via A(2a) adenosine receptors. *Glia, 39*, 133–147.

Norenberg, W., Wirkner, K., Assmann, H., Richter, M., & Illes, P. (1998). Adenosine A2A receptors inhibit the conductance of NMDA receptor channels in rat neostriatal neurons. *Amino Acids, 14*, 33–39.

Norenberg, W., Wirkner, K., & Illes, P. (1997). Effect of adenosine and some of its structural analogues on the conductance of NMDA receptor channels in a subset of rat neostriatal neurones. *British Journal of Pharmacology, 122*, 71–80.

Ohata, H., Cao, S., & Koehler, R. C. (2006). Contribution of adenosine A2A and A2B receptors and heme oxygenase to AMPA-induced dilation of pial arterioles in rats. *American Journal of Physiology. Regulatory, Integrative and Comparative Physiology, 291*, R728–R735.

Oishi, Y., Huang, Z. L., Fredholm, B. B., Urade, Y., & Hayaishi, O. (2008). Adenosine in the tuberomammillary nucleus inhibits the histaminergic system via A1 receptors and promotes non-rapid eye movement sleep. *Proceedings of the National Academy of Sciences of the United States of America, 105*, 19992–19997.

Ongini, E., Adami, M., Ferri, C., & Bertorelli, R. (1997). Adenosine A2A receptors and neuroprotection. *Annals of the New York Academy of Sciences, 825*, 30–48.

O'Regan, M. H., Simpson, R. E., Perkins, L. M., & Phillis, J. W. (1992). The selective A2 adenosine receptor agonist CGS 21680 enhances excitatory transmitter amino acid release from the ischemic rat cerebral cortex. *Neuroscience Letters, 138*, 169–172.

Othman, T., Yan, H., & Rivkees, S. A. (2003). Oligodendrocytes express functional A1 adenosine receptors that stimulate cellular migration. *Glia, 44*, 166–172.

Pak, M. A., Haas, H. L., Decking, U. K. M., & Schrader, J. (1994). Inhibition of adenosine kinase increases endogenous adenosine and depresses neuronal activity in hippocampal slices. *Neuropharmacology, 33*, 1049–1053.

Palmer, T. M., Gettys, T. W., & Stiles, G. L. (1995). Differential interaction with and regulation of multiple G-proteins by the rat A3 adenosine receptor. *The Journal of Biological Chemistry, 270*, 16895–16902.

Palmer, T. M., & Stiles, G. L. (2000). Identification of threonine residues controlling the agonist-dependent phosphorylation and desensitization of the rat A(3) adenosine receptor. *Molecular Pharmacology, 57*, 539–545.

Pankevych, H., Korkhov, V., Freissmuth, M., & Nanoff, C. (2003). Truncation of the A1 adenosine receptor reveals distinct roles of the membrane-proximal carboxyl terminus in receptor folding and G protein coupling. *The Journal of Biological Chemistry, 278*, 30283–30293.

Pascual, O., Casper, K. B., Kubera, C., Zhang, J., Revilla-Sanchez, R., Sul, J. Y., et al. (2005). Astrocytic purinergic signaling coordinates synaptic networks. *Science*, *310*, 113–116.

Peakman, M. C., & Hill, S. J. (1994). Adenosine A2B-receptor-mediated cyclic AMP accumulation in primary rat astrocytes. *British Journal of Pharmacology*, *111*, 191–198.

Pedata, F., Corsi, C., Melani, A., Bordoni, F., & Latini, S. (2001). Adenosine extracellular brain concentrations and role of A2A receptors in ischemia. *Annals of the New York Academy of Sciences*, *939*, 74–84.

Pereira, G. S., Rossato, J. I., Sarkis, J. J., Cammarota, M., Bonan, C. D., & Izquierdo, I. (2005). Activation of adenosine receptors in the posterior cingulate cortex impairs memory retrieval in the rat. *Neurobiology of Learning and Memory*, *83*, 217–223.

Phillis, J. W. (1995). The effects of selective A1 and A2a adenosine receptor antagonists on cerebral ischemic injury in the gerbil. *Brain Research*, *705*, 79–84.

Phillis, J. W. (2004). Adenosine and adenine nucleotides as regulators of cerebral blood flow: Roles of acidosis, cell swelling, and KATP channels. *Critical Reviews in Neurobiology*, *16*, 237–270.

Pin, J. P., Neubig, R., Bouvier, M., Devi, L., Filizola, M., Javitch, J. A., et al. (2007). International Union of Basic and Clinical Pharmacology. LXVII. Recommendations for the recognition and nomenclature of G protein-coupled receptor heteromultimers. *Pharmacological Reviews*, *59*, 5–13.

Pinhal-Enfield, G., Ramanathan, M., Hasko, G., Vogel, S. N., Salzman, A. L., Boons, G. J., et al. (2003). An angiogenic switch in macrophages involving synergy between Toll-like receptors 2, 4, 7, and 9 and adenosine A(2A) receptors. *The American Journal of Pathology*, *163*, 711–721.

Popoli, P., Blum, D., Domenici, M. R., Burnouf, S., & Chern, Y. (2008). A critical evaluation of adenosine A2A receptors as potentially "druggable" targets in Huntington's disease. *Current Pharmaceutical Design*, *14*, 1500–1511.

Popoli, P., Blum, D., Martire, A., Ledent, C., Ceruti, S., & Abbracchio, M. P. (2007). Functions, dysfunctions and possible therapeutic relevance of adenosine A(2A) receptors in Huntington's disease. *Progress in Neurobiology*, *81*, 331–348.

Popoli, P., Pintor, A., Domenici, M. R., Frank, C., Tebano, M. T., Pezzola, A., et al. (2002). Blockade of striatal adenosine A2A receptor reduces, through a presynaptic mechanism, quinolinic acid-induced excitotoxicity: Possible relevance to neuroprotective interventions in neurodegenerative diseases of the striatum. *The Journal of Neuroscience*, *22*, 1967–1975.

Prediger, R. D., Batista, L. C., & Takahashi, R. N. (2004). Adenosine A1 receptors modulate the anxiolytic-like effect of ethanol in the elevated plus-maze in mice. *European Journal of Pharmacology*, *499*, 147–154.

Prediger, R. D., Batista, L. C., & Takahashi, R. N. (2005). Caffeine reverses age-related deficits in olfactory discrimination and social recognition memory in rats. Involvement of adenosine A1 and A2A receptors. *Neurobiology of Aging*, *26*, 957–964.

Prediger, R. D., Fernandes, D., & Takahashi, R. N. (2005). Blockade of adenosine A2A receptors reverses short-term social memory impairments in spontaneously hypertensive rats. *Behavioural Brain Research*, *159*, 197–205.

Prediger, R. D., Pamplona, F. A., Fernandes, D., & Takahashi, R. N. (2005). Caffeine improves spatial learning deficits in an animal model of attention deficit hyperactivity disorder (ADHD)—The spontaneously hypertensive rat (SHR). *The International Journal of Neuropsychopharmacology*, *8*, 583–594.

Prediger, R. D., & Takahashi, R. N. (2005). Modulation of short-term social memory in rats by adenosine A1 and A(2A) receptors. *Neuroscience Letters*, *376*, 160–165.

Quiroz, C., Lujan, R., Uchigashima, M., Simoes, A. P., Lerner, T. N., Borycz, J., et al. (2009). Key modulatory role of presynaptic adenosine A2A receptors in cortical neurotransmission to the striatal direct pathway. *ScientificWorldJournal*, *9*, 1321–1344.

Rainnie, D. G., Grunze, H. C., McCarley, R. W., & Greene, R. W. (1994). Adenosine inhibition of mesopontine cholinergic neurons: Implications for EEG arousal. *Science, 263*, 689–692.
Ramkumar, V., Stiles, G. L., Beaven, M. A., & Ali, H. (1993). The A3 adenosine receptor is the unique adenosine receptor which facilitates release of allergic mediators in mast cells. *The Journal of Biological Chemistry, 268*, 16887–16890.
Rebola, N., Canas, P. M., Oliveira, C. R., & Cunha, R. A. (2005). Different synaptic and subsynaptic localization of adenosine A2A receptors in the hippocampus and striatum of the rat. *Neuroscience, 132*, 893–903.
Rebola, N., Pinheiro, P. C., Oliveira, C. R., Malva, J. O., & Cunha, R. A. (2003). Subcellular localization of adenosine A(1) receptors in nerve terminals and synapses of the rat hippocampus. *Brain Research, 987*, 49–58.
Rebola, N., Porciuncula, L. O., Lopes, L. V., Oliveira, C. R., Soares-da-Silva, P., & Cunha, R. A. (2005). Long-term effect of convulsive behavior on the density of adenosine A1 and A2A receptors in the rat cerebral cortex. *Epilepsia, 46*(Suppl. 5), 159–165.
Rebola, N., Rodrigues, R. J., Lopes, L. V., Richardson, P. J., Oliveira, C. R., & Cunha, R. A. (2005). Adenosine A1 and A2A receptors are co-expressed in pyramidal neurons and co-localized in glutamatergic nerve terminals of the rat hippocampus. *Neuroscience, 133*, 79–83.
Rebola, N., Sebastiao, A. M., de Mendonça, A., Oliveira, C. R., Ribeiro, J. A., & Cunha, R. A. (2003). Enhanced adenosine A2A receptor facilitation of synaptic transmission in the hippocampus of aged rats. *Journal of Neurophysiology, 90*, 1295–1303.
Rees, D. A., Lewis, B. M., Lewis, M. D., Francis, K., Scanlon, M. F., & Ham, J. (2003). Adenosine-induced IL-6 expression in pituitary folliculostellate cells is mediated via A2b adenosine receptors coupled to PKC and p38 MAPK. *British Journal of Pharmacology, 140*, 764–772.
Ren, G., Li, T., Lan, J. Q., Wilz, A., Simon, R. P., & Boison, D. (2007). Lentiviral RNAi-induced downregulation of adenosine kinase in human mesenchymal stem cell grafts: A novel perspective for seizure control. *Experimental Neurology, 208*, 26–37.
Ren, H., & Stiles, G. L. (1995). Separate promoters in the human A1 adenosine receptor gene direct the synthesis of distinct messenger RNAs that regulate receptor abundance. *Molecular Pharmacology, 48*, 975–980.
Ren, H., & Stiles, G. L. (1999). Dexamethasone stimulates human A1 adenosine receptor (A1AR) gene expression through multiple regulatory sites in promoter B. *Molecular Pharmacology, 55*, 309–316.
Ribeiro, J. A., Sebastiao, A. M., & de Mendonça, A. (2002). Adenosine receptors in the nervous system: Pathophysiological implications. *Progress in Neurobiology, 68*, 377–392.
Robledo, P., Ursu, G., & Mahy, N. (1999). Effects of adenosine and gamma-aminobutyric acid A receptor antagonists on N-methyl-D-aspartate induced neurotoxicity in the rat hippocampus. *Hippocampus, 9*, 527–533.
Rosin, D. L., Hettinger, B. D., Lee, A., & Linden, J. (2003). Anatomy of adenosine A2A receptors in brain: Morphological substrates for integration of striatal function. *Neurology, 61*, S12–S18.
Rosin, D. L., Robeva, A., Woodard, R. L., Guyenet, P. G., & Linden, J. (1998). Immunohistochemical localization of adenosine A2A receptors in the rat central nervous system. *The Journal of Comparative Neurology, 401*, 163–186.
Rudich, N., Ravid, K., & Sagi-Eisenberg, R. (2012). Mast cell adenosine receptors function: A focus on the A3 adenosine receptor and inflammation. *Frontiers in Immunology, 3*, 134.
Salvatore, C. A., Jacobson, M. A., Taylor, H. E., Linden, J., & Johnson, R. G. (1993). Molecular cloning and characterization of the human A3 adenosine receptor. *Proceedings of the National Academy of Sciences of the United States of America, 90*, 10365–10369.

Satoh, S., Matsumura, H., & Hayaishi, O. (1998). Involvement of adenosine A2A receptor in sleep promotion. *European Journal of Pharmacology*, *351*, 155–162.
Satoh, S., Matsumura, H., Koike, N., Tokunaga, Y., Maeda, T., & Hayaishi, O. (1999). Region-dependent difference in the sleep-promoting potency of an adenosine A2A receptor agonist. *The European Journal of Neuroscience*, *11*, 1587–1597.
Saura, J., Angulo, E., Ejarque, A., Casado, V., Tusell, J. M., Moratalla, R., et al. (2005). Adenosine A2A receptor stimulation potentiates nitric oxide release by activated microglia. *Journal of Neurochemistry*, *95*, 919–929.
Scammell, T. E., Gerashchenko, D. Y., Mochizuki, T., McCarthy, M. T., Estabrooke, I. V., Sears, C. A., et al. (2001). An adenosine A2a agonist increases sleep and induces Fos in ventrolateral preoptic neurons. *Neuroscience*, *107*, 653–663.
Schiffmann, S. N., Fisone, G., Moresco, R., Cunha, R. A., & Ferre, S. (2007). Adenosine A2A receptors and basal ganglia physiology. *Progress in Neurobiology*, *83*, 277–292.
Schiffmann, S. N., Jacobs, O., & Vanderhaeghen, J. J. (1991). Striatal restricted adenosine A2 receptor (RDC8) is expressed by enkephalin but not by substance P neurons: An in situ hybridization histochemistry study. *Journal of Neurochemistry*, *57*, 1062–1067.
Schiffmann, S. N., Libert, F., Vassart, G., & Vanderhaeghen, J. J. (1991). Distribution of adenosine A2 receptor mRNA in the human brain. *Neuroscience Letters*, *130*, 177–181.
Schulte, G., & Fredholm, B. B. (2002). Signaling pathway from the human adenosine A(3) receptor expressed in Chinese hamster ovary cells to the extracellular signal-regulated kinase 1/2. *Molecular Pharmacology*, *62*, 1137–1146.
Schulte, G., & Fredholm, B. B. (2003a). Signalling from adenosine receptors to mitogen-activated protein kinases. *Cellular Signalling*, *15*, 813–827.
Schulte, G., & Fredholm, B. B. (2003b). The G(s)-coupled adenosine A(2B) receptor recruits divergent pathways to regulate ERK1/2 and p38. *Experimental Cell Research*, *290*, 168–176.
Schwarzschild, M. A., Agnati, L., Fuxe, K., Chen, J. F., & Morelli, M. (2006). Targeting adenosine A2A receptors in Parkinson's disease. *Trends in Neurosciences*, *29*, 647–654.
Schwarzschild, M. A., Chen, J. F., & Ascherio, A. (2002). Caffeinated clues and the promise of adenosine A(2A) antagonists in PD. *Neurology*, *58*, 1154–1160.
Schwarzschild, M. A., Xu, K., Oztas, E., Petzer, J. P., Castagnoli, K., Castagnoli, N., Jr., et al. (2003). Neuroprotection by caffeine and more specific A2A receptor antagonists in animal models of Parkinson's disease. *Neurology*, *61*, S55–S61.
Sebastiao, A. M., & Ribeiro, J. A. (1996). Adenosine A2 receptor-mediated excitatory actions on the nervous system. *Progress in Neurobiology*, *48*, 167–189.
Sebastiao, A. M., & Ribeiro, J. A. (2009a). Adenosine receptors and the central nervous system. *Handbook of Experimental Pharmacology*, *193*, 471–534.
Sebastiao, A. M., & Ribeiro, J. A. (2009b). Triggering neurotrophic factor actions through adenosine A2A receptor activation: Implications for neuroprotection. *British Journal of Pharmacology*, *158*, 15–22.
Shen, H. Y., Coelho, J. E., Ohtsuka, N., Canas, P. M., Day, Y. J., Huang, Q. Y., et al. (2008). A critical role of the adenosine A2A receptor in extrastriatal neurons in modulating psychomotor activity as revealed by opposite phenotypes of striatum and forebrain A2A receptor knock-outs. *The Journal of Neuroscience*, *28*, 2970–2975.
Shen, W., Flajolet, M., Greengard, P., & Surmeier, D. J. (2008). Dichotomous dopaminergic control of striatal synaptic plasticity. *Science*, *321*, 848–851.
Sitaraman, S. V., Wang, L., Wong, M., Bruewer, M., Hobert, M., Yun, C. H., et al. (2002). The adenosine 2b receptor is recruited to the plasma membrane and associates with E3KARP and Ezrin upon agonist stimulation. *The Journal of Biological Chemistry*, *277*, 33188–33195.

Soriano, A., Ventura, R., Molero, A., Hoen, R., Casado, V., Cortes, A., et al. (2009). Adenosine A2A receptor-antagonist/dopamine D2 receptor-agonist bivalent ligands as pharmacological tools to detect A2A–D2 receptor heteromers. *Journal of Medicinal Chemistry, 52*, 5590–5602.

Studer, F. E., Fedele, D. E., Marowsky, A., Schwerdel, C., Wernli, K., Vogt, K., et al. (2006). Shift of adenosine kinase expression from neurons to astrocytes during postnatal development suggests dual functionality of the enzyme. *Neuroscience, 142*, 125–137.

Sun, C. N., Cheng, H. C., Chou, J. L., Lee, S. Y., Lin, Y. W., Lai, H. L., et al. (2006). Rescue of p53 blockage by the A2A adenosine receptor via a novel interacting protein, Translin-associated protein X. *Molecular Pharmacology, 70*, 454–466.

Sun, C. N., Chuang, H. C., Wang, J. Y., Chen, S. Y., Cheng, Y. Y., Lee, C. F., et al. (2010). The A2A adenosine receptor rescues neuritogenesis impaired by p53 blockage via KIF2A, a kinesin family member. *Developmental Neurobiology, 70*, 604–621.

Sun, Y., Duan, Y., Eisenstein, A. S., Hu, W., Quintana, A., Lam, W. K., et al. (2012). A novel mechanism of control of NFkappaB activation and inflammation involving A2B adenosine receptors. *Journal of Cell Science, 125*, 4507–4517.

Svenningsson, P., Hall, H., Sedvall, G., & Fredholm, B. B. (1997). Distribution of adenosine receptors in the postmortem human brain: An extended autoradiographic study. *Synapse, 27*, 322–335.

Svenningsson, P., Le Moine, C., Kull, B., Sunahara, R., Bloch, B., & Fredholm, B. B. (1997). Cellular expression of adenosine A2A receptor messenger RNA in the rat central nervous system with special reference to dopamine innervated areas. *Neuroscience, 80*, 1171–1185.

Tanaka, K. (2001). Alteration of second messengers during acute cerebral ischemia—Adenylate cyclase, cyclic AMP-dependent protein kinase, and cyclic AMP response element binding protein. *Progress in Neurobiology, 65*, 173–207.

Tebano, M. T., Martire, A., Potenza, R. L., Gro, C., Pepponi, R., Armida, M., et al. (2008). Adenosine A(2A) receptors are required for normal BDNF levels and BDNF-induced potentiation of synaptic transmission in the mouse hippocampus. *Journal of Neurochemistry, 104*, 279–286.

Thakkar, M. M., Winston, S., & McCarley, R. W. (2003). A1 receptor and adenosinergic homeostatic regulation of sleep–wakefulness: Effects of antisense to the A1 receptor in the cholinergic basal forebrain. *The Journal of Neuroscience, 23*, 4278–4287.

Thurner, P., Gsandtner, I., Kudlacek, O., Choquet, D., Nanoff, C., Freissmuth, M., et al. (2014). A two-state model for the diffusion of the A2A adenosine receptor in hippocampal neurons: Agonist-induced switch to slow mobility is modified by synapse-associated protein 102 (SAP102). *The Journal of Biological Chemistry, 289*, 9263–9274.

Ticho, S. R., & Radulovacki, M. (1991). Role of adenosine in sleep and temperature regulation in the preoptic area of rats. *Pharmacology, Biochemistry, and Behavior, 40*, 33–40.

Torvinen, M., Kozell, L. B., Neve, K. A., Agnati, L. F., & Fuxe, K. (2004). Biochemical identification of the dopamine D2 receptor domains interacting with the adenosine A2A receptor. *Journal of Molecular Neuroscience, 24*, 173–180.

Torvinen, M., Marcellino, D., Canals, M., Agnati, L. F., Lluis, C., Franco, R., et al. (2005). Adenosine A2A receptor and dopamine D3 receptor interactions: Evidence of functional A2A/D3 heteromeric complexes. *Molecular Pharmacology, 67*, 400–407.

Tozzi, A., Tscherter, A., Belcastro, V., Tantucci, M., Costa, C., Picconi, B., et al. (2007). Interaction of A2A adenosine and D2 dopamine receptors modulates corticostriatal glutamatergic transmission. *Neuropharmacology, 53*, 783–789.

Trincavelli, M. L., Costa, B., Tuscano, D., Lucacchini, A., & Martini, C. (2002). Up-regulation of A(2A) adenosine receptors by proinflammatory cytokines in rat PC12 cells. *Biochemical Pharmacology, 64*, 625–631.

Trincavelli, M. L., Tuscano, D., Marroni, M., Falleni, A., Gremigni, V., Ceruti, S., et al. (2002). A3 adenosine receptors in human astrocytoma cells: Agonist-mediated desensitization, internalization, and down-regulation. *Molecular Pharmacology*, *62*, 1373–1384.
Urade, Y., Eguchi, N., Qu, W. M., Sakata, M., Huang, Z. L., Chen, J. F., et al. (2003). Sleep regulation in adenosine A2A receptor-deficient mice. *Neurology*, *61*, S94–S96.
van Calker, D., Muller, M., & Hamprecht, B. (1979). Adenosine regulates via two different types of receptors, the accumulation of cyclic AMP in cultured brain cells. *Journal of Neurochemistry*, *33*, 999–1005.
Ventura, R., & Harris, K. M. (1999). Three-dimensional relationships between hippocampal synapses and astrocytes. *The Journal of Neuroscience*, *19*, 6897–6906.
Vidi, P. A., Chemel, B. R., Hu, C. D., & Watts, V. J. (2008). Ligand-dependent oligomerization of dopamine D(2) and adenosine A(2A) receptors in living neuronal cells. *Molecular Pharmacology*, *74*, 544–551.
Virus, R. M., Ticho, S., Pilditch, M., & Radulovacki, M. (1990). A comparison of the effects of caffeine, 8-cyclopentyltheophylline, and alloxazine on sleep in rats. Possible roles of central nervous system adenosine receptors. *Neuropsychopharmacology*, *3*, 243–249.
Von Lubitz, D. K., Lin, R. C., & Jacobson, K. A. (1995). Cerebral ischemia in gerbils: Effects of acute and chronic treatment with adenosine A2A receptor agonist and antagonist. *European Journal of Pharmacology*, *287*, 295–302.
Waltereit, R., & Weller, M. (2003). Signaling from cAMP/PKA to MAPK and synaptic plasticity. *Molecular Neurobiology*, *27*, 99–106.
Weaver, D. R. (1993). A2a adenosine receptor gene expression in developing rat brain. *Brain Research. Molecular Brain Research*, *20*, 313–327.
Wei, C. J., Augusto, E., Gomes, C. A., Singer, P., Wang, Y., Boison, D., et al. (2014). Regulation of fear responses by striatal and extrastriatal adenosine A2A receptors in forebrain. *Biological Psychiatry*, *75*, 855–863.
Wei, C. J., Li, W., & Chen, J. F. (2011). Normal and abnormal functions of adenosine receptors in the central nervous system revealed by genetic knockout studies. *Biochimica et Biophysica Acta*, *1808*, 1358–1379.
Wei, C. J., Singer, P., Coelho, J., Boison, D., Feldon, J., Yee, B. K., et al. (2011). Selective inactivation of adenosine A2A receptors in striatal neurons enhances working memory and reversal learning. *Learning & Memory*, *18*, 459–474.
Wiese, S., Jablonka, S., Holtmann, B., Orel, N., Rajagopal, R., Chao, M. V., et al. (2007). Adenosine receptor A2A-R contributes to motoneuron survival by transactivating the tyrosine kinase receptor TrkB. *Proceedings of the National Academy of Sciences of the United States of America*, *104*, 17210–17215.
Wirkner, K., Assmann, H., Koles, L., Gerevich, Z., Franke, H., Norenberg, W., et al. (2000). Inhibition by adenosine A(2A) receptors of NMDA but not AMPA currents in rat neostriatal neurons. *British Journal of Pharmacology*, *130*, 259–269.
Woods, A. S., Marcellino, D., Jackson, S. N., Franco, R., Ferre, S., Agnati, L. F., et al. (2008). How calmodulin interacts with the adenosine A(2A) and the dopamine D(2) receptors. *Journal of Proteome Research*, *7*, 3428–3434.
Wu, Y. C., Lai, H. L., Chang, W. C., Lin, J. T., Liu, Y. J., & Chern, Y. (2013). A novel Galphas-binding protein, Gas-2 like 2, facilitates the signaling of the A2A adenosine receptor. *Biochimica et Biophysica Acta*, *1833*, 3145–3154.
Xaus, J., Valledor, A. F., Cardo, M., Marques, L., Beleta, J., Palacios, J. M., et al. (1999). Adenosine inhibits macrophage colony-stimulating factor-dependent proliferation of macrophages through the induction of p27kip-1 expression. *Journal of Immunology*, *163*, 4140–4149.
Xiao, D., Bastia, E., Xu, Y. H., Benn, C. L., Cha, J. H., Peterson, T. S., et al. (2006). Forebrain adenosine A2A receptors contribute to L-3,4-dihydroxyphenylalanine-induced dyskinesia in hemiparkinsonian mice. *The Journal of Neuroscience*, *26*, 13548–13555.

Xu, K., Bastia, E., & Schwarzschild, M. (2005). Therapeutic potential of adenosine A(2A) receptor antagonists in Parkinson's disease. *Pharmacology & Therapeutics*, *105*, 267–310.
Xu, H. L., Mao, L., Ye, S., Paisansathan, C., Vetri, F., & Pelligrino, D. A. (2008). Astrocytes are a key conduit for upstream signaling of vasodilation during cerebral cortical neuronal activation in vivo. *American Journal of Physiology. Heart and Circulatory Physiology*, *294*, H622–H632.
Xu, H. L., & Pelligrino, D. A. (2007). ATP release and hydrolysis contribute to rat pial arteriolar dilatation elicited by neuronal activation. *Experimental Physiology*, *92*, 647–651.
Xu, F., Wu, H., Katritch, V., Han, G. W., Jacobson, K. A., Gao, Z. G., et al. (2011). Structure of an agonist-bound human A2A adenosine receptor. *Science*, *332*, 322–327.
Yu, L., Frith, M. C., Suzuki, Y., Peterfreund, R. A., Gearan, T., Sugano, S., et al. (2004). Characterization of genomic organization of the adenosine A2A receptor gene by molecular and bioinformatics analyses. *Brain Research*, *1000*, 156–173.
Yu, C., Gupta, J., Chen, J. F., & Yin, H. H. (2009). Genetic deletion of A2A adenosine receptors in the striatum selectively impairs habit formation. *The Journal of Neuroscience*, *29*, 15100–15103.
Yu, L., Huang, Z., Mariani, J., Wang, Y., Moskowitz, M., & Chen, J. F. (2004). Selective inactivation or reconstitution of adenosine A2A receptors in bone marrow cells reveals their significant contribution to the development of ischemic brain injury. *Nature Medicine*, *10*, 1081–1087.
Yu, L., Shen, H. Y., Coelho, J. E., Araujo, I. M., Huang, Q. Y., Day, Y. J., et al. (2008). Adenosine A2A receptor antagonists exert motor and neuroprotective effects by distinct cellular mechanisms. *Annals of Neurology*, *63*, 338–346.
Zezula, J., & Freissmuth, M. (2008). The A(2A)-adenosine receptor: A GPCR with unique features? *British Journal of Pharmacology*, *153*(Suppl. 1), S184–S190.
Zhang, Z., Chen, G., Zhou, W., Song, A., Xu, T., Luo, Q., et al. (2007). Regulated ATP release from astrocytes through lysosome exocytosis. *Nature Cell Biology*, *9*, 945–953.
Zhou, Q. Y., Li, C., Olah, M. E., Johnson, R. A., Stiles, G. L., & Civelli, O. (1992). Molecular cloning and characterization of an adenosine receptor: The A3 adenosine receptor. *Proceedings of the National Academy of Sciences of the United States of America*, *89*, 7432–7436.
Zhou, S. J., Zhu, M. E., Shu, D., Du, X. P., Song, X. H., Wang, X. T., et al. (2009). Preferential enhancement of working memory in mice lacking adenosine A(2A) receptors. *Brain Research*, *1303*, 74–83.

FURTHER READING

Cunha, R. A. (2008). Caffeine, adenosine receptors, memory and Alzheimer disease. *Medicina Clínica (Barcelona)*, *131*, 790–795.
Cunha, R. A., Ferre, S., Vaugeois, J. M., & Chen, J. F. (2008). Potential therapeutic interest of adenosine A2A receptors in psychiatric disorders. *Current Pharmaceutical Design*, *14*, 1512–1524.
Ferre, S., Quiroz, C., Woods, A. S., Cunha, R., Popoli, P., Ciruela, F., et al. (2008). An update on adenosine A2A–dopamine D2 receptor interactions: Implications for the function of G protein-coupled receptors. *Current Pharmaceutical Design*, *14*, 1468–1474.

CHAPTER TWO

Adenosine Receptor PET Imaging in Human Brain

Masahiro Mishina[*,†,1], **Kiich Ishiwata**[†]
[*]Department of Neurological Science, Graduate School of Medicine, Nippon Medical School, Tokyo, Japan
[†]Research Team for Neuroimaging, Tokyo Metropolitan Institute of Gerontology, Tokyo, Japan
[1]Corresponding author: e-mail address: mishina@nms.ac.jp

Contents

Abstract

Positron emission tomography (PET) is a nuclear medicine imaging technique that allows *in vivo* imaging of regional receptor-binding capacity. Advances in radiotracer chemistry have led to the development of novel imaging probes for adenosine receptors, especially adenosine A_1 and A_{2A} receptors. In this chapter, we discuss brain PET imaging for adenosine receptors and comparison of radioligands for PET imaging in health and diseases.

1. INTRODUCTION

Positron emission tomography (PET) is a nuclear medicine imaging technique that allows imaging and quantifying of cellular and molecular processes in humans. A small amount of radiopharmaceutical is introduced into a subject usually by intravenous injection. During its decay process, the radioisotope emits a positron (Fig. 2.1). After traveling a short distance,

International Review of Neurobiology, Volume 119
ISSN 0074-7742
http://dx.doi.org/10.1016/B978-0-12-801022-8.00002-7

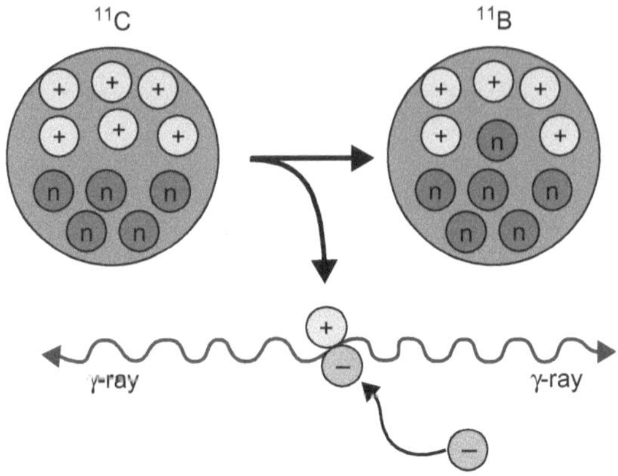

Figure 2.1 Decay process of ^{11}C. During its decay process, ^{11}C emits a positron, and changes to ^{11}B. The positron travels a short distance and annihilates on contact with an electron. The annihilation produces two γ-rays traveling in opposite directions.

the positron annihilates on contact with an electron from the surrounding environment. The annihilation produces two γ-rays traveling in opposite directions. The PET scanner detects the pairs of γ-rays emitted indirectly by positron-emitting radioisotopes. After an appropriate uptake period, a PET scanner measures the concentration of tracer in tissue. Thus, PET allows *in vivo* imaging of regional cerebral functions, including cerebral blood flow, molecular metabolism, and receptor-binding capacity.

Since 1995, a number of radioligands for mapping adenosine receptors by PET have been reported. Most of them are focused on the ligands for adenosine A_1 and A_{2A} receptors. Radiolabeled adenosine derivatives such as N^6-cyclohexyladenosine (Fastbom, Pazos, Probst, & Palacios, 1987) for adenosine A_1 receptor and 2-*p*-(2-carboxyethyl)-phenethylamino-5′-*N*-ethylcarboxamidoadenosine (CGS 21680) (Martinez-Mir, Probst, & Palacios, 1991) for adenosine A_{2A} receptor are used *in vitro*; however, these hydrophilic compounds are not appropriate for *in vivo* imaging adenosine receptors in the central nervous system. Lipophilic xanthine and non-xanthine derivatives are candidates for *in vivo* imaging (Table 2.1). A comprehensive overview of PET tracers for the different adenosine receptor subtypes has been presented in some recent reviews (Bauer & Ishiwata, 2009; Ishiwata, Kimura, de Vries, & Elsinga, 2007; Khanapur et al., 2013; Paul, Elsinga, Ishiwata, Dierckx, & van Waarde, 2011). The binding properties of PET ligands are used clinically and related ligands are summarized in Table 2.1.

Table 2.1 *In vitro* affinity of PET ligands for adenosine receptors

	Affinity (K_i nM)		Selectivity	Affinity (K_i nM)	Selectivity		References
	A_1	A_{2A}	A_{2A}/A_1 or A_1/A_{2A}	A_3	A_{2B}/A_{2A}	A_3/A_{2A}	
A_1 receptor ligand							
DPCPX	6.4 (***g***, ***1***)	590 (***g***, ***4***)	92				Shimada et al. (1991)
	0.17 (***b***, K_d)						Holschbach et al. (1998)
	2.58 (***h***, K_d)						Maemoto et al. (2004)
[^{11}C]KF15372	3.0 (***g***, ***1***)	430 (***g***, ***4***)	140				Shimada et al. (1991)
[^{11}C]MPDX	4.2 (***r***, ***1***)	>100 (***r***, ***5***)	>24				Noguchi et al. (1997)
[^{18}F]CPFPX	0.18 (***b***, ***2***)	812 (***r***, ***5***)	4500				Holschbach et al. (1998)
	0.63 (***r***, K_d)						Holschbach et al. (2002)
	1.26 (***h***, K_d)	940 (***h***, K_d)	>700				Holschbach et al. (2002)
[^{11}C]FR194921	4.96 (***r***, ***2***)	>100 (***r***, ***5***)					Maemoto et al. (2004)
	2.91 (***h***, ***2***)	>100 (***h***, ***5***)	>34	>100 (***h***, ***4***)			Maemoto et al. (2004)

Continued

Table 2.1 *In vitro* affinity of PET ligands for adenosine receptors—cont'd

	Affinity (K_i nM)		Selectivity	Affinity (K_i nM)	Selectivity		References
	A_1	A_{2A}	A_{2A}/A_1 or A_1/A_{2A}	A_3	A_{2B}/A_{2A}	A_3/A_{2A}	
A_{2A} receptor ligand							
[^{11}C]KF17837	62 (***r***, ***1***)	1.0 (***r***, ***5***)	62				Nonaka et al. (1994)
[^{11}C]TMSX ([^{11}C] KF18446)	1600 (***r***, ***1***)	5.9 (***r***, ***5***)	270				Ishiwata, Noguchi, et al. (2000)
[^{11}C]KF21213	>10,000 (***r***, ***1***)	3.0 (***r***, ***5***)	>3300				Wang et al. (2000)
[^{11}C]KW-6002	150 (***r***, ***1***)	2.2 (***r***, ***5***)	68				Hirani et al. (2001)
SCH 58261	121 (***r***, ***1***)	2.3 (***r***, ***5***)	53	>1000 (***r***, ***8***)			Zocchi, Ongini, Conti, et al. (1996)
[^{11}C]SCH 442416	1800 (***r***, ***2***)	0.50 (***r***, ***6***)	3600	>10,000 (***r***, ***8***)		>20,000	Todde et al. (2000)
	1111 (***h***, ***2***)	0.048 (***h***, ***6***)	23,000	>10,000 (***h***, ***8***)	>200,000	>200,000	Todde et al. (2000)
A_3 receptor ligand							
[^{18}F]FE@SUPPY[a]	11,500 (***r***, ***3***)	7330 (***r***, ***7***)		600 (***r***, ***8***)	19	12	Li et al. (1999)
				4.22 (***h***, ***8***)	2700		Li et al. (1999)

[a]Recently developed [^{18}F] 5-(2-fluoroethyl) 2,4-diethyl-3-(ethylsulfanylcarbonyl)-6-phenylpyridine-5-carboxylate (FE@SUPPY) may be a candidate radioligand for A_3 subtype (Wadsak et al., 2008) but has not been applied to humans and nonhuman primates.
Radioligands used for binding assay: ***1***, [^{3}H] N^6-cyclohexyladenosine; ***2***, [^{3}H]DPCPX; ***3***, [^{3}H] *R*-N^6-(phenylisopropyl)-adenosine (*R*-PIA); ***4***, [^{3}H]NECA; ***5***, [^{3}H]CGS 21680; ***6***, [^{3}H]SCH 58261; ***8***, [^{125}I] N^6-(4-amino-3-iodobenzyl)-5′-*N*-methylcarbamoyladenosine.
Receptor membrane source used for binding assay: ***b***, bovine; ***g***, guinea pig; ***h***, human; ***r***, rat.

PET studies of adenosine receptors in human brain are limited to adenosine A_1 and A_{2A} receptors, but not adenosine A_{2B} and A_3 receptors. In this chapter, therefore, we discuss brain PET imaging for adenosine A_1 and A_{2A} receptors and comparison of radioligands for PET imaging in health and diseases.

2. PET IMAGING OF ADENOSINE A_1 RECEPTORS

2.1. PET ligands for adenosine A_1 receptors

Two xanthine derivatives with high affinity and high selectivity for adenosine A_1 receptor, 8-cyclopentyl-1,3-dipropylxanthine (DPCPX) (Lohse et al., 1987) and its analog 8-dicyclopropylmethyl-1,3-dipropylxanthine (KF15372) (Shimada et al., 1991), were selected as lead compounds for *in vivo* imaging. First, KF15372 was labeled with positron-emitter ^{11}C in the propyl group (Ishiwata et al., 1995). Then, its ethyl and methyl derivatives, ^{11}C-8-dicyclopropylmethyl-1-ethyl-3-propylxanthine (^{11}C-EPDX) and ^{11}C-8-dicyclopropylmethyl-1-methyl-3-propylxanthine (^{11}C-MPDX; Fig. 2.2A), were developed and the latter was applied to humans (Fukumitsu et al., 2003). On the other hand, the replacement of the propyl group of DPCPX with fluoroalkyl groups was investigated (Holschbach et al., 2002), and ^{18}F-labeled fluoropropyl analog ^{18}F-8-cyclopentyl-3-(3-fluoropropyl)-1-propylxanthine (^{18}F-CPFPX; Fig. 2.2B) was applied to humans (Bauer et al., 2003). ^{18}F-CPFPX has a higher affinity for A_1R than ^{11}C-MPDX, but in humans, ^{11}C-MPDX was much more stable than ^{18}F-CPFPX (Bauer et al., 2003; Fukumitsu et al., 2005).

As a nonxanthine ligand, ^{11}C-2-(1-methyl-4-piperidinyl)-6-(2-phenylpyrazolo[1,5-*a*]pyridin-3-yl)-3(*2H*)-pyridazinone (^{11}C-FR194921) was evaluated in rats and nonhuman primates (Matsuya et al., 2005), but further studies have not been reported.

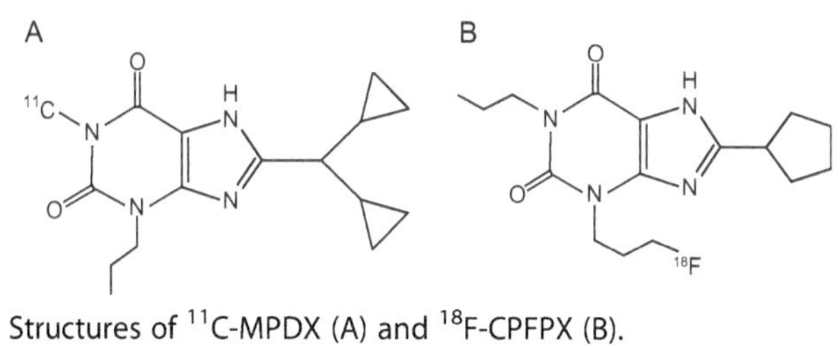

Figure 2.2 Structures of ^{11}C-MPDX (A) and ^{18}F-CPFPX (B).

In the search for single photon emission computed tomography (SPECT) ligands, radioiodinated DPCPX was evaluated but found to lack promising properties (Sihver et al., 2003).

2.2. Adenosine A_1 receptors in normal subjects

We successfully visualized adenosine A_1 receptors in living humans with ^{11}C-MPDX PET (Fukumitsu et al., 2003; Ishiwata et al., 2002). Kimura et al. (2004) confirmed that the nondisplaceable-binding potential (BP_{ND}) was suitable to evaluate the density of adenosine A_1 receptors for kinetic analysis for ^{11}C-MPDX in human brain. Figure 2.3 demonstrates that the density of adenosine A_1 receptors was large in the striatum and thalamus, moderate in the cerebral cortices and pons, and small in the cerebellum in the ^{11}C-MPDX PET image (Fukumitsu et al., 2005). Bauer et al. (2003) also successfully performed imaging of adenosine A_1 receptors in the human brain using ^{18}F-CPFPX PET. They also confirmed the specific binding of ^{18}F-CPFPX to adenosine A_1 receptors using unlabeled CPFPX (Meyer et al., 2006).

A study using autoradiography reported that the binding ability of adenosine A_1 receptor in the striatum was reduced in aged rats (Meerlo et al., 2004). Human PET studies reported that the binding ability of ^{18}F-CPFPX and ^{11}C-MPDX was negatively correlated with age, an effect that has previously been demonstrated with regard to dopamine D_1 and D_2 receptors (Ishibashi et al., 2009; Meyer et al., 2007; Mishina et al., 2012). As described

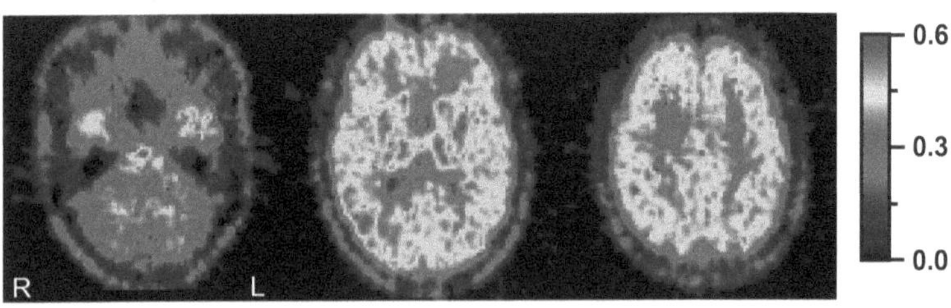

Figure 2.3 Brain images for 51-year-old man of distribution volume ratio (*DVR*) of ^{11}C-MPDX. Adenosine A_1 receptors are enriched in the striatum and thalamus as well as adenosine A_{2A} receptors. Unlike the adenosine A_{2A} receptors, however, the adenosine A_1 receptors are also widely distributed in the cerebral cortex. The pixel values for the [^{11}C]MPDX PET image are visualized as the *DVR*, because the brain anatomy is unclear in the nondisplaceable-binding potential ($BP_{ND} = DVR - 1.0$) images of [^{11}C]MPDX. Note that we use the values for the BP_{ND} in the kinetic analysis for [^{11}C]MPDX PET (Kimura et al., 2004).

below, we reported that age-related changes were different between adenosine A_1 receptors and adenosine A_{2A} receptors (Mishina et al., 2012).

A ^{18}F-CPFPX PET study showed that the density of adenosine A_1 receptors in the orbitofrontal cortex was increased in subjects who were deprived of sleep for 24 h compared with that in controls with regular sleep, and suggested that changes of expression of adenosine A_1 receptor contribute to homeostatic sleep regulation (Elmenhorst et al., 2007).

2.3. Adenosine A_1 receptors in brain diseases

Alzheimer's disease is the most frequent form of dementia. The pathological features are senile plaques composed of amyloid-β peptide fibrils, neurofibrillary tangles of hyperphosphorylated tau, and neurotransmitter deficits (Ferri et al., 2005). Several studies have been reported for adenosine receptors in Alzheimer's disease (Rahman, 2009). Postmortem studies in patients with Alzheimer's disease reported a reduced density of adenosine A_1 receptors in the hippocampus (Jaarsma, Sebens, & Korf, 1991; Jansen, Faull, Dragunow, & Synek, 1990; Kalaria, Sromek, Wilcox, & Unnerstall, 1990; Ulas, Brunner, Nguyen, & Cotman, 1993), although the reduction was also observed in vascular dementia (Deckert et al., 1998). Another postmortem study reported that the density of adenosine A_1 receptors in the striatum was also decreased in patients with Alzheimer's disease (Ikeda, Mackay, Dewar, & McCulloch, 1993). A PET study for Alzheimer's disease (Fukumitsu et al., 2008) showed that low density of adenosine A_1 receptors by ^{11}C-MPDX PET was observed in the temporal cortex and thalamus, while hypometabolism of glucose by 2-^{18}F-fluoro-2-deoxy-D-glucose (^{18}F-FDG) PET was observed in the parietotemporal cortex and posterior cingulate gyrus. The cerebral glucose metabolism is thought to reflect regional neuronal activities such as synaptic function (Magistretti & Pellerin, 1996; Pellerin & Magistretti, 1994; Tsacopoulos & Magistretti, 1996). Alteration of the density of adenosine A_1 receptors may be different with that of other neurotransmitter system in Alzheimer's disease.

Parkinson's disease is a progressive degenerative neurological disorder characterized clinically by resting tremor, bradykinesia, cogwheel rigidity, and postural instability (Lees, Hardy, & Revesz, 2009). These symptoms result primarily from the loss of dopaminergic neurons in the substantia nigra and can be reduced by levodopa and dopamine agonists. Adenosine A_1 receptors interact negatively with dopamine D_1 receptors in direct pathway neurons (Ferre et al., 1994; Yabuuchi et al., 2006) and are also presynaptic receptors that regulate the dopaminergic system (Yabuuchi et al., 2006). However, the roles of adenosine A_1 receptors remain unclear in patients with Parkinson's disease unlike adenosine A_{2A} receptors (Kelsey, Langelier, Oriel, & Reedy, 2009). In our ^{11}C-MPDX PET study, adenosine A_1 receptors seem monotonous in the putamen of Parkinson's disease compared with adenosine A_{2A} receptors (in preparation). In young patients with Parkinson's disease, chronic dopamine replacement therapy often leads to involuntary movements known as dyskinesia, which is one of the most inconvenient side effects. A recent study suggested that the dyskinesia might involve not only in adenosine A_{2A} receptors but also in adenosine A_1 receptors (Xiao et al., 2011).

The adenosine and its receptors have attracted attention as potential therapeutic targets for stroke (Williams-Karnesky & Stenzel-Poore, 2009). Although human PET studies were not available for adenosine A_1 receptors in patients with stroke, Nariai et al. (2003) has found that decreased ^{11}C-MPDX binding to adenosine A_1 receptors after reperfusion was a sensitive predictor of severe ischemic damage in an animal study.

A postmortem study found that the adenosine A_1 receptors were reduced in the epileptic temporal cortex in patients with temporal lobe epilepsy (Glass et al., 1996), although upregulation of adenosine A_1 receptor was found in the specimens of epileptogenic neocortex by surgical resection (Angelatou et al., 1993). In patients with temporal lobe epilepsy, binding of ^{11}C-MPDX was significantly decreased in the mesial temporal lobe of the focus side outside the hippocampus, whereas it was significantly increased in the frontal cortex of the focus side (in preparation). The regional abnormality observed by ^{11}C-MPDX PET was different from that observed by already established PET methods measuring central benzodiazepine receptor density by ^{11}C-flumazenil PET and glucose metabolism by ^{18}F-FDG PET.

In the improving patients with hemianopia caused by brain damage, a ^{11}C-MPDX PET study reported a compensatory increase in density of adenosine A_1 receptors in the injured portion of the primary visual cortex, while cerebral glucose metabolism and benzodiazepine receptor density were low in the primary visual cortex and visual association cortex (Suzuki et al., 2012).

An ^{18}F-CPFPX PET study reported that the density of adenosine A_1 receptors was decreased in the cerebral cortex of patients with liver cirrhosis and hepatic encephalopathy (Boy et al., 2008).

3. PET IMAGING OF ADENOSINE A_{2A} RECEPTORS

3.1. Ligands for adenosine A_{2A} receptors

Several ligands with high affinity and high selectivity for adenosine A_{2A} receptor are lead compounds for *in vivo* imaging radioligands: (*E*)-8-(3,4-dimethoxystyryl)-1,3-dipropyl-7-methylxanthine (KF17837) (Shimada, Suzuki, Nonaka, Ishii, & Ichikawa, 1992), 7-(2-phenylethyl)-5-amino-2-(2-furyl)-pyrazolo[4,3-*e*]-1,2,4- triazolo[1,5-*c*]pyrimidine (SCH 58261) (Zocchi, Ongini, Ferrara, Baraldi, & Dionisotti, 1996), and 4-(2-[7-amino-2-{2-furyl}1{1,2,4}triazolo{2,3-*a*}{1,3,5}triazin-5-yl- amino] ethyl)phenol (ZM241385) (Poucher et al., 1995). First, KF17837 with a xanthine structure was labeled with ^{11}C (Ishiwata et al., 1996; Stone-Elander, Thorell, Eriksson, Fredholm, & Ingvar, 1997). Thereafter, Ishiwata et al. (2005) prepared and evaluated several analogs of ^{11}C-KF17837, and ^{11}C-(*E*)-8-(3,4,5-trimethoxystyryl)-1,3,7-trimethylxanthine (^{11}C-TMSX, formally designated as ^{11}C-KF18446; Fig. 2.4A) was chosen for clinical application. A similar xanthine derivative, KW-6002 designated as istradefylline, is a nondopaminergic therapeutic agent for Parkinson's disease (Factor et al., 2010; Mizuno, Hasegawa, Kondo, Kuno, & Yamamoto, 2010). Istradefylline was approved for use of Parkinson's disease by the Ministry of Health, Labour, and Welfare, Japan, and has been available in Japan since March 2013. ^{11}C-KW-6002 PET (Fig. 2.4B) was applied for measuring

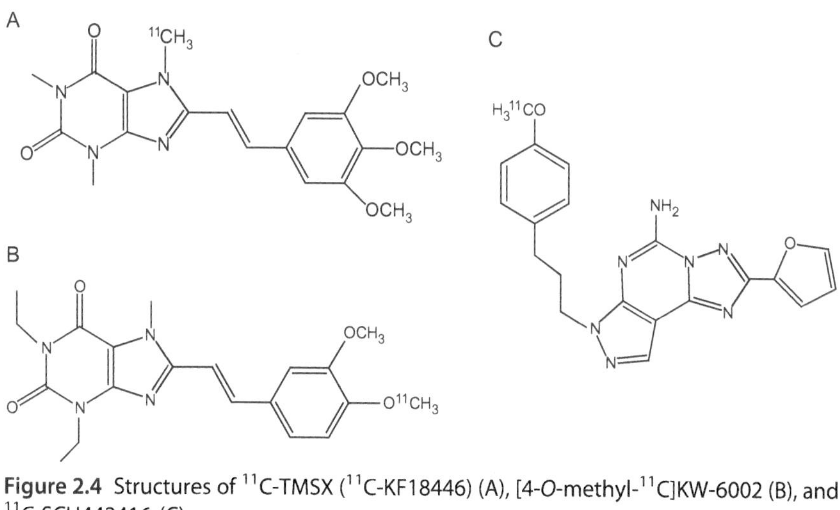

Figure 2.4 Structures of ^{11}C-TMSX (^{11}C-KF18446) (A), [4-*O*-methyl-^{11}C]KW-6002 (B), and ^{11}C-SCH442416 (C).

receptor occupancy by cold KW-6002 (Brooks et al., 2008). ^{11}C-KF21213 showed a slightly higher affinity and much greater selectivity than ^{11}C-TMSX (Wang et al., 2000); however, it has not been applied to human studies.

^{11}C-(*E*)-8-(3-bromostyryl)-3,7-dimethyl-1-propargylxanthine (^{11}C-BS-DMPX) and ^{11}C-(*E*)-3,7-dimethyl-8-(3-iodostyryl)-1-propargylxanthine (^{11}C-IS-DMPX) (Ishiwata, Shimada, et al., 2000) can potentially be labeled with longer half-life bromines (^{75}Br, half-life of 1.7 h, or ^{76}Br, half-life of 16.1 h) and iodines (^{124}I, half-life of 4.18 days, and ^{123}I, half-life of 13.3 h), but did not show preferable characteristics (Ishiwata, Shimada, et al., 2000) for PET or SPECT.

One of the disadvantages of xanthine-type radioligands such as TMSX and KW-6002 is photoisomerization. It was noted that photoisomerization occurred in the styryl group at the eight positions of xanthine-type adenosine A_{2A} receptor-selective ligands (Ishiwata, Wang, Kimura, Kawamura, & Ishii, 2003; Nonaka et al., 1993). Consequently, all procedures in PET studies were carried out under dim light until injection and also during plasma metabolite analysis (Ishiwata et al., 2005; Mishina et al., 2007, 2011).

From nonxanthine-type SCH 58261 as the lead, ^{11}C-5-amino-7-(3-(4-methoxyphenyl)propyl)-2-(2-furyl)-pyrazolo[4,3-*e*]-1,2,4-triazolo[1,5-*c*]pyrimidine (^{11}C-SCH 442416; Fig. 2.4C) was radiolabeled (Todde et al., 2000) and recently applied to humans (Ramlackhansingh et al., 2011). This ligand showed the highest affinity and selectivity among the PET ligands investigated so far, and it is noted that its affinity *in vitro* is 10 times higher in human adenosine receptors than in the receptors of rat. A ligand of ^{18}F-labeled fluoroethyl derivative of SCH 442416, ^{18}F-MRS5425, was also developed (Bhattacharjee et al., 2011).

For SPECT ligands, ^{123}I-7-(2-(4-(2-fluoro-4-iodophenyl)piperazin-1-yl) ethyl)-2-(furan-2-yl)-7*H*-pyrazolo[4,3-*e*][1,2,4]triazolo[1,5-*c*]pyrimidin-5-amine (^{123}I-MNI-420) was recently developed and used to measure an adenosine A_{2A} receptor occupancy by caffeine (Tavares, Batis, Barret, et al., 2013), and then the first human study was reported (Tavares, Batis, Papin, et al., 2013).

3.2. Adenosine A_{2A} receptors in normal subjects

We successfully visualized adenosine A_{2A} receptors in the human brain with PET and ^{11}C-TMSX (Bauer & Ishiwata, 2009; Ishiwata et al., 2007, 2010; Mishina et al., 2007). The specific binding of ^{11}C-TMSX to adenosine A_{2A} receptors was confirmed with a theophylline challenge

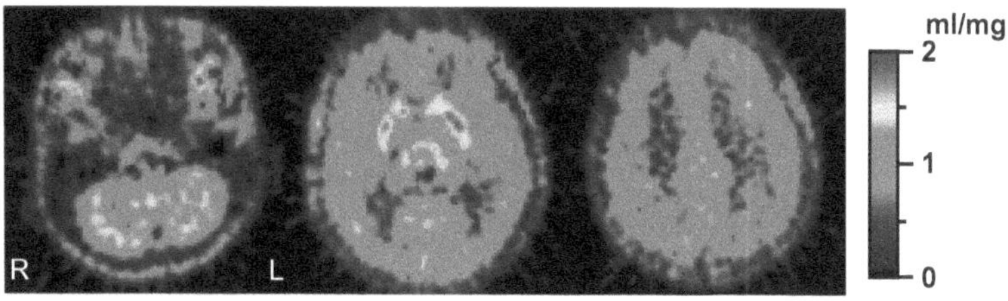

Figure 2.5 A parametric image for the distribution volume ratio (*DVR*) of ^{11}C-TMSX PET image for a 21-year-old man. The image demonstrates that density of adenosine A_{2A} receptors was high in the putamen and low in the cerebral cortex.

(Ishiwata et al., 2005). We performed test–retest studies and optimized the kinetics for ^{11}C-TMSX PET in normal subjects, thus confirming good reproducibility of ^{11}C-TMSX PET in the striatum (in preparation). In the ^{11}C-TMSX PET image (Fig. 2.5), the density of adenosine A_{2A} receptors is largest in the putamen, followed by the head of the caudate nucleus and the thalamus, but is low in the cerebral cortex, especially the frontal lobe (Ishiwata et al., 2005; Mishina et al., 2007). ^{11}C-TMSX PET has shown a large binding potential in the striatum where adenosine A_{2A} receptors are abundant, as found in postmortem and nonhuman studies, but the binding of ^{11}C-TMSX is larger in the human thalamus than in other mammals.

The effects of aging on adenosine A_1 and adenosine A_{2A} receptor may be different. Our ^{11}C-TMSX PET study did not demonstrate an effect of aging on levels of adenosine A_{2A} receptors in the human striatum (Mishina et al., 2012), although several studies reported that adenosine A_1 receptors decrease with age as mentioned above (Meerlo et al., 2004; Meyer et al., 2007; Mishina et al., 2012). On the other hand, an animal study reported that the bindings of the adenosine A_{2A} receptor agonist and antagonist in the cortical membranes were increased in aged rats compared with those in young rats (Lopes, Cunha, & Ribeiro, 1999).

3.3. Adenosine A_{2A} receptors in brain diseases

A postmortem study reported that the density of adenosine A_{2A} receptor-binding sites in Parkinson's disease was comparable to that found in normal subjects (Martinez-Mir et al., 1991). Using reverse transcription polymerase chain reaction and postmortem brain tissue, Hurley, Mash, and Jenner (2000) reported that mRNA levels for adenosine A_{2A} receptor of patients with Parkinson's disease were decreased in the caudate nucleus and anterior dorsal putamen, and were increased in the substantia nigra pars reticulata. Another postmortem study suggested that adenosine A_{2A} receptors were involved in the development of dyskinesia following long-term levodopa therapy in Parkinson's disease (Calon et al., 2004). Our ^{11}C-TMSX PET study with Parkinson's disease (Mishina et al., 2011) demonstrates that the putaminal density of adenosine A_{2A} receptors was increased in the patients with dyskinesia and that there was no significant difference in the striatal density of adenosine A_{2A} receptors between *de novo* patients and normal controls (Fig. 2.6). Another PET study with ^{11}C-SCH 44241 also reported that binding potentials for adenosine A_{2A} receptors were increased in patients with dyskinesia compared with those in patients without dyskinesia (Ramlackhansingh et al., 2011) In drawing attention to the asymmetrical symptoms in *de novo* patients, our study suggests that adenosine A_{2A} receptors were asymmetrically downregulated in the putamen but not in the head of the caudate nucleus. We speculate that the asymmetrical regulation of adenosine A_{2A} receptors was involved in compensation for the decrease in dopamine because the function of adenosine A_{2A} receptor is thought to be opposite to that of dopamine D_2 receptor (Fredholm & Svenningsson, 2003). We also found that the density of adenosine A_{2A} receptors was increased in the putamen after antiparkinsonian therapy in the *de novo* patients with Parkinson's disease. They did not developed dyskinesia during the period of this study. The finding may reflect alteration in compensation for the decreased dopamine by the antiparkinsonian therapy in the patients with Parkinson's disease. Our study suggested that the increase in putaminal adenosine A_{2A} receptors after antiparkinsonian therapy

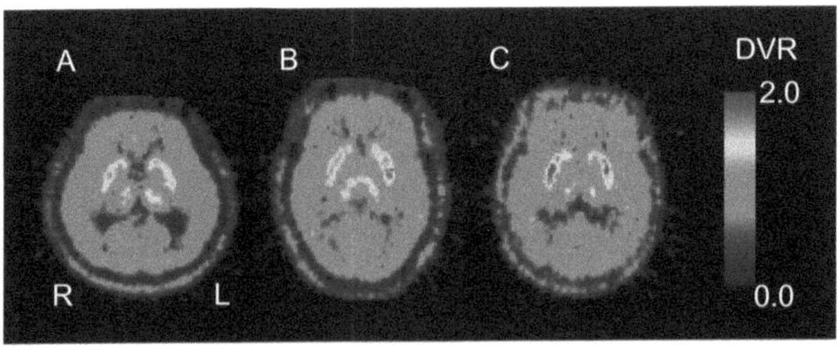

Figure 2.6 ^{11}C-TMSX PET images for a healthy man (A), a *de novo* patient with Parkinson's disease (B), and a patient with dyskinesia (C). The normal subject is a 56-year-old man (A). The *de novo* patient with Parkinson's disease is a 56-year-old man with left-dominant parkinsonism (B). The distribution volume ratio (*DVR*) of ^{11}C-TMSX was smaller in the right putamen than in the left. The patient with mild dyskinesia and Parkinson's disease is a 66-year-old **woman with left-dominant parkinsonism (C). Compared with the normal subject and *de novo* patient, the *DVR* of ^{11}C-TMSX in the striata was increased in the patient with dyskinesia.

preceded the development of dyskinesia in patients with Parkinson's disease (Mishina et al., 2011).

The density of adenosine A_{2A} receptors in the basal ganglia was lower in patients with Huntington's disease than in normal subjects (Martinez-Mir et al., 1991). The loss of adenosine A_{2A} receptors in the caudate nucleus, putamen, and external globus pallidus was more dominant than that of dopamine D_2 receptor binding in the patients with Huntington's disease (Glass, Dragunow, & Faull, 2000).

In patients with schizophrenia, postmortem studies using [^{3}H]CGS21680 reported that the adenosine A_{2A} receptors were increased in the striatum (Deckert et al., 2003; Kurumaji & Toru, 1998) and that the increase of the receptor density correlated with the dose of antipsychotic medication (Deckert et al., 2003).

4. CONCLUSIONS

Recently, adenosine receptors have attracted attention as potential therapeutic strategy (Jacobson & Gao, 2006). Although little information was available for adenosine receptors in the living human brain to date, molecular imaging for adenosine receptors was successful in several studies for developing PET ligands and is being applied to research on physiology and neurological disorders. The imaging techniques may be applied to various drug developments. Continued efforts to identify high-affinity and selective ligands should lead to PET probes suitable for these binding sites in the near future.

REFERENCES

Angelatou, F., Pagonopoulou, O., Maraziotis, T., Olivier, A., Villemeure, J. G., Avoli, M., et al. (1993). Upregulation of A_1 adenosine receptors in human temporal lobe epilepsy: A quantitative autoradiographic study. *Neuroscience Letters*, *163*(1), 11–14. http://dx.doi.org/10.1016/0304-3940(93)90217-9.

Bauer, A., Holschbach, M. H., Meyer, P. T., Boy, C., Herzog, H., Olsson, R. A., et al. (2003). In vivo imaging of adenosine A_1 receptors in the human brain with [^{18}F]CPFPX and positron emission tomography. *NeuroImage*, *19*(4), 1760–1769. http://dx.doi.org/10.1016/S1053-8119(03)00241-6.

Bauer, A., & Ishiwata, K. (2009). Adenosine receptor ligands and PET imaging of the CNS. *Handbook of Experimental Pharmacology*, *193*, 617–642. http://dx.doi.org/10.1007/978-3-540-89615-9_19.

Bhattacharjee, A. K., Lang, L., Jacobson, O., Shinkre, B., Ma, Y., Niu, G., et al. (2011). Striatal adenosine A_{2A} receptor-mediated positron emission tomographic imaging in 6-hydroxydopamine-lesioned rats using [^{18}F]-MRS5425. *Nuclear Medicine and Biology*, *38*(6), 897–906. http://dx.doi.org/10.1016/j.nucmedbio.2011.01.009.

Boy, C., Meyer, P. T., Kircheis, G., Holschbach, M. H., Herzog, H., Elmenhorst, D., et al. (2008). Cerebral A_1 adenosine receptors (A_1AR) in liver cirrhosis. *European Journal of Nuclear Medicine and Molecular Imaging, 35*(3), 589–597. http://dx.doi.org/10.1007/s00259-007-0586-z.

Brooks, D. J., Doder, M., Osman, S., Luthra, S. K., Hirani, E., Hume, S., et al. (2008). Positron emission tomography analysis of [^{11}C]KW-6002 binding to human and rat adenosine A_{2A} receptors in the brain. *Synapse, 62*(9), 671–681. http://dx.doi.org/10.1002/syn.20539.

Calon, F., Dridi, M., Hornykiewicz, O., Bedard, P. J., Rajput, A. H., & Di Paolo, T. (2004). Increased adenosine A_{2A} receptors in the brain of Parkinson's disease patients with dyskinesias. *Brain, 127*, 1075–1084. http://dx.doi.org/10.1093/brain/awh128.

Deckert, J., Abel, F., Kunig, G., Hartmann, J., Senitz, D., Maier, H., et al. (1998). Loss of human hippocampal adenosine A_1 receptors in dementia: Evidence for lack of specificity. *Neuroscience Letters, 244*(1), 1–4. http://dx.doi.org/10.1016/S0304-3940(98)00108-6.

Deckert, J., Brenner, M., Durany, N., Zochling, R., Paulus, W., Ransmayr, G., et al. (2003). Up-regulation of striatal adenosine A_{2A} receptors in schizophrenia. *Neuroreport, 14*(3), 313–316. http://dx.doi.org/10.1097/01.wnr.0000058242.21747.04.

Elmenhorst, D., Meyer, P. T., Winz, O. H., Matusch, A., Ermert, J., Coenen, H. H., et al. (2007). Sleep deprivation increases A_1 adenosine receptor binding in the human brain: A positron emission tomography study. *Journal of Neuroscience, 27*(9), 2410–2415. http://dx.doi.org/10.1523/JNEUROSCI.5066-06.2007.

Factor, S., Mark, M. H., Watts, R., Struck, L., Mori, A., Ballerini, R., et al. (2010). A long-term study of istradefylline in subjects with fluctuating Parkinson's disease. *Parkinsonism & Related Disorders, 16*(6), 423–426. http://dx.doi.org/10.1016/j.parkreldis.2010.02.014.

Fastbom, J., Pazos, A., Probst, A., & Palacios, J. M. (1987). Adenosine A_1 receptors in the human brain: A quantitative autoradiographic study. *Neuroscience, 22*(3), 827–839. http://dx.doi.org/10.1016/0306-4522(87)92962-9.

Ferre, S., Popoli, P., Gimenez-Llort, L., Finnman, U. B., Martinez, E., Scotti de Carolis, A., et al. (1994). Postsynaptic antagonistic interaction between adenosine A_1 and dopamine D_1 receptors. *Neuroreport, 6*(1), 73–76.

Ferri, C. P., Prince, M., Brayne, C., Brodaty, H., Fratiglioni, L., Ganguli, M., et al. (2005). Global prevalence of dementia: A Delphi consensus study. *Lancet, 366*(9503), 2112–2117. http://dx.doi.org/10.1016/S0140-6736(05)67889-0.

Fredholm, B. B., & Svenningsson, P. (2003). Adenosine-dopamine interactions: Development of a concept and some comments on therapeutic possibilities. *Neurology, 61*(Suppl.), S5–S9. http://dx.doi.org/10.1212/01.WNL.0000095204.89871.FF.

Fukumitsu, N., Ishii, K., Kimura, Y., Oda, K., Hashimoto, M., Suzuki, M., et al. (2008). Adenosine A_1 receptors using 8-dicyclopropylmethyl-1-[^{11}C]methyl-3-propylxanthine PET in Alzheimer's disease. *Annals of Nuclear Medicine, 22*(10), 841–847. http://dx.doi.org/10.1007/s12149-008-0185-5.

Fukumitsu, N., Ishii, K., Kimura, Y., Oda, K., Sasaki, T., Mori, Y., et al. (2003). Imaging of adenosine A_1 receptors in the human brain by positron emission tomography with [^{11}C]MPDX. *Annals of Nuclear Medicine, 17*(6), 511–515. http://dx.doi.org/10.1007/BF03006445.

Fukumitsu, N., Ishii, K., Kimura, Y., Oda, K., Sasaki, T., Mori, Y., et al. (2005). Adenosine A_1 receptor mapping of the human brain by PET with 8-dicyclopropylmethyl-1-^{11}C-methyl-3-propylxanthine. *Journal of Nuclear Medicine, 46*(1), 32–37.

Glass, M., Dragunow, M., & Faull, R. L. (2000). The pattern of neurodegeneration in Huntington's disease: A comparative study of cannabinoid, dopamine, adenosine and $GABA_A$ receptor alterations in the human basal ganglia in Huntington's disease. *Neuroscience, 97*(3), 505–519. http://dx.doi.org/10.1016/S0306-4522(00)00008-7.

Glass, M., Faull, R. L., Bullock, J. Y., Jansen, K., Mee, E. W., Walker, E. B., et al. (1996). Loss of A_1 adenosine receptors in human temporal lobe epilepsy. *Brain Research, 710*(1–2), 56–68. http://dx.doi.org/10.1016/0006-8993(95)01313-X.

Hirani, E., Gillies, J., Karasawa, A., Shimada, J., Kase, H., Opacka-Juffry, J., et al. (2001). Evaluation of [4-O-methyl-^{11}C]KW-6002 as a potential PET ligand for mapping central adenosine A_{2A} receptors in rats. *Synapse, 42*(3), 164–176. http://dx.doi.org/10.1002/syn.1110.

Holschbach, M. H., Fein, T., Krummeich, C., Lewis, R. G., Wutz, W., Schwabe, U., et al. (1998). A_1 adenosine receptor antagonists as ligands for positron emission tomography (PET) and single-photon emission tomography (SPET). *Journal of Medicinal Chemistry, 41*(4), 555–563. http://dx.doi.org/10.1021/jm9705465.

Holschbach, M. H., Olsson, R. A., Bier, D., Wutz, W., Sihver, W., Schuller, M., et al. (2002). Synthesis and evaluation of no-carrier-added 8-cyclopentyl-3-(3-[^{18}F]fluoropropyl)-1-propylxanthine ([^{18}F]CPFPX): A potent and selective A_1-adenosine receptor antagonist for in vivo imaging. *Journal of Medicinal Chemistry, 45*(23), 5150–5156. http://dx.doi.org/10.1021/jm020905i.

Hurley, M. J., Mash, D. C., & Jenner, P. (2000). Adenosine A_{2A} receptor mRNA expression in Parkinson's disease. *Neuroscience Letters, 291*(1), 54–58. http://dx.doi.org/10.1016/S0304-3940(00)01371-9.

Ikeda, M., Mackay, K. B., Dewar, D., & McCulloch, J. (1993). Differential alterations in adenosine A_1 and kappa 1 opioid receptors in the striatum in Alzheimer's disease. *Brain Research, 616*(1–2), 211–217. http://dx.doi.org/10.1016/0006-8993(93)90211-5.

Ishibashi, K., Ishii, K., Oda, K., Kawasaki, K., Mizusawa, H., & Ishiwata, K. (2009). Regional analysis of age-related decline in dopamine transporters and dopamine D_2-like receptors in human striatum. *Synapse, 63*(4), 282–290. http://dx.doi.org/10.1002/syn.20603.

Ishiwata, K., Furuta, R., Shimada, J., Ishii, S., Endo, K., Suzuki, F., et al. (1995). Synthesis and preliminary evaluation of [11C]KF15372, a selective adenosine A1 antagonist. *Applied Radiation and Isotopes, 46*(10), 1009–1013.

Ishiwata, K., Kimura, Y., de Vries, E. F. J., & Elsinga, P. H. (2007). PET tracers for mapping adenosine receptors as probes for diagnosis of CNS disorders. *Central Nervous System Agents in Medicinal Chemistry*, 7(1), 57–77. http://dx.doi.org/10.2174/187152407780059259.

Ishiwata, K., Kimura, Y., Oda, K., Ishii, K., Sakata, M., Kawasaki, K., et al. (2010). Development of PET radiopharmaceuticals and their clinical applications at the positron medical center. *Geriatrics & Gerontology International, 10*(Suppl. 1), S180–S196. http://dx.doi.org/10.1111/j.1447-0594.2010.00594.x.

Ishiwata, K., Mishina, M., Kimura, Y., Oda, K., Sasaki, T., & Ishii, K. (2005). First visualization of adenosine A_{2A} receptors in the human brain by positron emission tomography with [^{11}C]TMSX. *Synapse, 55*(2), 133–136. http://dx.doi.org/10.1002/syn.20099.

Ishiwata, K., Nariai, T., Kimura, Y., Oda, K., Kawamura, K., Ishii, K., et al. (2002). Preclinical studies on [^{11}C]MPDX for mapping adenosine A_1 receptors by positron emission tomography. *Annals of Nuclear Medicine, 16*(6), 377–382.

Ishiwata, K., Noguchi, J., Toyama, H., Sakiyama, Y., Koike, N., Ishii, S., et al. (1996). Synthesis and preliminary evaluation of [^{11}C]KF17837, a selective adenosine A_{2A} antagonist. *Applied Radiation and Isotopes*, 47(5–6), 507–511. http://dx.doi.org/10.1016/0969-8043(95)00295-2.

Ishiwata, K., Noguchi, J., Wakabayashi, S., Shimada, J., Ogi, N., Nariai, T., et al. (2000). ^{11}C-labeled KF18446: A potential central nervous system adenosine A_{2A} receptor ligand. *Journal of Nuclear Medicine, 41*(2), 345–354.

Ishiwata, K., Shimada, J., Wang, W. F., Harakawa, H., Ishii, S., Kiyosawa, M., et al. (2000). Evaluation of iodinated and brominated [^{11}C]styrylxanthine derivatives as in vivo radioligands mapping adenosine A_{2A} receptor in the central nervous system. *Annals of Nuclear Medicine, 14*(4), 247–253. http://dx.doi.org/10.1007/BF02988206.

Ishiwata, K., Wang, W. F., Kimura, Y., Kawamura, K., & Ishii, K. (2003). Preclinical studies on [^{11}C]TMSX for mapping adenosine A_{2A} receptors by positron emission tomography. *Annals of Nuclear Medicine, 17*(3), 205–211. http://dx.doi.org/10.1007/BF02990023.

Jaarsma, D., Sebens, J. B., & Korf, J. (1991). Reduction of adenosine A_1-receptors in the perforant pathway terminal zone in Alzheimer hippocampus. *Neuroscience Letters, 121*(1–2), 111–114. http://dx.doi.org/10.1016/0304-3940(91)90661-C.

Jacobson, K. A., & Gao, Z. G. (2006). Adenosine receptors as therapeutic targets. *Nature Reviews. Drug Discovery, 5*(3), 247–264. http://dx.doi.org/10.1038/nrd1983.

Jansen, K. L., Faull, R. L., Dragunow, M., & Synek, B. L. (1990). Alzheimer's disease: Changes in hippocampal N-methyl-D-aspartate, quisqualate, neurotensin, adenosine, benzodiazepine, serotonin and opioid receptors—An autoradiographic study. *Neuroscience, 39*(3), 613–627. http://dx.doi.org/10.1016/0306-4522(90)90246-Z.

Kalaria, R. N., Sromek, S., Wilcox, B. J., & Unnerstall, J. R. (1990). Hippocampal adenosine A_1 receptors are decreased in Alzheimer's disease. *Neuroscience Letters, 118*(2), 257–260. http://dx.doi.org/10.1016/0304-3940(90)90641-L.

Kelsey, J. E., Langelier, N. A., Oriel, B. S., & Reedy, C. (2009). The effects of systemic, intrastriatal, and intrapallidal injections of caffeine and systemic injections of A_{2A} and A_1 antagonists on forepaw stepping in the unilateral 6-OHDA-lesioned rat. *Psychopharmacology, 201*(4), 529–539. http://dx.doi.org/10.1007/s00213-008-1319-0.

Khanapur, S., Waarde, A., Ishiwata, K., Leenders, K. L., Dierckx, R. A., & Elsinga, P. H. (2013). Adenosine A_{2A} receptor antagonists as positron emission tomography (PET) tracers. *Current Medicinal Chemistry, 21*(3), 312–328.

Kimura, Y., Ishii, K., Fukumitsu, N., Oda, K., Sasaki, T., Kawamura, K., et al. (2004). Quantitative analysis of adenosine A_1 receptors in human brain using positron emission tomography and [1-methyl-^{11}C]8-dicyclopropylmethyl-1-methyl-3-propylxanthine. *Nuclear Medicine and Biology, 31*(8), 975–981. http://dx.doi.org/10.1016/j.nucmedbio.2004.06.005.

Kurumaji, A., & Toru, M. (1998). An increase in [^{3}H] CGS21680 binding in the striatum of postmortem brains of chronic schizophrenics. *Brain Research, 808*(2), 320–323. http://dx.doi.org/10.1016/S0006-8993(98)00840-3.

Lees, A. J., Hardy, J., & Revesz, T. (2009). Parkinson's disease. *Lancet, 373*(9680), 2055–2066. http://dx.doi.org/10.1016/S0140-6736(09)60492-X.

Li, A. H., Moro, S., Forsyth, N., Melman, N., Ji, X. D., & Jacobson, K. A. (1999). Synthesis, CoMFA analysis, and receptor docking of 3,5-diacyl-2, 4-dialkylpyridine derivatives as selective A_3 adenosine receptor antagonists. *Journal of Medicinal Chemistry, 42*(4), 706–721. http://dx.doi.org/10.1021/jm980550w.

Lohse, M. J., Klotz, K. N., Lindenborn-Fotinos, J., Reddington, M., Schwabe, U., & Olsson, R. A. (1987). 8-Cyclopentyl-1,3-dipropylxanthine (DPCPX)—A selective high affinity antagonist radioligand for A1 adenosine receptors. *Naunyn-Schmiedeberg's Archives of Pharmacology, 336*(2), 204–210. http://dx.doi.org/10.1007/BF00165806.

Lopes, L. V., Cunha, R. A., & Ribeiro, J. A. (1999). Increase in the number, G protein coupling, and efficiency of facilitatory adenosine A_{2A} receptors in the limbic cortex, but not striatum, of aged rats. *Journal of Neurochemistry, 73*(4), 1733–1738. http://dx.doi.org/10.1046/j.1471-4159.1999.731733.x.

Maemoto, T., Tada, M., Mihara, T., Ueyama, N., Matsuoka, H., Harada, K., et al. (2004). Pharmacological characterization of FR194921, a new potent, selective, and orally active antagonist for central adenosine A_1 receptors. *Journal of Pharmacological Sciences, 96*(1), 42–52.

Magistretti, P. J., & Pellerin, L. (1996). Cellular bases of brain energy metabolism and their relevance to functional brain imaging: Evidence for a prominent role of astrocytes. *Cerebral Cortex, 6*(1), 50–61. http://dx.doi.org/10.1093/cercor/6.1.50.

Martinez-Mir, M. I., Probst, A., & Palacios, J. M. (1991). Adenosine A_2 receptors: Selective localization in the human basal ganglia and alterations with disease. *Neuroscience*, *42*(3), 697–706. http://dx.doi.org/10.1016/0306-4522(91)90038-P.

Matsuya, T., Takamatsu, H., Murakami, Y., Noda, A., Ichise, R., Awaga, Y., et al. (2005). Synthesis and evaluation of [^{11}C]FR194921 as a nonxanthine-type PET tracer for adenosine A_1 receptors in the brain. *Nuclear Medicine and Biology*, *32*(8), 837–844. http://dx.doi.org/10.1016/j.nucmedbio.2005.06.008.

Meerlo, P., Roman, V., Farkas, E., Keijser, J. N., Nyakas, C., & Luiten, P. G. (2004). Ageing-related decline in adenosine A_1 receptor binding in the rat brain: An autoradiographic study. *Journal of Neuroscience Research*, *78*(5), 742–748. http://dx.doi.org/10.1002/jnr.20314.

Meyer, P. T., Elmenhorst, D., Boy, C., Winz, O., Matusch, A., Zilles, K., et al. (2007). Effect of aging on cerebral A_1 adenosine receptors: A [^{18}F]CPFPX PET study in humans. *Neurobiology of Aging*, *28*(12), 1914–1924. http://dx.doi.org/10.1016/j.neurobiolaging.2006.08.005.

Meyer, P. T., Elmenhorst, D., Matusch, A., Winz, O., Zilles, K., & Bauer, A. (2006). A_1 adenosine receptor PET using [^{18}F]CPFPX: Displacement studies in humans. *NeuroImage*, *32*(3), 1100–1105. http://dx.doi.org/10.1016/j.neuroimage.2006.04.202.

Mishina, M., Ishiwata, K., Kimura, Y., Naganawa, M., Oda, K., Kobayashi, S., et al. (2007). Evaluation of distribution of adenosine A_{2A} receptors in normal human brain measured with [^{11}C]TMSX PET. *Synapse*, *61*(9), 778–784. http://dx.doi.org/10.1002/syn.20423.

Mishina, M., Ishiwata, K., Naganawa, M., Kimura, Y., Kitamura, S., Suzuki, M., et al. (2011). Adenosine A_{2A} receptors measured with [^{11}C]TMSX PET in the striata of Parkinson's disease patients. *PLoS One*, *6*(2), e17338. http://dx.doi.org/10.1371/journal.pone.0017338.

Mishina, M., Kimura, Y., Naganawa, M., Ishii, K., Oda, K., Sakata, M., et al. (2012). Differential effects of age on human striatal adenosine A_1 and A_{2A} receptors. *Synapse*, *66*(9), 832–839. http://dx.doi.org/10.1002/syn.21573.

Mizuno, Y., Hasegawa, K., Kondo, T., Kuno, S., & Yamamoto, M. (2010). Clinical efficacy of istradefylline (KW-6002) in Parkinson's disease: A randomized, controlled study. *Movement Disorders*, *25*(10), 1437–1443. http://dx.doi.org/10.1002/mds.23107.

Nariai, T., Shimada, Y., Ishiwata, K., Nagaoka, T., Shimada, J., Kuroiwa, T., et al. (2003). PET imaging of adenosine A_1 receptors with ^{11}C-MPDX as an indicator of severe cerebral ischemic insult. *Journal of Nuclear Medicine*, *44*(11), 1839–1844.

Noguchi, J., Ishiwata, K., Furuta, R., Simada, J., Kiyosawa, M., Ishii, S., et al. (1997). Evaluation of carbon-11 labeled KF15372 and its ethyl and methyl derivatives as a potential CNS adenosine A_1 receptor ligand. *Nuclear Medicine and Biology*, *24*(1), 53–59. http://dx.doi.org/10.1016/S0969-8051(96)00161-8.

Nonaka, H., Ichimura, M., Takeda, M., Nonaka, Y., Shimada, J., Suzuki, F., et al. (1994). KF17837 ((E)-8-(3,4-dimethoxystyryl)-1,3-dipropyl-7-methylxanthine), a potent and selective adenosine A_2 receptor antagonist. *European Journal of Pharmacology*, *267*(3), 335–341. http://dx.doi.org/10.1016/0922-4106(94)90159-7.

Nonaka, Y., Shimada, J., Nonaka, H., Koike, N., Aoki, N., Kobayashi, H., et al. (1993). Photoisomerization of a potent and selective adenosine A_2 antagonist, (E)-1,3-dipropyl-8-(3,4-dimethoxystyryl)-7-methylxanthine. *Journal of Medicinal Chemistry*, *36*(23), 3731–3733. http://dx.doi.org/10.1021/jm00075a031.

Paul, S., Elsinga, P. H., Ishiwata, K., Dierckx, R. A., & van Waarde, A. (2011). Adenosine A_1 receptors in the central nervous system: Their functions in health and disease, and possible elucidation by PET imaging. *Current Medicinal Chemistry*, *18*(31), 4820–4835.

Pellerin, L., & Magistretti, P. J. (1994). Glutamate uptake into astrocytes stimulates aerobic glycolysis: A mechanism coupling neuronal activity to glucose utilization. *Proceedings of the National Academy of Sciences of the United States of America*, *91*(22), 10625–10629.

Poucher, S. M., Keddie, J. R., Singh, P., Stoggall, S. M., Caulkett, P. W., Jones, G., et al. (1995). The in vitro pharmacology of ZM 241385, a potent, non-xanthine A_{2A} selective adenosine receptor antagonist. *British Journal of Pharmacology*, *115*(6), 1096–1102. http://dx.doi.org/10.1111/j.1476-5381.1995.tb15923.x.

Rahman, A. (2009). The role of adenosine in Alzheimer's disease. *Current Neuropharmacology*, 7(3), 207–216. http://dx.doi.org/10.2174/157015909789152119.

Ramlackhansingh, A. F., Bose, S. K., Ahmed, I., Turkheimer, F. E., Pavese, N., & Brooks, D. J. (2011). Adenosine 2A receptor availability in dyskinetic and nondyskinetic patients with Parkinson disease. *Neurology*, *76*(21), 1811–1816. http://dx.doi.org/10.1212/WNL.0b013e31821ccce4.

Shimada, J., Suzuki, F., Nonaka, H., Ishii, A., & Ichikawa, S. (1992). (E)-1,3-dialkyl-7-methyl-8-(3,4,5-trimethoxystyryl)xanthines: Potent and selective adenosine A_2 antagonists. *Journal of Medicinal Chemistry*, *35*(12), 2342–2345. http://dx.doi.org/10.1021/jm00090a027.

Shimada, J., Suzuki, F., Nonaka, H., Karasawa, A., Mizumoto, H., Ohno, T., et al. (1991). 8-(Dicyclopropylmethyl)-1,3-dipropylxanthine: A potent and selective adenosine A_1 antagonist with renal protective and diuretic activities. *Journal of Medicinal Chemistry*, *34*(1), 466–469. http://dx.doi.org/10.1021/jm00105a072.

Sihver, W., Holschbach, M. H., Bier, D., Wutz, W., Schulze, A., Olsson, R. A., et al. (2003). Evaluation of radioiodinated 8-cyclopentyl-3-[(E)-3-iodoprop-2-en-1-yl]-1-propylxanthine ([*I]CPIPX) as a new potential A_1 adenosine receptor antagonist for SPECT. *Nuclear Medicine and Biology*, *30*(6), 661–668. http://dx.doi.org/10.1016/S0969-8051(03)00053-2.

Stone-Elander, S., Thorell, J. O., Eriksson, L., Fredholm, B. B., & Ingvar, M. (1997). In vivo biodistribution of [N-^{11}C-methyl]KF 17837 using 3-D-PET: Evaluation as a ligand for the study of adenosine A_{2A} receptors. *Nuclear Medicine and Biology*, *24*(2), 187–191. http://dx.doi.org/10.1016/S0969-8051(96)00216-8.

Suzuki, Y., Nariai, T., Kiyosawa, M., Mochizuki, M., Kimura, Y., Oda, K., et al. (2012). Increased adenosine A_1 receptor levels in hemianopia patients after cerebral injury: An application of PET using ^{11}C-8-dicyclopropylmethyl-1-methyl-3-propylxanthine. *Clinical Nuclear Medicine*, *37*(12), 1146–1151. http://dx.doi.org/10.1097/RLU.0b013e31826392a7.

Tavares, A. A., Batis, J., Barret, O., Alagille, D., Vala, C., Kudej, G., et al. (2013). In vivo evaluation of [^{123}I]MNI-420: A novel single photon emission computed tomography radiotracer for imaging of adenosine 2A receptors in brain. *Nuclear Medicine and Biology*, *40*(3), 403–409. http://dx.doi.org/10.1016/j.nucmedbio.2012.12.005.

Tavares, A. A., Batis, J. C., Papin, C., Jennings, D., Alagille, D., Russell, D. S., et al. (2013). Kinetic modeling, test-retest, and dosimetry of ^{123}I-MNI-420 in humans. *Journal of Nuclear Medicine*, *54*(10), 1760–1767. http://dx.doi.org/10.2967/jnumed.113.119933.

Todde, S., Moresco, R. M., Simonelli, P., Baraldi, P. G., Cacciari, B., Spalluto, G., et al. (2000). Design, radiosynthesis, and biodistribution of a new potent and selective ligand for in vivo imaging of the adenosine A_{2A} receptor system using positron emission tomography. *Journal of Medicinal Chemistry*, *43*(23), 4359–4362. http://dx.doi.org/10.1021/jm0009843.

Tsacopoulos, M., & Magistretti, P. J. (1996). Metabolic coupling between glia and neurons. *Journal of Neuroscience*, *16*(3), 877–885.

Ulas, J., Brunner, L. C., Nguyen, L., & Cotman, C. W. (1993). Reduced density of adenosine A_1 receptors and preserved coupling of adenosine A_1 receptors to G proteins in Alzheimer hippocampus: A quantitative autoradiographic study. *Neuroscience*, *52*(4), 843–854. http://dx.doi.org/10.1016/0306-4522(93)90533-L.

Wadsak, W., Mien, L. K., Shanab, K., Ettlinger, D. E., Haeusler, D., Sindelar, K., et al. (2008). Preparation and first evaluation of [^{18}F]FE@SUPPY: A new PET tracer for the adenosine A_3 receptor. *Nuclear Medicine and Biology*, *35*(1), 61–66. http://dx.doi.org/10.1016/j.nucmedbio.2007.09.004.

Wang, W. F., Ishiwata, K., Nonaka, H., Ishii, S., Kiyosawa, M., Shimada, J., et al. (2000). Carbon-11-labeled KF21213: A highly selective ligand for mapping CNS adenosine A_{2A} receptors with positron emission tomography. *Nuclear Medicine and Biology*, *27*(6), 541–546. http://dx.doi.org/10.1016/S0969-8051(00)00126-8.

Williams-Karnesky, R. L., & Stenzel-Poore, M. P. (2009). Adenosine and stroke: Maximizing the therapeutic potential of adenosine as a prophylactic and acute neuroprotectant. *Current Neuropharmacology*, 7(3), 217–227. http://dx.doi.org/10.2174/157015909789152209.

Xiao, D., Cassin, J. J., Healy, B., Burdett, T. C., Chen, J. F., Fredholm, B. B., et al. (2011). Deletion of adenosine A_1 or A_{2A} receptors reduces L-3,4-dihydroxyphenylalanine-induced dyskinesia in a model of Parkinson's disease. *Brain Research*, *1367*, 310–318. http://dx.doi.org/10.1016/j.brainres.2010.08.099.

Yabuuchi, K., Kuroiwa, M., Shuto, T., Sotogaku, N., Snyder, G. L., Higashi, H., et al. (2006). Role of adenosine A_1 receptors in the modulation of dopamine D_1 and adenosine A_{2A} receptor signaling in the neostriatum. *Neuroscience*, *141*(1), 19–25. http://dx.doi.org/10.1016/j.neuroscience.2006.04.047.

Zocchi, C., Ongini, E., Conti, A., Monopoli, A., Negretti, A., Baraldi, P. G., et al. (1996). The non-xanthine heterocyclic compound SCH 58261 is a new potent and selective A_{2A} adenosine receptor antagonist. *Journal of Pharmacology and Experimental Therapeutics*, *276*(2), 398–404.

Zocchi, C., Ongini, E., Ferrara, S., Baraldi, P. G., & Dionisotti, S. (1996). Binding of the radioligand [^{3}H]-SCH 58261, a new non-xanthine A_{2A} adenosine receptor antagonist, to rat striatal membranes. *British Journal of Pharmacology*, *117*(7), 1381–1386. http://dx.doi.org/10.1111/j.1476-5381.1996.tb15296.x.

CHAPTER THREE

An Overview of Adenosine A_{2A} Receptor Antagonists in Parkinson's Disease

Peter Jenner[1]

Neurodegenerative Diseases Research Group, Institute of Pharmaceutical Sciences, School of Biomedical Sciences, King's College, London, United Kingdom

[1]Corresponding author: e-mail address: peter.jenner@kcl.ac.uk

Contents

Abstract

Adenosine A_{2A} receptor antagonists represent a new way forward in the symptomatic treatment of Parkinson's disease (PD) through a non-dopaminergic mechanism. As a class, adenosine A_{2A} antagonists are effective in reversing motor deficits in haloperidol-treated rodents, 6-OHDA-lesioned rats, and MPTP-treated primates when combined with low doses of L-dopa or dopamine agonist drugs. Importantly, they improve motor function without worsening dyskinesia and they may prevent the onset of involuntary movements. Adenosine A_{2A} receptor antagonists are active in animal models of reduced cognition, anxiety, and depression and so this drug class may also be effective in controlling the neuropsychiatric components of nonmotor symptoms in PD. Preclinical evidence has shown that A_{2A} antagonists can prevent neuronal loss in experimental models of PD and their disease modifying activity needs to be explored in man.

Importantly, a number of A_{2A} antagonists have been studied in PD in clinical trial for their effects on motor function. So far, little evidence has emerged of an effect of monotherapy with adenosine antagonists in early PD. However, in later stage, patient populations already treated with dopaminergic drugs but exhibiting "wearing off," adenosine antagonists have been demonstrated to reduce "off" time without increasing

International Review of Neurobiology, Volume 119
ISSN 0074-7742
http://dx.doi.org/10.1016/B978-0-12-801022-8.00003-9

troublesome dyskinesia in phase IIB trials. However, in larger phase III evaluations, a consistent significant decrease in "off" time has proved more difficult to demonstrate—due in part to trial conduct. So far, only istradefylline has completed phase III development and it is now marketed for the treatment of PD.

Adenosine A_{2A} antagonists are the first non-dopaminergic approach to the treatment of PD to appear in the recent era. They represent a novel way of approaching therapy that will provide additional benefit to that achieved with dopaminergic medication, while avoiding common side effects and may in addition, improve some non-motor symptoms of PD that are currently poorly treated.

1. PROBLEMS IN THE TREATMENT OF PARKINSON'S DISEASE

Dopamine replacement therapy (L-dopa, dopamine agonists, decarboxylase inhibitors, COMT inhibitors, MAO-B inhibitors) has been successfully employed to control the motor symptoms of Parkinson's disease (PD). Over the past 50 years, individual drugs (L-dopa, dopamine agonists, MAO-B inhibitors) have been used to treat early PD with combinations of all the available dopaminergic agents, and a range of delivery forms (sustained release capsules, transdermal patches, infusion pumps) being employed in the later stages of the illness (Abbruzzese, Barone, Bonuccelli, Lopiano, & Antonini, 2012; Antonini, Ursino, Calandrella, Bernardi, & Plebani, 2010). However, the treatment of PD with dopaminergic drugs is associated with significant unmet clinical need in terms of both effect and side effect. Long-term treatment leads to the onset of motor fluctuations ("wearing off") and motor complications (dyskinesia) appearing as a consequence of both drug treatment and disease progression. Both have been associated with alterations in striatal output and changes in the function of GABAergic, glutamatergic, and cholinergic neurones among others (Obeso et al., 2008). While "wearing off" can be initially masked by using combinations of dopaminergic drugs, it is common for increasingly long periods of "off" time to persist even when medication is optimized. The involuntary movements that characterize dyskinesia affect a significant portion of the patient population and once established, appear following administration of each dose of dopaminergic medication (Guridi, Gonzalez-Redondo, & Obeso, 2012). In this situation, when drug doses need to be increased to improve movement with disease progression, dyskinesia can become troublesome and can limit therapy. As a consequence, new approaches to treatment are required that avoid further affecting dopaminergic function.

PD is now recognized as not only a disorder of movement but also in addition as having multiple nonmotor components (for example, constipation, REM sleep behavioral disorder, sensory change, anxiety, depression, and cognitive decline) which can precede the onset of motor symptoms, occur simultaneously with motor signs or appear subsequently (Chaudhuri, Odin, Antonini, & Martinez-Martin, 2011; Martinez-Martin, Rodriguez-Blazquez, Kurtis, Chaudhuri, & NMSS Validation Group, 2011; Rektorova, Aarsland, Chaudhuri, & Strafella, 2011). Nonmotor symptoms respond poorly to dopaminergic medication, they are often a more serious clinical problem than motor disability and the biggest determinant of disease outcome. Many nonmotor symptoms are recognized as originating from the progressive pathology occurring in PD outside of the basal ganglia and involving non-dopaminergic neurones. Consequently, there is a need to develop approaches to treatment that are non-dopaminergic in nature and that address this major unmet need.

2. NON-DOPAMINERGIC APPROACHES TO THE TREATMENT OF PD

Non-dopaminergic drug treatment of PD could be potentially effective at two levels. First, within the basal ganglia, output pathways beyond the damaged dopaminergic system are largely non-dopaminergic in nature and primarily controlled by GABA and glutamate transmission. However, these pathways have a large range of other neurotransmitter receptors on their surface that are potential targets for the drug manipulation of motor function. The available targets include 5-HT receptors, noradrenergic receptors, histamine receptors, nicotinic and muscarinic cholinergic receptors, ionotropic and metabotropic glutamate receptors, and adenosine receptors. Outside of the basal ganglia, the pathology of PD affects many other brain nuclei (for example, locus coeruleus, raphe nuclei, and dorsal motor nucleus of the vagus) and their target structures (for example, amygdala, hippocampus, and cortex) and a range of neurotransmitters other than dopamine (for example, 5-HT, noradrenaline, acetylcholine, GABA, and glutamate). Again, these provide a focus for using non-dopaminergic drugs to manipulate both the motor and motor symptoms of PD.

Non-dopaminergic approaches to the treatment of PD are already in routine use. Amantadine and anticholinergics are used to either control some motor symptoms or to suppress dyskinesia but their use is limited by efficacy and side effect profile. Other non-dopaminergic drugs are used on a symptomatic basis to control nonmotor symptoms of PD, for example atypical antipsychotics, anxiolytics, antidepressants, and cholinesterase inhibitors. Again, the effectiveness of these approaches in PD can be limited. More recently, a large range of non-dopaminergic drugs (including 5-HT agonists and antagonists, histamine agonist and antagonists, adrenergic antagonists, glutamate antagonists, cannabinoid and opiate agonists and antagonists, nicotinic agonists) have been assessed for their ability to alter motor function in experimental models of PD, most notably the 6-OHDA-lesioned rat and the MPTP-treated primate (Brotchie & Jenner, 2011; Johnston & Brotchie, 2006). They have followed numerous reports of compounds improving the motor benefit seen with L-dopa and/or the suppression of established dyskinesia. However, while a number of compounds are still undergoing clinical trial (Fox, Brotchie, & Lang, 2008), many of these approaches have failed during evaluation in PD due to a lack of efficacy, a worsening of motor function or unacceptable side effect profiles and narrow therapeutic windows.

An exception appears to be the effects of adenosine A_{2A} receptor antagonists on the motor symptoms of PD and their potential effects on nonmotor symptoms, most notably neuropsychiatric events such as depression, anxiety, and cognitive decline. The following sections present an overview of the preclinical and clinical evidence of the effectiveness of this compound class in the treatment of PD. This should be seen as a brief prelude to the more detailed chapters making up this volume that follow in which the available data is explored in depth.

3. ADENOSINE A_{2A} RECEPTOR ANTAGONISTS AND MOTOR FUNCTION

The selective localization of A_{2A} receptors to the indirect output pathway from the striatum (Brooks et al., 2008) and their ability to control GABAergic, glutamatergic, and cholinergic function makes them a prime target for the development of adenosine antagonists that reverse the altered motor function occurring in PD (Kase et al., 2003; Mori & Shindou, 2003; Shindou, Mori, Kase, & Ichimura, 2001; Shindou et al., 2002; Shindou, Richardson, Mori, Kase, & Ichimura, 2003). A number of selective xanthine and non-xanthine A_{2A} receptor antagonists (istradefylline, preladenant, tozadenant, vipadenant, and ST 1535) have been assessed for their effects in the same range of preclinical models of PD (haloperidol-treated mouse, 6-OHDA-lesioned rat, MPTP-treated primate) and subsequently studied in clinical trial.

3.1. Improvement in motor disability

The effects observed in models of decreased motor function have been more or less the same for all selective adenosine A_{2A} receptor antagonists studied (Morelli, Carta, & Jenner, 2009). They reverse catalepsy induced by the administration of haloperidol to rodents (Shiozaki et al., 1999). While this is the least relevant model to idiopathic PD (as it does not involve nigral damage or a reduction in dopamine levels), their effectiveness perhaps indicates an ability to alter motor function beyond the blocked striatal postsynaptic dopamine receptor. In unilateral 6-OHDA-lesioned rats, the administration of adenosine A_{2A} antagonists alone can induce a mild degree of ipsilateral rotation indicative of a predominant effect on the intact basal ganglia in the unlesioned hemisphere (Rose, Ramsay Croft, & Jenner, 2007). This might be indicative of some mild symptomatic action relevant to early PD but interestingly, this is not observed with all A_{2A} receptor antagonists tested for reasons which remain unclear. When combined with L-dopa or a dopamine agonist, adenosine A_{2A} receptor antagonists enhance the rate of contralateral rotation that occurs with the dopaminergic drug alone (Hodgson et al., 2009; Koga, Kurokawa, Ochi, Nakamura, & Kuwana, 2000; Kuwana et al., 1999; Pinna, Volpini, Cristalli, & Morelli, 2005; Tronci et al., 2007). This is indicative of an ability to potentiate the antiparkinsonian actions of dopaminergic medication seen in PD. An important facet of this effect is that the action of the adenosine A_{2A} antagonists is most prominent when combined with a low dose of L-dopa with little or no additional effect being observed when used with high L-dopa doses.

The most predictive model of drug effect in PD in man is the MPTP-treated primate. All adenosine A_{2A} antagonists tested in this model have shown effects on motor function (Grondin et al., 1999; Hodgson et al., 2010; Kanda, Jackson et al., 1998; Kanda, Tashiro, Kuwana, & Jenner, 1998; Kuwana et al., 1999; Rose et al., 2006). Some such as istradefylline when administered alone, produce a significant but modest decrease in motor disability and an increase in locomotor activity but others, such as ST 1535 do not. As for the 6-OHDA-lesioned rat, the reasons for this are not known but may relate to intrinsic activity at A_{2A} receptors or to the relative selectivity of molecules for A_{2A} receptors compared to A_1 sites.

It again suggests that some A_{2A} antagonists might exert a mild symptomatic effect in PD if used as monotherapy. In conjunction with L-dopa or dopamine agonist drugs, A_{2A} receptor antagonists as a class, enhance the reversal of motor disability and increase in locomotor activity seen with dopaminergic agents alone. This strongly predicts an effect of adenosine A_{2A} receptor antagonists as adjunct therapy in those patients with PD already receiving dopaminergic therapy. Again, the most significant effects of the adenosine A_{2A} antagonists are seen when they are combined with low or threshold doses of L-dopa. This suggests that their greatest clinical effect should be seen in early PD where L-dopa sparing strategies could be utilized or where there is no wish to increase the dose of L-dopa or dopamine agonist being used for monotherapy. There is one other facet of the effects of adenosine A_{2A} receptor antagonists in the MPTP-treated primate that is fascinating. There appears to be a sensitization to the effects of dopaminergic drugs after a single administration of istradefylline that lasts for 2–3 days and well beyond the presence of the drug in plasma or brain (Kanda et al., 2000). This is seen with both L-dopa and dopamine agonists and it may reflect a long lasting adaptive neuromodulatory effect induced by acute blockade of A_{2A} receptors. It also suggests that another potential use of A_{2A} receptor antagonists might be in combination with other non-dopaminergic classes of drugs that modulate basal ganglia function through different mechanisms, such as amantadine or zonisamide but this remains to be tested.

3.2. Improved motor function without dyskinesia

Avoidance of dyskinesia is a key objective for novel approaches to the treatment of PD and a strong association between A_{2A} receptor expression in imaging and postmortem studies in PD and in experimental models of PD has identified the receptor as an important target (Calon et al., 2004; Ramlackhansingh et al., 2011; Xiao et al., 2011). Indeed, in preclinical studies, adenosine A_{2A} receptor antagonists have been shown to have a low propensity to provoke dyskinesia. The most extensive studies have examined the effects of istradefylline in the MPTP-treated primate. In common marmosets primed with L-dopa to express dyskinesia, istradefylline administered alone does not provoke involuntary movements on either acute or repeated treatment (Kanda, Jackson, et al., 1998; Uchida et al., 2014). When combined with L-dopa, istradefylline does not cause an increase in dyskinesia despite improving motor disability and increasing locomotor activity. In fact, on repeated administration of istradefylline with L-dopa there is a trend for the intensity of dyskinesia to diminish. All these data suggest, that as an adjunct to dopaminergic therapy in later-stage PD, the use of istradefylline would improve motor function with no significant impact on existing dyskinesia. This contrasts with the effects of increasing dopaminergic medication in this population where an exacerbation of dyskinesia inevitably occurs.

Once dyskinesia is induced in PD by dopaminergic therapy, it is persistent, if not permanent and reversal is problematic except through the use of invasive drug infusion (apomorphine, DuoDopa) or by surgery (DBS, pallidotomy). As a consequence, the avoidance of dyskinesia induction is a key objective of therapy. There is some evidence to suggest that the administration of istradefylline with dopaminergic therapy can prevent or delay the priming process that underlies dyskinesia induction. Thus, in MPTP-treated primates, initiating apomorphine treatment in conjunction with istradefylline, prevented the rapid onset of dyskinesia that occurred on repeated administration of apomorphine alone (Bibbiani et al., 2003). If true, this would suggest that adenosine A_{2A} receptor antagonists may have a role in the early treatment of PD not only for symptomatic improvement but also for dyskinesia prevention. However, in 6-OHDA-lesioned rats, istradefylline did not prevent the development of abnormal involuntary movements in response to L-dopa treatment (Lundblad, Vaudano, & Cenci, 2003).

4. ADENOSINE A_{2A} RECEPTOR ANTAGONISTS AND NONMOTOR SYMPTOMS OF PD

While the majority of adenosine A_{2A} receptors are located within the basal ganglia, some are also found in brain regions associated with higher function, such as the nucleus accumbens and olfactory tubercle. This suggests that adenosine A_{2A} receptors might play a role in the neuropsychiatric nonmotor symptoms occurring in PD (for example, anxiety, depression, and cognitive impairment). There has been relatively little investigation of this possibility until recently but an association with A_{2A} receptors had been established using A_{2A} receptor knock-out mice and pharmacological manipulation of A_{2A} receptor function (El Yacoubi, Costentin, & Vaugeois, 2003; El Yacoubi et al., 2001). This is potentially important as in PD, anxiety and depression are not well controlled by classical antidepressant and anxiolytic drugs and changes in cognition only show a small improvement in response to cholinesterase inhibitors which can, in some cases, worsen motor features.

Recently, studies undertaken using istradefylline has demonstrated more clearly a role for adenosine A_{2A} receptor antagonists in the control of neuropsychiatric components of PD. Istradefylline was shown to be active in the tail suspension test and forced swim test and on learned helplessness suggesting antidepressant potential (Yamada, Kobayashi, Mori, Jenner, & Kanda, 2013; Yamada et al., 2014). Positive effects in all three paradigms suggest a high predictive value and the possibility that would istradefylline be effective not only against anxiety in PD but also against anxiety syndromes affecting the general population. In a similar manner, istradefylline showed activity in the rat elevated plus maze and suppressed anxiety and enhanced fear extinction in the rat conditioned fear stress test that are both highly suggestive of an anxiolytic effect of the drug. In relation to cognitive impairment in PD, cognitive performance in rats with a 6-OHDA lesion in prefrontal cortex was improved by istradefylline treatment (Kadowaki Horita, Kobayashi, Mori, Jenner, & Kanda, 2013).

The wide range of actions seen with istradefylline suggests that A_{2A} adenosine receptor antagonists have a greater role to play in the treatment of PD over and above a simple improvement of motor function. The ability to improve both motor and nonmotor features of PD provides a strong additional reason for employing this drug class in PD as it provides benefit not seen with dopaminergic medications and it would limit the necessity for administration of multiple drug treatments in an elderly population where compliance with medication can be low.

5. A_{2A} ADENOSINE RECEPTOR ANTAGONISTS AND NEUROPROTECTION IN PD

So far, this overview of the effects of adenosine A_{2A} receptor antagonists in PD has focused on the symptomatic benefit that might ensue from their actions. However, another issue highly relevant to the treatment of PD requires some discussion, namely the ability of adenosine A_{2A} receptor antagonists to be disease modifying or neuroprotective. To date, no drug for the treatment of PD is labeled as disease modifying and there have been numerous failures of potential neuroprotectants in clinical trial (AlDakheel, Kalia, & Lang, 2014). This has occurred despite promising data from experimental models of PD involving nigral dopaminergic cell degeneration. There may be many reasons for this—the cause of neuronal cell death remains unclear, a lack of predictability of clinical effect from current models of PD and the heterogeneous nature of PD itself. However, a major clue has come from epidemiological studies that have repeatedly shown that caffeine intake is related to a decrease risk of developing PD (Palacios et al., 2012). Among its many actions, caffeine also acts as an adenosine A_{2A} receptor antagonist and this has led to investigation of the potential of A_{2A} receptor manipulation as a means of controlling disease progression in PD.

Initial studies demonstrated the ability of caffeine to protect against MPTP toxicity in mice and linked its effects to the A_{2A} adenosine receptor through pharmacological and genetic manipulation of both A_{2A} and A_1 receptors (Chen et al., 2001). Selective deletion of A_{2A} receptors in the forebrain neurones did not attenuate MPTP toxicity in mice suggesting that the protective effect is generated elsewhere and perhaps involved glial cells (Yu et al., 2008). Notably with respect to the known pathology of PD, the knockout of A_{2A} receptors in mice protected against neuronal degeneration induced by the expression of A53T and A30P mutant α-synuclein (Kachroo & Schwarzschild, 2012).

From a pharmacological perspective, the important discovery was that the effects of caffeine in protecting against MPTP-induced dopaminergic neuron loss, were mimicked a range of selective A_{2A} adenosine receptor antagonists, including istradefylline and SCH 58261 (Chen et al., 2001). In subsequent experiments, istradefylline was shown to protect against dopaminergic neuronal loss in 6-OHDA-lesioned rats and against both nigral cell body and striatal dopaminergic terminal loss in MPTP-treated mice as well as inhibiting the associated microglia activation (Ikeda, Kurokawa, Aoyama, & Kuwana, 2002; Pierri, Vaudano, Sager, & Englund, 2005).

While all of these data should be interpreted with caution with respect to translation to clinical effect in PD in man, the overall picture from epidemiological study through genetic manipulation of adenosine receptors and pharmacological intervention is consistent with a protective effect mediated through adenosine A_{2A} receptors and that translates into a functional response to the administration of selective A_{2A} receptor antagonists.

6. A_{2A} RECEPTOR ANTAGONISTS AND CLINICAL OUTCOMES IN PD

To date, at least four xanthine and non-xanthine selective adenosine A_{2A} receptor antagonists have been studied in clinical trial in PD—istradefylline, preladenant, vipadenant, and tozadenant—and to a point, the results obtained have been remarkably similar (Perez-Lloret & Merello, 2014; Pinna, 2009, 2014; Salamone, 2010). All have been studied in patients with mid- to late-stage PD treated with optimal dopaminergic therapy (L-dopa, dopamine agonists, MAO-B inhibitors, COMT inhibitors) but exhibiting significant "wearing off". In phase IIB studies of efficacy in PD, all adenosine A_{2A} receptor antagonists so far investigated have resulted in a significant reduction in daily "off" time in the region of 1 h compared to placebo but with either no alteration in dyskinesia duration and intensity or with only a small increase in "nontroublesome" dyskinesia (Factor et al., 2013; Hauser et al., 2011; LeWitt et al., 2008; Stacy et al., 2008). This seems to be robustly indicative of an additional increase in "on" time in this patient population that could not be achieved using dopaminergic medication.

Two of these compounds have proceeded to full phase III development in the same patient population—istradefylline and preladenant. Vipadenant was terminated because of concerns over the outcome of preclinical toxicology testing (see http://vernalis.com/media-centre/latest-releases/584-vernalis-announces-a2a-receptor-antagonist-programme-for-parkinsons-disease-continues-with-next-generation-compound) and tozadenant remains on going but still to be evaluated in phase III investigations. However, it is in phase III that the results of clinical trials started to diverge. In USA-based clinical studies, istradefylline was found to significantly decrease "off" time compared to placebo, while not increasing "troublesome" dyskinesia in one study but was negative in a second investigation (Hauser et al., 2008; Pourcher et al., 2012). However, in subsequent phase IIB and phase III investigations in Japan, the ability of istradefylline to significantly reduce "off" time in patients, 90% of whom had already been treated with L-dopa and other dopaminergic medications, without exacerbating dyskinesia was confirmed in two studies (Mizuno et al., 2010; Mizuno, Kondo, & Japanese Istradefylline Study Group, 2013). This has led to the registration and marketing of istradefylline in Japan as the first adenosine A_{2A} receptor antagonist recognized for the treatment of PD. Further clinical evaluation of istradefylline in the USA and Europe is now underway.

In contrast, the phase III development of preladenant has failed (see http://www.mercknewsroom.com/press-release/research-and-development-news/merck-provides-update-phase-iii-clinical-program-prelade). No significant decrease in "off" time was detected compared to placebo in any of the phase III investigations (and some parallel phase II investigations). This was unexpected and largely it remains unexplained. The reasons for the failure to detect the effects of adenosine A_{2A} receptor antagonists in phase III clinical trials, needs some discussion. For example, the decrease in "off" time could be lost in the noise associated with multicenter, multinational large-scale clinical investigations. This seems unlikely as both COMT inhibitors (entacapone, tolcapone) and MAO-B inhibitors (selegiline, rasagiline) that produce similar size effects have been demonstrated to be effective in similar patient populations. The definition of "optimized" dopaminergic therapy as an entry criteria to the trials may not have been adequately standardized leading to differences between centers that affected active drug outcomes.

Perhaps significantly, the design of clinical trials used to evaluate the effects of adenosine A_{2A} receptor antagonists in PD complies with the requirements of regulatory authorities to show benefit over and above that which can be achieved by the use of existing medications—i.e., dopaminergic drugs. This makes the window of opportunity for seeing a significant effect small but importantly, it does not reflect the known pharmacology of the increased effect obtained by combining A_{2A} receptor antagonists with L-dopa. All the preclinical experience shows that adenosine A_{2A} receptor antagonists show maximal effect when combined with low or threshold doses of L-dopa and that effect is difficult to see when high L-dopa doses are employed. In one preladenant study, rasagiline was used as an active comparator but no effect of the drug was seen compared to placebo (Hauser et al., 2014—presented at the American Academy of Neurology). Since rasagiline is already approved and marketed for use in "wearing off" in PD as shown by the LARGO and PRESTO studies (Parkinson Study Group, 2005; Rascol et al., 2005), the underlying reasons for this failure may reflect on why preladenant also appeared inactive. In general, it seems that the more diverse phase III clinical trials become with respect to numbers of patients, centers, and territories, the more difficult it becomes to see a consistent effect of adenosine A_{2A} antagonists in PD. This may be the reason behind the success of the Japanese phase III studies of istradefylline where a relative small number of centers were used within one territory in a carefully conducted and monitored investigation. A similar approach may be the way forward in subsequent A_{2A} receptor antagonist developments.

The role of caffeine intake in the negative outcome of the preladenant phase III studies has been debated. It could be argued that a larger and more variable population of patients with differing caffeine intakes would potentially have an effect on trial endpoints if caffeine contributed to the decrease in "off" time. However, many smaller phase II studies of a range of adenosine A_{2A} receptor antagonists have been successful despite unrestricted caffeine consumption. Caffeine intake presumably would have been the same at entry into the studies as on completion so being reflected in basal motor disability scores as judged by the UPDRS. However, there is so far little evidence that caffeine exerts anything but a modest antiparkinsonian action and does not seem to improve day time somnolence (Postuma et al., 2012).

Certainly, caffeine causes hyperactivity when administered to animals and this may affect behavior in models of PD (Fenu & Morelli, 1998). But caffeine is pharmacologically different from the selective A_{2A} receptor antagonists currently under development for PD. Caffeine is a nonselective antagonist of both A_1 and A_{2A} adenosine receptors and also a phosphodiesterase inhibitor among other actions. It is estimated that the caffeine intake from a cup of coffee would only lead to 20% occupation of A_{2A} receptors (Fredholm, Battig, Holmen, Nehlig, & Zvartau, 1999). At this time, the argument that the A_{2A} receptor antagonists used to treat PD are merely the equivalent of "a cup of coffee" has no merit.

While the role of adenosine A_{2A} receptor antagonists as an adjunct to dopaminergic therapy to treat "wearing off" in PD seems established, these drugs may have more use and efficacy in earlier treatment strategies. So far, there is no evidence of an effect of either istradefylline or preladenant as monotherapy in early PD, although this has not been thoroughly studied. However, the synergy seen between istradefylline and low doses of both L-dopa and dopamine agonist drugs in experimental models of PD suggests two further possibilities. First, istradefylline might be used in an L-dopa sparing strategy to reduce L-dopa dosage in patients where unacceptable side effects, such as dyskinesia, are occurring but a further improvement in motor function is required. Second, istradefylline might be used in patients who show insufficient improvement from low dose L-dopa or dopamine agonist monotherapy to avoid any further increase in dopaminergic medication so that the onset of motor fluctuations and motor complications can be avoided. Both potential indications should be explored in subsequent clinical investigations.

7. CONCLUSIONS

Adenosine A_{2A} receptor antagonists offer an exciting new approach to the symptomatic treatment of PD. Preclinical studies have provided abundant evidence of their potential to improve motor symptoms and to reduce the risk of dyskinesia without the usual dopaminergic side effects associated with current treatment. In addition, they may provide a more effective means of addressing some of the nonmotor, as well as motor symptoms of PD so avoiding a polypharmacy approach to patient management. The indication that there may be a neuroprotective element to their actions is exciting but needs further exploration. What is required now is a greater clinical experience of the use of adenosine A_{2A} receptor antagonists in the treatment of PD and further clinical investigation of their full range of effects in both early- and late-stage PD.

REFERENCES

Abbruzzese, G., Barone, P., Bonuccelli, U., Lopiano, L., & Antonini, A. (2012). Continuous intestinal infusion of levodopa/carbidopa in advanced Parkinson's disease: Efficacy, safety and patient selection. *Functional Neurology, 27*, 147–154.

AlDakheel, A., Kalia, L. V., & Lang, A. E. (2014). Pathogenesis-targeted, disease-modifying therapies in Parkinson disease. *Neurotherapeutics: The Journal of the American Society for Experimental NeuroTherapeutics, 11*, 6–23.

Antonini, A., Ursino, G., Calandrella, D., Bernardi, L., & Plebani, M. (2010). Continuous dopaminergic delivery in Parkinson's disease. *Journal of Neurology, 257*, S305–S308.

Bibbiani, F., Oh, J. D., Petzer, J. P., Castagnoli, N., Jr., Chen, J. F., Schwarzschild, M. A., et al. (2003). A_{2A} antagonist prevents dopamine agonist-induced motor complications in animal models of Parkinson's disease. *Experimental Neurology, 184*, 285–294.

Brooks, D. J., Doder, M., Osman, S., Luthra, S. K., Hirani, E., Hume, S., et al. (2008). Positron emission tomography analysis of [11C]KW-6002 binding to human and rat adenosine A_{2A} receptors in the brain. *Synapse, 62*, 671–681.

Brotchie, J., & Jenner, P. (2011). New approaches to therapy. *International Review of Neurobiology, 98*, 123–150.

Calon, F., Dridi, M., Hornykiewicz, O., Bedard, P. J., Rajput, A. H., & Di Paolo, T. (2004). Increased adenosine A_{2A} receptors in the brain of Parkinson's disease patients with dyskinesias. *Brain: A Journal of Neurology, 127*, 1075–1084.

Chaudhuri, K. R., Odin, P., Antonini, A., & Martinez-Martin, P. (2011). Parkinson's disease: The non-motor issues. *Parkinsonism & Related Disorders, 17*, 717–723.

Chen, J. F., Xu, K., Petzer, J. P., Staal, R., Xu, Y. H., Beilstein, M., et al. (2001). Neuroprotection by caffeine and A_{2A} adenosine receptor inactivation in a model of Parkinson's disease. *Journal of Neuroscience: The Official Journal of the Society for Neuroscience, 21*, RC143.

El Yacoubi, M., Costentin, J., & Vaugeois, J. M. (2003). Adenosine A_{2A} receptors and depression. *Neurology, 61*, S82–S87.

El Yacoubi, M., Ledent, C., Parmentier, M., Bertorelli, R., Ongini, E., Costentin, J., et al. (2001). Adenosine A_{2A} receptor antagonists are potential antidepressants: Evidence based on pharmacology and A_{2A} receptor knockout mice. *British Journal of Pharmacology, 134*, 68–77.

Factor, S. A., Wolski, K., Togasaki, D. M., Huyck, S., Cantillon, M., Ho, T. W., et al. (2013). Long-term safety and efficacy of preladenant in subjects with fluctuating Parkinson's disease. *Movement Disorders: Official Journal of the Movement Disorder Society, 28*, 817–820.

Fenu, S., & Morelli, M. (1998). Motor stimulant effects of caffeine in 6-hydroxydopamine-lesioned rats are dependent on previous stimulation of dopamine receptors: A different role of D1 and D2 receptors. *European Journal of Neuroscience, 10*, 1878–1884.

Fox, S. H., Brotchie, J. M., & Lang, A. E. (2008). Non-dopaminergic treatments in development for Parkinson's disease. *Lancet Neurology, 7*, 927–938.

Fredholm, B. B., Battig, K., Holmen, J., Nehlig, A., & Zvartau, E. E. (1999). Actions of caffeine in the brain with special reference to factors that contribute to its widespread use. *Pharmacological Reviews, 51*, 83–133.

Grondin, R., Bedard, P. J., Hadj Tahar, A., Gregoire, L., Mori, A., & Kase, H. (1999). Antiparkinsonian effect of a new selective adenosine A_{2A} receptor antagonist in MPTP-treated monkeys. *Neurology, 52*, 1673–1677.

Guridi, J., Gonzalez-Redondo, R., & Obeso, J. A. (2012). Clinical features, pathophysiology, and treatment of levodopa-induced dyskinesias in Parkinson's disease. *Parkinson's Disease, 2012*, 943159.

Hauser, R. A., Cantillon, M., Pourcher, E., Micheli, F., Mok, V., Onofrj, M., et al. (2011). Preladenant in patients with Parkinson's disease and motor fluctuations: A phase 2, double-blind, randomised trial. *Lancet Neurology, 10*, 221–229.

Hauser, R. A., Shulman, L. M., Trugman, J. M., Roberts, J. W., Mori, A., Ballerini, R., et al. (2008). Study of istradefylline in patients with Parkinson's disease on levodopa with motor fluctuations. *Movement Disorders: Official Journal of the Movement Disorder Society, 23*, 2177–2185.
Hauser, R., Stocchi, F., Rascol, O., Huyck, S., Ha, X., Capece, R., et al. (2014). Phase-3 clinical trials of adjunctive therapy with preladenant, an adenosine 2a antagonist, in patients with Parkinson's disease (P7.087). *Neurology, 82*(10 Suppl.), P7.087.
Hodgson, R. A., Bedard, P. J., Varty, G. B., Kazdoba, T. M., Di Paolo, T., Grzelak, M. E., et al. (2010). Preladenant, a selective A_{2A} receptor antagonist, is active in primate models of movement disorders. *Experimental Neurology, 225*, 384–390.
Hodgson, R. A., Bertorelli, R., Varty, G. B., Lachowicz, J. E., Forlani, A., Fredduzzi, S., et al. (2009). Characterization of the potent and highly selective A_{2A} receptor antagonists preladenant and SCH 412348 [7-[2-[4–2,4-difluorophenyl]-1-piperazinyl]ethyl]-2-(2-furanyl)-7H-pyrazolo[4,3-e][1,2,4]triazolo[1,5-c]pyrimidin-5-amine] in rodent models of movement disorders and depression. *Journal of Pharmacology and Experimental Therapeutics, 330*, 294–303.
Ikeda, K., Kurokawa, M., Aoyama, S., & Kuwana, Y. (2002). Neuroprotection by adenosine A_{2A} receptor blockade in experimental models of Parkinson's disease. *Journal of Neurochemistry, 80*, 262–270.
Johnston, T. H., & Brotchie, J. M. (2006). Drugs in development for Parkinson's disease: An update. *Current Opinion in Investigational Drugs, 7*, 25–32.
Kachroo, A., & Schwarzschild, M. A. (2012). Adenosine A_{2A} receptor gene disruption protects in an alpha-synuclein model of Parkinson's disease. *Annals of Neurology, 71*, 278–282.
Kadowaki Horita, T., Kobayashi, M., Mori, A., Jenner, P., & Kanda, T. (2013). Effects of the adenosine A_{2A} antagonist istradefylline on cognitive performance in rats with a 6-OHDA lesion in prefrontal cortex. *Psychopharmacology, 230*, 345–352.
Kanda, T., Jackson, M. J., Smith, L. A., Pearce, R. K., Nakamura, J., Kase, H., et al. (1998). Adenosine A_{2A} antagonist: A novel antiparkinsonian agent that does not provoke dyskinesia in parkinsonian monkeys. *Annals of Neurology, 43*, 507–513.
Kanda, T., Jackson, M. J., Smith, L. A., Pearce, R. K., Nakamura, J., Kase, H., et al. (2000). Combined use of the adenosine A_{2A} antagonist KW-6002 with L-DOPA or with selective D1 or D2 dopamine agonists increases antiparkinsonian activity but not dyskinesia in MPTP-treated monkeys. *Experimental Neurology, 162*, 321–327.
Kanda, T., Tashiro, T., Kuwana, Y., & Jenner, P. (1998). Adenosine A_{2A} receptors modify motor function in MPTP-treated common marmosets. *Neuroreport, 9*, 2857–2860.
Kase, H., Aoyama, S., Ichimura, M., Ikeda, K., Ishii, A., Kanda, T., et al. (2003). Progress in pursuit of therapeutic A_{2A} antagonists the adenosine A_{2A} receptor selective antagonist KW6002: Research and development toward a novel nondopaminergic therapy for Parkinson's disease. *Neurology, 61*, S97–S100.
Koga, K., Kurokawa, M., Ochi, M., Nakamura, J., & Kuwana, Y. (2000). Adenosine A_{2A} receptor antagonists KF17837 and KW-6002 potentiate rotation induced by dopaminergic drugs in hemi-Parkinsonian rats. *European Journal of Pharmacology, 408*, 249–255.
Kuwana, Y., Shiozaki, S., Kanda, T., Kurokawa, M., Koga, K., Ochi, M., et al. (1999). Antiparkinsonian activity of adenosine A_{2A} antagonists in experimental models. *Advances in Neurology, 80*, 121–123.
LeWitt, P. A., Guttman, M., Tetrud, J. W., Tuite, P. J., Mori, A., Chaikin, P., et al. (2008). Adenosine A_{2A} receptor antagonist istradefylline (KW-6002) reduces "off" time in Parkinson's disease: A double-blind, randomized, multicenter clinical trial (6002-US-005). *Annals of Neurology, 63*, 295–302.
Lundblad, M., Vaudano, E., & Cenci, M. A. (2003). Cellular and behavioural effects of the adenosine A_{2a} receptor antagonist KW-6002 in a rat model of L-DOPA-induced dyskinesia. *Journal of Neurochemistry, 84*, 1398–1410.
Martinez-Martin, P., Rodriguez-Blazquez, C., Kurtis, M. M., & Chaudhuri, K. R., SNMSS Validation Group. (2011). The impact of non-motor symptoms on health-related quality

of life of patients with Parkinson's disease. *Movement Disorders: Official Journal of the Movement Disorder Society, 26*, 399–406.
Mizuno, Y., Hasegawa, K., Kondo, T., Kuno, S., Yamamoto, M., & Japanese Istradefylline Study Group (2010). Clinical efficacy of istradefylline (KW-6002) in Parkinson's disease: A randomized, controlled study. *Movement Disorders: Official Journal of the Movement Disorder Society, 25*, 1437–1443.
Mizuno, Y., Kondo, T., & Japanese Istradefylline Study Group. (2013). Adenosine A2A receptor antagonist istradefylline reduces daily OFF time in Parkinson's disease. *Movement Disorders: Official Journal of the Movement Disorder Society, 28*, 1138–1141.
Morelli, M., Carta, A. R., & Jenner, P. (2009). Adenosine A_{2A} receptors and Parkinson's disease. *Handbook of Experimental Pharmacology, 193*, 589–615.
Mori, A., & Shindou, T. (2003). Modulation of GABAergic transmission in the striatopallidal system by adenosine A_{2A} receptors: A potential mechanism for the antiparkinsonian effects of A_{2A} antagonists. *Neurology, 61*, S44–S48.
Obeso, J. A., Rodriguez-Oroz, M. C., Benitez-Temino, B., Blesa, F. J., Guridi, J., Marin, C., et al. (2008). Functional organization of the basal ganglia: Therapeutic implications for Parkinson's disease. *Movement Disorders: Official Journal of the Movement Disorder Society, 23*(Suppl. 3), S548–S559.
Palacios, N., Gao, X., McCullough, M. L., Schwarzschild, M. A., Shah, R., Gapstur, S., et al. (2012). Caffeine and risk of Parkinson's disease in a large cohort of men and women. *Movement Disorders: Official Journal of the Movement Disorder Society, 27*, 1276–1282.
Parkinson Study Group. (2005). A randomized placebo-controlled trial of rasagiline in levodopa-treated patients with Parkinson disease and motor fluctuations: The PRESTO study. *Archives of Neurology, 62*, 241–248.
Perez-Lloret, S., & Merello, M. (2014). Two new adenosine receptor antagonists for the treatment of Parkinson's disease: Istradefylline versus tozadenant. *Expert Opinion on Pharmacotherapy, 15*, 1097–1107.
Pierri, M., Vaudano, E., Sager, T., & Englund, U. (2005). KW-6002 protects from MPTP induced dopaminergic toxicity in the mouse. *Neuropharmacology, 48*, 517–524.
Pinna, A. (2009). Novel investigational adenosine A_{2A} receptor antagonists for Parkinson's disease. *Expert Opinion on Investigational Drugs, 18*, 1619–1631.
Pinna, A. (2014). Adenosine A_{2A} receptor antagonists in Parkinson's disease: Progress in clinical trials from the newly approved istradefylline to drugs in early development and those already discontinued. *CNS Drugs, 28*, 455–474.
Pinna, A., Volpini, R., Cristalli, G., & Morelli, M. (2005). New adenosine A_{2A} receptor antagonists: Actions on Parkinson's disease models. *European Journal of Pharmacology, 512*, 157–164.
Postuma, R. B., Lang, A. E., Munhoz, R. P., Charland, K., Pelletier, A., Moscovich, M., et al. (2012). Caffeine for treatment of Parkinson disease: A randomized controlled trial. *Neurology, 79*, 651–658.
Pourcher, E., Fernandez, H. H., Stacy, M., Mori, A., Ballerini, R., & Chaikin, P. (2012). Istradefylline for Parkinson's disease patients experiencing motor fluctuations: Results of the KW-6002-US-018 study. *Parkinsonism & Related Disorders, 18*, 178–184.
Ramlackhansingh, A. F., Bose, S. K., Ahmed, I., Turkheimer, F. E., Pavese, N., & Brooks, D. J. (2011). Adenosine 2A receptor availability in dyskinetic and nondyskinetic patients with Parkinson disease. *Neurology, 76*, 1811–1816.
Rascol, O., Brooks, D. J., Melamed, E., Oertel, W., Poewe, W., Stocchi, F., et al. (2005). Rasagiline as an adjunct to levodopa in patients with Parkinson's disease and motor fluctuations (LARGO, Lasting effect in adjunct therapy with rasagiline given once daily, study): A randomised, double-blind, parallel-group trial. *Lancet, 365*, 947–954.
Rektorova, I., Aarsland, D., Chaudhuri, K. R., & Strafella, A. P. (2011). Nonmotor symptoms of Parkinson's disease. *Parkinson's Disease, 2011*, 351461.

Rose, S., Jackson, M. J., Smith, L. A., Stockwell, K., Johnson, L., Carminati, P., et al. (2006). The novel adenosine A_{2a} receptor antagonist ST1535 potentiates the effects of a threshold dose of L-DOPA in MPTP treated common marmosets. *European Journal of Pharmacology, 546*, 82–87.

Rose, S., Ramsay Croft, N., & Jenner, P. (2007). The novel adenosine A_{2a} antagonist ST1535 potentiates the effects of a threshold dose of L-dopa in unilaterally 6-OHDA-lesioned rats. *Brain Research, 1133*, 110–114.

Salamone, J. D. (2010). Preladenant, a novel adenosine A_{2A} receptor antagonist for the potential treatment of parkinsonism and other disorders. *IDrugs: The Investigational Drugs Journal, 13*, 723–731.

Shindou, T., Mori, A., Kase, H., & Ichimura, M. (2001). Adenosine A_{2A} receptor enhances GABA(A)-mediated IPSCs in the rat globus pallidus. *Journal of Physiology, 532*, 423–434.

Shindou, T., Nonaka, H., Richardson, P. J., Mori, A., Kase, H., & Ichimura, M. (2002). Presynaptic adenosine A_{2A} receptors enhance GABAergic synaptic transmission via a cyclic AMP dependent mechanism in the rat globus pallidus. *British Journal of Pharmacology, 136*, 296–302.

Shindou, T., Richardson, P. J., Mori, A., Kase, H., & Ichimura, M. (2003). Adenosine modulates the striatal GABAergic inputs to the globus pallidus via adenosine A_{2A} receptors in rats. *Neuroscience Letters, 352*, 167–170.

Shiozaki, S., Ichikawa, S., Nakamura, J., Kitamura, S., Yamada, K., & Kuwana, Y. (1999). Actions of adenosine A_{2A} receptor antagonist KW-6002 on drug-induced catalepsy and hypokinesia caused by reserpine or MPTP. *Psychopharmacology, 147*, 90–95.

Stacy, M., Silver, D., Mendis, T., Sutton, J., Mori, A., Chaikin, P., et al. (2008). A 12-week, placebo-controlled study (6002-US-006) of istradefylline in Parkinson disease. *Neurology, 70*, 2233–2240.

Tronci, E., Simola, N., Borsini, F., Schintu, N., Frau, L., Carminati, P., et al. (2007). Characterization of the antiparkinsonian effects of the new adenosine A_{2A} receptor antagonist ST1535: Acute and subchronic studies in rats. *European Journal of Pharmacology, 566*, 94–102.

Uchida, S., Tashiro, T., Kawai-Uchida, M., Mori, A., Jenner, P., & Kanda, T. (2014). The adenosine A_{2A}-receptor antagonist istradefylline enhances the motor response of L-DOPA without worsening dyskinesia in MPTP-treated common marmosets. *Journal of Pharmacological Sciences, 124*, 480–485.

Xiao, D., Cassin, J. J., Healy, B., Burdett, T. C., Chen, J. F., Fredholm, B. B., et al. (2011). Deletion of adenosine A(1) or A((2)A) receptors reduces L-3,4-dihydroxyphenylalanine-induced dyskinesia in a model of Parkinson's disease. *Brain Research, 1367*, 310–318.

Yamada, K., Kobayashi, M., Mori, A., Jenner, P., & Kanda, T. (2013). Antidepressant-like activity of the adenosine A_{2A} receptor antagonist, istradefylline (KW-6002), in the forced swim test and the tail suspension test in rodents. *Pharmacology, Biochemistry, and Behavior, 114–115*, 23–30.

Yamada, K., Kobayashi, M., Shiozaki, S., Ohta, T., Mori, A., Jenner, P., et al. (2014). Antidepressant activity of the adenosine A2A receptor antagonist, istradefylline (KW-6002) on learned helplessness in rats. *Psychopharmacology, 231*(14), 2839–2849.

Yu, L., Shen, H. Y., Coelho, J. E., Araujo, I. M., Huang, Q. Y., Day, Y. J., et al. (2008). Adenosine A_{2A} receptor antagonists exert motor and neuroprotective effects by distinct cellular mechanisms. *Annals of Neurology, 63*, 338–346.

CHAPTER FOUR

Mode of Action of Adenosine A_{2A} Receptor Antagonists as Symptomatic Treatment for Parkinson's Disease

Akihisa Mori[1]

Strategic Product Portfolio Department, Kyowa Hakko Kirin Co., Ltd., Tokyo, Japan
[1]Corresponding author: e-mail address: akihisa.mori@kyowa-kirin.co.jp

Contents

Abstract

Adenosine A_{2A} receptor antagonists are classified to be a recent new therapeutic strategy for the symptomatic treatment of Parkinson's disease, a hypokinetic movement disorder. First, this chapter addresses how adenosine A_{2A} receptors are involved with brain motor control via the basal ganglia–thalamocortical circuit, considering anatomical and ultrastructural localization of the receptor in critical areas/neurons of the circuit. Then, based on the understanding of the functional significance of the receptor in the circuit, the mode of action of adenosine A_{2A} receptor antagonists is explained by dynamism of the circuit and possible cellular mechanisms, highlighting the importance of the pathophysiological difference proposed between normal and disease state.

International Review of Neurobiology, Volume 119
ISSN 0074-7742
http://dx.doi.org/10.1016/B978-0-12-801022-8.00004-0

1. INTRODUCTION

Among the many challenges for drug discovery related to adenosine receptor science, there is a growing body of evidence that adenosine A_{2A} receptor antagonists have finally become a new treatment class for Parkinson's disease (PD). The first-in-class adenosine A_{2A} receptor antagonist Istradefylline (KW-6002) was launched in Japan in 2013. This innovation resulted not only from nonclinical studies with animal models (see Chapter 3 by Peter Jenner) but also from clinical studies (Hauser et al., 2008; LeWitt et al., 2008; Mizuno et al., 2010; Mizuno, Kondo, & the Japanese Istradefylline Study Group, 2013; Stacy et al., 2008). Recently, the major mechanism for adenosine A_{2A} receptor function in motor control is considered to be in the basal ganglia–thalamocortical circuit in the brain, due to the high density of the receptor localization and also the physiological significance of the circuit for motor control. This chapter addresses several hypotheses proposed for the functional/anatomical role of the adenosine A_{2A} receptors and describes how/why the blockade of the receptors by A_{2A} antagonists brings motor improvement to PD patients, considering the pathophysiology of the disease.

2. THE BASAL GANGLIA–THALAMOCORTICAL CIRCUIT AND PATHOPHYSIOLOGY OF PD

Over the past two decades, recent research has resulted in new insights into the structure and function of basal ganglia and into the pathophysiological basis of PD (DeLong, 1990; Wichmann & DeLong, 2003). Figure 4.1 shows a simplified diagram of the basal ganglia–thalamocortical circuitry. The striatum has two GABAergic output pathways, one of which is the striatopallidal pathway projecting to the external segment of globus pallidus (GPe), initiating the "indirect pathway," which passes first to the GPe, then from the GPe to the subthalamic nucleus (STN), and finally to the internal segment of the globus pallidus (GPi) and the substantia nigra pars reticulata ([SNr], the GPi/SNr complex); and the other of which is the striatonigral pathway projecting to the GPi/SNr complex, the "direct pathway." One of the key points to remember is that the nigrostriatal dopaminergic projection from the substantia nigra pars compacta (SNc) onto the striatum provides two opposite influences on those striatal two output pathways. The striatopallidal pathway receives an inhibitory influence via dopamine D2

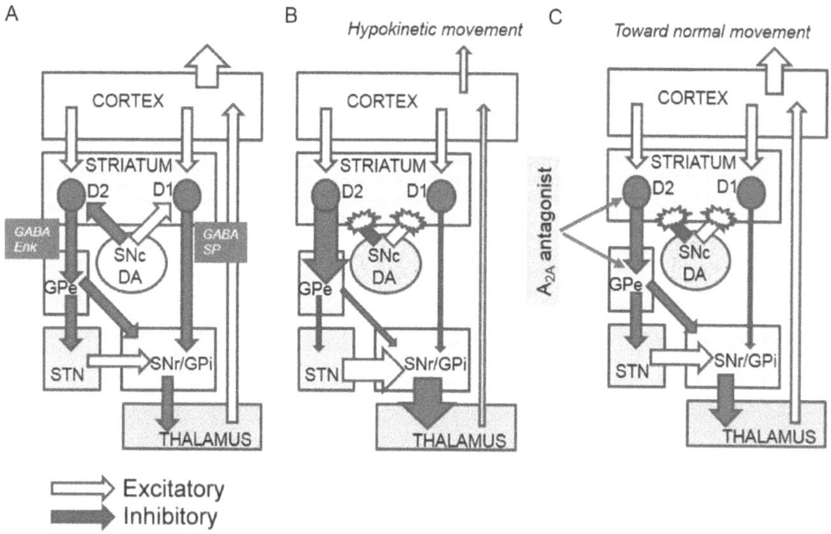

Figure 4.1 Simplified schematic diagram of the basal ganglia–thalamocortical circuit for three states with a proposed mechanism of adenosine A_{2A} receptor antagonist as symptomatic treatment for Parkinson's disease. (A) Normal state. (B) Parkinson's disease state (PD state). Excessive activation of the striatopallidal pathway induced by loss of D2 receptor-mediated inhibition, being followed by augmented excitatory output from STN, leads to exaggerated activation of the SNr/GPi, which translates as increased inhibition of the thalamus. (C) Treatment for Parkinson's disease with adenosine A_{2A} receptor antagonist (PD + A_{2A} therapy). Reduced activity of the striatopallidal pathway by A_{2A} receptor antagonists can recover an entire coordination of the circuit, via decreased inhibition of the thalamus, toward normal control. GABA, gamma-aminobutyric acid; D1, dopamine D1 receptors; D2, dopamine D2 receptors; ENK, enkephalin; SP, substance P. *Adapted from Alexander and Crutcher (1990), DeLong (1990), Kase, Mori, and Jenner (2004), and Xu, Batia, and Schwarzchild (2005).*

receptors, and the striatonigral pathway receives an excitatory influence via dopamine D1 receptors (see Fig. 4.1A). Therefore, once nigral dopaminergic neurons are degenerated (like in PD), the inhibitory and excitatory regulation of the striatum mediated via dopamine receptors becomes lost, resulting in both an increased excitation of the striatopallidal pathway and a decreased activity of the striatonigral pathway. Such an imbalance in activity between the two major striatal output pathways causes an excessive inhibition of the thalamocortical pathway, partly via the GPe reducing GABAergic inhibition of the STN and then via the STN inducing an increased glutamatergic input onto the GPi/SNr complex. Subsequently, the thalamocortical activity, suppressed by GABAergic inhibition from the GPi/SNr complex, drives an abnormal output from the cortex (see Fig. 4.1B), resulting in hypokinetic movement. The indirect pathway including the striatopallidal pathway has been considered to be more significant, than the direct pathway, in the expression of motor disability in PD (Obeso et al., 2004). In fact, clinically used dopamine D2 receptor agonists for PD therapy are considered to work on the indirect pathway. Furthermore, with recent advances in understanding of the basal ganglia–thalamocortical circuit, high-frequency deep brain stimulation (DBS) targeting the STN (STN–DBS), resulting in neuronal inhibition of subthalamic output, has been introduced as a popular neurosurgical treatment of PD (Krack, Hariz, Baunez, Guridi, & Obeso, 2010).

Adenosine A_{2A} receptors are known to strictly express in the striatopallidal medium spiny neurons (MSNs) (Section 3). Since the MSNs are the primary neuronal target of adenosine A_{2A} receptors and considering that the proposed PD pathophysiology is the striatopallidal pathway, it is expected that the A_{2A} receptor may regulate the activity of the pathway.

3. STRIATAL MSNs

One of the key anatomical structures of the basal ganglia is the stratum (caudate nucleus and putamen). Prior to presenting the detailed anatomical and ultrastructural discussion for adenosine A_{2A} receptor localization in the basal ganglia (Section 4), this section briefly introduces the striatal neuronal population.

The principal neurons in the striatum are the GABAergic MSNs, constituting roughly 90% of all striatal neurons in most mammals (Kawaguchi, 1997). The remaining interneurons are classified into large cholinergic neurons and three distinct GABAergic interneurons (Tapper, Koós, & Wilson, 2004). These striatal MSNs receive massive glutamatergic inputs from the cerebral cortex and thalamus, and various intrastriatal input from interneurons.

MSNs as principal neurons are further divided into striatopallidal and striatonigral MSNs, based on which area of the brain they pass through as striatal outputs. The striatopallidal MSNs pass through the GPe from the striatum, initiating the "indirect pathway." The MSNs have specific markers which are dopamine D2 receptors and enkephalin. The striatonigral MSNs themselves compose the "direct pathway" in the basal ganglia–thalamocortical circuit, and their specific markers are dopamine D1 receptors and dynorphin/substance P (SP) (Alexander & Crutcher, 1990; Gërfen, 2004; Gërfen et al., 1990; Graybiel, 1990). Recently, it has further been confirmed that D1 and D2 MSNs differ in their intrinsic excitability and dendritic morphology (Surmeier et al., 2010).

4. LOCALIZATION OF ADENOSINE A_{2A} RECEPTORS ON/AROUND STRIATAL MSNs

4.1. Regional and cellular anatomy of adenosine A_{2A} receptors

A_{2A} receptor messenger ribonucleic acid (mRNA) has been richly detected in the caudate nucleus, putamen, nucleus accumbens, and olfactory tubercles in rodents, primates, and humans (Augood & Emson, 1994; Dixon, Gubitz, Sirinathsinghji, Richardson, & Freeman, 1996; Fink et al., 1992; Johansson, Georgiev, Parkinson, & Fredholm, 1993; Schiffmann, Jacobs, & Vanderhaeghen, 1991; Schiffmann, Libert, Vassart, & Vanderhaeghen, 1991; Svenningsson et al., 1998, 1997; Xu et al., 2005) and has further been demonstrated to be expressed in the cortex (Lee et al., 2003), hippocampus (Cunha et al., 1994; Lee et al., 2003), cerebellar Purkinje cells (Svenningsson, Le Moine, et al., 1997), and olfactory bulb (Kaelin-Lang, Lauterburg, & Burgunder, 1999) as well. Recent consensus is that the brain regions expressing high levels of the adenosine A_{2A} receptor molecule are the caudate nucleus, putamen, nucleus accumbens, olfactory tubercles, and GPe and have been reported in the rat and human brains (DeMet & Chicz-DeMet, 2002; Fredholm, Lindstrom, Dionisotti, & Ongini, 1998; Jarvis & Williams, 1989; Martinez-Mir, Probst, & Palacios, 1991; Nonaka et al., 1994; Parkinson & Fredholm, 1990; Svenningsson, Hall, Sedvall, & Fredholm, 1997). Low levels of immunoreactivity of the receptor have been found in cortex, hippocampus, thalamus, and cerebellum (Lee et al., 2003; Rosin, Robeva, Woodard, Guyenet, & Linden, 1998). Although in the globus pallidus, A_{2A} receptor transcript is not detected (Augood & Emson, 1994; Fink et al., 1992; Schiffmann, Jacobs, et al., 1991), immunohistochemical study in the rat brain revealed that nerve terminals located in the region had the A_{2A} receptor molecule, which was estimated at the axon terminal of the striatopallidal neurons (Rosin, Hettinger, Lee, & Linden, 2003). Triggered by PD pathophysiology drawn with the basal ganglia–thalamocortical circuit, there has been a growing body of interest in A_{2A} receptors in the striatum and GPe, both of which are very crucial portions in the circuit for constructing the striatopallidal pathway. It has been revealed that the A_{2A} receptor mRNA is highly and specifically expressed in the GABAergic striatopallidal MSNs, but not in the striatonigral MSNs; the A_{2A} receptor mRNA colocalizes with the D2 receptor mRNA (Fink et al., 1992; Schiffmann & Vanderhaeghen, 1993; Svenningsson et al., 1998) as well as with the preproenkephalin mRNA, although the A_{2A} mRNA (Schiffmann, Jacobs, et al., 1991; Svenningsson et al., 1998) had very little coexpression with D1 receptor and SP mRNA (Svenningsson et al., 1998).

4.2. Ultrastructural aspect of adenosine A_{2A} receptors

To identify morphological substrates for the integration of striatal function, an ultrastructural analysis of A_{2A} receptors on/around striatal MSNs, using the vesicular glutamate transporters as makers of glutamatergic terminals, has been conducted by Rosin et al. (1998, 2003). The results are summarized below:

- 83% of all A_{2A} receptors containing synaptic contacts were postsynaptic A_{2A} receptors forming asymmetric synapses, suggesting a site of excitatory input onto striatopallidal MSNs.
- 12% of all A_{2A} receptors containing synaptic contacts were presynaptic A_{2A} receptors on excitatory terminals forming asymmetric synapses, suggesting glutamatergic synaptic contacts into MSNs.
- A_{2A} receptors were found postsynaptically to inhibitory symmetric input, suggesting synaptic contacts from GABAergic interneurons, cholinergic neurons, and local axon collaterals from MSNs.
- 3% of all A_{2A} receptors containing synaptic contacts were presynaptic A_{2A} receptors found on terminals forming symmetric synapses with GABA-labeled or unlabeled soma or dendritic profiles, suggesting GABAergic terminals of recurrent axon collaterals.

Also, Hettinger, Lee, Linden, and Rosin (2001) demonstrated that of the 714 A_{2A}-immunoreactive profiles examined in rat striatum, 37% were apposed to GABA-labeled profiles. The most common appositions were A_{2A}-labeled dendrites apposed to GABA-labeled dendrites (18%), axon terminal (4%), and soma (1%); and A_{2A}-labeled axons apposed to GABA-labeled dendrites (8%), axon terminals (2%), and somata (1%) (Xu et al., 2005). Those A_{2A} receptors are considered to be on spiny neurons, but not on GABAergic interneurons, since the A_{2A} receptor mRNA has not been detected in GABAergic interneurons of rodents or primates (Augood & Emson, 1994; Schiffmann, Jacobs, et al., 1991; Schiffmann, Libert, et al., 1991) and there was little evidence for A_{2A} receptor immunoreactivity in the soma of GABAergic interneurons in the striatum (Hettinger et al., 2001).

Very interestingly, there were no striatal nerve terminals colabeling A_{2A} receptor and tyrosine hydroxylase, suggesting less possibility that nigrostriatal dopaminergic neurons have adenosine A_{2A} receptors (Hettinger et al., 2001).

Whether striatal cholinergic large aspiny interneurons express significant levels of A_{2A} receptors is still a question. Although the A_{2A} receptor mRNA was detected in these neurons using a modification of 3′ end amplification polymerase chain reaction (PCR) (Preston et al., 2000; Richardson et al., 2000) and a mouse immunofluorescence analysis showed that choline acetyltransferase-positive neurons were labeled by A_{2A} receptors (Tozzi et al., 2011), most *in situ* hybridization studies did not detect the expression in cholinergic neurons in rodents or primates (Augood & Emson, 1994; Fink et al., 1992; Schiffmann, Jacobs, et al., 1991; Schiffmann, Libert, et al., 1991; Svenningsson et al., 1998).

From those cellular and ultrastructural analysis results for adenosine A_{2A} receptor localization in the striatum, the author attempted to make a schematic diagram of A_{2A} receptors on/around striatopallidal MSNs (see Fig. 4.2A).

Section 5 provides several insights/aspects regarding hypotheses of the mode of action in adenosine A_{2A} receptor antagonist-induced antiparkinsonian and how A_{2A} receptors contribute to the regulation of the striatopallidal pathway and other functions of the basal ganglia.

5. PROPOSED MECHANISM OF ADENOSINE A_{2A} RECEPTOR FUNCTION AND MODE OF ACTION OF A_{2A} RECEPTOR ANTAGONISTS ON MOTOR CONTROL VIA THE BASAL GANGLIA

5.1. A_{2A} receptor-induced dual excitatory modulation of striatopallidal GABAergic system

Striatal and pallidal A_{2A} receptor-mediated excitatory modulation of striatopallidal system has been proposed to be the physiological function of adenosine via A_{2A} receptors in the basal ganglia (Mori & Shindou, 2003). The following sections summarize the intrastriatal and pallidal mechanisms of the modulation and an implication to the mode of action of A_{2A} receptor antagonists as PD therapy.

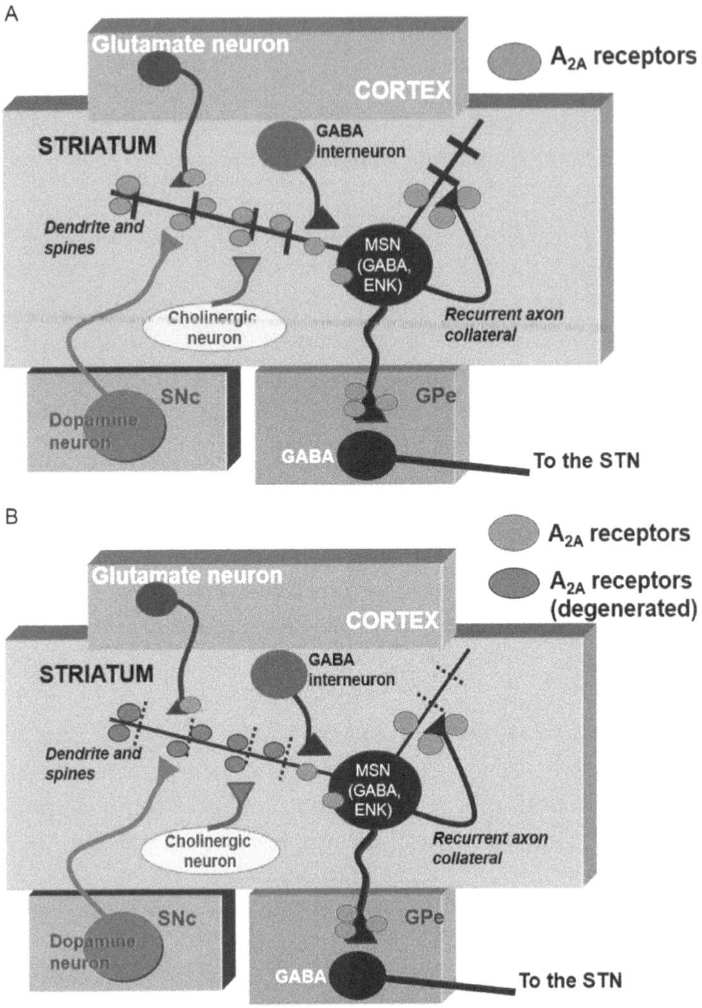

Figure 4.2 Schematic diagram of a striatopallidal medium spiny neuron (MSN) possessing adenosine A_{2A} receptors and neurons constituting synaptic circuits onto the cell. (A) A striatopallidal MSN in normal state. (B) A striatopallidal MSN in Parkinson's disease state exerting selective loss of dendritic spines, resulting in degeneration of some of adenosine A_{2A} receptors localizing at spines (see text).

5.1.1 The A_{2A} receptor-mediated modulation of MSNs in the striatum

First, A_{2A} receptor-mediated inhibition of GABAergic synaptic transmission onto the striatal MSN has been found in an *in vitro* electrophysiological study using a whole-cell patch clamp method applied to striatal spiny neurons of rat striatal slices (Mori, Shindou, Ichimura, Nonaka, & Kase, 1996). In the study, inhibitory postsynaptic currents (IPSCs) recorded from a spiny projection neuron, evoked by intrastriatal field stimulation, were investigated under conditions in which excitatory inputs were blocked. The average amplitude of IPSCs was significantly suppressed by A_{2A} agonist CGS21680, and the suppression was subsequently antagonized by A_{2A} antagonist KF17837. Also, through analysis of spontaneous miniature IPSCs (mIPSCs), the A_{2A} receptor-mediated suppression of IPSCs was found to be attributable to presynaptic mechanism. These results demonstrated the existence of presynaptic A_{2A} receptor-mediated suppression of striatal GABAergic synaptic transmission onto spiny projection neurons. These electrophysiological findings were reproduced by Chergui, Bouron, Normand, and Mulle (2000). The results were consistent with neurochemical studies, using striatal synaptosomal preparation, showing that GABA release from striatal nerve terminals was reduced by A_{2A} receptor stimulation, being blocked by the A_{2A} antagonist KF17837 (Kirk & Richardson, 1994; Kurokawa, Kirk, Kirkpatrick, Kase, & Richardson, 1994).

In the striatum, other than the glutamatergic components from the cortex, the other major factor to determine the membrane excitability of MSN is intrastriatal GABAergic input, which is divided into feedback inhibition via recurrent axon collaterals of spiny neurons themselves and feed-forward inhibition from GABAergic interneurons. Because the reversal potential of chloride operated by $GABA_A$ receptors is close to the spike threshold potentials driven by excitatory inputs, GABAergic inputs act mainly to shunt the glutamatergic inputs. Striatal A_{2A} receptor modulation of either of both feedback and feed-forward inhibition systems could weaken the GABAergic shunting effects on excitatory inputs to spiny neurons, leading an excitation of the neurons, resulting in an increase of striatopallidal excitability (Mori & Shindou, 2003).

5.1.2 The A_{2A} receptor-mediated modulation in the GP

In contrast to the striatal modulation onto MSNs, pallidal A_{2A} receptor activation has been found to facilitate GABAergic IPSCs onto GP neurons in the rat pallidal slice preparation (Shindou, Mori, Kase, & Ichimura, 2001). The facilitation of average amplitude of IPSCs onto GP neurons was dose dependently induced by the A_{2A} agonist CGS21680, and the CGS21680-induced enhancement of IPSCs was completely antagonized by the A_{2A} antagonists KF17837 and ZM241385. Paired-pulse facilitation and mIPSC analysis revealed that the A_{2A} receptor-mediated facilitation was via the presynaptic mechanism, like striatal modulation. The results are consistent with neurochemical studies demonstrating that activation of the A_{2A} receptor, which was stimulated by CGS21680 and blocked by the A_{2A} antagonist KF17837, enhanced the electrically stimulated release of GABA from pallidal slices (Mayfield, Suzuki, & Zahniser, 1993). Shindou, Richardson, Mori, Kase, and Ichimura (2003) also revealed that the A_{2A} receptor-mediated facilitation was caused by an action on the striatopallidal spiny neuron terminals, not on the GP neuron intranuclear axon collaterals (Kita, 1994; Nambu & Llinas, 1997) which were electrophysiologically (Kita, Chang, & Kitai, 1983; Nakanishi, Kita, & Kitai, 1987) and immunochemically (Oertel & Mugnaini, 1984; Smith, Parent, Seguela, & Descarries, 1987) identified as GABAergic neurons. Also, the facilitating modulation by A_{2A} receptors on GABAergic transmission onto GP neurons involved the sequential activation of the A_{2A} receptor, adenylyl cyclase, and then cyclic-AMP-dependent protein kinase (Shindou et al., 2002). Separated from the striatal A_{2A} receptor modulation onto MSNs, the results provided evidence for the existence of isolated presynaptic A_{2A} receptor-mediated modulation of GABA release from the pallidal terminal of striatopallidal spiny neurons. This facilitation of GABAergic transmission onto GP neurons is considered to directly suppress the excitability of GP neurons projecting to the STN, thus causing increased neuronal activity in the STN (Mori & Shindou, 2003).

5.1.3 The A_{2A} receptor-mediated modulation of the output from the entire striatopallidal pathway

The electrophysiological study to investigate GABAergic IPSCs, recorded from GP neurons evoked by striatal simulation (i.e., striatopallidal IPSCs) in rat striatopallidal slices (see Fig. 4.3A), demonstrated that the striatopallidal IPSCs were enhanced by adenosine (100 μM) and the A_{2A} receptor agonist CGS21680. The enhancement of striatopallidal IPSCs was antagonized by A_{2A} antagonists ZM241385 and KF17837 (Fig. 4.3B–D). Also, it has been shown that pallidal GABAergic transmission affected by A_{2A} receptors had no contribution of axon collaterals of GP neurons. The single-cell RT-PCR analysis did not detect A_{2A} receptor mRNA in GP cells by electrophysiologically methods (Shindou et al., 2003). This was the first evidence of A_{2A} receptor-mediated modulation of the entire striatopallidal pathway, demonstrated by *in vitro* electrophysiological and gene expression studies. Also, the enhancement of averaged mean pallidal IPSC amplitude was approximately 30% higher than those observed in the isolated GP slice study (Section 5.1.2). Thus, the activation of A_{2A} receptors in both the striatum

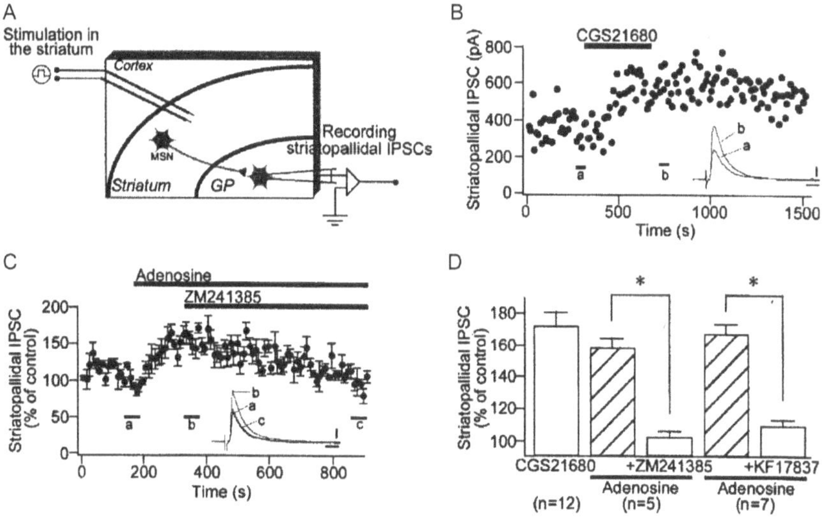

Figure 4.3 Adenosine A_{2A} receptor-mediated enhancement of striatopallidal IPSCs in the GP. (A) Oblique slice preparation includes the striatum and GP, and striatopallidal GABAergic projection fibers. Stimulation and recording electrodes are shown. (B) The time course of the amplitude of evoked IPSCs during the application of the A_{2A} agonist, CGS21680 (1 μM). Inset: superimposed traces of an average of consecutive striatopallidal IPSCs (eight traces) before and during application of CGS21680, taken at indicated time points. (C) Enhancement of striatopallidal IPSCs by adenosine (100 μM) via A_{2A} receptors. The A_{2A} antagonist ZM241385 (1 μM) reversed the potentiation of the striatopallidal IPSCs caused by adenosine ($n = 5$). Insets: superimposed traces taken at the indicated time points. Calibration in insets of B, C: *x*-axis, 20 ms; *y*-axis, 100 pA. (D) Summary of pharmacological characterization of the adenosine A_{2A} receptor-mediated modulation. The selective A_{2A} antagonists ZM241385 (1 μM) and KF17837 (1 μM) significantly blocked the adenosine-induced potentiation of striatopallidal IPSCs. $^{*}P < 0.01$ by paired *t*-test. *From Shindou et al. (2003).*

and the GP is more effective than the sole activation of pallidal A_{2A} receptors for the modulation of striatopallidal GABAergic outputs (Shindou et al., 2003). Also, Ochi et al. (2000) provided evidence of the modulation via both the striatum and GP, with an *in vivo* microdialysis study measuring pallidal GABA levels in free-moving rats. Either intrastriatal microinjection or intrapallidal infusion of the A_{2A} antagonist CGS21680 showed significantly increased pallidal GABA levels (see Fig. 4.4A and B). The *in vivo* study examining pallidal GABA levels indicated an *in vivo* existence of dual modulation of A_{2A} receptors via both the striatum and GP, which is leading upregulation of the striatopallidal pathway. Also, in a rat PD model (6-hydroxydopamine-lesioned rats), basal pallidal GABA levels were

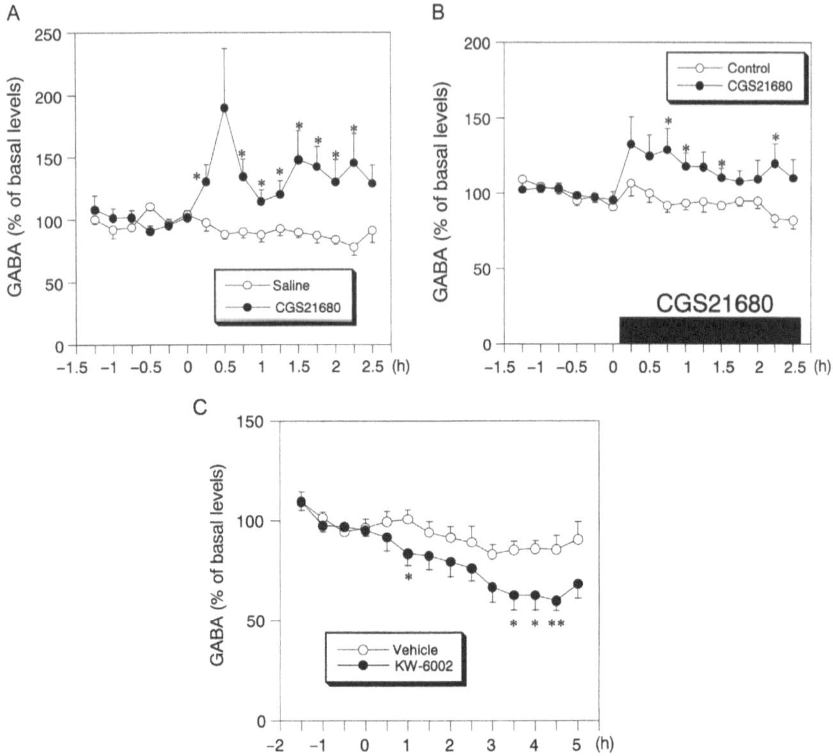

Figure 4.4 Adenosine A_{2A} receptor-regulated pallidal GABA levels evaluated by *in vivo* microdialysis method in rats. (A) Effects of intrastriatal microinjection of the A_{2A} agonist CGS21680 (0.05 μg/μL; $n=9$) or saline ($n=10$) on pallidal GABA levels of normal rats. (B) Effects of intrapallidal infusion of CGS21680 (10 μmol/L) via dialysis probe on pallidal GABA levels in normal rats. $n=11$ (control) or 13 (CGS21680). (C) Effects of KW-6002 (3 mg/kg, p.o.) on pallidal GABA levels in 6-OHDA-lesioned rats. Data point represents vehicle ($n=13$) and KW-6002 ($n=12$). *From Ochi et al. (2000).*

significantly higher than those of nonlesioned rats, consistent with the pathophysiology of the activated striatopallidal pathway in PD. Interestingly, the upregulated pallidal GABA levels in the rat PD models were ameliorated by the A_{2A} receptor antagonist KW-6002 (Istradefylline) (see Fig. 4.4C) at same dosages that achieved antiparkinsonian effects in the animals (Koga, Kuroawa, Ochi, Nakamura, & Kuwana, 2000).

In vitro and *in vivo* studies on the entire striatopallidal system demonstrated existence of the dual modulation via A_{2A} receptors located at both the striatum and GP. It is interpreted that when A_{2A} receptors of either area are activated, the striatopallidal pathway excitability is driven to be

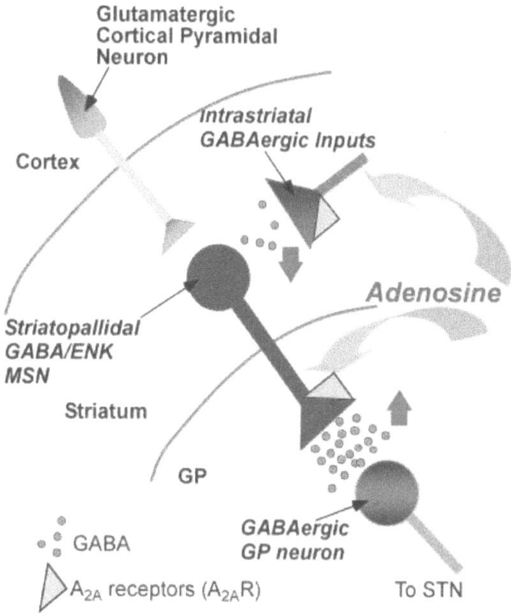

Figure 4.5 Schematic diagram of the corticostriatal and striatopallidal pathways, representing the A_{2A} receptor-mediated dual modulation of GABAergic synaptic transmission in the striatopallidal system. Adenosine via striatal A_{2A} receptor causes disinhibitory modulation of the excitability of striatopallidal MSN via suppression of GABA input onto the MSNs. Adenosine via pallidal A_{2A} receptors activates the nerve terminals of striatopallidal MSNs to increase GABA release. *From Mori and Shindou (2003).*

enhanced, which translates to excessive inhibition on the GP principal neurons projecting to the STN (see Fig. 4.5) (Mori & Shindou, 2003).

5.1.4 Models for the synaptic connection of MSNs affected by presynaptic A_{2A} receptor modulation

As described earlier, the proposed dual modulation by A_{2A} receptors on the striatopallidal pathway includes opposite effects on GABA release by A_{2A} receptors between the striatal and pallidal terminals of same spiny neurons. The pallidal modulation has shown consistent results among various studies, with a mechanism of second messenger cascade of A_{2A} receptors. However, considering the complexity of the intrastriatal network, the following three models of intrastriatal synaptic contacts affected by A_{2A} receptors have been proposed (Mori & Shindou, 2003; Xu et al., 2005) (see Fig. 4.6). In the first

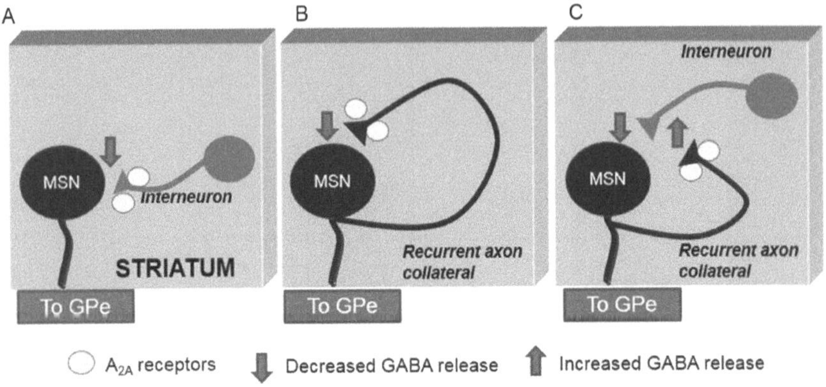

Figure 4.6 Three considerable models for inhibitory GABAergic synaptic contacts onto the striatopallidal MSNs, which can be affected by adenosine via presynaptic A_{2A} receptors to induce disinhibition of MSNs. (A) Synaptic contact from interneurons onto MSNs. (B) Synaptic contact via recurrent axon collaterals. (C) Synaptic contact from interneurons onto MSNs, which presynaptically suppressed by increased GABAergic inputs via activation of A_{2A} receptors of recurrent axon collaterals.

model, the A_{2A} receptor-modulated feed-forward system via GABAergic interneurons (Fig. 4.6A) requires A_{2A} receptor expression on these neurons, yet there has been little evidence of the receptors detected neither *in situ* nor in ultrastructural studies. Therefore, a presynaptic inhibitory model of recurrent axon collaterals is more reliable since the receptors have been conformed to express on the structure (see Fig. 4.6B). There are some arguments regarding if the recurrent feedback is critical or not to determine if a spiny neuron can be exited (Tapper et al., 2004; Wickens, Arbuthnott, & Shindou, 2007). One critical point of the model is if the opposite modulation on the same transmitter (i.e., GABA) via the same receptors at different locations on the same neuron can coexist.

Therefore, Xu et al. (2005) described another model suggesting that facilitation of GABA release by A_{2A} receptors (the same as pallidal modulation) can be generated at synapses contacted at presynaptic sites of a GABAergic interneuron (or other spiny neuron). The suppression of GABAergic contact from an interneuron (or other spiny neuron) onto a spiny neuron causes disinhibition on the neuron (see Fig. 4.6C). This model can partly be supported by the finding of double patch clamp recording between neighboring MSNs, demonstrating that A_{2A} activation enhances recurrent inputs (Shindou, Arbuthnott, & Wickens, 2008).

Interestingly, it has been reported that the transient potassium channel in medium spiny projection neurons received two opposite muscarinic modulations, depending upon two different resting membrane potentials range-driven by cortical input. This result suggests that acetylcholine (ACh) stabilizes the state of the neuron, rather than simply excite or inhibit it (Akins, Surmeier, & Kitai, 1990). Considering this dual modulation model for the potassium channel discovered in MSNs, the existence of A_{2A} receptor-mediated opposite modulation of GABA release in MSNs at different nerve terminals located in the striatum and GPe cannot be excluded (see Fig. 4.6B).

Concerning the complexity of the GABAergic network in the striatum and various modes of the excitability state of each neuron, a mixture of models is possible (see both Fig. 4.6B and C), including recurrent modulation, and may work for synaptic contacts affected by striatal A_{2A} receptor-mediated modulation. The synaptic models for the striatal modulation via A_{2A} receptors still remains to be investigated with future consideration/understanding of the neuronal processing mechanism in the striatum. The interpretation of GABAergic circuits in the basal ganglia is especially challenging (Wickens et al., 2007).

In summary, recent consensus confirmed by several studies is that A_{2A} receptors regulate striatal GABAergic synapses onto MSNs, and the activation of striatal A_{2A} receptors provides an increased GABAergic output from striatopallidal pathway in the GPe.

5.1.5 Physiological consideration of the A_{2A} receptor-mediated dual excitatory modulation of the striatopallidal pathway for the basal ganglia circuit (see Figs. 4.1C and 4.7)

How the modulation contributes to motor control and creates an aspect for the antiparkinsonian mechanism by adenosine A_{2A} receptor antagonist is shown in Fig. 4.7, which is a schematic diagram focusing on the activity of the striatopallidal pathway as a key factor for the entire coordination of basal ganglia–thalamocortical circuit (Section 2). The functional significance of A_{2A} receptors and the mode of action of the A_{2A} antagonist can be explained in the following text.

In the normal state, A_{2A} receptor-mediated excitatory modulation onto the pathway is well balanced with dopamine D2 receptor-mediated inhibitory regulation on the striatopallidal pathway (see Fig. 4.7A). In the PD state, when the D2 receptor system is significantly damaged due to loss of dopamine, the A_{2A} receptor-mediated modulation becomes at least relatively dominant to regulate the pathway, resulting in an increased excitation of the pathway (see Fig. 4.7B). This induces an entire disturbance of basal

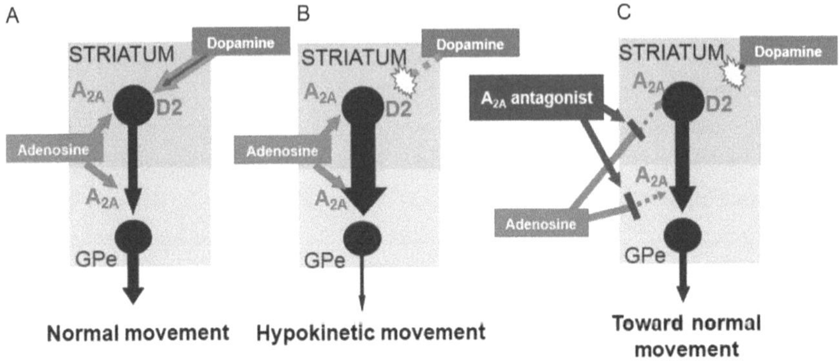

Figure 4.7 Schematic diagram focusing on regulation of the excitability of the striatopallidal pathway in three states, based on the A_{2A} receptor-mediated dual modulation of GABAergic synaptic transmission in the striatopallidal system (See Fig. 4.5). (A) Normal state. (B) Parkinson's disease state (PD state). (C) Treatment for Parkinson's disease with adenosine A_{2A} receptor antagonist (PD+A_{2A} therapy) (see text and Fig. 4.1).

ganglia–thalamocortical circuit as described in the PD state, which causes hypokinetic motor dysfunction (see Fig. 4.1B). Furthermore, recently, an increased density of A_{2A} receptors in the basal ganglia has been reported in PD patients, in comparison with normal subjects (Calon et al., 2004; Mishina et al., 2011; Morelli et al., 2007; Ramlackhansingh et al., 2011). This suggests that in addition to decreased influence via D2 receptors, an increase of A_{2A} receptor-mediated excitation may be occurred to drive the pathway overexcited in PD. Therefore, when the therapy with A_{2A} antagonist for PD is applied to block the A_{2A} receptor-mediated dual excitatory modulation in both the striatum and GPe on the striatopallidal pathway, the excessive excitation of the pathway is reduced, resulting in the entire basal ganglia–thalamocortical balance to shift toward normalization, even with the loss of dopamine (see Figs. 4.1C and 4.7C) (Kase et al., 2004; Xu et al., 2005).

5.2. Functional/physiological interaction hypotheses of adenosine A_{2A} receptors with other receptors

Several receptor–receptor interaction hypotheses between A_{2A} receptors and other receptors have been proposed mainly to explain more precisely A_{2A} receptor cellular and intermembrane mechanisms on neurons that directly or indirectly contribute to the regulation of the striatopallidal pathway.

5.2.1 Interaction with dopamine D2 receptors

The dopaminergic system seems to be a "center player" in PD. Thus, many investigations have been performed to learn how adenosine receptors work through or with dopamine receptors. In particular, dopamine D2 receptors (A_{2A}–D2 interaction) have been studied, as adenosine A_{2A} receptors colocalize with dopamine D2 receptors in the striatopallidal MSNs. Since the early 1990s, many studies and proposals have been made regarding the functional interaction between A_{2A} and D2 receptors in the striatum. One of first evidences was that A_{2A} receptor activation in the rat striatal membrane reduced the affinity of the D2 receptor for its agonist (Ferré, von Euler, Johansson, Fredholm, & Fuxe, 1991), which was followed by the reduction of G-protein coupling of the D2 receptor (Ferré, Snaprud, & Fuxe, 1993). As a result of those studies, the interaction has been interpreted as (i) A_{2A}–D2 intramembrane receptor interaction and (ii) A_{2A}–D2 receptor interaction at the second messenger level (Ferré et al., 2008; Xu et al., 2005). A_{2A}–D2 receptor heterodimerization demonstrated in mammalian transfected cells (Canals et al., 2004, 2003; Kamiya, Saitoh, Yoshioka, & Nakamura, 2003) has supported the concept of the intramembrane interaction. The interaction is considered to determine A_{2A} receptor-mediated regulation of the D2 receptor function. As A_{2A} and D2 receptors couple with G proteins to stimulate or inhibit adenylyl cyclase and the cAMP-PKA signaling pathways, respectively, a reciprocal interaction of both receptors at the second messenger level has been proposed, demonstrating that A_{2A} receptor-mediated signaling was antagonized by D2 receptor activation (Fuxe et al., 2001). Similar types of interactions have also been observed in several cells transfected with both receptors (Dasgupta et al., 1996; Kull et al., 1999; Salim et al., 2000), and human striatal sections (Díaz-Cabiale et al., 2001). Hillion et al. (2002) have shown that D2 receptor activation suppressed cAMP production via A_{2A} receptor stimulation in a human neuroblastoma cell line. From a striatal functional perspective, the electrophysiological study of spiny neurons in rat corticostriatal slices demonstrated that although spontaneous excitatory postsynaptic currents (sEPSCs) were affected neither by the A_{2A} receptor antagonists nor by the D2 agonists, coapplication of both A_{2A} antagonists and D2 agonists significantly reduced sEPCSs and is considered to be a presynaptic mechanism. This suggests that A_{2A} and D2 receptors converge in the presynaptic control of corticostriatal glutamatergic transmission by exerting an opposite function; however, the precise mechanisms of action are less understood (Tozzi et al., 2007).

Whether the A_{2A}–D2 intramembrane interaction is the mode of action for adenosine A_{2A} receptor antagonist resulting in symptomatic improvement of PD is still an open question. Motor dysfunction exerted in dopamine D2 receptor knockout (KO) mice was completely ameliorated by a single administration of the selective adenosine A_{2A} receptor antagonist KW-6002 (Istradefylline), with reversed change in enkephalin and SP expression (Aoyama, Kase, & Borrelli, 2000). It has also been demonstrated that adenosine A_{2A} receptor agonists reduced locomotor activity of D2 KO, as well as wild-type mice (Chen et al., 2001). These findings clearly indicate that there is a mechanism other than the receptor–receptor interaction to explain A_{2A} receptor-related motor control and the antiparkinsonian mode of action seen with the use of adenosine A_{2A} receptor antagonists.

In some cases, rather than acute symptomatic phenomena in motor control, the interaction between A_{2A} and D2 receptors may be related to a more chronic state via phosphorylation and other intercellular mechanisms for striatal neuronal plasticity, neurodegeneration, etc.

5.2.2 Synergistic interaction of A_{2A} receptors with metabotropic glutamate receptors

A functional/physical interaction between the A_{2A} receptor and the metabotropic glutamate (mGlu) receptor mGlu5 has been suggested to clarify modulation of striatopallidal MSNs. This is based on several findings: (i) the synergistic reduction of the affinity of dopamine receptors by A_{2A} and mGlu5 receptor agonists (Ferré et al., 1999; Popoli et al., 2001); (ii) *in vitro* successful membrane preparation of both A_{2A} and mGlu5 receptors with heteromeric receptor complexes (Ferré et al., 2002); (iii) the synergistic induction of c-fos level by coactivation of A_{2A} and mGlu5 receptors (Ferré et al., 2002); (iv) activation of both receptors induced an increase of extracellular GABA levels in the ventral pallidum, indicating coactivation of striatopallidal pathway (Díaz-Cabiale et al., 2002); (v) synergistic mGlu5 and A_{2A} antagonist-induced motor stimulation; and (vii) attenuation of mGlu5 antagonist-induced motor stimulation in forebrain-conditioned A_{2A} receptor KO mice (Kachroo et al., 2005).

Most of the interactions noted in the preceding text have been proposed as postsynaptic interactions, which can be generated on dendritic spines/spine heads receiving corticostriatal glutamatergic input onto striatal MSNs. Although a molecular basis for this functional interaction via the phosphorylation of DARPP-32 (dopamine- and cAMP-regulated phosphoprotein of 32 kDa) has been suggested (Nishi et al., 2003; Shen et al., 2013), the mechanism by which the receptors counteract one another on the neuronal surface must be further investigated.

In addition, the presynaptic interaction between the A_{2A} receptor and mGlu5 to facilitate the presynaptic glutamate release has recently been proposed (Pintor, Pezzola, Reggio, Quarta, & Popoli, 2000; Rodrigues, Alfaro, Rebola, Oliveria, & Cunha, 2005). Both receptors were found to be colocalized in corticostriatal glutamatergic nerve terminal, which also supports presynaptic interaction between the receptors (Rodrigues et al., 2005). However, considering that the expression of the A_{2A} receptor mRNA is detected much less in the cortex than in other significant regions like the stratum, the contribution of these presynaptic A_{2A} receptors in corticostriatal synapses associated with motor control via the basal ganglia remains to be further investigated and clarified.

Significant immunoreactivity of one of the group III metabotropic glutamate receptors, the mGlu4 receptor mRNA, has also been found in the striatum and their localization is confirmed in the GPe (Bradley et al., 1999; Matsui & Kita, 2003). Hence, an interaction of the receptors with the A_{2A} system in regulating inhibitory tone at the levels of the GPe has been proposed (Jones et al., 2012). Valenti et al. (2003) reported that mGlu4 receptor activation could presynaptically inhibit striatopallidal transmission, evaluated by IPSCs in GP neurons. This was same target with opposite direction of the presynaptic A_{2A} receptor-mediated excitatory action of the terminals of the striatopallidal pathway located in the GPe (Shindou et al., 2001, 2002, 2003). The emergence of correlated oscillatory activity in the subthalamopallidal circuit (not shown in Fig. 4.1) has been noted after the destruction of dopaminergic neurons in PD (Terman, Rubin, Yew, & Wiloson, 2002). In addition to the A_{2A} receptor mechanism in the striatopallidal pathway, if the physiological interaction of A_{2A} receptors with mGlu4 receptors occurs in this circuit, it is very important to take this into consideration in the symptomatic improvement of PD by A_{2A} receptor antagonists.

5.2.3 A_{2A} receptor-related modulations of ACh system in the striatum

A_{2A} receptor-mediated presynaptic regulation of striatal ACh release has been shown (Kurokawa et al., 1994; Kurokawa, Koga, Kase, Nakamura, & Kuwana, 1996). It has been documented by *in vitro* striatal synaptosome and *in vivo* microdialysis that the A_{2A} receptor antagonist KF17837 blocked enhanced ACh release stimulated by the A_{2A} receptor agonist CGS21680. This function is interesting to consider relative to the antiparkinsonian effect of A_{2A} receptor antagonists. However, it is still questionable whether striatal cholinergic large spiny interneurons express significant levels of A_{2A} receptors, as discussed in Section 4.2. Further investigation into this area is still needed.

5.2.4 Presynaptic A_{2A}–A_1 receptor interaction

Adenosine A_{2A} and A_1 receptor heterodimer-induced reciprocal interaction has recently been discussed in the regulation of corticostriatal glutamatergic transmission onto MSNs (Schwarzchild, Agnati, Fuxe, Chen, & Morelli, 2006). This was suggested from data indicating A_{2A}–A_1 receptor heteromers in the cell surface of cotransfected cells, as well as colocalization of both receptors in the same striatal glutamatergic terminals (Ciruela et al., 2006; Ferré, Aganati, et al., 2007; Ferré, Ciruela, et al., 2007). Adenosine A_{2A} and A_1 receptors work in opposite directions to regulate the second messenger system via adenylyl cyclase activation by the difference in adenosine concentrations. The functional evidence and physiological necessity of the heteromer of two different adenosine-operated receptor subtypes, with an extracellular adenosine concentration-dependent switching mechanism, to regulate glutamate release are still unknown (Ciruela et al., 2006; Ferré, Aganati, et al., 2007; Ferré, Ciruela, et al., 2007). The selective adenosine A_{2A} receptor antagonist KW-6002 (Istradefylline) was effective in postnatal forebrain-specific conditional A_{2A} receptor KO mice in acute MPTP neurotoxicity (Yu et al., 2008). This may suggest that A_{2A} receptors in the brain control motor activity independent of the A_{2A}–A_1 heteromer interaction mechanism. Further research will verify if such a receptor interaction using same endogenous ligand (i.e., adenosine) works physiologically, like the receptor conformational analysis of adenosine A_{2A} receptors producing a cellular response (Hino et al., 2012).

6. NEW ASPECT FOR THE PATHOPHYSIOLOGICAL CHANGE TO STRIATOPALLIDAL MSNs IN PD

The significant pathophysiological change of striatopallidal MSNs, as distinguished from striatonigral neurons, has recently been discovered. Although the dopaminergic denervation and degeneration of nigrostriatal pathway to the striatum is well known, it has been very unclear how the loss of dopamine affects striatal neurons, leading to PD symptoms. Day et al. (2006) have demonstrated both anatomical and physiological evidence of a selective loss of glutamatergic synapses in striatopallidal neurons after dopamine depletion, more precisely in the speed, selectivity, and magnitude of the loss than reported in previous studies (Dunah et al., 2000; Ingham, Hood, Taggart, & Arbuthnott, 1998; Ingham, Hood, van Maldegem, Weenink, & Arbuthnott, 1993; McNeill, Sally, Rafols, & Shoulson, 1998; Stephens et al., 2005). Bacterial artificial chromosome transgenic mice, which enhanced green fluorescent protein to distinguish between striatopallidal and striatonigral neurons, were utilized to show that dopamine depletion by reserpine treatment caused a reduction in spine density and number of spines in D2 receptor expression, but not D1 receptor expression MSNs. Dendritic length and branching were also markedly reduced in dopamine-depleted striatopallidal neurons. In parallel with the loss of spines, miniature EPSC frequency in striatopallidal neurons was decreased, which indicates that presynaptic functional components of the glutamatergic synapse had been eliminated (Day et al., 2006). This selective degeneration of dendritic spines was also observed in spines of neurons that lacked D1 receptor labeling (indirectly meaning to be the MSNs expressing D2 receptors), in 6-hydroxydopamine-treated condition as another PD models. Even though dopamine presynaptically regulates glutamatergic synapses in MSNs (Bamford et al., 2004; Cepeda et al., 2001), there is no evidence that this regulation specifically affects the synapses that are formed on the striatopallidal neurons (Day et al., 2006). To clarify a part of the mechanism, it was revealed that dendrites of striatopallidal neurons were more excitable than those of striatonigral neurons. The subsequent dopamine depletion increased calcium entry through the dendritic voltage-dependent L-type calcium channels associated with the backpropagation of action potentials in striatopallidal neurons (Day, Wokosin, Plotkin, Tian, & Surmeier, 2008). Reserpine treatment results in depletion of dopamine and produces akinesia and muscle rigidity, which morphologically resemble the symptoms of PD (Davis et al., 1979). How and when during the progression of human PD the selective degeneration of dendritic structures of striatopallidal MSNs occurs is unknown. The method of dopamine depletion employed in Day's studies (i.e., reserpine treatment) did not result in the loss of the dopamine neuron. This suggests that the phenomena may occur at a relatively early stage of PD.

The author tried to adapt this new and interesting pathophysiological aspect into the schematic diagram shown in Fig. 4.2B. Although regarding A_{2A} receptor ultrastructural localization on striatopallidal MSNs, a majority (about 80%) of all of A_{2A} receptor density is condensed in the dendritic spines (see Section 4.2) in the normal state, and most of them are destroyed in PD due to the loss of dopamine resulting in the denervation of the dendritic spines of the striatopallidal MSNs. Therefore in the PD state, it is speculated that most prominent A_{2A} receptors working on a striatopallidal MSN are those expressed in both the presynaptic GABAergic synapses and the postsynaptic site of the GABA synapses, not on the dendritic spines/spine heads. Those surviving A_{2A} receptors in the PD state are considered to regulate intrastriatal GABAergic contacts, not coordinating the interaction with D2 and mGlu receptors. Therefore, the likely target of A_{2A} receptor antagonists in PD therapy is interpreted to be those receptors controlling the GABAergic input onto MSNs.

As described earlier, when any mechanism related to the A_{2A} receptors on striatopallidal MSNs is considered, it is notable to distinguish between the normal state and the disease state (PD), with respect to depletion/denervation of dopamine/dopamine neurons. For example, a behavioral pharmacology study of D2 KO mice has shown that the A_{2A} antagonist KW-6002 (Istradefylline) had the same degree of locomotor stimulant effect between wild-type mice and D2 KO mice (Aoyama et al., 2000). It is assumed that the mechanism of locomotor stimulant effect of the A_{2A} antagonist in either wild-type mice or D2 KO mice may occur due to the action of A_{2A} receptors including those on the dendritic spines of striatopallidal MSNs. Some of this may be due to an interaction with glutamate receptors, since D2 KO mice maintain their dopaminergic cells, which is different than the PD models used for Day's pathophysiology study. However, KW-6002 (Istradefylline) attenuated motor dysfunction in a reserpine-induced catalepsy model (Shiozaki et al., 1999). The mechanism of action is via A_{2A} receptors surviving on other cellular sites of the striatopallidal neurons, not via the A_{2A} receptors of dendritic spines, which is different from the D2 KO mice model.

7. CONCLUDING REMARKS

Recently, it has been reported from a clinical study in 21 PD patients with cerebral blood flow (CBF) imaging that the adenosine A_{2A} receptor antagonist SYN115 has produced a highly significant decrease in thalamic CBF, consistent with reduced pallidothalamic inhibition via the indirect pathway (Black, Koller, Campbell, Gusnard, & Bandak, 2010). This is direct evidence that the mode of action of the A_{2A} receptor antagonist occurred via deactivation of the indirect pathway as predicted by the hypothesis mentioned earlier in this chapter. Also, in a rat *in vivo* subthalamic DBS study, administration of the adenosine A_{2A} receptor antagonist MSX-3 has reduced the frequency and intensity parameters of DBS needed to attenuate tremulous jaw movements, a pharmacological model of tremor in rodents. These results suggest that A_{2A} receptor antagonism acts to enhance the suppressive effect of STN–DBS on tremor. Additional studies are needed to determine whether the ability of A_{2A} receptor antagonists to enhance sensitivity to DBS occurs in human PD patients (Collins-Praino et al., 2013).

The function and anatomical architecture of the basal ganglia–thalamocortical circuit has been providing great insights and ideas into exploration of the mode of action of adenosine A_{2A} receptor antagonist in PD. Now, the target of A_{2A} receptor antagonists is clearly the A_{2A} receptor-mediated excitatory modulation of the striatopallidal pathway. However, the precise cellular and receptor level mechanism of the modulation and further insight into the intrastriatal network require investigation. Recent pathophysiological studies have revealed changes from and differences between the normal state and PD. This reveals a clear gap between the various states of PD. Any exploration of the mechanism of receptor function seeking new therapies, like that of adenosine A_{2A} receptors, should consider the pathophysiological changes that occur with disease progression, as seen in PD.

REFERENCES

Akins, P. T., Surmeier, D. J., & Kitai, S. T. (1990). Muscarinic modulation of transient K^+ conductance in rat neostriatal neurons. *Nature*, *344*, 240–242.

Alexander, G. E., & Crutcher, M. D. (1990). Functional architecture of basal ganglia circuits: Neuronal substrates of processing. *Trends in Neurosciences*, *13*, 266–271.

Aoyama, S., Kase, H., & Borrelli, E. (2000). Rescue of locomotor impairment in dopamine D2 receptor deficient mice by an adenosine A_{2A} receptor antagonist. *The Journal of Neuroscience*, *20*, 5848–5852.

Augood, S. J., & Emson, P. C. (1994). Adenosine A_{2A} receptor mRNA in expressed by enkephalin cells but not by somatostatin cells in rat striatum: A co-expression study. *Brain Research. Molecular Brain Research*, *22*, 204–210.

Bamford, N. S., Zhang, H., Schmitz, Y., Wu, N. P., Cepeda, C., Levine, M. S., et al. (2004). Heterosynaptic dopamine neurotransmission selects sets of corticostriatal terminals. *Neuron*, *42*, 653–663.

Black, K. J., Koller, J. M., Campbell, M. C., Gusnard, D. A., & Bandak, S. I. (2010). Quantification of indirect pathway inhibition by the adenosine A_{2A} antagonist SYN115 in Parkinson's disease. *The Journal of Neuroscience*, *30*, 16284–16292.

Bradley, S. R., Standaert, D. G., Rhodes, K. J., Rees, H. D., Testa, C. M., Levey, A. I., et al. (1999). Immunohistochemical localization of subtype 4a metabotropic glutamate receptors in the rat and mouse basal ganglia. *The Journal of Comparative Neurology*, *402*, 33–56.

Calon, F., Dridi, M., Homykiewicz, O., Bedard, P. J., Rajput, A. H., & Di Paolo, T. (2004). Increased adenosine A_{2A} receptors in the brain of Parkinson's disease patients with dyskinesia. *Brain*, *127*, 1075–1084.

Canals, M., Burgueño, J., Marcellino, D., Cabello, N., Canela, E. I., Mallol, J., et al. (2004). Homodimerization of adenosine A_{2A} receptors: Qualitative and quantitative assessment by fluorescence and bioluminescence energy transfer. *Journal of Neurochemistry, 88*, 726–734.

Canals, M., Marcellino, D., Fanelli, F., Ciruela, F., de Bendetti, P., Goldberg, S. R., et al. (2003). Adenosine A2A -dopamine D2 receptor-receptor heteromerization: Qualitative and quantitative assessment by fluorescence and bioluminescence energy transfer. *Journal of Biological Chemistry, 278*, 46741–46749.

Cepeda, C., Hurst, R. S., Altemus, K. L., Flores-Hernandez, J., Calvert, C. R., Jokel, E. S., et al. (2001). Facilitated glutamatergic transmission in the striatum of D2 dopamine receptor-deficient mice. *Journal of Neurophysiology, 85*, 659–670.

Chen, J.-F., Moratella, R., Impagnatiello, F., Grandy, D. K., Cuellar, B., Rubinstein, M., et al. (2001). The role of the D2 dopamine receptor (D2R) in A_{2A} adenosine receptor (A_{2A} R)-mediated behavioral and cellular response in as revealed by A_{2A} and D2 receptor knockout mice. *Proceedings of the National Academy of Sciences of the United States of America, 98*, 1970–1975.

Chergui, K., Bouron, A., Normand, E., & Mulle, C. (2000). Functional GluR6 kinate receptors in the striatum: Indirect downregulation of synaptic transmission. *The Journal of Neuroscience, 20*, 2175–2182.

Ciruela, F., Casadó, V., Rodrigues, R. J., Luján, R., Burgueño, J., Canals, M., et al. (2006). Presynaptic control of striatal glutamatergic neurotransmission by adenosine A1–A_{2A} receptor heteromers. *The Journal of Neuroscience, 26*, 2080–2087.

Collins-Praino, L. E., Paul, N. E., Ledgard, F., Podurgiel, S. J., Kovner, R., Baqi, Y., et al. (2013). Deep brain stimulation of the subthalamic nucleus reverses oral tremor in pharmacological models of parkinsonism: Interaction with the effects of adenosine A_{2A} antagonism. *European Journal of Neuroscience, 38*, 2183–2191.

Cunha, R. A., Johansson, B., van der Ploeg, I., Sebastião, A. M., Ribeiro, J. A., & Fredholm, B. B. (1994). Evidence for functionally important adenosine A_{2A} receptor in the rat hippocampus. *Brain Research, 649*, 208–216.

Dasgupta, S., Ferré, S., Kull, B., Hedlund, P. B., Finnman, U. B., Ahlberg, S., et al. (1996). Adenosine A_{2A} receptors modulate the binding characteristics of dopamine D2 receptors in stably cotransfected fibroblast cells. *European Journal of Pharmacology, 316*, 325–331.

Davis, C. G., Willams, A. C., Markey, S. P., Ebert, M. H., Caine, E. D., Reichert, C. M., et al. (1979). Chronic parkinsonism secondary to intravenous injection of meperidine analogues. *Psychiatry Research, 1*, 249–254.

Day, M., Wang, Z., Ding, J., An, X., Ingham, C. A., Shering, A. F., et al. (2006). Selective elimination of glutamatergic synapses on striatopallidal neurons in Parkinson disease models. *Nature Neuroscience, 9*, 251–259.

Day, M., Wokosin, D., Plotkin, J. L., Tian, X., & Surmeier, D. J. (2008). Differential excitability and modulation of striatal medium spiny neuron dendrites. *The Journal of Neuroscience, 28*, 11603–11614.

DeLong, M. (1990). Primate models of movement disorders of basal ganglia origin. *Trends in Neurosciences, 13*, 281–285.

DeMet, E. M., & Chicz-DeMet, A. (2002). Localization of adenosine A_{2A} receptors in rat brain with [^{3}H]ZM-241385. *Naunyn-Schmiedeberg's Archives of Pharmacology, 366*, 478–481.

Díaz-Cabiale, Z., Hurd, Y., Guidolin, D., Finnman, U. B., Zoli, M., Agnati, L. F., et al. (2001). Adenosine A_{2A} agonist CGS21680 decrease the affinity of dopamine D2 receptors for dopamine in human stratum. *Neuroreport, 12*, 1831–1834.

Díaz-Cabiale, Z., Vivó, M., Del Arco, A., O'Conner, W. T., Harte, M. K., Müller, E. C., et al. (2002). Metabotropic glutamate mGlu5 receptor-mediated modulation of the ventral striatopallidal GABA pathway in rats. Interactions with adenosine A_{2A} and dopamine D2 receptors. *Neuroscience Letters, 324*, 154–158.

Dixon, A. K., Gubitz, A. K., Sirinathsinghji, D. J., Richardson, P. J., & Freeman, T. C. (1996). Tissue distribution of adenosine receptor mRNAs in the rat. *British Journal of Pharmacology, 118*, 1461–1468.

Dunah, A. W., Wang, Y., Yasuda, R. P., Kameyama, K., Huganir, R. L., Wolfe, B. B., et al. (2000). Alterations in subunit expression, composition, and phosphorylation of striatal N-methyl-D-aspartate glutamate receptors in a rat 6-hydroxy dopamine model of Parkinson's disease. *Molecular Pharmacology, 57*, 342–352.

Ferré, S., Aganati, L. F., Ciruela, F., Lluis, C., Woods, A. S., Fuxe, K., et al. (2007). Neurotransmitter receptor heteromers and their integrative role in local modules: The striatal spine module. *Brain Research Reviews, 55*, 55–67.

Ferré, S., Ciruela, F., Quiroz, C., Luján, R., Popoli, P., Cunha, R. A., et al. (2007). Adenosine receptor heteromers and their integrative role in striatal function. *Scientific World Journal*, 7, 74–85.

Ferré, S., Karcz-Kubicha, M., Hope, B. T., Popoli, P., Burgueno, J., Gutierrez, M., et al. (2002). Synergistic interaction between adenosine A_{2A} and glutamate mGlu5 receptors: Implications for striatal neuronal function. *Proceedings of the National Academy of Sciences of the United States of America, 99*, 11940–11945.

Ferré, S., Popoli, P., Rimondini, R., Reggio, R., Kehr, J., & Fuxe, K. (1999). Adenosine A_{2A} and group I metabotropic glutamate receptors synergistically modulate the binding characteristics of dopamine D2 receptors in the rat striatum. *Neuropharmacology, 38*, 129–140.

Ferré, S., Quiroz, C., Woods, A. C., Cunha, R., Popoli, P., Ciruela, F., et al. (2008). An update on adenosine A_{2A}-dopamine D2 receptor interactions. Implications for the function of G protein-coupled receptors. *Current Pharmaceutical Design, 14*, 1468–1474.

Ferré, S., Snaprud, P., & Fuxe, K. (1993). Opposing actions of an adenosine A_{2A} receptor agonist and a GTP analogue on the regulation of dopamine D2 receptors in rat neostriatal membranes. *European Journal of Pharmacology, 244*, 3110315.

Ferré, S., von Euler, G., Johansson, B., Fredholm, B. B., & Fuxe, K. (1991). Stimulation of high-affinity adenosine A2 receptors decrease the affinity of dopamine D2 receptors in rat striatal membranes. *Proceedings of the National Academy of Sciences of the United States of America, 88*, 7238–7241.

Fink, J. S., Weaver, D. R., Rivkees, S. A., Peterfreund, R. A., Pollack, A. E., Adler, E. M., et al. (1992). Molecular cloning of the rat A_{2A} adenosine receptors: Selective co-expression with D2 dopamine receptors in rat striatum. *Molecular Brain Research, 14*, 186–195.

Fredholm, B. B., Lindstrom, K., Dionisotti, S., & Ongini, E. (1998). [^{3}H]SCH58261, a selective adenosine A_{2A} receptor antagonist, is a useful ligand in autoradiographic studies. *Journal of Neurochemistry, 70*, 1210–1216.

Fuxe, K., Stromberg, I., Popoli, P., Rimondini-Giorgini, R., Torvinen, M., Ogren, S. O., et al. (2001). Adenosine receptors and Parkinson's disease. Relevance of antagonistic adenosine and dopamine receptor interactions in the striatum. *Advances in Neurology, 86*, 345–353.

Gërfen, C. R. (2004). Basal ganglia. In G. Paxinos (Ed.), *The rat nervous system* (pp. 445–508). Amsterdam: Elsevier Academic Press.

Gërfen, C. R., Engber, T. M., Mahan, L. C., Susel, Z., Chase, T. N., Monsma, F. J., et al. (1990). D1 and D2 dopamine receptor-regulated gene expression of striatonigral and striatopallidal neurons. *Science, 250*, 1429–1432.

Graybiel, A. M. (1990). Neurotransmitter and neuromodulators in the basal ganglia. *Trends in Neurosciences, 13*, 244–254.

Hauser, R. A., Shulman, L. M., Trugman, J. M., Roberts, J. W., Mori, A., Ballerini, R., et al. (2008). Study of istradefylline in patients with Parkinson's disease on levodopa with motor fluctuations. *Movement Disorders, 23*, 2177–2185.

Hettinger, B. D., Lee, A., Linden, J., & Rosin, D. L. (2001). Ultrastructural localization of adenosine A_{2A} receptors suggests multiple cellular sites for modulation of GABAergic neurons in rat striatum. *The Journal of Comparative Neurology*, *431*, 331–346.

Hillion, J., Canals, M., Torvinen, M., Casado, V., Scott, R., Terasmaa, A., et al. (2002). Coaggregation, cointernalization, and codesensitization of adenosine A_{2A} receptors and dopamine D2 receptors. *Journal of Biological Chemistry*, *277*, 18091–18097.

Hino, T., Arakawa, T., Iwanari, H., Yurugi-Kobayashi, T., Ikeda-Suno, C., Nakada-Nakura, Y., et al. (2012). G-protein-coupled receptor inactivation by an allosteric inverse-agonist antibody. *Nature*, *482*, 237–241.

Ingham, C. A., Hood, S. H., Taggart, P., & Arbuthnott, G. W. (1998). Plasticity of synapses in the rat neostriatum after unilateral lesion of the nigrostriatal dopaminergic pathway. *The Journal of Neuroscience*, *18*, 4732–4743.

Ingham, C. A., Hood, S. H., van Maldegem, A., Weenink, A., & Arbuthnott, G. W. (1993). Morphological changes in the rat neostriatum after unilateral 6-hydroxydopamine injections into the nigrostriatal pathway. *Experimental Brain Research*, *93*, 17–27.

Jarvis, M. F., & Williams, M. (1989). Direct autoradiographic localization of adenosine A2A receptors in rat brain using the A2A -selective agonist, [3H]CGS21680. *European Journal of Pharmacology*, *168*, 243–246.

Johansson, B., Georgiev, V., Parkinson, F. E., & Fredholm, B. B. (1993). The binding of the adenosine A_{2A} receptor selective agonist [^{3}H]CGS21680 to rat cortex differs from its binding to rat striatum. *European Journal of Pharmacology*, *247*, 103–110.

Jones, C. K., Bubser, M., Thompson, A. D., Dickerson, J. W., Turle-Lorenzo, N., Amelric, M., et al. (2012). The metabotropic glutamate receptor 4-positive allosteric modulator VU0364770 produces efficacy alone and in combination with L-DOPA or an adenosine 2A antagonist in preclinical models of Parkinson's disease. *The Journal of Pharmacology and Experimental Therapeutics*, *340*, 404–421.

Kachroo, A., Orlando, L. R., Grandy, D. K., Chen, J.-F., Young, A. B., & Schwarzschild, M. (2005). Interactions between metabotropic glutamate 5 and adenosine A_{2A} receptors in normal and parkinsonian mice. *The Journal of Neuroscience*, *25*, 10414–10419.

Kaelin-Lang, A., Lauterburg, T., & Burgunder, J. M. (1999). Expression of adenosine A_{2A} receptors gene in the olfactory bulb and spinal cord of rat and mouse. *Neuroscience Letters*, *261*, 189–191.

Kamiya, T., Saitoh, O., Yoshioka, K., & Nakamura, H. (2003). Oligomerization of adenosine A_{2A} and dopamine D2 receptors in living cells. *Biochemical and Biophysical Research Communications*, *306*, 544–549.

Kase, H., Mori, A., & Jenner, P. (2004). Adenosine A_{2A} receptor antagonists: Beyond dopaminergic therapies for Parkinson's disease. *Drug Discovery Today: Therapeutic Strategies*, *1*, 51–57.

Kawaguchi, Y. (1997). Neostriatal cell subtypes and their functional roles. *Neuroscience Research*, *27*, 1–8.

Kirk, I. P., & Richardson, P. J. (1994). Adenosine A_{2A} receptor-mediated modulation of striatal [^{3}H]GABA and [^{3}H]acetylcholine release. *Journal of Neurochemistry*, *62*, 960–966.

Kita, H. (1994). Parvalbumin immunopositive neurons in rat globus pallidus: A light and electron microscope study. *Brain Research*, *657*, 31–41.

Kita, H., Chang, H. T., & Kitai, S. T. (1983). Pallidal inputs to subthalamus intracellular analysis. *Brain Research*, *264*, 255–265.

Koga, K., Kuroawa, M., Ochi, M., Nakamura, J., & Kuwana, Y. (2000). Adenosine A(2A) receptor antagonist KF17837 and KW-6002 potentiate rotation induced by dopaminergic drugs in hemi-parkinsonian rats. *European Journal of Pharmacology*, *408*, 249–255.

Krack, P., Hariz, M. I., Baunez, C., Guridi, J., & Obeso, J. A. (2010). Deep brain stimulation: From neurology to psychiatry? *Trends in Neurosciences*, *33*, 474–484.

Kull, B., Ferré, S., Arslan, G., Svenningsson, P., Fuxe, K., Owman, C., et al. (1999). Reciprocal interactions between adenosine A_{2A} and dopamine D2 receptors in Chinese hamster ovary cells co-transfected with two receptors. *Biochemical Pharmacology, 58*, 1035–1045.

Kurokawa, M., Kirk, I. P., Kirkpatrick, K. A., Kase, H., & Richardson, P. (1994). Inhibition of KF17837 of adenosine A_{2A} receptor-mediated modulation of striatal GABA and ACh release. *British Journal of Pharmacology, 113*, 43–48.

Kurokawa, M., Koga, K., Kase, H., Nakamura, J., & Kuwana, Y. (1996). Adenosine A_{2A} receptor-mediated modulation of striatal acetylcholine release in vivo. *Journal of Neurochemistry, 66*, 1882–1888.

Lee, Y. C., Chien, C. L., Sun, C. N., Huang, C. L., Huang, N. K., Chiang, M. C., et al. (2003). Characterization of the rat A_{2A} adenosine receptor gene: A 4.8-kb promoter-proximal DNA fragment confers selective expression in the central nervous system. *European Journal of Neuroscience, 18*, 1786–1796.

LeWitt, P. A., Guttman, M., Tetrud, J. W., Tuite, P. J., Mori, A., Chaikin, P., et al. (2008). Adenosine A_{2A} receptor antagonist Istradefylline (KW-6002) reduces "off" time in Parkinson' s disease: A double-blind, randomized, multicenter clinical trial (6002-US-005). *Annals of Neurology, 63*, 295–302.

Martinez-Mir, M. I., Probst, A., & Palacios, J. M. (1991). Adenosine A_{2A} receptors: Selective localization in the human ganglia and alterations with disease. *Neuroscience, 42*, 697–706.

Matsui, T., & Kita, H. (2003). Activation of group III metabotropic glutamate receptors presynaptically reduces both GABAergic and glutamatergic transmission in the rat globus pallidus. *Neuroscience, 122*, 727–737.

Mayfield, R. D., Suzuki, F., & Zahniser, N. R. (1993). Adenosine A_{2A} receptor modulation of electrically evoked endogenous GABA release from slices of rat globus pallidus. *Journal of Neurochemistry, 60*, 2334–2337.

McNeill, T. H., Sally, A. B., Rafols, J. A., & Shoulson, I. (1998). Atrophy of medium spiny I striatal dendrites in advanced Parkinson's disease. *Brain Research, 455*, 148–152.

Mishina, M., Ishiwata, K., Naganawa, M., Kimura, Y., Kitamura, S., Suzuki, M., et al. (2011). Adenosine A_{2A} receptors measured with [^{11}C] TMSX PET in the striata of Parkinson's disease patients. *PLoS One, 6*, e17338.

Mizuno, Y., Hasegawa, K., Kondo, S., Kuno, S., Yamamoto, M., & Japanese Istradefylline Study Group. (2010). Clinical efficacy of Istradefylline (KW-6002) in Parkinson's disease: A randomized, controlled study. *Movement Disorders, 25*, 1437–1443.

Mizuno, Y., Kondo, T., & the Japanese Istradefylline Study Group (2013). Adenosine A_{2A} receptor antagonist Istradefylline reduces daily off time in Parkinson's disease. *Movement Disorders, 28*, 1138–1141.

Morelli, M., DiPaolo, T., Wardas, J., Calon, F., Xiao, D., & Schwarzschild, M. (2007). Role of adenosine A_{2A} receptors in parkinsonian motor impairment and L-DOPA-induced motor complications. *Progress in Neurobiology, 83*, 293–309.

Mori, A., & Shindou, T. (2003). Modulation of GABAergic transmission in the striatopallidal system by adenosine A_{2A} receptors. *Neurology, 61*, S44–S48.

Mori, A., Shindou, T., Ichimura, M., Nonaka, H., & Kase, H. (1996). The role of adenosine A_{2A} receptors in regulating GABAergic synaptic transmission in striatal medium spiny neurons. *The Journal of Neuroscience, 16*, 605–611.

Nakanishi, H., Kita, G., & Kitai, S. T. (1987). Electrical membrane properties of rat subthalamic neurons in an in vivo slice preparation. *Brain Research, 437*, 35–44.

Nambu, A., & Llinas, R. (1997). Morphology of globus pallidus neurons: Its correlation with electrophysiology in guinea pig brain slices. *Journal of Comparative Neurology, 377*, 85–94.

Nishi, A., Liu, F., Matsuyama, S., Hamada, M., Higashi, H., Nain, A. C., et al. (2003). Metabotropic mGlu5 receptors regulate adenosine A_{2A} receptor signaling. *Proceedings of the National Academy of Sciences of the United States of America, 100*, 1322–1327.

Nonaka, H., Mori, A., Ichimura, M., Shindou, T., Yanagawa, K., Shimada, J., et al. (1994). Binding of [^{3}H]KF17837S, a selective adenosine A_{2A} receptor antagonist, to rat brain membranes. *Molecular Pharmacology, 46*, 817–822.

Obeso, J. A., Rodriguez-Oroz, M., Martin, C., Alonoso, F., Zamarbide, I., Lancigo, J. L., et al. (2004). The origin of motor fluctuations in Parkinson's disease: Importance of dopaminergic innervation and basal ganglia circuits. *Neurology, 62*, S17–S30.

Ochi, M., Koga, K., Kurokawa, M., Kase, H., Nakamura, J., & Kuwana, Y. (2000). Systemic administration of adenosine A_{2A} receptor antagonist reverses increased GABA release in the globus pallidus of unilateral 6-hydroxy dopamine-lesioned rats: A microdialysis study. *Neuroscience, 100*, 53–62.

Oertel, W. H., & Mugnaini, E. (1984). Immunocytochemical studies of GABAergic neurons in rat basal ganglia and their relations to other neuronal systems. *Neuroscience Letters, 47*, 233–238.

Parkinson, F. E., & Fredholm, B. B. (1990). Autoradiographic evidence for G-protein coupled A_{2A}-receptor in rat neostriatum using [3H]-CGS21680 as a ligand. *Naunyn-Schmiedeberg's Archives of Pharmacology, 342*, 85–89.

Pintor, A., Pezzola, A., Reggio, R., Quarta, D., & Popoli, P. (2000). The mGlu5 receptor agonist CHPG stimulates striatal glutamate release: Possible involvement of A_{2A} receptors. *Neuroreport, 11*, 3611–3614.

Popoli, P., Pezzola, A., Torvinen, M., Reggio, R., Pintor, A., Scarchilli, L., et al. (2001). The selective mGlu(5) receptor agonist CHPG inhibits quinpirole-induced turning in 6-hydroxydopamine-induced rats and modulates the binding characteristics of dopamine D(2) receptors in the rat striatum: Interactions with adenosine A(2a) receptors. *Neuropharmacology, 25*, 505–513.

Preston, Z., Lee, K., Widdowson, L., Freeman, T. C., Dixson, A. K., & Richardson, P. J. (2000). Adenosine receptor expression and function in rat striatal cholinergic interneurons. *British Journal of Pharmacology, 130*, 886–890.

Ramlackhansingh, A. F., Bose, S. K., Ahmed, I., Trukheimer, F. E., Ravese, N., & Brooks, J. (2011). Adenosine 2A receptor availability in dyskinetic and nondyskinetic patients with Parkinson's disease. *Neurology, 76*, 1811–1816.

Richardson, P. J., Dixon, A. K., Lee, K., Bell, M. I., Cox, P. J., Williams, R., et al. (2000). Correlating physiology with gene expression in striatal cholinergic neurons. *Journal of Neurochemistry, 74*, 839–846.

Rodrigues, R. J., Alfaro, T. M., Rebola, N., Oliveria, C. R., & Cunha, R. A. (2005). Co-localization and functional interaction between A_{2A} and metabotropic group 5 receptors in glutamatergic nerve terminals of the rat striatum. *Journal of Neurochemistry, 92*, 433–441.

Rosin, D. L., Hettinger, B. D., Lee, A., & Linden, J. (2003). Anatomy of adenosine A_{2A} receptors in brain. *Neurology, 61*, S12–S18.

Rosin, D. L., Robeva, A., Woodard, R. L., Guyenet, P. G., & Linden, J. (1998). Immunochemical localization of adenosine A_{2A} receptors in the rat central nervous system. *The Journal of Comparative Neurology, 401*, 163–186.

Salim, H., Ferré, S., Dalal, A., Peterfreund, R. A., Fuxe, K., Vincent, J. D., et al. (2000). Activation of adenosine A_1 and A_{2A} receptors modulates dopamine D2 receptor-induced responses in stably transfected human neuroblastoma cells. *Journal of Neurochemistry, 74*, 432–439.

Schiffmann, S. N., Jacobs, O., & Vanderhaeghen, J. J. (1991). Striatal restricted adenosine A_{2A} receptor (RDC8) is expressed by enkephalin but not by substance P neurons: An in situ hybridization histochemistry study. *Journal of Neurochemistry, 57*, 1062–1067.

Schiffmann, S. N., Libert, F., Vassart, G., & Vanderhaeghen, J. J. (1991). Distribution of adenosine A_{2A} receptor mRNA in the human brain. *Neuroscience Letters, 130*, 177–181.

Schiffmann, S. N., & Vanderhaeghen, J. J. (1993). Adenosine A_{2A} receptors regulate the gene expression of striatopallidal and striatonigral neurons. *The Journal of Neuroscience, 13*, 1080–1087.

Schwarzchild, M. A., Agnati, L., Fuxe, K., Chen, J.-F., & Morelli, M. (2006). Targeting adenosine A_{2A} receptors in Parkinson's disease. *Trends in Neurosciences, 29*, 647–654.

Shen, H.-Y., Canas, P. M., Garcia-Sanz, P., Lan, J.-Q., Boison, D., Moratalla, R., et al. (2013). Adenosine A_{2A} receptors in striatal glutamatergic terminals and GABAergic neurons oppositely modulate pyschostimulant action and DARPP-32 phosphorylation. *PLoS One, 8*, 1–11.

Shindou, T., Arbuthnott, G. W., & Wickens, J. R. (2008). Actions of adenosine A_{2A} receptors on synaptic connections of spiny projection neurons in the neostriatal inhibitory network. *Journal of Neurophysiology, 99*, 1884–1889.

Shindou, T., Mori, A., Kase, H., & Ichimura, M. (2001). Adenosine A_{2A} receptor enhances $GABA_A$-mediated IPSCs in the rat globus pallidus. *Journal of Physiology (London), 532*, 423–434.

Shindou, T., Nonaka, H., Richardson, P. J., Mori, A., Kase, H., & Ichimura, M. (2002). Presynaptic adenosine A_{2A} receptors enhance GABAergic synaptic transmission via a cyclic AMP dependent mechanism in the rat globus pallidus. *British Journal of Pharmacology, 136*, 296–302.

Shindou, T., Richardson, P. J., Mori, A., Kase, H., & Ichimura, M. (2003). Adenosine modulates the striatal GABAergic inputs to the globus pallidus via adenosine A_{2A} receptors in rats. *Neuroscience Letters, 352*, 167–170.

Shiozaki, S., Ischikawa, S., Nakamura, J., Kitamura, S., Yamada, K., & Kuwana, Y. (1999). Actions of adenosine receptor antagonist KW-6002 on drug-induced catalepsy and hypokinesia caused by reserpine or MPTP. *Psychopharmacology, 147*, 90–95.

Smith, Y., Parent, A., Seguela, P., & Descarries, L. (1987). Distribution of GABA immunoreactive neurons in the basal ganglia of the squirrel monkey. *Journal of Comparative Neurology, 259*, 50–64.

Stacy, M., Sliver, D., Mendis, T., Sutton, J., Mori, A., Chakin, P., et al. (2008). 12-Week, placebo-controlled study (6002-US-006) of Istradefylline in Parkinson's disease. *Neurology, 70*, 2233–2240.

Stephens, B., Mueller, A., Shering, A. F., Hood, S. H., Taggart, P., Arbuthnott, G. W., et al. (2005). Evidence of a breakdown of corticostriatal connections in Parkinson's disease. *Neuroscience, 132*, 741–754.

Surmeier, D. J., Shen, W., Day, M., Gertler, T., Chan, S., Tian, X., et al. (2010). The role of dopamine in modulating the structure and function of striatal circuits. In A. Bjorklund & M. A. Cenci (Eds.), *Progress in brain research: Vol. 183* (pp. 149–167). Amsterdam: Elsevier B.V.

Svenningsson, P., Hall, H., Sedvall, G., & Fredholm, B. B. (1997). Distribution of adenosine receptors in the postmortem human brain: An extended autographic study. *Synapse, 27*, 322–335.

Svenningsson, P., Le Moine, C., Aubert, I., Burbaud, P., Fredholm, B. B., & Bloch, B. (1998). Cellular distribution of adenosine A_{2A} receptor mRNA in the primate striatum. *The Journal of Comparative Neurology, 399*, 229–240.

Svenningsson, P., Le Moine, C., Kull, B., Sunahara, R., Bloch, B., & Fredholm, B. B. (1997). Cellular expression of adenosine A_{2A} receptor messenger RNA in the rat central nervous system with special reference to dopamine innervated areas. *Neuroscience, 80*, 1171–1185.

Tapper, J. M., Koós, T., & Wilson, C. J. (2004). GABAergic microcircuits in the neostriatum. *Trends in Neurosciences, 11*, 662–669.

Terman, D., Rubin, J. E., Yew, A. C., & Wiloson, C. J. (2002). Activity patterns in a model for the subthalamopallidal network of the basal ganglia. *The Journal of Neuroscience*, *22*, 2963–2976.

Tozzi, A., de Iure, A., Di Filippo, M., Tantucci, M., Costa, C., Borsini, F., et al. (2011). The distinct role of medium spiny neurons and cholinergic interneurons in the D2/A_{2A} receptor interaction in the striatum: Implications for Parkinson's disease. *The Journal of Neuroscience*, *31*, 1850–1862.

Tozzi, A., Tscherter, A., Balcastro, V., Tantucci, M., Costa, C., Picconi, B., et al. (2007). Interaction of A_{2A} adenosine and D2 dopamine receptors modulates corticostriatal glutamatergic transmission. *Neuropharmacology*, *53*, 783–789.

Valenti, O., Mariono, M. J., Wittmann, M., Lis, E., DiLella, A. G., Kinney, G. G., et al. (2003). Group III metabotropic glutamate receptor-mediated modulation of the striatopallidal synapse. *The Journal of Neuroscience*, *23*, 7218–7226.

Wichmann, T., & DeLong, M. R. (2003). Functional neuroanatomy of the basal ganglia in Parkinson's disease. In A. Gordin, S. Kaakkola, & H. Teravainen (Eds.), *Parkinson's disease: Advances in neurology* (pp. 9–18). Philadelphia: Lippincott Williams & Wilkins.

Wickens, J. R., Arbuthnott, G. W., & Shindou, T. (2007). Stimulation of GABA function in the basal ganglia: Computational models of GABAergic mechanisms in basal ganglia function. In J. M. Tepper, E. D. Abercrombile, & J. P. Bolam (Eds.), *Progress in brain research: Vol. 160* (pp. 313–329). Amsterdam: Elsevier B.V.

Xu, K., Batia, E., & Schwarzchild, M. (2005). Therapeutic potential of adenosine A_{2A} receptor antagonists in Parkinson's disease. *Pharmacology & Therapeutics*, *105*, 267–310.

Yu, L., Shen, H. Y., Coelho, J. E., Araújo, I. M., Huang, Q. Y., Day, Y. J., et al. (2008). Adenosine A_{2A} receptors antagonists exert motor and neuroprotective effects by distinct cellular mechanism. *Annals of Neurology*, *63*, 338–346.

CHAPTER FIVE

Adenosine Receptors and Dyskinesia in Pathophysiology

Masahiko Tomiyama[1]
Department of Neurology, Aomori Prefectural Central Hospital, Aomori, Japan
[1]Corresponding author: e-mail address: masahiko_tomiyama@med.pref.aomori.jp

Contents

Abstract

First, the recent progress in the pathogenesis of levodopa-induced dyskinesia was described. Serotonin neurons play an important role in conversion from levodopa to dopamine and in the release of converted dopamine into the striatum in the Parkinsonian state. Since serotonin neurons lack buffering effects on synaptic dopamine concentration, the synaptic dopamine markedly fluctuates depending on the fluctuating levodopa concentration in the serum after taking levodopa. The resultant pulsatile stimulation makes the striatal direct-pathway neurons get potential that releases excessive GABA into the output nuclei of the basal ganglia. When levodopa is administered, the stored GABA is released, the output nuclei become hypoactive, and then dyskinesias emerge. Second, effects of adenosine A_{2A} receptor antagonists on dyskinesia were described. It has been demonstrated that the expression of adenosine A_{2A} receptors is increased in Parkinson's disease (PD) patients with dyskinesias, suggesting that blockade of A_{2A} receptors is beneficial for dyskinesias. Preclinical studies have shown that A_{2A} receptor antagonists reduce liability of dyskinesias in PD models. Clinical trials have demonstrated that A_{2A} antagonists increase functional ON-time (ON without troublesome dyskinesia) in PD patients suffering from wearing-off phenomenon, although they may increase dyskinesia in patients with advanced PD.

1. PATHOGENESIS OF LEVODOPA-INDUCED DYSKINESIA

The current model of basal ganglia circuit (Albin, Young, & Penney, 1989; DeLong, 1990) postulates that stimulation of the direct and indirect striatal efferent pathways provokes motor activation and motor inhibition,

International Review of Neurobiology, Volume 119
ISSN 0074-7742
http://dx.doi.org/10.1016/B978-0-12-801022-8.00005-2

respectively, and that smooth motor performance results from the balanced influence of these two pathways on the neural activity of the basal ganglia output nuclei (Kravitz et al., 2010). In Parkinson's disease (PD), the balance is lost due to the depletion in striatal dopamine caused by degeneration of dopaminergic neurons in the substantia nigra (Obeso et al., 2008).

The L-3,4-dihydroxyphenylalanine (levodopa) is the most effective drug to improve the PD motor symptoms as this compound is converted to dopamine in the brain (Hornykiewicz, 2010; Rascol, Lozano, Stern, & Poewe, 2011). However, its chronic use is associated with the development of motor complications, including levodopa-induced dyskinesia (LID). LID is characterized by jerking and writhing movements that diminish patients' quality of life and over 90% of patients with PD develop LID at the 15-years follow-up (Hely, Morris, Reid, & Trafficante, 2005). Alterations of neuronal activity, both within the basal ganglia and between circuits crucial to the coordination of movement, have been suggested to be associated with appearance of dyskinesias (Brotchie, Lee, & Venderova, 2005), although the mechanisms underlying LID are not completely understood. LID is thought to result from the combined effects of nigrostriatal dopaminergic denervation and levodopa treatment (Cenci, 2007). In short, in the dopamine-denervated striatum levodopa is converted to dopamine and the converted dopamine is released by serotonin neurons instead of devastated dopamine neurons (Tanaka et al., 1999). Since serotonin neurons lack dopamine transporters and presynaptic dopamine D2 receptors both of which regulate dopamine concentration in the synaptic clefts, the dopamine concentration fluctuates markedly depending on levodopa concentration in the serum after levodopa administration (Zhang et al., 2003). The pulsatile unphysiological stimulation of dopamine receptors in the dopamine-denervated striatum following levodopa intake appears to be a primary cause of LID (de la Fuente-Fernández et al., 2004; Olanow & Obeso, 2000; Pavese et al., 2006). There are two physiological patterns of firing of dopamine neurons: one is the tonic firing and the other is phasic firing. Such an unphysiological fluctuation of dopamine concentration in the dopamine-denervated striatum after levodopa intake could lead to phasic-like stimulation principally on dopamine D1 receptors (Dreyer, Herrik, Berg, & Hounsgaard, 2010). Phasic stimulation of dopamine receptors contributes to neural plasticity (Dreyer et al., 2010). Accordingly, levodopa-induced phasic-like dopamine release from the serotonin neurons could result in abnormal plastic changes in the direct-pathway neurons bearing dopamine D1 receptors. It has been demonstrated that the rat model of LID lacks both depotentiation and long-term depression at the corticostriatal synapses, indicating that the synapses where dopamine and glutamate informations are integrated become hypersensitive to excitatory stimuli in the dyskinesia-primed state (Picconi et al., 2003, 2011). We have shown that dendritic spines of the striatal direct-pathway neurons become enlarged in the LID model (Nishijima et al., 2013, 2014). Enlargement of spines is a morphological hallmark of synaptic hypersensitivity (Matsuzaki, Honkura, Ellis-Davies, & Kasai, 2004). Thus, electrophysiological and morphological results suggest that pulsatile dopaminergic stimulation to the rat PD model makes the striatal direct-pathway neurons hypersensitive to glutamate transmission from the cortex. This hypersensitivity of corticostriatal synapses caused by repeated levodopa treatment results in increased signal transductions in the direct-pathway neurons (Cenci & Konradi, 2010) and as a result, the neurons get to have potentials to release excessive GABA into the output nuclei of the basal ganglia, substantia nigra pars reticulata, and globus pallidus interna (Tomiyama, Mori, et al., 2004; Yamamoto, Pierce, & Soghomonian, 2006). We have shown enlargement of the globus pallidus interna of a rat model of LID (Tomiyama, Mori, et al., 2004). When levodopa is administered, excessive GABA is released into the output nuclei (Yamamoto et al., 2006) and the nuclei become hypoactive, resulting in dyskinesia (Papa, Desimone, Fiorani, & Oldfield, 1999). It can be stated that the mechanisms to gain the ability of release the excessive GABA into the output nuclei of the basal ganglia is so-called "priming."

Other than the striatum, transcranial magnetic stimulation studies have suggested that the motor cortex from PD patients with LID also get hypersensitivity since the patients present a lack of depotentiation-like cortical plasticity (Huang, Rothwell, Lu, Chuang, & Chen, 2011). We have demonstrated enlargement of dendritic spines of the motor cortex neurons in the rat LID model and support the hypothesis (Ueno et al., 2014).

2. ADENOSINE A_{2A} RECEPTORS AND LEVODOPA-INDUCED DYSKINESIA

Thus, levodopa is a highly effective drug but the treatment has the nonsolved problems, LID, in the advanced stage of PD. Alternatives to current treatments that avoid LID may lie in nondopaminergic manipulations of the indirect pathway of the basal ganglia. One possible approach is through pharmacological antagonism of adenosine A_{2A} receptors. Efficacy of A_{2A} receptor antagonists for Parkinsonian symptoms is described in detail elsewhere. Although the effects of the A_1 receptor on LID is not well understood, the role of the A_{2A} receptor associated with LID has been extensively examined (Morelli et al., 2007; Schwarzschild, Agnati, Fuxe, Chen, & Morelli, 2006; Pinna, Fenu & Morelli, 2001). The A_{2A} receptor is expressed in axon terminals of the striatopallidal indirect pathway neurons and terminals of their collaterals to striatal projection neurons (Mori & Shindou, 2003). Accordingly, it is hypothesized that the mechanisms by which A_{2A} receptor antagonists exert their effects on motor function result from their inhibitory actions on the indirect pathway neurons expressing both A_{2A} and D2 receptors (Martinez-Mir, Probst, & Palacios, 1991; Richardson, Kase, & Jenner, 1997). In other words, blocking A_{2A} receptors on striopallidal neurons of the indirect pathway may produce parallel behavior activation by mimicking the motor stimulant actions of colocalized D2 receptors on these neurons (Ferré, Fredholm, Morelli, Popoli, & Fuxe, 1997). Inhibitory D2 receptor activity is negatively influenced through the A_{2A} receptor's role in second messenger systems or through the formation of receptor heterodimers with cannabinoid receptors; conversely, activation of A_{2A} receptors disrupts the inhibitory actions of D2 within the indirect pathway leading to downstream effects like those elicited through administration of D2 receptor agonists (Bonaventura et al., 2014; Morelli et al., 2007; Pinna et al., 2014).

Preclinical evidence points to the reduced dyskinesia liability of A_{2A} receptor antagonists. Unilaterally 6-hydroxydopamine-lesioned rats rotate contralateral to the lesion site when treated with levodopa. This phenomenon increases with chronic administration of levodopa. This behavioral sensitization is thought to be predictive of the dyskinetic liability (Morelli, Fenu, Pinna, & Di Chiara, 1994). When low doses of levodopa are administered to 6-hydroxydopamine-lesioned rats or A_{2A} knockout mice in conjunction with an A_{2A} antagonist, contralateral turning behavior is observed without sensitization (Fredduzzi et al., 2002; Hodgson et al., 2009; Morelli et al., 2007). These results obtained from rat PD models are consistent with those from primate PD models. Kanda et al. (1998) demonstrated that istradefylline reversed MPTP-induced motor deficits in marmosets in a dose-dependent fashion. In addition, these studies revealed that chronic istradefylline administration produced no dyskinetic behavior in animals in which levodopa reliably induced dyskinesias. Moreover, Hodgson et al. (2009) reported that when levodopa-sensitized models were treated with preladenant at a dose that significantly reduced their Parkinsonian symptoms, there was no induction of LID.

In addition to data from these behavioral examinations, autoradiographic and *in situ* hybridization studies have demonstrated that the expression of adenosine A_{2A} receptor and the mRNA are increased in the striatum from the rodent model of LID and PD patients with LID (Calon et al., 2004; Tomiyama, Kimura, et al., 2004). Recently, two PET studies have reported higher levels of A_{2A} receptor binding in PD patients with LID compared with PD patients without LID (Mishina et al., 2011; Ramlackhansingh et al., 2011). These findings offer strong support for the hypothesis that A_{2A} receptor antagonists may have a beneficial antidyskinetic potential and are unlikely to exacerbate dyskinesia in patients with PD. In an animal experiment, Bibbiani et al. (2003) has shown that A_{2A} receptor antagonists have prevention effects on LID. However, a recent animal study has demonstrated that A_{2A} receptor antagonists are likely to have a reduced dyskinetic liability relative to levodopa but do not block dyskinesias when coadministered with levodopa (Jones, Bleickardt, Mullins, Parker, & Hodgson, 2013).

Monotherapy of an adenosine A_{2A} receptor antagonist, istradefylline, for drug-naïve PD patients improves Parkinsonian symptoms without provoking dyskinesia (Fernandez et al., 2010). In patients with PD, a recent meta-analysis of istradefylline reveals a significant difference in dyskinesia compared to istradefylline 40 mg with placebo (Chen, Wang, Wei, Gu, & Wei, 2013). On the other hand, there was no difference in case of 20 mg istradeffyline (Chen et al., 2013). However, it should be kept in mind that troublesome dyskinesia does not increase in clinical trials of A_{2A} receptor antagonists in PD patients with wearing-off (Factor et al., 2010, 2013; Hauser, Hubble, Truong, & Istradefylline US-001 Study Group, 2003; Hauser et al., 2008, 2011; LeWitt et al., 2008; Mizuno, Kondo, & Japanese Istradefylline Study Group, 2013; Mizuno et al., 2010; Ngwuluka et al., 2010; Pourcher et al., 2012; Stacy et al., 2008). All the trials were conducted in the manner of additional administration of A_{2A} receptor antagonists to previous antiparkinsonian treatment. This means that istradeffyline increases functional ON-time (ON without troublesome dyskinesia) in PD patients suffering from OFF phenomenon. In our small open-label study of istradefylline, 20 mg/day for 8 weeks, for PD patients with wearing-off but with low UPDRS-III score during the best ON phase (less than 20), there was apparent improvements in best-on UPDRS score and in duration of OFF time (Table 5.1). No patients newly developed LID, although some patients with LID at the entry of the study complained of worsening of LID (Table 5.1). These results suggest that istradeffyline does not provoke dyskinesia in patients with PD without dyskinesia, but it

Table 5.1 Patient's characteristics and results of an open-label trial of Istradefylline (20 mg/day) for Parkinson's disease patients with wearing-off and with low UPDRS-III score during the best ON phase (less than 20) at the entry of the study

	Entry	8 weeks
Patients (F/M)	14 (8/6)	
Age (years)	61.7 ± 11.8	
Disease duration (years)	11.2 ± 2.7	
Number of antiparkinsonian drugs	3.5	
Levodopa (mg/day)	525	
MMSE score	29.1 ± 0.7	
Hoehn–Yahr score	2.4 ± 0.8	2.3 ± 0.7
UPDRS-III score (best-on)	9.8 ± 5.8	6.6 ± 4.4*
Off (h/day)	6.0 ± 2.1	4.0 ± 2.2*
Nontroublesome dyskinesia (%)	36	36
Troublesome dyskinesia (%)	21	29

* $P < 0.01$ (compared with scores at the entry).

possibly worsens existing dyskinesia. Thus, clinical studies of Istradefylline suggest that A_{2A} receptor antagonists may not improve dyskinesia when coadministered with levodopa (Factor et al., 2010, 2013; Hauser et al., 2003, 2008, 2011; LeWitt et al., 2008; Mizuno et al., 2010, 2013; Ngwuluka et al., 2010; Pourcher et al., 2012; Stacy et al., 2008). When A_{2A} receptor antagonists are administered as an adjuvant to levodopa, it should be kept in mind that they may worsen existing dyskinesia. However, they apparently improve Parkinsonian motor symptoms. Animal studies have demonstrated that A_{2A} receptor antagonists lower the dose of levodopa required to get significant antiparkinsonian efficacy in models of PD (Hauser et al., 2003; Hodgson et al., 2009). Accordingly, in the clinical use of them, it is plausible that levodopa dose-sparing potentials of A_{2A} receptor antagonists improve LID.

REFERENCES

Albin, R. L., Young, A. B., & Penney, J. B. (1989). The functional anatomy of basal ganglia disorders. *Trends in Neurosciences*, *12*, 366–375.

Bibbiani, F., Oh, J. D., Petzer, J. P., Castagnoli, N., Jr., Chen, J. F., Schwarzschild, M. A., et al. (2003). A2A antagonist prevents dopamine agonist-induced motor complications in animal models of Parkinson's disease. *Experimental Neurology*, *184*, 285–294.

Bonaventura, J., Rico, A. J., Moreno, E., Sierra, S., Sánchez, M., Luquin, N., et al. (2014). L-DOPA-treatment in primates disrupts the expression of A(2A) adenosine-CB(1) cannabinoid-D(2) dopamine receptor heteromers in the caudate nucleus. *Neuropharmacology*, *79*, 90–100.
Brotchie, J. M., Lee, J., & Venderova, K. (2005). Levodopa-induced dyskinesia in Parkinson's disease. *Journal of Neural Transmission*, *112*, 359–391.
Calon, F., Dridi, M., Hornykiewicz, O., Bédard, P. J., Rajput, A. H., & Di Paolo, T. (2004). Increased adenosine A2A receptors in the brain of Parkinson's disease patients with dyskinesias. *Brain*, *127*, 1075–1084.
Cenci, M. A. (2007). Dopamine dysregulation of movement control in L-DOPA-induced dyskinesia. *Trends in Neurosciences*, *30*, 236–243.
Cenci, M. A., & Konradi, C. (2010). Maladaptive striatal plasticity in L-DOPA-induced dyskinesia. *Progress in Brain Research*, *183*, 209–233.
Chen, W., Wang, H., Wei, H., Gu, S., & Wei, H. (2013). Istradefylline, an adenosine A_2A receptor antagonist, for patients with Parkinson's disease: A meta-analysis. *Journal of Neurological Sciences*, *324*, 21–28.
de la Fuente-Fernández, R., Sossi, V., Huang, Z., Furtado, S., Lu, J. Q., Calne, D. B., et al. (2004). Levodopa-induced changes in synaptic dopamine levels increase with progression of Parkinson's disease: Implications for dyskinesias. *Brain*, *127*, 2747–2754.
DeLong, M. R. (1990). Primate models of movement disorders of basal ganglia origin. *Trends in Neurosciences*, *13*, 281–285.
Dreyer, J. K., Herrik, K. F., Berg, R. W., & Hounsgaard, J. D. (2010). Influence of phasic and tonic dopamine release on receptor activation. *Journal of Neuroscience*, *30*, 14273–14283.
Factor, S., Mark, M. H., Watts, R., Struck, L., Mori, A., Ballerini, R., et al. (2010). A long-term study of istradefylline in subjects with fluctuating Parkinson's disease. *Parkinsonism & Related Disorders*, *16*, 423–426.
Factor, S. A., Wolski, K., Togasaki, D. M., Huyck, S., Cantillon, M., Ho, T. W., et al. (2013). Long-term safety and efficacy of preladenant in subjects with fluctuating Parkinson's disease. *Movement Disorders*, *28*, 817–820.
Fernandez, H. H., Greeley, D. R., Zweig, R. M., Wojcieszek, J., Mori, A., Sussman, N. M., et al. (2010). Istradefylline as monotherapy for Parkinson disease: Results of the 6002-US-051 trial. *Parkinsonism & Related Disorders*, *16*, 16–20.
Ferré, S., Fredholm, B. B., Morelli, M., Popoli, P., & Fuxe, K. (1997). Adenosine-dopamine receptor–receptor interactions as an integrative mechanism in the basal ganglia. *Trends in Neurosciences*, *20*, 482–487.
Fredduzzi, S., Moratalla, R., Monopoli, A., Cuellar, B., Xu, K., Ongini, E., et al. (2002). Persistent behavioral sensitization to chronic L-DOPA requires A2A adenosine receptors. *Journal of Neuroscience*, *22*, 1054–1062.
Hauser, R. A., Cantillon, M., Pourcher, E., Micheli, F., Mok, V., Onofrj, M., et al. (2011). Preladenant in patients with Parkinson's disease and motor fluctuations: A phase 2, double-blind, randomised trial. *Lancet Neurology*, *10*, 221–229.
Hauser, R. A., Hubble, J. P., Truong, D. D., & Istradefylline US-001 Study Group. (2003). Randomized trial of the adenosine A(2A) receptor antagonist istradefylline in advanced PD. *Neurology*, *61*, 297–303.
Hauser, R. A., Shulman, L. M., Trugman, J. M., Roberts, J. W., Mori, A., Ballerini, R., et al. (2008). Study of istradefylline in patients with Parkinson's disease on levodopa with motor fluctuations. *Movement Disorders*, *23*, 2177–2185.
Hely, M. A., Morris, J. G., Reid, W. G., & Trafficante, R. (2005). Sydney Multicenter Study of Parkinson's disease: Non-L-DOPA-responsive problems dominate at 15 years. *Movement Disorders*, *20*, 190–199.
Hodgson, R. A., Bertorelli, R., Varty, G. B., Lachowicz, J. E., Forlani, A., Fredduzzi, S., et al. (2009). Characterization of the potent and highly selective A2A receptor antagonists

preladenant and SCH 412348 [7-[2-[4-2,4-difluorophenyl]-1-piperazinyl]ethyl]-2-(2-furanyl)-7H-pyrazolo[4,3-e][1,2,4]triazolo[1,5-c]pyrimidin-5-amine] in rodent models of movement disorders and depression. *Journal of Pharmacology and Experimental Therapeutics, 330*, 294–303.

Hornykiewicz, O. (2010). A brief history of levodopa. *Journal of Neurology, 257*(Suppl. 2), S249–S252.

Huang, Y. Z., Rothwell, J. C., Lu, C. S., Chuang, W. L., & Chen, R. S. (2011). Abnormal bidirectional plasticity-like effects in Parkinson's disease. *Brain, 134*, 2312–2320.

Jones, N., Bleickardt, C., Mullins, D., Parker, E., & Hodgson, R. (2013). A2A receptor antagonists do not induce dyskinesias in drug-naive or L-DOPA sensitized rats. *Brain Research Bulletin, 98*, 163–169.

Kanda, T., Jackson, M. J., Smith, L. A., Pearce, R. K., Nakamura, J., Kase, H., et al. (1998). Adenosine A2A antagonist: A novel antiparkinsonian agent that does not provoke dyskinesia in parkinsonian monkeys. *Annals of Neurology, 43*, 507–513.

Kravitz, A. V., Freeze, B. S., Parker, P. R., Kay, K., Thwin, M. T., Deisseroth, K., et al. (2010). Regulation of parkinsonian motor behaviours by optogenetic control of basal ganglia circuitry. *Nature, 466*, 622–626.

LeWitt, P. A., Guttman, M., Tetrud, J. W., Tuite, P. J., Mori, A., Chaikin, P., et al. (2008). Adenosine A2A receptor antagonist istradefylline (KW-6002) reduces "off" time in Parkinson's disease: A double-blind, randomized, multicenter clinical trial (6002-US-005). *Annals of Neurology, 63*, 295–302.

Martinez-Mir, M. I., Probst, A., & Palacios, J. M. (1991). Adenosine A2 receptors: Selective localization in the human basal ganglia and alterations with disease. *Neuroscience, 42*, 697–706.

Matsuzaki, M., Honkura, N., Ellis-Davies, G. C., & Kasai, H. (2004). Structural basis of long-term potentiation in single dendritic spines. *Nature, 429*, 761–766.

Mishina, M., Ishiwata, K., Naganawa, M., Kimura, Y., Kitamura, S., Suzuki, M., et al. (2011). Adenosine A(2A) receptors measured with [C]TMSX PET in the striata of Parkinson's disease patients. *PLoS One, 6*, e17338.

Mizuno, Y., Hasegawa, K., Kondo, T., Kuno, S., Yamamoto, M., & Japanese Istradefylline Study Group (2010). Clinical efficacy of istradefylline (KW-6002) in Parkinson's disease: A randomized, controlled study. *Movement Disorders, 25*, 1437–1443.

Mizuno, Y., Kondo, T., & Japanese Istradefylline Study Group. (2013). Adenosine A2A receptor antagonist istradefylline reduces daily OFF time in Parkinson's disease. *Movement Disorders, 28*, 1138–1141.

Morelli, M., Di Paolo, T., Wardas, J., Calon, F., Xiao, D., & Schwarzschild, M. A. (2007). Role of adenosine A2A receptors in parkinsonian motor impairment and L-DOPA-induced motorcomplications. *Progress in Neurobiology, 83*, 293–309.

Morelli, M., Fenu, S., Pinna, A., & Di Chiara, G. (1994). Adenosine A2 receptors interact negatively with dopamine D1 and D2 receptors in unilaterally 6-hydroxydopamine-lesioned rats. *European Journal of Pharmacology, 251*, 21–25.

Mori, A., & Shindou, T. (2003). Modulation of GABAergic transmission in the striatopallidal system by adenosine A2A receptors: A potential mechanism for the antiparkinsonian effects of A2A antagonists. *Neurology, 61*(Suppl. 6), S44–S48.

Ngwuluka, N., Pillay, V., Du Toit, L. C., Ndesendo, V., Choonara, Y., Modi, G., et al. (2010). Levodopa delivery systems: Advancements in delivery of the gold standard. *Expert Opinion on Drug Delivery*, 7, 203–224.

Nishijima, H., Arai, A., Kimura, T., Mori, F., Yamada, J., Migita, K., et al. (2013). Drebrin immunoreactivity in the striatum of a rat model of levodopa-induced dyskinesia. *Neuropathology, 33*, 391–396.

Nishijima, H., Suzuki, S., Kon, T., Funamizu, Y., Ueno, T., Haga, R., et al. (2014). Morphologic changes of dendritic spines of striatal neurons in the levodopa-induced dyskinesia model. *Movement Disorders, 29*, 336–343.

Obeso, J. A., Marin, C., Rodriguez-Oroz, C., Blesa, J., Benitez-Temiño, B., Mena-Segovia, J., et al. (2008). The basal ganglia in Parkinson's disease: Current concepts and unexplained observations. *Annals of Neurology, 64*(Suppl. 2), S30–S46.

Olanow, C. W., & Obeso, J. A. (2000). Pulsatile stimulation of dopamine receptors and levodopa-induced motor complications in Parkinson's disease: Implications for the early use of COMT inhibitors. *Neurology, 55*(Suppl. 4), S72–S77.

Papa, S. M., Desimone, R., Fiorani, M., & Oldfield, E. H. (1999). Internal globus pallidus discharge is nearly suppressed during levodopa-induced dyskinesias. *Annals of Neurology, 46*, 732–738.

Pavese, N., Evans, A. H., Tai, Y. F., Hotton, G., Brooks, D. J., Lees, A. J., et al. (2006). Clinical correlates of levodopa-induced dopamine release in Parkinson disease: A PET study. *Neurology, 67*, 1612–1617.

Picconi, B., Bagetta, V., Ghiglieri, V., Paillè, V., Di Filippo, M., Pendolino, V., et al. (2011). Inhibition of phosphodiesterases rescues striatal long-term depression and reduces levodopa-induced dyskinesia. *Brain, 134*, 375–387.

Picconi, B., Centonze, D., Håkansson, K., Bernardi, G., Greengard, P., Fisone, G., et al. (2003). Loss of bidirectional striatal synaptic plasticity in L-DOPA-induced dyskinesia. *Nature Neuroscience, 6*, 501–506.

Pinna, A., Bonaventura, J., Farré, D., Sánchez, M., Simola, N., Mallol, J., et al. (2014). L-DOPA disrupts adenosine A(2A)-cannabinoid CB(1)-dopamine D(2) receptor heteromer cross-talk in the striatum of hemiparkinsonian rats: Biochemical and behavioral studies. *Experimental Neurology, 253*, 180–191.

Pinna, A., Fenu, S., & Morelli, M. (2001). Motor stimulant effects of the adenosine A(2A) receptor antagonist SCH 58261 do not develop tolerance after repeated treatments in 6-hydroxydopamine-lesioned rats. *Synapse, 39*, 233–238.

Pourcher, E., Fernandez, H. H., Stacy, M., Mori, A., Ballerini, R., & Chaikin, P. (2012). Istradefylline for Parkinson's disease patients experiencing motor fluctuations: Results of the KW-6002-US-018 study. *Parkinsonism & Related Disorders, 18*, 178–184.

Ramlackhansingh, A. F., Bose, S. K., Ahmed, I., Turkheimer, F. E., Pavese, N., & Brooks, D. J. (2011). Adenosine 2A receptor availability in dyskinetic and nondyskinetic patients with Parkinson disease. *Neurology, 76*, 1811–1816.

Rascol, O., Lozano, A., Stern, M., & Poewe, W. (2011). Milestones in Parkinson's disease therapeutics. *Movement Disorders, 26*, 1072–1082.

Richardson, P. J., Kase, H., & Jenner, P. G. (1997). Adenosine A2A receptor antagonists as new agents for the treatment of Parkinson's disease. *Trends in Pharmacological Sciences, 18*, 338–344.

Schwarzschild, M. A., Agnati, L., Fuxe, K., Chen, J. F., & Morelli, M. (2006). Targeting adenosine A2A receptors in Parkinson's disease. *Trends in Neurosciences, 29*, 647–654.

Stacy, M., Silver, D., Mendis, T., Sutton, J., Mori, A., Chaikin, P., et al. (2008). A 12-week, placebo-controlled study (6002-US-006) of istradefylline in Parkinson disease. *Neurology, 70*, 2233–2240.

Tanaka, H., Kannari, K., Maeda, T., Tomiyama, M., Suda, T., & Matsunaga, M. (1999). Role of serotonergic neurons in L-DOPA-derived extracellular dopamine in the striatum of 6-OHDA-lesioned rats. *Neuroreport, 10*, 631–634.

Tomiyama, M., Kimura, T., Maeda, T., Tanaka, H., Kannari, K., & Baba, M. (2004). Upregulation of striatal adenosine A2A receptor mRNA in 6-hydroxydopamine-lesioned rats intermittently treated with L-DOPA. *Synapse, 52*, 218–222.

Tomiyama, M., Mori, F., Kimura, T., Ichinohe, N., Wakabayashi, K., Matsunaga, M., et al. (2004). Hypertrophy of medial globus pallidus and substantia nigra reticulata in 6-hydroxydopamine-lesioned rats treated with L-DOPA: Implication for L-DOPA-induced dyskinesia in Parkinson's disease. *Neuropathology, 24*, 290–295.

Ueno, T., Yamada, J., Nishijima, H., Arai, A., Migita, K., Baba, M., et al. (2014). Morphological and electrophysiological changes in intratelencephalic-typepyramidal neurons in the motor cortex of a rat model of levodopa-induced dyskinesia. *Neurobiology of Disease*, *64*, 142–149.
Yamamoto, N., Pierce, R. C., & Soghomonian, J. J. (2006). Subchronic administration of L-DOPA to adult rats with a unilateral 6-hydroxydopamine lesion of dopamine neurons results in a sensitization of enhanced GABA release in the substantia nigra, pars reticulata. *Brain Research*, *1123*, 196–200.
Zhang, J., Qu, F. R., Nakatsuka, A., Nomura, T., Nagai, M., & Nomoto, M. (2003). Pharmacokinetics of L-DOPA in plasma and extracellular fluid of striatum in common marmosets. *Brain Research*, *993*, 54–58.

CHAPTER SIX

Clinical/Pharmacological Aspect of Adenosine A_{2A} Receptor Antagonist for Dyskinesia

Tomoyuki Kanda[1], Shin-ichi Uchida
Central Nervous System R&D Unit, R&D Division, Kyowa Hakko Kirin Co., Ltd., Shizuoka, Japan
[1]Corresponding author: e-mail address: tomoyuki.kanda@kyowa-kirin.co.jp

Contents

Abstract

Dopamine replacement therapy using the dopamine precursor, L-3,4-dihydroxyphenylalanine (L-DOPA), with a peripheral dopa decarboxylase inhibitor is the most effective treatment currently available for the symptoms of Parkinson's disease (PD). However, the long-term use of dopaminergic therapies for PD is often limited by the development of motor response complications, such as dyskinesia. Adenosine A_{2A} receptors are a promising nondopaminergic target for the treatment of PD. The treatment of motor response complications involves combinations of regular and controlled release L-DOPA, perhaps with the addition of a COMT inhibitor or the use of a longer-acting dopamine agonist. However, when dyskinesia is already established, the increase in dopaminergic load produced by the addition of a dopamine agonist can result in an increase in the severity and duration of dyskinesia. Currently, there are no well-tolerated antidyskinesia agents available. Amantadine, which may exert its effects through the inhibition of *N*-methyl-D-aspartate (NMDA) receptors, shows some effects on established dyskinesia. Dyskinesia has a negative impact on the quality of life of patients,

International Review of Neurobiology, Volume 119
ISSN 0074-7742
http://dx.doi.org/10.1016/B978-0-12-801022-8.00006-4

sometimes being more disabling than PD itself. Although some patients prefer experiencing dyskinesia than being in the OFF state and unable to move, alternative, more effective therapies are still required for severe disabling dyskinesia to afford patients an improved quality of life while in the ON state. The mechanisms causing and maintaining the dyskinesia have not been clarified. The application of a non-dopaminergic approach to modify the basal ganglial activity would be helpful to better understand and treat dyskinesia. The use of an adenosine A_{2A} receptor may provide one such approach. In this literature review, we will summarize the current knowledge from both clinical and nonclinical studies on the effects of adenosine A_{2A} receptor blockade on dyskinesia.

1. INTRODUCTION

Parkinson's disease (PD) is a chronic, gradually progressive neurodegenerative disorder characterized by motor dysfunctions, such as bradykinesia and akinesia (Fahn, 2003; Jellinger, 1986). Most motor symptoms of the disease result from the progressive degeneration of dopaminergic neurons in the substantia nigra, which eventually leads to reduced stimulation of the striatal dopamine receptors (Barzilai & Melamed, 2003; Dauer & Przedborski, 2003; Hunot & Hirsch, 2003). Dopamine replacement therapy using a dopamine precursor, L-3,4-dihydroxyphenylalanine (L-DOPA), with a peripheral dopa decarboxylase inhibitor (DCI) is the most effective treatment currently available for the symptoms of PD (Fahn, 2008). However, the onset of dyskinesias and/or psychosis and a loss of drug efficacy can limit this pharmacological intervention (Ahlskog & Muenter, 2001; Rascol et al., 2003). Most patients experience motor response fluctuations within 5 year of starting treatment, and dyskinesias then frequently appear (Ahlskog & Muenter, 2001).

The dyskinesia can present in a variety of clinical forms, such as dystonia and chorea, which occur at different time points after drug treatment. When the dyskinesia is already established, the increase in dopaminergic load produced by the addition of a dopamine agonist can result in an increase in the severity and duration of the dyskinesia (Bonuccelli, Colzi, & Del Dotto, 2002).

The mechanisms responsible for the induction of dyskinesia have not been fully elucidated. However, the activities of basal ganglia are thought to be involved in promoting the development of the dyskinesias. It was suggested that dyskinesia results from an imbalance between the indirect striopallidal output pathway and the direct striopallidal/-nigral pathway.

In particular, the indirect pathway has been implicated as becoming dysfunctional, and thus indirectly causing alterations in the pallidothalamic output. On the other hand, the physiological mechanisms underlying the development of dyskinesia and already existing dyskinesia may be different. Both the degeneration of the dopamine neurons and the chronic and intermittent dopaminergic receptor stimulation by dopaminergic treatments are thought to be important pathophysiological factors associated with the induction and development of dyskinesia.

The treatment of the response fluctuations currently involves combinations of regular and controlled release L-DOPA, perhaps with the addition of a catechol *O*-methyl transferase (COMT) inhibitor or the use of a long-duration dopamine agonist (Antonini et al., 2010; Marsala, Gioulis, Ceravolo, & Tinazzi, 2012; Matheson & Spencer, 2000). Little can be done to reverse an existing dyskinesia, although amantadine may be effective by inhibiting NMDA receptors (Olanow, Stern, & Sethi, 2009; Thomas et al., 2004). An alternative approach would be to target nondopaminergic receptors, which contribute a selective site for the modulation of basal ganglial output pathways. Adenosine A_{2A} receptor antagonists may provide one such possibility for treatment.

Adenosine receptors are classified into three major types: A_1, A_2 and A_3, with adenosine A_2 receptors being further subdivided into two subtypes, A_{2A} and A_{2B} (Bruns, Lu, & Pugsley, 1986; Daly, Butts-Lamb, & Padgett, 1983, Fredholm, IJzerman, Jacobson, Klotz, & Linden, 2001). Adenosine A_{2A} receptors are highly localized to the caudate–putamen, nucleus accumbens, and olfactory tubercle (Jarvis & Williams, 1991; Schiffmann, Libert, Vassart, & Vanderhaeghen, 1989). Within the caudate–putamen, adenosine A_{2A} receptors are selectively localized on the GABA- and encephalin-containing medium size spiny neurons which form the indirect output pathway from the striatum to the globus pallidus external segment, and which bear dopamine D_2 receptors on their surface (Schiffmann, Jacobs, & Vanderhaeghen, 1991).

The indirect pathway has been implicated in the onset of parkinsonian disability and the genesis of L-DOPA-induced dyskinesia (LID) (Obeso et al., 2004). Hence, modulation of A_{2A} receptors may have a profound influence on the motor function, and there is considerable recent evidence to support such a role (Jenner et al., 2009). Istradefylline is a potent and selective adenosine A_{2A} receptor antagonist. This compound is the highest development stage compound in this class. Istradefylline ((*E*)-1,3-diethyl-8-(3,4-dimethoxystyryl)-7-methyl-3,7-dihydo-1*H*-purine-2,6-dione, formerly known as KW-6002) was approved in Japan for adjunct treatment for PD with L-DOPA in 2013. A Phase III clinical study of this compound is also ongoing in the United States. Istradefylline showed selectivity as an adenosine A_{2A} receptor antagonist relative to its affinity for other adenosine receptor subtypes and showed specificity based on the lack of observed affinity for other neurotransmitter receptors, such as dopamine and acetylcholine receptors and transporters *in vitro* (Saki, Yamada, Koshimura, Sasaki, & Kanda, 2013). Istradefylline also does not affect the dopamine metabolic enzymes, such as monoamine oxidase (MAO)-A, MAO-B and COMT (Saki et al., 2013). Thus, istradefylline thought to be a pure adenosine A_{2A} antagonist. The results of the effects of istradefylline on preclinical and clinical studies will be helpful to understand the correlation between adenosine A_{2A} receptors and dyskinesia.

The pathophysiology of adenosine A_{2A} receptor function and dyskinesia is summarized and discussed in Chapter 5. Thus the following sections will summarize the current knowledge based on both clinical and nonclinical studies of the effects of adenosine A_{2A} receptor blockade and dyskinesia.

2. NONCLINICAL STUDIES OF THE EFFECTS OF ADENOSINE A_{2A} RECEPTOR ANTAGONISTS ON DYSKINESIA

Several clinical studies of adenosine A_{2A} antagonist drugs, such as istradefylline, have been published regarding its use for the treatment of PD (Hauser et al., 2008; LeWitt et al., 2008, Mizuno et al., 2010; Mizuno, Kondo, & the Japanese Istradefylline Study Group, 2013; Pourcher et al., 2012; Stacy et al., 2008), as have studies of preladenant (Factor et al., 2013; Hauser et al., 2011). However, only a few studies have examined the potential of adenosine A_{2A} antagonists for treating dyskinesia. On the other hand, the effects of adenosine A_{2A} antagonists on dyskinesia have been studied using nonhuman primates or rodent models of dyskinesia caused by repeated L-DOPA treatment. The current knowledge based on these nonclinical findings will summarize in section 2.1 and 2.2.

2.1. The effects of adenosine A_{2A} receptor antagonism in 6-OHDA-lesioned hemiparkinsonian rodents

Given that LID correlates with the neuroplasticity, the neural circuits and structure of the basal ganglia are thought to be important to the development and maintenance of the LID. In this regard, rodent models of PD are associated with some weaknesses. For example, the anatomical differences observed in the rodent basal ganglia structures compared with nonhuman primates and humans. For example, the striatal components in rodents are not divided in the caudate nucleus and putamen, while those in humans and nonhuman primates are clearly divided. In addition, there are also species differences in the structures of the basal ganglia, especially the globus pallidus, between rodents and human and nonhuman primates. The globus pallidus of humans and nonhuman primates are divided within the internal (GPi) and external (GPe) segments, while the rodents do not exhibit this division (Parent & Hazarti, 1995). However, rodent models of PD have some advantages in nonclinical research for LID, such as their ease in handling and reproducibility.

In hemiparkinsonian rodents, the nigrostriatal dopamine pathway is unilaterally lesioned by a dopamine neurotoxin, 6-hydroxy-dopamine (6-OHDA), and these rodents represent a well-known model of PD (Ungerstedt, 1976). Unilateral 6-OHDA-lesioned rodents show a contralateral rotational response to dopaminergic agents, such as L-DOPA and/or dopamine agonist drugs (Ungerstedt, 1976). Therefore, a contralateral rotational response in this animal is thought to be an indicator of the antiparkinsonian activities of the therapeutic compounds. In addition, dopaminergic priming is required to produce the full effects of dopaminergic treatment in this animal (Henry, Crossman, & Brotchie, 1998). The effective dose of L-DOPA, that induced the rotational response, induced rotational sensitization in this animal model (Pinna, Fenu, & Morelli, 2001). Sensitization of the contralateral rotational response by repeated L-DOPA treatment is considered to be a motor response complication in rodents, especially those with LID.

The administration of an adenosine A_{2A} antagonist, such as istradefylline, increased the rotational response when coadministered with a dopaminergic drug (Koga, Kurokawa, Ochi, Nakamura, & Kuwana, 2000). Thus, adenosine A_{2A} antagonists can reduce the dosage of dopaminergic drug required to induce a response, while inducing the same magnitude of action. Pinna et al. (2001) also reported that chronic use of the combined treatment of an adenosine A_{2A} antagonist and L-DOPA did not produce rotational response sensitization. Thus, they suggested that A_{2A} receptor antagonism either directly blocks an A_{2A} receptor-dependent neuroplasticity underlying the sensitized response to repeated L-DOPA, or indirectly prevents sensitization by simply avoiding the higher doses of L-DOPA that may be required to induce sensitization. They also suggested that adenosine A_{2A} antagonism might reduce the development of LID by exerting L-DOPA-sparing effects during adjunctive therapy, especially in the early stage of the disease.

The direct contribution of adenosine A_{2A} antagonists to the rotational response sensitization caused by repeated L-DOPA administration was also studied using adenosine A_{2A} receptor-deficient animals (Fredduzzi et al., 2002). The rotational response sensitization by repeated L-DOPA administration was prevented by A_{2A} receptor deletion. These results suggested that a blockade of adenosine A_{2A} receptors prevented the development of LID in addition to L-DOPA-sparing effects. Unfortunately, the mechanism by which adenosine A_{2A} receptor blockade prevents rotational response sensitization has not been fully clarified.

Contralateral rotation induced by dopaminergic treatment in unilateral 6-OHDA-lesioned rodents is generally interpreted to indicate antiparkinsonian efficacy. Thus, enhancement of the rotational response by the coadministration of L-DOPA or a dopamine agonist drug and other drugs is considered to be due to additive and/or synergistic effects of the concomitant agents. Furthermore, it was difficult to divide the combined effects and sensitization effects of the coadministered drugs. However, the results of studies using the rotational sensitization model in unilateral 6-OHDA-lesioned rodents suggested that the adenosine A_{2A} receptor function may be involved in developing and/or maintaining the LID.

As described above, the contralateral rotational responses to L-DOPA in hemiparkinsonian rodents are generally interpreted as a measure of desirable antiparkinsonian efficacy. Indeed, enhancement of the contralateral rotational responses to L-DOPA by A_{2A} antagonists provided an initial indication of their therapeutic potential as drugs that could ameliorate the symptoms of PD. An alternative method and/or measurement would be to use a more intuitive and relevant model of LID.

The abnormal involuntary movements (AIMs) in 6-OHDA-lesioned hemiparkinsonian rodents may provide such a model. Cenci, Lee, and Bjorklund (1998) have characterized the induction of AIMs after the repeated administration of L-DOPA in 6-OHDA-lesioned hemiparkinsonian rodents. They summarized that the AIMs in 6-OHDA-lesioned hemiparkinsonian rodents are thought to correlate with LID. The AIMs scale is the rate determined using the forelimb, axial, and orolingual subscales that assess the choreoathetoid movements that bear a relatively close resemblance to the LID of PD patients. The AIMs rating scale uses subjective behavioral changes, limb dyskinesia characterized by repetitive and rhythmic movements or dystonic posturing of the forelimb on the side contralateral to the lesion, axial dyskinesia characterized by lateral flexion and axial rotation/torsion affecting the neck and the upper trunk toward the side contralateral to the lesion, and orolingual dyskinesia affecting the orofacial musculature, including chewing movements, tongue protrusions, and jaw tremors (Cenci et al., 2002).

The role of A_{2A} antagonists have also been investigated in AIMs, as well as rotational sensitization models, of 6-OHDA-lesioned hemiparkinsonian animals. Lundblad, Vaudano, and Cenci (2003) reported the effects of an adenosine A_{2A} antagonist, KW-6002, and examined whether it could modulate the dyskinetic effects of L-DOPA in 6-OHDA-lesioned hemiparkinsonian rats with AIMs due to previous exposure to chronic L-DOPA. They found that KW-6002 did not elicit any AIMs on its own, but failed to reduce the severity of dyskinesia when coadministered with L-DOPA. They also reported that chronic treatment with KW-6002 only produced a significant relief of motor disability in the rotarod test in the absence of any AIMs. Combined treatment with L-DOPA and KW-6002 improved the rotarod performance to a significantly higher degree than did each of the two drugs alone. However, this combined treatment induced dyskinesia to about the same degree as did L-DOPA alone.

In addition to behavioral assessments, they carried out histochemical assessments of the effects of KW-6002 alone or in combination with L-DOPA. The striatal levels of preproenkephalin mRNA were reduced to approximately 20% of the control level by KW-6002 alone, but neither the coadministration of KW-6002 and L-DOPA nor L-DOPA alone significantly altered the expression of this transcript in the dopamine-denervated striatum. Alone or in combination with L-DOPA, KW-6002 did not have any significant effect on the prodynorphin mRNA expression or FosB/DFosB-like immunoreactivity in the dopamine-denervated striatum.

Tronci et al. (2007) also reported that another adenosine A_{2A} antagonist, ST1535 (2-butyl-9-methyl-8-(2*H*-1,2,3-triazol-2-yl)-9*H*-purin-6-ylamine), did not modify the AIMs when given in combination with L-DOPA. Thus, they concluded that treatment with an adenosine A_{2A} receptor antagonist alone can relieve motor disability without inducing behavioral or cellular signs of dyskinesia in 6-OHDA-lesioned hemiparkinsonian rats and suggested that cotreatment with an adenosine A_{2A} antagonist and L-DOPA potentiates the therapeutic effects, but not the dyskinesiogenic potential, of the latter drug. However, in the mouse AIMs model with unilateral 6-OHDA lesions, conditional knockout of the forebrain A_{2A} receptor significantly attenuated the AIMs induced by daily L-DOPA treatment (Xiao et al., 2006). Thus, they suggested that the A_{2A} receptor deficit in the forebrain of animals was important for the development and maintenance of the AIMs. However, the above findings based on rotational response sensitization and AIMs are inconsistent with the findings for adenosine A_{2A} receptor blockade and the development and/or maintenance of the LID. The hemiparkinsonian rodents with unilateral 6-OHDA lesions did not display the same dyskinesias observed in nonhuman primate PD models or in PD patients following chronic L-DOPA treatment. Thus, there are some difficulties in drawing conclusions based on the effects of adenosine A_{2A} receptor antagonists using rodent models of LID. Nevertheless, the existence of some correlation between the adenosine A_{2A} receptor function and the development or maintenance of the dyskinesia was suggested by the rodent studies.

2.2. Adenosine A_{2A} receceptor antagonism in nonhuman primate models of dyskinesia

The dopaminergic neurotoxin (1-methyl-4-phenyl-1,2,3,6-tetrahydropyridine (MPTP))-treated nonhuman primate is the most appropriate pharmacological model of Parkinson's disease that is currently available. The behavioral changes and cellular loss closely mimic those seen in PD patients. The response to antiparkinsonian agents and the doses required in this primate model are also similar to those in human subjects (Grondin et al., 1999; Kanda et al., 1998; Löschmann et al., 1992). In MPTP-treated nonhuman primates, the adverse effects of dopaminergic drugs that were observed included vomiting, stereotyped behavior, and L-DOPA or levodopa/carbidopa-induced dyskinesia (Grondin et al., 1999; Pearce, Jackson, Smith, Jenner, & Marsden, 1995). In the case of MPTP-treated common marmosets, which were orally dosed with L-DOPA with DCI twice daily for 4 weeks with an optimal dose (L-DOPA; 10 mg/kg), dyskinesia developed as a response to therapy. Several studies have described the effects of adenosine A_{2A} antagonists in MPTP-treated nonhuman primate models, including the effects on LID.

We first reported the effects of the adenosine A_{2A} antagonist istradefylline on MPTP-treated common marmosets in the late 1990s (Kanda et al., 1998). In that study, we demonstrated that the oral administration of istradefylline reversed the motor disability in MPTP-treated common marmosets in a dose-dependent manner. However, istradefylline only modestly increased the overall locomotor activity and did not cause abnormal movements, such as stereotypy. The subsequent oral administration of istradefylline (10 mg/kg) for 21 days induced little or no dyskinesia. A further single oral administration of istradefylline (10 mg/kg), 1 week after the chronic dosing (Day 28), again induced little or no dyskinesia. These data were reproduced using the same experimental procedure. The results were in line with those of our previous investigation (Fig. 6.1). These data described that istradefylline alone induced little or no dyskinesia in MPTP-treated primates previously primed to exhibit dyskinesia by prior exposure to L-DOPA. Thus, istradefylline alone did not induce dyskinesia at doses that reversed parkinsonian symptoms in L-DOPA-primed MPTP-lesioned common marmosets.

We also investigated whether istradefylline, in combination with L-DOPA, could enhance the antiparkinsonian activity and exacerbate LID in MPTP-treated common marmosets (Kanda et al., 2000). Three

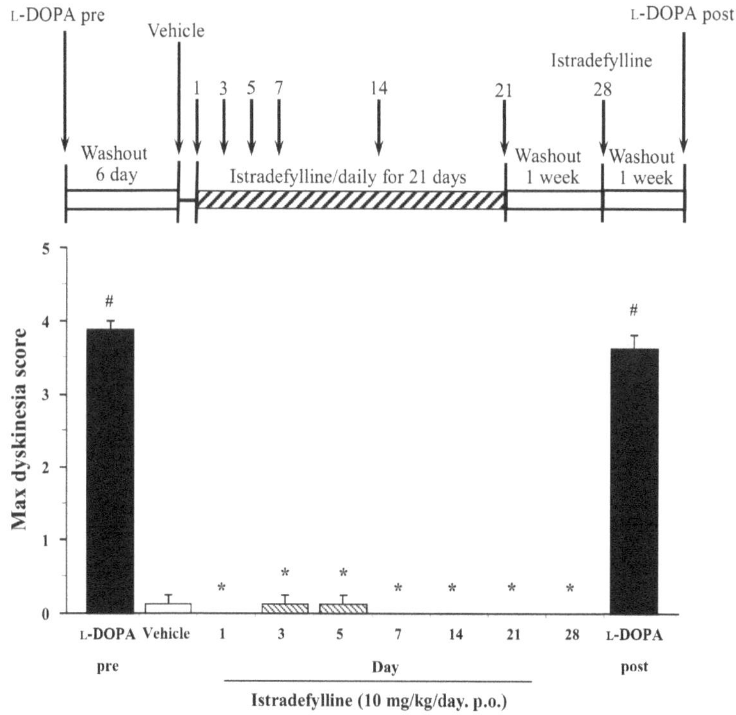

Figure 6.1 Effect of istradefylline (10 mg/kg p.o.) administered daily for 21 days on dyskinesia in MPTP-treated marmosets primed with L-DOPA. The animals previously received 28 days of L-DOPA (10 mg/kg plus benserazide 2.5 mg/kg p.o.) twice daily for induction of dyskinesia. Each column represents the mean maximal dyskinesia score (Max dyskinesia score) + SEM for eight animals. #$p < 0.05$ compared with control (vehicle-treated group). *$p < 0.05$ compared with L-DOPA pre (L-DOPA-treated group).

weeks of chronic administration of L-DOPA (12.5 mg/kg p.o. twice/day) plus carbidopa (12.5 mg/kg p.o. twice/day) induced dyskinesia, which was reproduced following the administration of each dose of L-DOPA. The administration of a threshold dose of L-DOPA (2.5 mg/kg p.o.) induced a moderate degree of dyskinesia. Coadministration of istradefylline (10 mg/kg/day) daily for 5 days with this threshold dose of L-DOPA (2.5 mg/kg) also produced mild dyskinesia, but this was not different from that produced by L-DOPA alone. The intensity of dyskinesia appeared to decrease over the treatment period, but this effect did not reach statistical significance.

Thus, we evaluated the effects of 3 weeks of combination treatment with istradefylline and suboptimal L-DOPA treatment on established dyskinesia (Uchida et al., 2014). When L-DOPA (10 mg/kg) and DCI-primed MPTP-treated common marmosets were treated with L-DOPA (2.5 mg/kg), mild dyskinesia was induced. Although the dose used in the L-DOPA control (L-DOPA: 2.5 mg/kg p.o.) is the suboptimal dose in animal models of parkinsonism, it produced significant dyskinesia compared to vehicle-treated L-DOPA-primed common marmosets ($p < 0.05$). However, the dyskinetic response to the suboptimal dose of L-DOPA (2.5 mg/kg p.o.) was significantly lower ($p < 0.01$) than that in the L-DOPA pretreatment group [L-DOPA (10 mg/kg p.o.)]. When 10 mg/kg of istradefylline was coadministered orally with a suboptimal dose of L-DOPA (2.5 mg/kg p.o.) for 21 days, the amplitude of the involuntary movements (dyskinesia) decreased significantly on Day 21 compared with that observed with the suboptimal dose of L-DOPA alone (Fig. 6.2). Dyskinesia was observed following the single oral administration of istradefylline (10 mg/kg) with a suboptimal dose of L-DOPA (2.5 mg/kg) after 1 week (Day 28) of chronic dosing, but these findings were significantly milder ($p < 0.05$) than those induced by a suboptimal dose of L-DOPA (2.5 mg/kg p.o.) alone (L-DOPA control group). These results indicated that the chronic coadministration of istradefylline with a suboptimal dose of L-DOPA did not worsen the dyskinesia observed with the suboptimal dose of L-DOPA alone; furthermore, the dyskinesia associated with the combination treatment was significantly decreased by Day 21. Again, the dyskinetic response to the optimal dose of L-DOPA (10 mg/kg, L-DOPA posttreatment) was unchanged by chronic treatment with istradefylline. These data demonstrated that in the L-DOPA-primed MPTP-treated marmosets, istradefylline alone itself does not induce dyskinesia, and also suggested that chronic coadministration of istradefylline with a suboptimal dose of L-DOPA causes no worsening of dyskinesia.

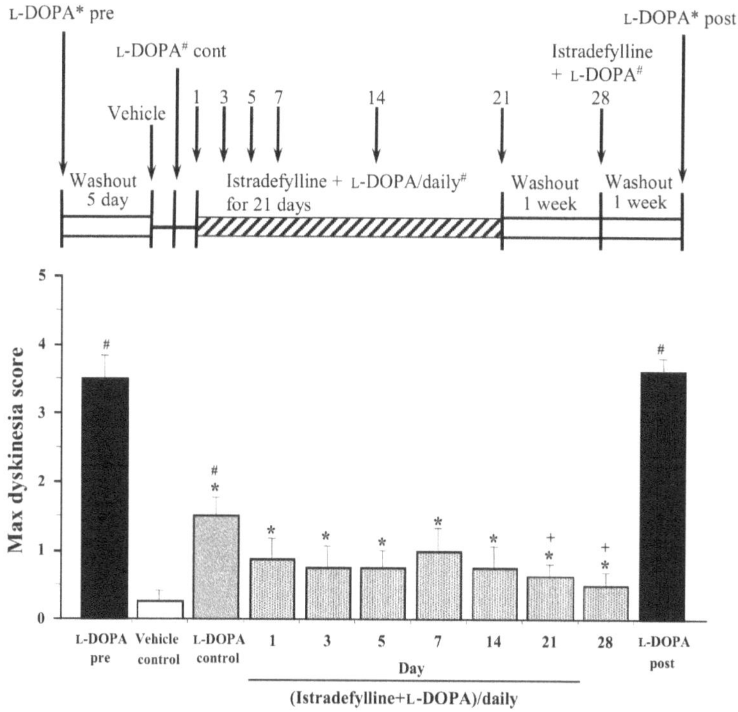

Figure 6.2 Effects of coadministration of istradefylline (10 mg/kg po) and L-DOPA (2.5 mg/kg plus benserazide 0.625 mg/kg po) administered daily for 21 days on dyskinesia in MPTP-treated common marmosets primed with L-DOPA. Upper panel represents experimental protocol. * L-DOPA (10 mg/kg plus benserazide 2.5 mg/kg po). # L-DOPA (2.5 mg/kg plus benserazide 0.625 mg/kg po). pre, pretreatment; cont, control treatment; post, posttreatment. Lower panel represents the maximum dyskinesia scores. The animals previously received 28 days of L-DOPA (10 mg/kg plus benserazide 2.5 mg/kg po) twice daily for induction of dyskinesia. Each column represents the mean (+SEM) of the maximal dyskinesia score (Max dyskinesia score) for eight animals. #$p < 0.05$ compared with vehicle control; *$p < 0.05$ compared with L-DOPA pre (10 mg/kg); +$p < 0.05$ compared with L-DOPA control (2.5 mg/kg). *This figure is from figures 1 and 3 in Uchida et al. (2014) with permission.*

Grondin et al. (1999) further explored the antiparkinsonian effects of istradefylline, as well as its effects on dyskinesia, in cynomolgus monkeys. Istradefylline alone, at doses ranging from 60 to 90 mg/kg, reversed the motor disability in MPTP-treated cynomolgus monkeys. The magnitude of the effect obtained with 90 mg/kg of istradefylline was comparable to that observed with an optimal dose of L-DOPA (50 mg). Unlike L-DOPA, however, istradefylline did not produce dyskinesia. When istradefylline (60 to 90 mg/kg) was added to 50 mg of L-DOPA, the dyskinesia was not increased. These data demonstrate that, as was observed in the MPTP-treated common marmosets, the selective adenosine A_{2A} receptor antagonist, istradefylline, improved the motor disability without inducing dyskinesia in the MPTP-treated cynomolgus monkey.

These results from MPTP-treated common marmosets and cynomolgus monkeys can be summarized as follows: (1) Istradefylline itself does not induce dyskinesia in MPTP-treated nonhuman primates previously primed to exhibit dyskinesia by prior exposure to L-DOPA. (2) The chronic coadministration of istradefylline with L-DOPA did not worsen the severity of existing dyskinesia. (3) The severity of dyskinesia tended to be reduced in animals with chronic coadministration of istradefylline with L-DOPA. These findings from MPTP-treated common marmosets and cynomolgus monkeys showed consistent effects of the adenosine A_{2A} antagonist, istradefylline.

In addition to the above findings, Bibbiani et al. (2003) found that chronic coadministration of istradefylline completely prevented the induction of dyskinesia that is normally seen in MPTP-lesioned cynomolgus monkeys after 1–2 weeks of daily administration of a dopamine receptor agonist. Istradefylline showed modest antiparkinsonian activity when given alone. Once-daily coadministration of istradefylline with apomorphine prevented the development of dyskinesias, which appeared in control animals 7–10 days after initiating apomorphine treatment. Animals initially given apomorphine plus istradefylline for 3 weeks did not begin to manifest apomorphine-induced dyskinesias until 10–12 days after discontinuing the A_{2A} antagonist. These findings suggest that istradefylline can attenuate the induction, as well as the expression, of motor response alterations due to chronic dopaminergic stimulation in parkinsonian animals, possibly by blocking A_{2A} receptor-stimulated signaling pathways. They also strengthen the rationale for developing A_{2A} antagonists as an early treatment strategy for PD.

3. CLINICAL OUTCOMES OF ADENOSINE A_{2A} RECEPTOR ANTAGONIST ON DYSKINESIA

The first clinical report describing the effects of an adenosine A_{2A} antagonist was published by Bara-Jimenez et al. (2003). They found that istradefylline alone or in combination with a steady-state intravenous infusion of each patient's optimal L-DOPA dose had no effect on the parkinsonian severity. At a low dose of L-DOPA, however, istradefylline (80 mg) potentiated the antiparkinsonian response by 36%, but with 45% less dyskinesia compared with that induced by the optimal dose L-DOPA alone. All cardinal parkinsonian signs improved, especially resting tremors. In addition, istradefylline prolonged the effective half-life of L-DOPA by an average of 47 min (76%). No medically important drug toxicity occurred. They suggested that the A_{2A} receptor-related mechanisms contributed to symptom production in PD, and that drugs able to selectively block these receptors may help palliate symptoms in L-DOPA-treated patients with PD.

Based on preclinical data suggesting a beneficial effect of istradefylline on PD symptoms, this adenosine A_{2A} receptor antagonist has undergone extensive clinical evaluation. However, only a few reports have mentioned the effects of istradefylline on dyskinesia. A 12-week, double-blind, randomized, placebo-controlled study in levodopa-treated PD subjects with motor fluctuations and peak-dose dyskinesias was reported by Hauser et al. (2008). Another 12-week, double-blind, placebo-controlled study in levodopa-treated PD subjects with motor fluctuations was reported by LeWitt et al. (2008). Two Japanese 12-week, double-blind, placebo-controlled studies in levodopa-treated PD subjects with motor fluctuations were reported by Mizuno et al. (2010, 2013). The clinical trials involving istradefylline are summarized in Table 6.1.

Hauser et al. (2008) evaluated the efficacy, safety, and tolerability of istradefylline (20 mg) once daily versus placebo as an adjunct to L-DOPA in subjects with PD who had motor fluctuations. Their 12-week, multicenter, double-blind, placebo-controlled, randomized, Phase 3 study of istradefylline was conducted in subjects experiencing an average daily OFF time of at least 3 h (116 randomized to istradefylline; 115 to placebo). All were on stable L-DOPA regimens; 90% were also on stable regimens of other anti-Parkinson's medications. The istradefylline-treated subjects had significant placebo-corrected reductions in daily OFF time from baseline to the endpoint: 4.6% and 0.7 h. With regard to ON time with troublesome dyskinesia, the changes between istradefylline and placebo were not significant. Istradefylline was well tolerated, with 6 (5.2%) istradefylline-treated and 7 (6.1%) placebo-treated subjects withdrawing from the study because of adverse events (AEs). Dyskinesia, lightheadedness, tremors, constipation, and weight loss were reported more often with istradefylline than placebo. The authors of that study concluded that istradefylline slightly exacerbated the dyskinesia frequency, but not the severity. However, severity of increased dyskinesia is almost always limited to nontroublesome dyskinesia. Therefore, they concluded that istradefylline is well tolerated

Table 6.1 Summary of clinical trials with istradefylline

Study	Design	Dosage (mg/day)	Clinical findings
Bara-Jimenez et al. (2003)	Six-week, dose-escalation study in PD subjects with optimal and suboptimal levodopa dosing	40 80	When given istradefylline with suboptimal doses of levodopa, istradefylline 80 mg/day improved motor function to a level similar to that seen in subjects optimally treated with levodopa, but with fewer dyskinesias. The efficacy half-time of levodopa was prolonged with istradefylline
Hauser, Hubble, Truong, and Istradefylline US-001 Study Group (2003)	Twelve-week, double-blind, randomized, placebo-controlled study in levodopa-treated PD subjects with motor fluctuations and peak-dose dyskinesias	20 40	Change in off time significantly reduced in the istradefylline-treated group compared to placebo. Dyskinesia severity was unchanged, but "ON" time with dyskinesia increased in the istradefylline group compared with the placebo group
Stacy et al. (2008)	Twelve-week, double-blind, placebo-controlled study in levodopa-treated PD subjects with motor complications	20 60	Change in percentage of off time significantly reduced in both 20 mg/day and 60 mg/day istradefylline treatment groups compared to placebo. The most frequently reported treatment-related adverse events were dyskinesia (23.9% in subjects receiving 20 mg/day istradefylline, 22.6% in subjects receiving 60 mg/day istradefylline, and 14.3% in subjects receiving placebo). Dyskinesia was generally mild or moderate in intensity

Hauser et al. (2008)	Twelve-week, double-blind, placebo-controlled study in levodopa-treated PD subjects with motor fluctuations	20	Change in off time significantly reduced in the istradefylline-treated group compared to placebo. The most frequently reported adverse events with istradefylline were dyskinesia. Dyskinesia was generally mild or moderate in intensity. No significant changes of ON time with troublesome dyskinesia were observed in istradefylline treatment group
LeWitt et al. (2008)	Twelve-week, double-blind, placebo-controlled study in levodopa-treated PD subjects with motor fluctuations	40	Change in awake off time significantly reduced in the istradefylline-treated group compared to placebo. The most frequently reported drug-related events, dyskinesia occurred more often for istradefylline (30.2%) than for placebo subjects (15.2%). Treatment-emergent dyskinesia was mostly mild or moderate in intensity. Severe dyskinesia was reported for two subjects in each group (istradefylline, 1.6%; placebo, 3.0%)
Mizuno et al. (2010)	Twelve-week, double-blind, placebo-controlled study in levodopa-treated PD subjects with motor fluctuations	20 40	The daily OFF time reduced from baseline by 1.31 h for 20 mg/day istradefylline, 1.58 h for 40 mg/day istradefylline, and 0.66 h for placebo. Istradefylline significantly reduced the daily OFF time compared with placebo. The UPDRS Part III subscale score (ON state) reduced by 5.7 at endpoint in both istradefylline groups

Continued

Table 6.1 Summary of clinical trials with istradefylline—cont'd

Study	Design	Dosage (mg/day)	Clinical findings
			The most commonly reported drug-related treatment-emergent adverse event (TEAE) was dyskinesia. No dose-dependent increase was found in the incidence of dyskinesia. The intensity was mild to moderate in all cases and no subjects experienced severe dyskinesia
Pourcher et al. (2012)	Twelve-week, double-blind, placebo-controlled, randomized, parallel-group study in PD subjects on levodopa therapy with motor fluctuations	10 20 40	No difference in the amount and percentage of time spent in the OFF state between treatment and placebo groups. Improvement in UPDRS motor subscale was observed in the ON state for 40 mg/day
Mizuno et al. (2013)	Twelve-week, double-blind, placebo-controlled study in levodopa-treated PD subjects with motor fluctuations	20 40	The change in daily OFF time was significantly reduced in the istradefylline group compared with placebo. The most common adverse event was dyskinesia (placebo, 4.0%; istradefylline 20 mg/day, 13.0%; istradefylline 40 mg/day, 12.1%). All occurrences were mild or moderate in severity and were not dose dependent. Istradefylline 40 mg/day significantly improved UPDRS Part II (OFF state) and Part III (ON state) scores compared with placebo

and significantly reduces the OFF time when used as an adjunct to L-DOPA in PD subjects with motor fluctuations.

LeWitt et al. (2008) also evaluated the efficacy, safety, and tolerability of istradefylline (40 mg once daily) versus placebo as an adjunct to levodopa in subjects with PD who had motor fluctuations. The results of that study were similar to those of the previous report (Hauser et al., 2008). Compared with the baseline findings, the decrease of daily awake OFF time for istradefylline was $-10.8 \pm 16.6\%$, while that for placebo was $-4.0 \pm 15.7\%$. This effect corresponded to changes from baseline in the total daily awake OFF time of -1.8 ± 2.8 h for istradefylline and -0.6 ± 2.7 h for placebo. The study treatment was generally well tolerated, and the incidence of all treatment-emergent AEs was similar for both groups (istradefylline, 89.1%; placebo, 86.4%). Drug-related, treatment-emergent, AEs were slightly more common for subjects receiving istradefylline (66.7%) than placebo (57.6%). Overall, the most frequently reported events were dyskinesia, dizziness, insomnia, nausea, and accidents involving falls. Of the most frequently reported drug-related events, dyskinesia occurred more often for the istradefylline-treated patients (30.2%) than for the placebo-treated subjects (15.2%), while accidents involving falls were more common in the placebo group (9.1%) than in the istradefylline group (3.1%). The treatment-emergent dyskinesia was mostly mild or moderate in intensity. Severe dyskinesia was reported for two subjects in each group (istradefylline, 1.6%; placebo, 3.0%).

Mizuno et al. (2010, 2013) evaluated the efficacy of istradefylline at an oral dose of 20 mg or 40 mg once daily for 12 weeks in Japanese PD patients with motor complications on levodopa therapy based on the change in the daily OFF time compared with placebo, and to assess the safety at these doses. The first Mizuno et al. study reported that at the end point, the daily OFF time was reduced from baseline by 1.31 h by 20 mg/day istradefylline, 1.58 h by 40 mg/day istradefylline, and 0.66 h by placebo; istradefylline significantly reduced the daily OFF time compared with placebo. The Unified Parkinson's Disease Rating Scale (UPDRS) Part III subscale score (Motor examination) in the ON state was reduced by 5.7 at the end point in the 20 and 40 mg/day istradefylline groups and 3.7 in the placebo group. The most commonly reported drug-related, treatment-emergent AE was dyskinesia, which occurred in 2.5% (3/119) of subjects receiving placebo, 8.5% (10/118) receiving 20 mg/day istradefylline, and 6.4% (8/125) receiving 40 mg/day istradefylline. No dose-dependent increase was found in the incidence of dyskinesia. The intensity was mild to moderate in all cases, and no subjects experienced severe dyskinesia.

The second study by Mizuno et al. reported that the change in daily OFF time was significantly reduced in the istradefylline 20 mg/day (−0.99 h) and istradefylline 40 mg/day (−0.96 h) groups compared with the placebo group (−0.23 h). The most common AE was dyskinesia (placebo, 4.0%; istradefylline 20 mg/day, 13.0%; istradefylline 40 mg/day, 12.1%). All occurrences were mild or moderate in severity, and were not dose-dependent. The clinical outcomes of istradefylline treatment were fairly consistent. Istradefylline reduced the OFF time without worsening the severity of dyskinesia.

A clinical study of another adenosine A_{2A} receptor antagonist, preladenant, has also been reported (Hauser et al., 2011). They reported that the mean daily OFF time from baseline to week 12 was reduced versus placebo in patients on 5 mg preladenant (difference −1.0 h) and 10 mg preladenant (−1.2 h). The changes in the mean daily OFF time versus placebo were not significant for 1 mg preladenant (0.2 h) or 2 mg preladenant (−0.7 h). The most common AEs in the combination preladenant groups versus placebo were a worsening of Parkinson's disease (22 [11%] vs. 4 [9%]), somnolence (20 [10%] vs. 3 [6%]), dyskinesia (18 [9%] vs. 6 [13%]), nausea (17 [9%] vs. 5 [11%]), constipation (15 [8%] vs. 1 [2%]), and insomnia (15 [8%] vs. 4 [9%]).

Based on the results of these clinical studies, adenosine A_{2A} receptor antagonists, such as istradefylline and preladenant, can increase the duration of ON time without troublesome dyskinesia (good ON time). In addition, adenosine A_{2A} antagonism significantly reduced the awake OFF time in PD patients with motor fluctuations. Importantly, this reduction in OFF time occurred without increasing the ON time with "troublesome" dyskinesia. Adenosine A_{2A} antagonists such as istradefylline provided these clinically meaningful benefits for PD subjects treated with conventional doses of L-DOPA and who, in most instances, were also receiving adjunctive medications.

4. COMPARISON BETWEEN NONCLINICAL AND CLINICAL FINDINGS ON ADENOSINE A_{2A} RECEPTOR BLOCKADE AND DYSKINESIA

In the case of istradefylline, it is possible to compare the effects on rodents, nonhuman primates, and patients with PD. Insufficient information is currently available for other agents. The results from rodent studies using the sensitization of the rotational response and AIMs in hemiparkinsonian rodents were inconsistent. It is therefore difficult to draw conclusions about these findings which described the effects of istradefylline on dyskinesia in rodent models of LID. However, despite the sometimes conflicting findings, rodent studies have some advantage for understanding the effects of adenosine A_{2A} antagonism on dyskinesia, mainly that they are easy to handle, allowing a variety of experimental procedures to be performed. Interestingly, the results from rodent studies did conclusively suggest that adenosine A_{2A} antagonists may prevent the development of dyskinesia. This possibility should be examined in a clinical trial in the future.

Table 6.2 presents the comparison of the results from MPTP-treated nonhuman primates and the clinical findings of istradefylline on dyskinesia in humans. There are some differences between nonclinical and clinical studies, such as the items evaluated, the L-DOPA dosage, and other concomitant drugs used. This makes the interpretation and translation of the results of nonclinical studies, especially MPTP-treated nonhuman primates, difficult. Dyskinesia is a common complication of chronic treatment with L-DOPA, but the mechanism(s) responsible for the induction of dyskinesia have not been fully elucidated. LID has been associated with a sequence of

Table 6.2 Comparison between nonclinical and clinical studies of the effects of istradefylline on dyskinesia in L-DOPA adjunct therapy

	Nonclinical (MPTP-treated nonhuman primates)	**Clinical study on PD patients**
Evaluation item	Maximum dyskinesia score Severity of dyskinesia	Frequency and severity of dyskinesia
L-DOPA dosage	Suboptimal dosage	Optimal dosage
Other concomitant drug	None	More than 80% patient treated with other dopamine agonist, MAO-B I, COMT I, etc.
Findings	No worsen severity of dyskinesia Otherwise slightly reduced severity	Increased frequency of dyskinesia No worsen severity of dyskinesia
Conclusion	Adenosine A_{2A} antagonist does not exacerbate LID	Adenosine A_{2A} antagonist increases frequency of dyskinesia associated with increase ON time with nontroublesome dyskinesia

events, including intermittent stimulation of dopamine receptors, downstream changes in protein and gene expression levels and abnormalities in nondopaminergic transmitter systems (Jenner, 2008). These events combine to produce the complex changes in the firing patterns in signaling pathways from the basal ganglia to the cortex. Based on recent electrophysiological studies, both excessive decreases in the GPi firing frequency and modifications of the firing pattern in the subthalamic nucleus are thought to play an important role in the pathogenesis of LID (Boraud, Bezard, Bioulac, & Gross, 2001; Gilmour et al., 2011).

It is well known that once PD patients are primed to exhibit dyskinesia by L-DOPA, they exhibit marked dyskinesia with every dopaminergic medication. This suggests that chronic dopamine receptor stimulation is important for the induction of dyskinesia in L-DOPA-primed patients. Istradefylline is a selective adenosine A_{2A} receptor antagonist without any stimulatory effects on dopamine receptors. This may explain why istradefylline alone did not directly cause dyskinesia in the dyskinesogenic condition. Interestingly, an increase in adenosine A_{2A} signaling was found in the CPu of 6-OHDA-lesioned rats, and in MPTP-treated primates, as well as in PD patients chronically treated with L-DOPA (Calon et al., 2004; Pinna et al., 2002; Tomiyama et al., 2004). These results suggest that an increase in adenosine A_{2A} signaling is associated with either the development or the expression of dyskinesia. Therefore, the selective adenosine A_{2A} receptor antagonists may prevent the development of dyskinesia.

The data from clinical studies of istradefylline, which involved 12–16 weeks of treatment in PD patients with motor response complications induced by L-DOPA, demonstrated that the frequency of dyskinesia as a part of AEs (objective scale) was increased by istradefylline treatment, but with fewer cases rated as "severe." In the clinical studies, dyskinesia was also evaluated by a patient 24-h diary, which asked patients to record their ON time with/without troublesome dyskinesia in 30-min intervals for awake time (subjective scale). As a result, some clinical studies have shown that istradefylline treatment caused a statistically significant increase in the amount of ON time without troublesome dyskinesia (LeWitt et al., 2008; Mizuno et al., 2013). Taken together, the clinical outcomes were well translated from the nonclinical studies discussed in Section 2.2 about L-DOPA-primed MPTP-treated marmosets, these include the following: (1) Istradefylline cannot prevent or diminish dyskinesia itself in the dyskinesiogenic condition associated with L-DOPA combination therapy. (2) Istradefylline never worsens the severity of dyskinesia. There is no subjective scale for assessing dyskinesia in marmosets, but the severity rating scale used in nonclinical studies can be interpreted to evaluate whether there is "troublesome dyskinesia," and if the score is rated 3 or more, it can be considered that there was an increase in the severity of the dyskinesia. When evaluated this way, both the marmoset data and clinical data indicated that istradefylline does not increase the severity of the dyskinesia. In addition, the combination of istradefylline and suboptimal L-DOPA, which induced mild dyskinesia, had antiparkinsonian activity without worsening the dyskinesia, which suggests that lowering the L-DOPA dose and adding the adenosine A_{2A} receptor antagonist may be able to maintain the antiparkinsonian effects with a reduction of dyskinesia.

5. CONCLUSIONS

The current knowledge about both clinical and nonclinical studies on the effects of adenosine A_{2A} receptor blockade and dyskinesia was summarized in this review. The results from studies of MPTP-treated nonhuman primates have revealed some of the effects of adenosine A_{2A} antagonism using istradefylline in L-DOPA-primed MPTP-treated nonhuman primates, and these correlated will with the human clinical evidence. This confirmed that adenosine A_{2A} receptor antagonist therapy (represented by istradefylline) appears to be a useful new therapeutic option for PD patients with motor complications. The mechanism(s) and genesis of the dyskinesia are not fully understood, but targeting the adenosine A_{2A} receptor can open a new door to exploring the causes and progression of dyskinesia.

REFERENCES

Ahlskog, J. E., & Muenter, M. D. (2001). Frequency of levodopa-related dyskinesias and motor fluctuations as estimated from the cumulative literature. *Movement Disorders*, *16*, 448–458.

Antonini, A., Barone, P., Ceravolo, R., Fabbrini, G., Tinazzi, M., & Abbruzzese, G. (2010). Role of pramipexole in the management of Parkinson's disease. *CNS Drugs*, *24*, 829–841.

Bara-Jimenez, W., Sherzai, A., Dimitrova, T., Favit, A., Bibbiani, F., Gillespie, M., et al. (2003). Adenosine A_{2A} receptor antagonist treatment of Parkinson's disease. *Neurology*, *61*, 293–296.

Barzilai, A., & Melamed, E. (2003). Molecular mechanisms of selective dopaminergic neuronal death in Parkinson's disease. *Trends in Molecular Medicine*, *9*(3), 126–132.

Bibbiani, F., Oh, J. D., Petzer, J. P., Castagnoli, N., Jr., Chen, J.-F., Schwarzschild, M. A., et al. (2003). A_{2A} antagonist prevents dopamine agonist-induced motor complications in animal models of Parkinson's disease. *Experimental Neurology*, *184*, 285–294.

Bonuccelli, U., Colzi, A., & Del Dotto, P. (2002). Pergolide in the treatment of patients with early and advanced Parkinson's disease. *Clinical Neuropharmacology*, *25*, 1–10.

Boraud, T., Bezard, E., Bioulac, B., & Gross, C. E. (2001). Dopamine agonist-induced dyskinesias are correlated to both firing pattern and frequency alterations of pallidal neurones in the MPTP-treated monkey. *Brain, 124*, 546–557.

Bruns, R. F., Lu, G. H., & Pugsley, T. A. (1986). Characterization of the A2 adenosine receptor labeled by [3H]NECA in rat striatal membranes. *Molecular Pharmacology, 29*, 331–346.

Calon, F., Dridi, M., Hornykiewicz, O., Bédard, P. J., Rajput, A. H., & Di Paolo, T. (2004). Increased adenosine A_{2A} receptors in the brain of Parkinson's disease patients with dyskinesias. *Brain, 127*, 1075–1084.

Cenci, M. A., Lee, C. S., & Bjorklund, A. (1998). L-DOPA-induced dyskinesia in rat is associated with striatal overexpression of prodynorphin and glutamic acid decarboxylase mRNA. *European Journal of Neuroscience, 10*, 2694–2706.

Cenci, M. A., Whishaw, I. Q., & Schallert, T. (2002). Animal models of neurological deficits: how relevant is the rat? *Nature Reviews Neuroscience, 3*, 574–579.

Daly, J. W., Butts-Lamb, P., & Padgett, W. (1983). Subclasses of adenosine receptors in the central nervous system: Interaction with caffeine and related methylxanthines. *Cellular and Molecular Neurobiology, 3*, 69–74.

Dauer, W., & Przedborski, S. (2003). Parkinson's disease mechanisms and models. *Neuron, 39*(6), 889–909.

Factor, S. A., Wolski, K., Togasaki, D. M., Huyck, S., Cantillon, M., Ho, T. W., et al. (2013). Long-term safety and efficacy of preladenant in subjects with fluctuating Parkinson's disease. *Movement Disorders, 28*(6), 817–820.

Fahn, S. (2003). Description of Parkinson's disease as a clinical syndrome. *Annals of the New York Academy of Sciences, 991*, 1–14.

Fahn, S. (2008). The history of dopamine and levodopa in the treatment of Parkinson's disease. *Movement Disorders, 23*(Suppl. 3), S497–S508.

Fredduzzi, S., Moratalla, R., Monopoli, A., Cuellar, B., Xu, K., Ongini, E., et al. (2002). Persistent behavioral sensitization to chronic L-DOPA requires A_{2A} adenosine receptors. *Journal of Neuroscience, 22*, 1054–1062.

Fredholm, B. B., IJzerman, A. P., Jacobson, K. A., Klotz, K. N., & Linden, J. (2001). International Union of Pharmacology. XXV. Nomenclature and classification of adenosine receptors. *Pharmacological Reviews, 53*, 527–552.

Gilmour, T. P., Lieu, C. A., Nolt, M. J., Piallat, B., Deogaonkar, M., & Subramanian, T. (2011). The effects of chronic levodopa treatments on the neuronal firing properties of the subthalamic nucleus and substantia nigra reticulata in hemiparkinsonian rhesus monkeys. *Experimental Neurology, 228*, 53–58.

Grondin, R., Bédard, P. J., Tahar, A. H., Grégoire, L., Mori, A., & Kase, H. (1999). Antiparkinsonian effect of a new selective adenosine A_{2A} receptor antagonist in MPTP-treated monkeys. *Neurology, 52*, 1673–1677.

Hauser, R. A., Cantillon, M., Pourcher, E., Micheli, F., Mok, V., Onofrj, M., et al. (2011). Preladenant in patients with Parkinson's disease and motor fluctuations: A phase 2, double-blind, randomised trial. *Lancet Neurology, 10*, 221–229.

Hauser, R. A., Hubble, J. P., Truong, D. D., & Istradefylline US-001 Study Group. (2003). Randomized trial of the adenosine A_{2A} receptor antagonist istradefylline in advanced PD. *Neurology, 61*(3), 297–303.

Hauser, R. A., Shulman, L. M., Trugman, J. M., Roberts, J. W., Mori, A., Ballerini, R., et al. (2008). Study of istradefylline in patients with Parkinson's disease on levodopa with motor fluctuations. *Movement Disorders, 23*(15), 2177–2185.

Henry, B., Crossman, A. R., & Brotchie, J. M. (1998). Characterization of enhanced behavioral responses to L-DOPA following repeated administration in the 6-hydroxydopamine-lesioned rat model of Parkinson's disease. *Experimental Neurology, 151*(2), 334–342.

Hunot, S., & Hirsch, E. (2003). Neuroinflammatory processes in Parkinson's disease. *Annals of Neurology, 53*(Suppl. 3), S49–S58 (discussion S58–S60).

Jarvis, M. F., & Williams, M. (1991). Direct autoradiographic localization of adenosine A2 receptors in the rat brain using the A2-selective agonist, [3H]CGS 21680. *European Journal of Pharmacology, 168*, 243–246.

Jellinger, K. (1986). Overview of morphological changes in Parkinson's disease. *Advances in Neurology, 45*, 1–18.

Jenner, P. (2008). Molecular mechanisms of L-DOPA-induced dyskinesia. *Nature Reviews. Neuroscience, 9*, 665–677.

Jenner, P., Mori, A., Hauser, R., Morelli, M., Fredholm, B. B., & Chen, J.-F. (2009). Adenosine, adenosine A_{2A} antagonists, and Parkinson's disease. *Parkinsonism and Related Disorders, 15*, 406–413.

Kanda, T., Jackson, M. J., Smith, L. A., Pearce, R. K. B., Nakamura, J., Kase, H., et al. (1998). Adenosine A_{2A} antagonist: A novel antiparkinsonian agent that does not provoke dyskinesia in parkinsonian monkeys. *Annals of Neurology, 43*, 507–513.

Kanda, T., Jackson, M. J., Smith, L. A., Pearce, R. K. B., Nakamura, J., Kase, H., et al. (2000). Combined use of the adenosine A_{2A} antagonist KW-6002 with L-DOPA or with selective D1 or D2 dopamine agonists increases antiparkinsonian activity but not dyskinesia in MPTP-treated monkeys. *Experimental Neurology, 162*, 321–327.

Koga, K., Kurokawa, M., Ochi, M., Nakamura, J., & Kuwana, Y. (2000). Adenosine A_{2A} receptor antagonists KF17837 and KW-6002 potentiate rotation induced by dopaminergic drugs in hemi-parkinsonian rats. *European Journal of Pharmacology, 408*(3), 249–255.

LeWitt, P. A., Guttman, M., Tetrud, J. W., Tuite, P. J., Mori, A., Chaikin, P., et al. (2008). Adenosine A2A receptor antagonist istradefylline (KW-6002) reduces "Off" time in Parkinson's disease: A double-blind, randomized, multicenter clinical trial (6002-US-005). *Annals of Neurology, 63*, 295–302.

Löschmann, P. A., Smith, L. A., Lange, K. W., Jähnig, P., Jenner, P., & Marsden, C. D. (1992). Motor activity following the administration of selective D-1 and D-2 dopaminergic drugs to MPTP-treated common marmosets. *Psychopharmacology, 109*, 49–56.

Lundblad, M., Vaudano, E., & Cenci, M. A. (2003). Cellular and behavioural effects of the adenosine A_{2A} receptor antagonist KW-6002 in a rat model of L-DOPA-induced dyskinesia. *Journal of Neurochemistry, 84*, 1398–1410.

Marsala, S. Z., Gioulis, M., Ceravolo, R., & Tinazzi, M. (2012). A systematic review of catechol-o-methyltransferase inhibitors: Efficacy and safety in clinical practice. *Clinical Neuropharmacology, 35*, 185–190.

Matheson, A. J., & Spencer, C. M. (2000). Ropinirole: A review of its use in the management of Parkinson's disease. *Drugs, 60*, 115–137.

Mizuno, Y., Hasegawa, K., Kondo, T., Kuno, S., Yamamoto, M., & Japanese Istradefylline Study Group. (2010). Clinical efficacy of istradefylline (KW-6002) in Parkinson's disease: A randomized, controlled study. *Movement Disordors, 25*, 1437–1443.

Mizuno, Y., Kondo, T., & the Japanese Istradefylline Study Group. (2013). Adenosine A_{2A} receptor antagonist istradefylline reduces daily OFF time in Parkinson's disease. *Movement Disorders, 28*(8), 1138–1141.

Obeso, J. A., Rodriguez-Oroz, M., Marin, C., Alonso, F., Zamarbide, I., Lanciego, J. L., et al. (2004). The origin of motor fluctuations in Parkinson's disease: Importance of dopaminergic innervation and basal ganglia circuits. *Neurology, 62*, S17–S30.

Olanow, C. W., Stern, M. B., & Sethi, K. (2009). The scientific and clinical basis for the treatment of Parkinson disease. *Neurology, 72*(Suppl. 4), S1–S136.

Parent, A., & Hazarti, L. N. (1995). Functional anatomy of the basal ganglia. II. The place of subthalamic nucleus and external pallidum in basal ganglia circuitry. *Brain Research Review, 20*, 128–154.

Pearce, R. K. B., Jackson, M., Smith, L., Jenner, P., & Marsden, C. D. (1995). Chronic L-DOPA administration induces dyskinesias in the 1-methyl-4-phenyl-1,2,3,6-tetrahydropyridine-treated common marmoset (Callithrix jacchus). *Movement Disorders*, *10*, 731–740.

Pinna, A., Corsi, C., Carta, A. R., Valentini, V., Pedata, F., & Morelli, M. (2002). Modification of adenosine extracellular levels and adenosine A_{2A} receptor mRNA by dopamine denervation. *European Journal of Pharmacology*, *446*, 75–82.

Pinna, A., Fenu, S., & Morelli, M. (2001). Motor stimulant effects of the adenosine A_{2A} receptor antagonist SCH 58261 do not develop tolerance after repeated treatments in 6-hydroxydopamine-lesioned rats. *Synapse*, *39*(3), 233–238.

Pourcher, M., Fernandez, H. H., Stacy, M., Mori, A., Ballerini, R., & Chaikin, P. (2012). Istradefylline for Parkinson's disease patients experiencing motor fluctuations: Results of the KW-6002-US-018 study. *Parkinsonism and Related Disorders*, *18*, 178–184.

Rascol, O., Payoux, P., Ory, F., Ferreira, J. J., Brefel-CourbonM, C., & Montastruc, J. L. (2003). Limitations of current Parkinson's disease therapy. *Annals of Neurology*, *53*(Suppl. 3), S3–S12.

Saki, M., Yamada, K., Koshimura, E., Sasaki, K., & Kanda, T. (2013). In vitro pharmacological profile of the A_{2A} receptor antagonist istradefylline. *Naunyn-Schmiedeberg's Archives of Pharmacology*, *386*, 963–972.

Schiffmann, S. N., Jacobs, O., & Vanderhaeghen, J. J. (1991). Striatal restricted adenosine A2 receptor (RDC8) is expressed by enkephalin but not by substance P neurons: An in situ hybridization histochemistry study. *Journal of Neurochemistry*, *57*, 1062–1067.

Schiffmann, S. N., Libert, F., Vassart, G., & Vanderhaeghen, J. J. (1989). Distribution of adenosine A2 receptor mRNA in the human brain. *Neuroscience Letters*, *130*, 177–181.

Stacy, M., Silver, D., Mendis, T., Sutton, J., Mori, A., Chaikin, P., et al. (2008). A 12-week, placebo-controlled study (6002-US-006) of istradefylline in Parkinson disease. *Neurology*, *70*, 2233–2240.

Thomas, A., Iacono, D., Luciano, A. L., Armellino, K., Di Iorio, A., & Onofrj, M. (2004). Duration of amantadine benefit on dyskinesia of severe Parkinson's disease. *Journal of Neurology, Neurosurgery, and Psychiatry*, *75*, 141–143.

Tomiyama, M., Kimura, T., Maeda, T., Tanaka, H., Kannari, K., & Baba, M. (2004). Upregulation of striatal adenosine A_{2A} receptor mRNA in 6-hydroxydopamine-lesioned rats intermittently treated with L-DOPA. *Synapse*, *52*, 218–222.

Tronci, E. 1., Simola, N., Borsini, F., Schintu, N., Frau, L., Carminati, P., et al. (2007). Characterization of the antiparkinsonian effects of the new adenosine A_{2A} receptor antagonist ST1535: acute and subchronic studies in rats. *European Journal of Pharmacology*, *566*, 94–102.

Uchida, S., Tashiro, T., Kawai-Uchida, M., Mori, A., Jenner, P., & Kanda, T. (2014). The adenosine A_{2A} receptor antagonist, istradefylline enhances the motor response of L-DOPA without worsening dyskinesia in MPTP-treated common marmosets. *Journal of Pharmacological Sciences*, *124*, 480–485.

Ungerstedt, U. (1976). 6-Hydroxydopamine-induced degeneration of the nigrostriatal dopamine pathway: The turning syndrome. *Pharmacology & Therapeutics. Part B: General & Systematic Pharmacology*, *2*, 37–40.

Xiao, K., Bastia, E., Xu, Y. H., Benn, C. L., Cha, J. H., Peterson, T. S., et al. (2006). Forebrain adenosine A_{2A} receptors contribute to L-3,4-dihydroxyphenylalanine-induced dyskinesia in hemiparkinsonian mice. *Journal of Neuroscience*, *26*, 13548–13555.

CHAPTER SEVEN

Interaction of Adenosine Receptors with Other Receptors from Therapeutic Perspective in Parkinson's Disease

Nicolas Morin*,†, **Thérèse Di Paolo***,†,1
*Neuroscience Research Unit, Centre de recherche du CHU de Québec, Quebec, Quebec, Canada
†Faculty of Pharmacy, Laval University, Quebec, Quebec, Canada
1Corresponding author: e-mail address: therese.dipaolo@crchul.ulaval.ca

Contents

Abstract

Altered dopaminergic neurotransmission in the basal ganglia is observed in Parkinson's disease (PD) and L-3,4-dihydroxyphenylalanine (L-DOPA)-induced dyskinesias (LID). An attractive alternative for treating LID is to use adjunct drugs to modulate non-dopaminergic neurotransmitter systems in the basal ganglia. For example, adenosine receptors have received attention over the past years for the treatment of PD and LID. Adenosine interacts closely with dopamine and plays an important role in the function of striatal GABAergic efferent neurons. Excitatory glutamatergic neurotransmission is also modulated by adenosine in the striatum. Hence, based on the unique cellular and

International Review of Neurobiology, Volume 119
ISSN 0074-7742
http://dx.doi.org/10.1016/B978-0-12-801022-8.00007-6

regional distribution of this system, adenosine neurotransmission could have an important implication for the development of new therapeutic strategies targeting the basal ganglia disorders. Indeed, A_{2A} adenosine receptor antagonists were shown to improve motor deficits in PD and to reduce the severity of LID. A_{2A} receptor subtypes are selectively found on striatopallidal neurons and can couple with receptors of interest in PD, such as D2 dopamine and metabotropic glutamate receptor type 5 (mGlu5) receptors, and form functional heteromeric complexes. This chapter will review relevant studies investigating the role and contribution of adenosine receptor subtypes in pathophysiology of PD and LID. The interactions of adenosine receptors, especially A_1 and A_{2A} receptor subtypes, with other receptors implicated in the pathophysiology of PD and LID such as dopaminergic and glutamatergic receptors will be reviewed. The implication of these interactions in the development and expression of PD symptoms and LID needs further investigation to find novel drug targets.

ABBREVIATIONS

DA dopamine
GABA γ-aminobutyric acid
GP globus pallidus
GPe external globus pallidus
L-DOPA L-3,4-dihydroxyphenylalanine
LID L-DOPA-induced dyskinesias
mGlu metabotropic glutamate
MPEP 2-methyl-6-(phenylethynyl)pyridine
MPTP 1-methyl-4-phenyl-1,2,3,6-tetrahydropyridine
NMDA *N*-methyl-D-aspartate
PD Parkinson's disease
PKA protein kinase A
PP-1 protein phosphatase-1
SNc substantia nigra pars compacta
STN subthalamic nucleus

1. INTRODUCTION

Parkinson's disease (PD) is the most common neurodegenerative movement disorder and is likely to increase due to the aging population (Siderowf & Stern, 2003). PD is characterized by tremor, rigidity, bradykinesia, and instability in postural reflexes that is principally attributed to the death of dopamine (DA) neurons in the substantia nigra *pars compacta* (SNc). At the moment, there is no cure for PD. Neuroprotection or disease modification, defined as an intervention that protects or rescues vulnerable neurons in order to slow, stop, or reverse disease progression is not yet available for PD but laboratory studies have found promising agents (Olanow, Stern, & Sethi, 2009). The treatment of motor symptoms with the DA precursor, L-3,4-dihydroxyphenylalanine (L-DOPA), introduced 50 years ago still remains the gold standard therapy for PD. However, involuntary movements termed L-DOPA-induced dyskinesias (LID) appear in the vast majority of PD patients after chronic treatment and may become disabling (Mones, Elizan, & Siegel, 1971; Olanow & Koller, 1998). Once dyskinesias appear, even if treatment is stopped for several weeks, the first dose will trigger them again, showing that L-DOPA has permanently or persistently modified the brain response to DA.

Though investigated in numerous studies, the mechanisms involved in the occurrence of PD and LID are still unknown. While L-DOPA and DA agonists are effective at reversing the motor symptoms of the disease, little they do to combat the progressive underlying degeneration of DA neurons in the SNc. The important decrease of striatal DA is responsible for the major Parkinsonian syndromes, but other neurotransmitters and neuromodulators are also affected and implicated in motor and nonmotor PD symptoms and in the development of motor complications. More attention has therefore been placed on finding alternative nondopaminergic drugs that could resolve some or all these problems. To modulate basal ganglia DA neurotransmission, a recent attractive strategy is to treat LID with adjunct drugs to nondopaminergic neurotransmitter systems such as glutamate and adenosine (Blanchet et al., 1999; Brotchie, 1998, 2003; Calon & Di Paolo, 2002; Grondin et al., 1999; Henry, Fox, Crossman, & Brotchie, 2001).

Adenosine, a purinergic messenger, plays a crucial role in many physiological processes and is released by many cells including neurons and glia (Dunwiddie & Masino, 2001; Ribeiro, Sebastiao, & de Mendonca, 2002). In the basal ganglia, adenosine interacts closely with DA and is involved in the function of striatal γ-aminobutyric acid (GABA)ergic striatopallidal neurons projecting from the caudate nucleus and the putamen, mainly to the external segment of the globus pallidus (GPe) (Ferre et al., 2004; Kase, 2001; Martinez-Mir, Probst, & Palacios, 1991; Morelli et al., 2007; Schiffmann, Libert, Vassart, & Vanderhaeghen, 1991; Svenningsson et al., 1998). The implication of adenosine to regulate the excessive glutamate neurotransmission observed in PD and LID is also demonstrated (Blandini & Armentero, 2012; Domenici et al., 2004;

Ferre et al., 2002). In this regard, adenosine has recently received increasing attention because its interaction with DA and glutamate receptors could have major implications for the development of new pharmacological targets for the treatment of PD and LID. Over the past few years, several behavioral analyses have shown the potential efficacy of adenosine receptor modulators in the treatment of PD and LID, such as A_{2A} receptor antagonist (Calon et al., 2004; Chase, Bibbiani, Bara-Jimenez, Dimitrova, & Oh-Lee, 2003; Grondin et al., 1999; Kanda et al., 2000; Morelli & Pinna, 2001)

Much remains to be learned about the biochemical processes that underlie the development of PD and LID, how dopaminergic and non-dopaminergic drugs can be used to avoid the initiation of dyskinesias in early PD, to prevent or inhibit their expression in later stages of the disease through a normalization of the basal ganglia functions and multiple neurotransmitter interactions. This chapter reviews relevant studies investigating the role and the contribution of adenosine receptor subtypes in pathophysiology of PD and LID with a focus on the interactions of adenosine receptors with other receptors of interest in PD and LID, such as dopaminergic, and glutamatergic receptors.

2. ADENOSINE RECEPTORS IN THE BASAL GANGLIA AND SIGNAL TRANSDUCTION

Adenosine is considered as an extracellular signal molecule, because this neurotransmitter is neither stored in synaptic vesicles nor released as a classical neurotransmitter as DA (Ribeiro et al., 2002). Indeed, there are no specific synapses where adenosine is the principal neurotransmitter (Ribeiro et al., 2002). Adenosine binds to four classes of specific G-protein-coupled receptor subtypes named A_1, A_{2A}, A_{2B}, and A_3 (Fredholm, AP, Jacobson, Klotz, & Linden, 2001). Activation of A_1 and A_3 receptors, which couple to G-proteins Gi and Go, is associated with inhibition, whereas activation of A_{2A} and A_{2B} receptors, which couple to G-protein Gs, stimulates adenylate cyclase (Yaar, Jones, Chen, & Ravid, 2005). Interestingly, adenosine receptor subtypes A_1 and A_{2A} are mainly localized in the basal ganglia, more precisely in the striatum, as compared to A_{2B} and A_3 adenosine receptor subtypes that are widely distributed in the brain (Rosin, Hettinger, Lee, & Linden, 2003; Rosin, Robeva, Woodard, Guyenet, & Linden, 1998).

2.1. Localization, organization, and functioning of A_1 adenosine receptor subtype

The A_1 adenosine receptor subtype is highly expressed in the central nervous system and is also localized at both the pre- and postsynaptic sites. Interestingly, the medium spiny neurons dynorphin-positive that express D1 DA receptors and project to the substantia nigra pars reticulata and medial globus pallidus (GP), which constitute the direct pathway, express mainly the A_1 receptor subtype and have low levels of A_{2A} receptor (Jenner et al., 2009; Morelli, Carta, & Jenner, 2009). In the basal ganglia, A_1 receptors are mainly present in the striatonigral GABAergic and corticostriatal glutamatergic neurons, suggesting that A_1 receptors are localized in both D1 and D2 dopaminergic receptor-containing neurons (Ferre, Fredholm, Morelli, Popoli, & Fuxe, 1997).

As mentioned previously, activation of A_1 receptors leads to inhibition of adenylyl cyclase and voltage-dependent calcium ion channels, whereas this receptor can stimulate phospholipases C and D, and can activate potassium ion channels (Ribeiro et al., 2002). The stimulation of A_1 receptors generally leads to inhibition of neurotransmitter release, such as the inhibition of glutamate release at corticostriatal terminals in the striatum (Dunwiddie & Masino, 2001; Svenningsson, Le Moine, Fisone, & Fredholm, 1999). Hence, with the reduction of glutamatergic activation, the adenosine A_1 receptor can indirectly control the function of GABAergic interneurons (Ribeiro et al., 2002). Besides its direct neuromodulatory effects, activation of A_1 receptors could lead to the reduction of excitability and firing rate as well as the hyperpolarization of the resting membrane potential (Dunwiddie & Masino, 2001).

2.2. Localization, organization, and functioning of A_{2A} adenosine receptor subtype

In contrast to the widespread distribution of the other adenosine receptor subtypes in the brain, the A_{2A} receptor is mainly limited to the striatum, nucleus accumbens, and olfactory tubercle (Jarvis & Williams, 1989; Martinez-Mir et al., 1991; Olah & Stiles, 1995; Peterfreund, MacCollin, Gusella, & Fink, 1996; Schiffmann, Libert, Vassart, Dumont, & Vanderhaeghen, 1990; Schiffmann, Libert, et al., 1991). A_{2A} receptors are homogeneously distributed throughout the lateral, medial, and ventromedial areas of the striatum, which represents the motor, associative, and limbic portions, respectively (Rosin et al., 2003). The A_{2A} receptor is also coexpressed at postsynaptic sites of the same medium spiny neurons as those bearing DA D2 receptors (Fink et al., 1992; Pollack, Harrison, Wooten, & Fink, 1993), containing enkephalin (Augood & Emson, 1994; Schiffmann, Jacobs, & Vanderhaeghen, 1991), and projecting to the GPe, which constitute the indirect pathway. Hence, the A_{2A} receptor subtype modulates the input and output activity of GABAergic medium spiny projection neurons (Hettinger, Lee, Linden, & Rosin, 2001; Rosin et al., 2003). The restricted and specific distribution of A_{2A} receptors to the indirect pathway of the basal ganglia provide specificity that could lead to a reduced incidence of adverse effects; the A_{2A} receptor is considered as a promising nondopaminergic therapeutic target to treat PD and to prevent the development of LID.

Presynaptic and glial A_{2A} receptors are also localized in the caudate nucleus and the putamen (Rosin et al., 2003). The A_{2A} receptors are found on presynaptic glutamatergic corticostriatal terminals, and these receptors can modulate positively the glutamatergic cortical input by stimulating glutamate release (Rosin et al., 2003). In addition, A_{2A} receptors are located on the presynaptic terminals of cholinergic interneurons, where they represent 5% of striatal neurons (Jin, Johansson, & Fredholm, 1993), and striatal axon collaterals (Hettinger et al., 2001). This specific localization suggests that these receptors could also modulate cholinergic and GABAergic inputs to medium spiny neurons (Mori & Shindou, 2003).

Indeed, the presence of pre- and postsynaptic A_{2A} receptors on corticostriatal glutamatergic and striatopallidal GABAergic neurons, respectively, suggests that A_{2A} receptors can increase the excitability of medium spiny neurons. Consequently, there is evidence that A_{2A} receptors are largely involved in the regulation of synaptic plasticity (Svenningsson & Fredholm, 2003). Moreover, in the GPe at striatopallidal GABAergic terminals, presynaptic A_{2A} receptors are expressed and they can increase the release of GABA (Mori & Shindou, 2003; Rosin et al., 2003). The increase of GABA suppresses the excitability of GPe projection neurons and leads to the disinhibition of subthalamic nucleus (STN) and its overactivity, resulting in the development of the Parkinsonian syndrome (Mori & Shindou, 2003; Rosin et al., 2003).

The A_{2A} receptor subtype is coupled to an abundant protein called Golf in the striatum, which can activate adenylyl cyclase (Kull, Svenningsson, & Fredholm, 2000). Stimulation of adenylyl cyclase increases the phosphorylation of DARPP-32 at Thr-34 while decreasing the phosphorylation of DARPP-32 at Thr-75 (Lindskog et al., 2002). Phosphorylation of DARPP-32 at Thr-34 and Thr-75 converts this protein into an inhibitor of protein phosphatase-1 (PP-1) and protein kinase A (PKA), respectively

(Greengard, 2001). Interestingly, PP-1 activity decreases the phosphorylation state of receptors of interests such as glutamate receptor subunits and transcription factors such as c-fos occurring after dopaminergic denervation in PD (Chase et al., 2003; Ferre et al., 2004; Ribeiro et al., 2002). When the A_{2A} receptor is blocked with specific antagonists, the PKA activation is reduced and therefore relieves the inhibition of PP-1 by DARPP-32.

In addition to the interactions observed with other receptors of interest in PD, the A_{2A} receptor seems to play a crucial and functional role in the modulation of motor behavior and can also reduce LID or motor complications. Indeed, adenosine A_{2A} antagonists can reduce the excessive striatopallidal and STN neuronal activity and they are considered as a novel and promising approach in PD therapy (Blandini & Armentero, 2012; Chase et al., 2003; Kase, 2001; Morelli et al., 2007; Mori & Shindou, 2003). Several behavioral studies in animal models and in Parkinsonian patients have showed the beneficial effects of A_{2A} receptor antagonism in the treatment of PD and its related motor complications, including LID (Calon et al., 2004; Chase et al., 2003; Fredduzzi et al., 2002; Grondin et al., 1999; Kanda et al., 2000; Morelli & Pinna, 2001). The relevant studies on the effects of A_{2A} adenosine receptor antagonists and their involvement in PD and LID are discussed in Chapters 3 to 6 of the present textbook.

3. A_1 AND A_{2A} ADENOSINE RECEPTOR INTERACTIONS AND HETERODIMERIZATION

Presynaptically, electron microscopy has shown that A_{2A} receptors can colocalize with A_1 receptors in corticostriatal afferents, where they act together to modulate and regulate glutamate release (Ciruela, Casado, et al., 2006). Hence, experiments using isolated striatal nerve terminal preparations have shown that most of the striatal glutamatergic terminals contain both A_1 and A_{2A} receptors (Ciruela, Casado, et al., 2006). Moreover, experiments conducted with cotransfected cells using bioluminescence resonance energy transfer techniques have shown the existence of A_1–A_{2A} adenosine receptor heteromers (Ciruela, Ferre, et al., 2006; Schiffmann, Fisone, Moresco, Cunha, & Ferre, 2007). In response to variation in adenosine concentrations, functional studies in striatal glutamatergic terminals have shown that the A_1–A_{2A} adenosine receptor heteromer provides a "switch mechanism" that can produce opposite effects on glutamate release (Ciruela, Casado, et al., 2006; Ciruela, Ferre, et al., 2006; Schiffmann et al., 2007).

4. ADENOSINE AND DOPAMINE RECEPTOR INTERACTIONS AND HETERODIMERIZATION

4.1. A_1 adenosine receptor subtype

In addition to its concrete effect on neurotransmission and as a neuromodulator, adenosine interacts also with DA receptors in the basal ganglia (Franco et al., 2000; Samadi, Rouillard, Bedard, & Di Paolo, 2007). The existence of functional interactions between DA and adenosine receptors is supported by the anatomic localization of these receptors in striatal projection neurons (Morelli et al., 2007). Indeed, A_1 receptors are mainly coexpressed with D1 DA receptors on striatal neurons that project to the internal globus pallidus and the substantia nigra (Samadi et al., 2007).

D_1 and adenosine A_1 receptors are known to form functionally interacting complexes, the heteromer D1/A_1, in cortical neurons and basal ganglia (Fuxe, Ferré, Genedani, Franco, & Agnati, 2007; Fuxe, Ferré, Zoli, & Agnati, 1998). Simultaneous pretreatment with A_1 and D1 receptors agonists in D1/A_1 cells was shown to decrease D1 receptor adenylyl cyclase signaling (Ginés et al., 2000). DA via D1 receptor activation increases the activity of A_1 receptors by potentiating *N*-methyl-D-aspartate (NMDA)-mediated adenosine release (Harvey & Lacey, 1997). Inversely, the formation of A_1/D1 heterodimers can also lead to the decrease of uncoupling of D1 receptors to the G-protein or D1 receptors in the high-affinity state (Fuxe et al., 1998).

4.2. A_{2A} adenosine receptor subtype

The specific and unique localization of A_{2A} receptors offers the basis for interesting interactions and modulations of this receptor with dopaminergic neurotransmission in the basal ganglia (Morelli et al., 2007). D2 and adenosine A_{2A} receptors are known to form functional hetero-oligomers (Canals et al., 2003; Fuxe et al., 2007; Hillion et al., 2002). Long-term administration of A_{2A} or D2 agonists induces an internalization and desensitization of the D2/A_{2A} complex (Canals et al., 2003; Hillion et al., 2002), whereas D2 antagonists trigger an increase in D2 and A_{2A} immunoreactivity (Torvinen et al., 2005). Moreover, an A_{2A} receptor agonist was shown to reduce D2 receptor affinity, and this change was related to conformational changes in the binding site of D2 receptors (Salim et al., 2000).

An increase of striatal c-fos levels is observed when D2 receptor signaling is reduced, and this might be related to the activation of A_{2A} receptors

(Pinna, Wardas, Cozzolino, & Morelli, 1999). It has been proposed that the antiparkinsonian and antidyskinetic actions of A_{2A} antagonists result from blocking the A_{2A}–D2 hetero-oligomer activity (Salim et al., 2000). A_{2A} antagonists might lead to the enhancement of D2 receptor signaling and to the blockade of increased A_{2A} receptor signaling in the striatum (Salim et al., 2000).

5. ADENOSINE AND GLUTAMATE RECEPTOR INTERACTIONS AND HETERODIMERIZATION

5.1. A_1 adenosine receptor subtype

Molecular and functional interactions between the metabotropic glutamate receptor type 1 (mGlu1) receptor and the adenosine A_1 receptors in cerebellum, primary cortical neurons, and heterologous transfected cells have been observed (Ciruela et al., 2001). Coimmunoprecipitation experiments showed a close and subtype-specific interaction between mGlu1 and A_1 receptors in rat cerebellar synaptosomes and in cotransfected HEK-293 cells (Ciruela et al., 2001). It has been suggested that activation of the mGlu1 receptor can suppress responses mediated via adenosine A_1 receptors, probably by activating protein kinase C (Shahraki & Stone, 2003). Hence, there is evidence that A_1 receptors can interact with both metabotropic and ionotropic glutamate receptors (Adams, Cowen, Short, & Lawrence, 2008; Ciruela et al., 2001; de Novellis et al., 2002; Hobson et al., 2013; Ismayilova, Crossman, Verkhratsky, & Brotchie, 2004; Shahraki & Stone, 2003; Wang, 2007).

5.2. A_{2A} adenosine receptor subtype

Besides adenosine A_{2A} and DA D2 receptors, mGlu5 receptors are also known to interact and colocalize postsynaptically in the striatopallidal GABAergic efferent neurons (Bogenpohl, Ritter, Hall, & Smith, 2012; Conn, Battaglia, Marino, & Nicoletti, 2005; Ferre et al., 2002). Interestingly, this colocalization provides a structural framework for the existence of multiple functional interactions of A_{2A}, D2, and mGlu5 receptors (Conn et al., 2005). Recent experiments using optical sectioning techniques found that A_{2A} and mGlu5 receptors are also colocalized in rat striatal cultures (Fuxe et al., 2003).

A_{2A} receptors can increase the phosphorylation of DARPP-32 at Thr-34 via an extracellular signal-regulated kinase pathway and the induction of c-fos expression is also increased in striatopallidal neurons when A_{2A} and mGlu5 receptors are coactivated (Ferre et al., 2002; Nishi et al., 2003). An important increase in the adenosine release was observed when glutamatergic neurotransmission becomes overactive (Ferre & Fuxe, 2000; Nash & Brotchie, 2000). Hence, the mGlu5 receptor-mediated effect in the striatum was abolished by the blockade of A_{2A} receptors (Domenici et al., 2004).

A_{2A} and mGlu5 receptors were also shown to reduce the affinity of D2 receptors in the striatum (Ferre et al., 1999). In previous studies from our laboratory, we reported that development of LID over a month of treatment was lower by overall ~70% with addition of 2-methyl-6-(phenylethynyl)pyridine (MPEP), an mGlu5 receptor antagonist, to the L-DOPA treatment in *de novo* 1-methyl-4-phenyl-1,2,3,6-tetrahydropyridine (MPTP) monkeys (Morin, Gregoire, et al., 2013) and this was associated with a normalization of glutamate (Morin, Morissette, et al., 2013) and DA neurotransmission (Morin, Jourdain, Morissette, Gregoire, & Di Paolo, 2014). The mechanism underlying the antidyskinetic activity of mGlu5 receptor antagonist is associated with the blockade of mGlu5 receptors and its interactions with other receptors that overall restore normal glutamate and dopaminergic neurotransmissions. This effect may be caused by desensitization of the A_{2A}–mGlu5 heteromeric complex, leading to removal of the blockade of D2 receptor-induced signaling effect (Fuxe et al., 2003). Interestingly, the density of the DA D2 receptor was only decreased in the dyskinetic monkeys, whereas the addition of MPEP to L-DOPA prevented this increase (Morin et al., 2014). These findings suggest that the striatal A_{2A}–D2–mGlu5 multimeric receptor complexes are involved in the striatal plasticity and could be relevant for the management of PD and LID (Ferre et al., 2002; Fuxe et al., 2003; Morelli et al., 2007).

6. DISCUSSION

A_{2A} receptor antagonists, such as Preladenant and KW-6002 (istradefylline), are reported to reduce the severity of LID in primate models of movement disorders (Bara-Jimenez et al., 2003; Grondin et al., 1999; Hodgson et al., 2010; Salamone, 2010; Varty et al., 2008). Moreover, in the basal ganglia of *postmortem* PD patients with motor complications, changes of glutamate neurotransmission and its receptors such as NMDA, α-amino-3-hydroxy-5-methyl-4-isoxazolepropionic acid, mGlu2 and mGlu5 receptors were reported (Calon, Rajput, Hornykiewicz, Bedard, & Di Paolo, 2003; Ouattara et al., 2011; Samadi et al., 2009). In these same patients, A_{2A} receptor levels were measured and they were increased in the basal ganglia and associated with LID (Calon et al., 2004). These observations in human brains do not provide a causal link between glutamate and adenosine receptors in motor complications. Nevertheless, they showed impaired neurotransmission of adenosine and glutamate receptors, suggesting possible interactions between these receptors. Moreover, in a primate model of PD and LID, a chronic treatment with L-DOPA and a NMDA receptor antagonist CI-1041, that prevented LID, was shown to normalize basal ganglia A_{2A} receptors as well as NMDA receptors suggesting the close link between these neurotransmitters in dyskinesias (Morissette et al., 2006).

L-DOPA-induced supersensitivity of A_{2A} adenosine receptors might represent one plausible mechanism of LID, but there is conflicting data in the literature. Postmortem studies have shown that basal ganglia adenosine receptors particularly of the A_{2A} subtype and its mRNA were increased in PD patients (Calon et al., 2004), while others reported a decrease of its mRNA levels (Hurley, Mash, & Jenner, 2000). Moreover, an increase of the A_{2A} adenosine receptors and its mRNA levels were increased in PD patients with LID as compared to PD patients without LID and controls (Calon et al., 2004). In MPTP-lesioned monkey and 6-hydroxydopamine rat models, A_{2A} receptor mRNA were shown to be increased or unchanged following the DA denervation, DA denervation with an L-DOPA treatment and in the presence of LID (Kaelin-Lang, Liniger, Probst, Lauterburg, & Burgunder, 2000; Morissette et al., 2006; Pinna et al., 2002; Tel et al., 2002; Tomiyama et al., 2004). As observed in the *postmortem* human brains, A_{2A} receptor levels were only increased in dyskinetic monkeys as compared to controls and DA-denervated monkeys treated or not with L-DOPA (Morelli, Fenu, Pinna, & Di Chiara, 1994; Morissette et al., 2006). These changes suggest that both striatal A_{2A} receptors and their mRNA levels are modulated by dyskinesiogenic L-DOPA treatment and might be related to the close interactions with DA and glutamate systems in the striatum.

7. CONCLUSION

The brain distribution of adenosine receptors, their pharmacological characteristics as well as their interactions with other neurotransmitters reviewed here show interesting activities. While A_1 adenosine receptors remain an interesting target, more studies are needed to better understand the implication of A_1 adenosine receptor and the role of its interactions with other receptors in the basal ganglia. A_{2A} receptor antagonists have shown antidyskinetic activity without worsening the antiparkinsonian effect of L-DOPA and have beneficial effects on specific symptoms of PD, such as tremors and rigidity. Hence, the A_{2A} receptor may be a target to modulate basal ganglia dopaminergic neurotransmission while avoiding the issues associated with direct dopaminergic drugs, such as motor complications. The interactions of adenosine receptors, especially A_{2A} receptor subtypes, with dopaminergic and glutamatergic receptors represent an interesting area of research. In the light of these interactions, new therapeutic approaches could include the combination of an A_{2A} receptor antagonist with dopaminomimetic or antiglutamatergic drugs. This combination could lead to the use of lower doses of each drug, especially L-DOPA, and could have an impact on PD symptoms and the development of motor complications. For example, A_{2A} receptor antagonists could have additive or synergistic activity when combined to mGlu5 receptor antagonists. The implication of these interactions in the development and expression of PD symptoms and LID needs further investigation to find new targets and ultimately, novel pharmacological treatments.

ACKNOWLEDGMENTS

This work was supported by a grant from the Canadian Institutes of Health Research to T. D. P. N. M. received a professional health care studentship from the Fonds de la recherche en santé du Québec.

REFERENCES

Adams, C. L., Cowen, M. S., Short, J. L., & Lawrence, A. J. (2008). Combined antagonism of glutamate mGlu5 and adenosine A2A receptors interact to regulate alcohol-seeking in rats. *The International Journal of Neuropsychopharmacology, 11*, 229–241.

Augood, S. J., & Emson, P. C. (1994). Adenosine A2a receptor mRNA is expressed by enkephalin cells but not by somatostatin cells in rat striatum: A co-expression study. *Brain Research. Molecular Brain Research, 22*, 204–210.

Bara-Jimenez, W., Sherzai, A., Dimitrova, T., Favit, A., Bibbiani, F., Gillespie, M., et al. (2003). Adenosine A(2A) receptor antagonist treatment of Parkinson's disease. *Neurology, 61*, 293–296.

Blanchet, P. J., Konitsiotis, S., Whittemore, E. R., Zhou, Z. L., Woodward, R. M., & Chase, T. N. (1999). Differing effects of N-methyl-D-aspartate receptor subtype selective antagonists on dyskinesias in levodopa-treated 1-methyl-4-phenyl- tetrahydropyridine monkeys. *The Journal of Pharmacology and Experimental Therapeutics, 290*, 1034–1040.

Blandini, F., & Armentero, M. T. (2012). New pharmacological avenues for the treatment of L-DOPA-induced dyskinesias in Parkinson's disease: Targeting glutamate and adenosine receptors. *Expert Opinion on Investigational Drugs, 21*, 153–168.

Bogenpohl, J. W., Ritter, S. L., Hall, R. A., & Smith, Y. (2012). Adenosine A2A receptor in the monkey basal ganglia: Ultrastructural localization and colocalization with the metabotropic glutamate receptor 5 in the striatum. *Journal of Comparative Neurology, 520*, 570–589.

Brotchie, J. M. (1998). Adjuncts to dopamine replacement: A pragmatic approach to reducing the problem of dyskinesia in Parkinson's disease. *Movement Disorders, 13*, 871–876.
Brotchie, J. M. (2003). CB1 cannabinoid receptor signalling in Parkinson's disease. *Current Opinion in Pharmacology, 3*, 54–61.
Calon, F., & Di Paolo, T. (2002). Levodopa response motor complications -GABA receptors and preproenkephalin expression in human brain. *Parkinsonism & Related Disorders, 8*, 449–454.
Calon, F., Dridi, M., Hornykiewicz, O., Bedard, P. J., Rajput, A. H., & Di Paolo, T. (2004). Increased adenosine A2A receptors in the brain of Parkinson's disease patients with dyskinesias. *Brain, 127*, 1075–1084.
Calon, F., Rajput, A. H., Hornykiewicz, O., Bedard, P. J., & Di Paolo, T. (2003). Levodopa-induced motor complications are associated with alterations of glutamate receptors in Parkinson's disease. *Neurobiology of Disease, 14*, 404–416.
Canals, M., Marcellino, D., Fanelli, F., Ciruela, F., de Benedetti, P., Goldberg, S. R., et al. (2003). Adenosine A2A-dopamine D2 receptor-receptor heteromerization: Qualitative and quantitative assessment by fluorescence and bioluminescence energy transfer. *The Journal of Biological Chemistry, 278*, 46741–46749.
Chase, T. N., Bibbiani, F., Bara-Jimenez, W., Dimitrova, T., & Oh-Lee, J. D. (2003). Translating A2A antagonist KW6002 from animal models to parkinsonian patients. *Neurology, 61*, S107–S111.
Ciruela, F., Casado, V., Rodrigues, R. J., Lujan, R., Burgueno, J., Canals, M., et al. (2006). Presynaptic control of striatal glutamatergic neurotransmission by adenosine A1-A2A receptor heteromers. *The Journal of Neuroscience, 26*, 2080–2087.
Ciruela, F., Escriche, M., Burgueno, J., Angulo, E., Casado, V., Soloviev, M. M., et al. (2001). Metabotropic glutamate 1alpha and adenosine A1 receptors assemble into functionally interacting complexes. *The Journal of Biological Chemistry, 276*, 18345–18351.
Ciruela, F., Ferre, S., Casado, V., Cortes, A., Cunha, R. A., Lluis, C., et al. (2006). Heterodimeric adenosine receptors: A device to regulate neurotransmitter release. *Cellular and Molecular Life Sciences, 63*, 2427–2431.
Conn, P. J., Battaglia, G., Marino, M. J., & Nicoletti, F. (2005). Metabotropic glutamate receptors in the basal ganglia motor circuit. *Nature Reviews Neuroscience, 6*, 787–798.
de Novellis, V., Marabese, I., Uliano, R., Palazzo, E., Scafuro, A., sca Rossi, F., et al. (2002). Type I and II metabotropic glutamate receptors modulate periaqueductal grey glycine release: Interaction between mGlu2/3 and A1 adenosine receptors. *Neuropharmacology, 43*, 1061–1069.
Domenici, M. R., Pepponi, R., Martire, A., Tebano, M. T., Potenza, R. L., & Popoli, P. (2004). Permissive role of adenosine A2A receptors on metabotropic glutamate receptor 5 (mGluR5)-mediated effects in the striatum. *Journal of Neurochemistry, 90*, 1276–1279.
Dunwiddie, T. V., & Masino, S. A. (2001). The role and regulation of adenosine in the central nervous system. *Annual Review of Neuroscience, 24*, 31–55.
Ferre, S., Ciruela, F., Canals, M., Marcellino, D., Burgueno, J., Casado, V., et al. (2004). Adenosine A2A-dopamine D2 receptor-receptor heteromers. Targets for neuropsychiatric disorders. *Parkinsonism & Related Disorders, 10*, 265–271.
Ferre, S., Fredholm, B. B., Morelli, M., Popoli, P., & Fuxe, K. (1997). Adenosine-dopamine receptor-receptor interactions as an integrative mechanism in the basal ganglia. *Trends in Neurosciences, 20*, 482–487.
Ferre, S., & Fuxe, K. (2000). Adenosine as a volume transmission signal. A feedback detector of neuronal activation. *Progress in Brain Research, 125*, 353–361.
Ferre, S., Karcz-Kubicha, M., Hope, B. T., Popoli, P., Burgueno, J., Gutierrez, M. A., et al. (2002). Synergistic interaction between adenosine A2A and glutamate mGlu5 receptors: Implications for striatal neuronal function. *Proceedings of the National Academy of Sciences of the United States of America, 99*, 11940–11945.

Ferre, S., Popoli, P., Rimondini, R., Reggio, R., Kehr, J., & Fuxe, K. (1999). Adenosine A2A and group I metabotropic glutamate receptors synergistically modulate the binding characteristics of dopamine D2 receptors in the rat striatum. *Neuropharmacology*, *38*, 129–140.

Fink, J. S., Weaver, D. R., Rivkees, S. A., Peterfreund, R. A., Pollack, A. E., Adler, E. M., et al. (1992). Molecular cloning of the rat A2 adenosine receptor: Selective co-expression with D2 dopamine receptors in rat striatum. *Brain Research. Molecular Brain Research*, *14*, 186–195.

Franco, R., Ferre, S., Agnati, L., Torvinen, M., Gines, S., Hillion, J., et al. (2000). Evidence for adenosine/dopamine receptor interactions: Indications for heteromerization. *Neuropsychopharmacology*, *23*, S50–S59.

Fredduzzi, S., Moratalla, R., Monopoli, A., Cuellar, B., Xu, K., Ongini, E., et al. (2002). Persistent behavioral sensitization to chronic L-DOPA requires A2A adenosine receptors. *The Journal of Neuroscience*, *22*, 1054–1062.

Fredholm, B. B., AP, I. J., Jacobson, K. A., Klotz, K. N., & Linden, J. (2001). International Union of Pharmacology. XXV. Nomenclature and classification of adenosine receptors. *Pharmacological Reviews*, *53*, 527–552.

Fuxe, K., Agnati, L. F., Jacobsen, K., Hillion, J., Canals, M., Torvinen, M., et al. (2003). Receptor heteromerization in adenosine A2A receptor signaling: Relevance for striatal function and Parkinson's disease. *Neurology*, *61*, S19–S23.

Fuxe, K., Ferré, S., Genedani, S., Franco, R., & Agnati, L. F. (2007). Adenosine receptor-dopamine receptor interactions in the basal ganglia and their relevance for brain function. *Physiology & Behavior*, *92*, 210–217.

Fuxe, K., Ferré, S., Zoli, M., & Agnati, L. F. (1998). Integrated events in central dopamine transmission as analyzed at multiple levels. Evidence for intramembrane adenosine A2A/dopamine D2 and adenosine A1/dopamine D1 receptor interactions in the basal ganglia. *Brain Research. Brain Research Reviews*, *26*, 258–273.

Ginés, S., Hillion, J., Torvinen, M., Le Crom, S., Casado, V., Canela, E. I., et al. (2000). Dopamine D1 and adenosine A1 receptors form functionally interacting heteromeric complexes. *Proceedings of the National Academy of Sciences of the United States of America*, *97*, 8606–8611.

Greengard, P. (2001). The neurobiology of slow synaptic transmission. *Science*, *294*, 1024–1030.

Grondin, R., Bedard, P. J., Hadj Tahar, A., Gregoire, L., Mori, A., & Kase, H. (1999). Antiparkinsonian effect of a new selective adenosine A2A receptor antagonist in MPTP-treated monkeys. *Neurology*, *52*, 1673–1677.

Harvey, J., & Lacey, M. G. (1997). A postsynaptic interaction between dopamine D1 and NMDA receptors promotes presynaptic inhibition in the rat nucleus accumbens via adenosine release. *The Journal of Neuroscience*, *17*, 5271–5280.

Henry, B., Fox, S. H., Crossman, A. R., & Brotchie, J. M. (2001). Mu- and delta-opioid receptor antagonists reduce levodopa-induced dyskinesia in the MPTP-lesioned primate model of Parkinson's disease. *Experimental Neurology*, *171*, 139–146.

Hettinger, B. D., Lee, A., Linden, J., & Rosin, D. L. (2001). Ultrastructural localization of adenosine A2A receptors suggests multiple cellular sites for modulation of GABAergic neurons in rat striatum. *The Journal of Comparative Neurology*, *431*, 331–346.

Hillion, J., Canals, M., Torvinen, M., Casadó, V., Scott, R., Terasmaa, A., et al. (2002). Coaggregation, cointernalization, and codesensitization of adenosine A2A receptors and dopamine D2 receptors. *The Journal of Biological Chemistry*, *277*, 18091–18097.

Hobson, B. D., O'Neill, C. E., Levis, S. C., Monteggia, L. M., Neve, R. L., Self, D. W., et al. (2013). Adenosine A1 and dopamine d1 receptor regulation of AMPA receptor phosphorylation and cocaine-seeking behavior. *Neuropsychopharmacology*, *38*, 1974–1983.

Hodgson, R. A., Bedard, P. J., Varty, G. B., Kazdoba, T. M., Di Paolo, T., Grzelak, M. E., et al. (2010). Preladenant, a selective A(2A) receptor antagonist, is active in primate models of movement disorders. *Experimental Neurology*, *225*, 384–390.

Hurley, M. J., Mash, D. C., & Jenner, P. (2000). Adenosine A(2A) receptor mRNA expression in Parkinson's disease. *Neuroscience Letters, 291*, 54–58.

Ismayilova, N., Crossman, A., Verkhratsky, A., & Brotchie, J. (2004). Effects of adenosine A1, dopamine D1 and metabotropic glutamate 5 receptors-modulating agents on locomotion of the reserpinised rats. *European Journal of Pharmacology, 497*, 187–195.

Jarvis, M. F., & Williams, M. (1989). Direct autoradiographic localization of adenosine A2 receptors in the rat brain using the A2-selective agonist, [3H]CGS 21680. *European Journal of Pharmacology, 168*, 243–246.

Jenner, P., Mori, A., Hauser, R., Morelli, M., Fredholm, B. B., & Chen, J. F. (2009). Adenosine, adenosine A 2A antagonists, and Parkinson's disease. *Parkinsonism & Related Disorders, 15*, 406–413.

Jin, S., Johansson, B., & Fredholm, B. B. (1993). Effects of adenosine A1 and A2 receptor activation on electrically evoked dopamine and acetylcholine release from rat striatal slices. *The Journal of Pharmacology and Experimental Therapeutics, 267*, 801–808.

Kaelin-Lang, A., Liniger, P., Probst, A., Lauterburg, T., & Burgunder, J. M. (2000). Adenosine A2A receptor gene expression in the normal striatum and after 6-OH-dopamine lesion. *Journal of Neural Transmission, 107*, 851–859.

Kanda, T., Jackson, M. J., Smith, L. A., Pearce, R. K., Nakamura, J., Kase, H., et al. (2000). Combined use of the adenosine A(2A) antagonist KW-6002 with L-DOPA or with selective D1 or D2 dopamine agonists increases antiparkinsonian activity but not dyskinesia in MPTP-treated monkeys. *Experimental Neurology, 162*, 321–327.

Kase, H. (2001). New aspects of physiological and pathophysiological functions of adenosine A2A receptor in basal ganglia. *Bioscience, Biotechnology, and Biochemistry, 65*, 1447–1457.

Kull, B., Svenningsson, P., & Fredholm, B. B. (2000). Adenosine A(2A) receptors are colocalized with and activate g(olf) in rat striatum. *Molecular Pharmacology, 58*, 771–777.

Lindskog, M., Svenningsson, P., Pozzi, L., Kim, Y., Fienberg, A. A., Bibb, J. A., et al. (2002). Involvement of DARPP-32 phosphorylation in the stimulant action of caffeine. *Nature, 418*, 774–778.

Martinez-Mir, M. I., Probst, A., & Palacios, J. M. (1991). Adenosine A2 receptors: Selective localization in the human basal ganglia and alterations with disease. *Neuroscience, 42*, 697–706.

Mones, R. J., Elizan, T. S., & Siegel, G. J. (1971). Analysis of L-dopa induced dyskinesias in 51 patients with Parkinsonism. *Journal of Neurology, Neurosurgery and Psychiatry, 34*, 668–673.

Morelli, M., Carta, A. R., & Jenner, P. (2009). Adenosine A2A receptors and Parkinson's disease. *Handbook of Experimental Pharmacology, 193*, 589–615.

Morelli, M., Di Paolo, T., Wardas, J., Calon, F., Xiao, D., & Schwarzschild, M. A. (2007). Role of adenosine A2A receptors in parkinsonian motor impairment and l-DOPA-induced motor complications. *Progress in Neurobiology, 83*, 293–309.

Morelli, M., Fenu, S., Pinna, A., & Di Chiara, G. (1994). Adenosine A2 receptors interact negatively with dopamine D1 and D2 receptors in unilaterally 6-hydroxydopamine-lesioned rats. *European Journal of Pharmacology, 251*, 21–25.

Morelli, M., & Pinna, A. (2001). Interaction between dopamine and adenosine A2A receptors as a basis for the treatment of Parkinson's disease. *Neurological Sciences, 22*, 71–72.

Mori, A., & Shindou, T. (2003). Modulation of GABAergic transmission in the striatopallidal system by adenosine A2A receptors: a potential mechanism for the antiparkinsonian effects of A2A antagonists. *Neurology, 61*, S44–S48.

Morin, N., Gregoire, L., Morissette, M., Desrayaud, S., Gomez-Mancilla, B., Gasparini, F., et al. (2013). MPEP, an mGlu5 receptor antagonist, reduces the development of L-DOPA-induced motor complications in de novo parkinsonian monkeys: Biochemical correlates. *Neuropharmacology, 66*, 355–364.

Morin, N., Jourdain, V. A., Morissette, M., Gregoire, L., & Di Paolo, T. (2014). Long-term treatment with l-DOPA and an mGlu5 receptor antagonist prevents changes in brain basal ganglia dopamine receptors, their associated signaling proteins and neuropeptides in parkinsonian monkeys. *Neuropharmacology, 79*, 688–706.

Morin, N., Morissette, M., Gregoire, L., Gomez-Mancilla, B., Gasparini, F., & Di Paolo, T. (2013). Chronic treatment with MPEP, an mGlu5 receptor antagonist, normalizes basal ganglia glutamate neurotransmission in L-DOPA-treated parkinsonian monkeys. *Neuropharmacology, 73*, 216–231.

Morissette, M., Dridi, M., Calon, F., Hadj Tahar, A., Meltzer, L. T., Bedard, P. J., et al. (2006). Prevention of dyskinesia by an NMDA receptor antagonist in MPTP monkeys: Effect on adenosine A2A receptors. *Synapse, 60*, 239–250.

Nash, J. E., & Brotchie, J. M. (2000). A common signaling pathway for striatal NMDA and adenosine A2a receptors: Implications for the treatment of Parkinson's disease. *The Journal of Neuroscience, 20*, 7782–7789.

Nishi, A., Liu, F., Matsuyama, S., Hamada, M., Higashi, H., Nairn, A. C., et al. (2003). Metabotropic mGlu5 receptors regulate adenosine A2A receptor signaling. *Proceedings of the National Academy of Sciences of the United States of America, 100*, 1322–1327.

Olah, M. E., & Stiles, G. L. (1995). Adenosine receptor subtypes: Characterization and therapeutic regulation. *Annual Review of Pharmacology and Toxicology, 35*, 581–606.

Olanow, C. W., & Koller, W. C. (1998). An algorithm (decision tree) for the management of Parkinson's disease: Treatment guidelines. *Neurology, 50*, S1–S57.

Olanow, C. W., Stern, M. B., & Sethi, K. (2009). The scientific and clinical basis for the treatment of Parkinson disease. *Neurology, 72*, S1–S136.

Ouattara, B., Gregoire, L., Morissette, M., Gasparini, F., Vranesic, I., Bilbe, G., et al. (2011). Metabotropic glutamate receptor type 5 in levodopa-induced motor complications. *Neurobiology of Aging, 32*, 1286–1295.

Peterfreund, R. A., MacCollin, M., Gusella, J., & Fink, J. S. (1996). Characterization and expression of the human A2a adenosine receptor gene. *Journal of Neurochemistry, 66*, 362–368.

Pinna, A., Corsi, C., Carta, A. R., Valentini, V., Pedata, F., & Morelli, M. (2002). Modification of adenosine extracellular levels and adenosine A(2A) receptor mRNA by dopamine denervation. *European Journal of Pharmacology, 446*, 75–82.

Pinna, A., Wardas, J., Cozzolino, A., & Morelli, M. (1999). Involvement of adenosine A2A receptors in the induction of c-fos expression by clozapine and haloperidol. *Neuropsychopharmacology, 20*, 44–51.

Pollack, A. E., Harrison, M. B., Wooten, G. F., & Fink, J. S. (1993). Differential localization of A2a adenosine receptor mRNA with D1 and D2 dopamine receptor mRNA in striatal output pathways following a selective lesion of striatonigral neurons. *Brain Research, 631*, 161–166.

Ribeiro, J. A., Sebastiao, A. M., & de Mendonca, A. (2002). Adenosine receptors in the nervous system: Pathophysiological implications. *Progress in Neurobiology, 68*, 377–392.

Rosin, D. L., Hettinger, B. D., Lee, A., & Linden, J. (2003). Anatomy of adenosine A2A receptors in brain: Morphological substrates for integration of striatal function. *Neurology, 61*, S12–S18.

Rosin, D. L., Robeva, A., Woodard, R. L., Guyenet, P. G., & Linden, J. (1998). Immunohistochemical localization of adenosine A2A receptors in the rat central nervous system. *The Journal of Comparative Neurology, 401*, 163–186.

Salamone, J. D. (2010). Preladenant, a novel adenosine A(2A) receptor antagonist for the potential treatment of parkinsonism and other disorders. *IDrugs, 13*, 723–731.

Salim, H., Ferre, S., Dalal, A., Peterfreund, R. A., Fuxe, K., Vincent, J. D., et al. (2000). Activation of adenosine A1 and A2A receptors modulates dopamine D2 receptor-induced responses in stably transfected human neuroblastoma cells. *Journal of Neurochemistry, 74*, 432–439.

Samadi, P., Rajput, A., Calon, F., Gregoire, L., Hornykiewicz, O., Rajput, A. H., et al. (2009). Metabotropic glutamate receptor II in the brains of Parkinsonian patients. *Journal of Neuropathology and Experimental Neurology, 68*, 374–382.

Samadi, P., Rouillard, C., Bedard, P. J., & Di Paolo, T. (2007). Functional neurochemistry of the basal ganglia. *Handbook of Clinical Neurology, 83*, 19–66.

Schiffmann, S. N., Fisone, G., Moresco, R., Cunha, R. A., & Ferre, S. (2007). Adenosine A2A receptors and basal ganglia physiology. *Progress in Neurobiology, 83*, 277–292.

Schiffmann, S. N., Jacobs, O., & Vanderhaeghen, J. J. (1991). Striatal restricted adenosine A2 receptor (RDC8) is expressed by enkephalin but not by substance P neurons: An in situ hybridization histochemistry study. *Journal of Neurochemistry, 57*, 1062–1067.

Schiffmann, S. N., Libert, F., Vassart, G., Dumont, J. E., & Vanderhaeghen, J. J. (1990). A cloned G protein-coupled protein with a distribution restricted to striatal medium-sized neurons. Possible relationship with D1 dopamine receptor. *Brain Research, 519*, 333–337.

Schiffmann, S. N., Libert, F., Vassart, G., & Vanderhaeghen, J. J. (1991). Distribution of adenosine A2 receptor mRNA in the human brain. *Neuroscience Letters, 130*, 177–181.

Shahraki, A., & Stone, T. W. (2003). Interactions between adenosine and metabotropic glutamate receptors in the rat hippocampal slice. *British Journal of Pharmacology, 138*, 1059–1068.

Siderowf, A., & Stern, M. (2003). Update on Parkinson disease. *Annals of Internal Medicine, 138*, 651–658.

Svenningsson, P., & Fredholm, B. B. (2003). Exciting news about A2A receptors. *Neurology, 61*, S10–S11.

Svenningsson, P., Le Moine, C., Aubert, I., Burbaud, P., Fredholm, B. B., & Bloch, B. (1998). Cellular distribution of adenosine A2A receptor mRNA in the primate striatum. *The Journal of Comparative Neurology, 399*, 229–240.

Svenningsson, P., Le Moine, C., Fisone, G., & Fredholm, B. B. (1999). Distribution, biochemistry and function of striatal adenosine A2A receptors. *Progress in Neurobiology, 59*, 355–396.

Tel, B. C., Zeng, B. Y., Cannizzaro, C., Pearce, R. K., Rose, S., & Jenner, P. (2002). Alterations in striatal neuropeptide mRNA produced by repeated administration of L-DOPA, ropinirole or bromocriptine correlate with dyskinesia induction in MPTP-treated common marmosets. *Neuroscience, 115*, 1047–1058.

Tomiyama, M., Kimura, T., Maeda, T., Tanaka, H., Kannari, K., & Baba, M. (2004). Upregulation of striatal adenosine A2A receptor mRNA in 6-hydroxydopamine-lesioned rats intermittently treated with L-DOPA. *Synapse, 52*, 218–222.

Torvinen, M., Torri, C., Tombesi, A., Marcellino, D., Watson, S., Lluís, C., et al. (2005). Trafficking of adenosine A2A and dopamine D2 receptors. *Journal of Molecular Neuroscience, 25*, 191–200.

Varty, G. B., Hodgson, R. A., Pond, A. J., Grzelak, M. E., Parker, E. M., & Hunter, J. C. (2008). The effects of adenosine A2A receptor antagonists on haloperidol-induced movement disorders in primates. *Psychopharmacology, 200*, 393–401.

Wang, S. J. (2007). Caffeine facilitation of glutamate release from rat cerebral cortex nerve terminals (synaptosomes) through activation protein kinase C pathway: An interaction with presynaptic adenosine A1 receptors. *Synapse, 61*, 401–411.

Yaar, R., Jones, M. R., Chen, J. F., & Ravid, K. (2005). Animal models for the study of adenosine receptor function. *Journal of Cellular Physiology, 202*, 9–20.

CHAPTER EIGHT

Effects of the Adenosine A_{2A} Receptor Antagonist on Cognitive Dysfunction in Parkinson's Disease

Shin-ichi Uchida[1], Takako Kadowaki-Horita, Tomoyuki Kanda
Central Nervous System Research Laboratories, Research & Development Division, Kyowa Hakko Kirin Co., Ltd., Sunto-gun, Shizuoka, Japan
[1]Corresponding author: e-mail address: shinichi.uchida@kyowa-kirin.co.jp

Contents

Abstract

Parkinson's disease (PD) is primarily characterized by motor abnormalities, but cognitive changes also occur in the early and late stages of the disease process. In PD patients, cognitive dysfunction is associated with reduced quality of life, as well as increased morbidity and mortality, resulting in increases in caregiver burden, and health-related costs. Therefore, safe and effective approaches are needed to treat cognitive dysfunction in PD patients. The underlying pathophysiology of cognitive dysfunction is complex and not fully understood, however. α-Synuclein, amyloid-related proteins, and cholinergic deficits have been reported to partially contribute to cognitive dysfunction. Changes in cortical dopamine (DA) content may also be responsible for early cognitive changes in patients with PD. Certainly, dopaminergic afferents to the frontal cortex degenerate in PD, and there is a reduction of DA content in the prefrontal cortex (PFC). It has also

International Review of Neurobiology, Volume 119
ISSN 0074-7742
http://dx.doi.org/10.1016/B978-0-12-801022-8.00008-8

been reported that PFC dopaminergic input plays an important role in working memory performance. Moreover, PFC DA levels and working memory performance are significantly reduced by a 6-hydroxydopamine lesion in the PFC of a rat. Recent findings in the areas of pharmacological manipulation and genetic ablation suggest that the adenosine A_{2A} receptor is also related to cognitive functions, especially working memory. In addition, the blockade of adenosine A_{2A} receptors reverses cognitive dysfunction in PFC-lesioned rats, and this blocking effect may be due to an increase in PFC DA content. Therefore, adenosine A_{2A} receptor antagonists not only improve motor performance, but they may also lead to improved cognitive function in those with PD.

1. INTRODUCTION

Cognitive dysfunction is one of the most common and important non-motor features of Parkinson's disease (PD), given its effect on the patient's physical functioning and quality of life (QOL) (Schrag, Jahanshahi, & Quinn, 2000). Most PD patients will eventually develop dysfunction in cognitive abilities such as attention, executive function, and memory. However, dementia in PD is only recognized and acted upon in 27% of patients with current symptoms (Hu et al., 2011). Thus, researchers have been trying to improve our understanding of PD-related cognitive dysfunction and dementia, clarifying the involved biology, developing more accurate diagnostic criteria, and investigating drug candidates for treating the cognitive disability. The underlying mechanisms of cognitive dysfunction in PD are only partially known, and mechanism-based treatments are limited. The dysmetabolism of α-synuclein and amyloid-related proteins, and cholinergic deficits have been reported to contribute to cognitive dysfunction in PD (Irwin, Lee, & Trojanowski, 2013).

Treatment with L-DOPA or dopamine (DA) agonists and cholinesterase inhibitors has produced some reductions in cognitive dysfunction during the early stages of PD (Svenningsson, Westman, Ballard, & Aarsland, 2012). However, dopaminergic therapy at high concentrations can cause cognitive deterioration. Rivastigmine, a cholinesterase inhibitor, is the only medication licensed for the treatment of mild-to-moderate dementia associated with PD, but rivastigmine also has a risk of worsening motor symptoms in PD patients. As a result, novel approaches are needed to treat cognitive dysfunction because the QOL of PD patients is strongly affected by their level of cognitive functioning (Schrag et al., 2000). There is increasing evidence to suggest that, in cognitively impaired PD patients, profound DA depletion not only occurs in the striatum but also in the prefrontal cortex (PFC) (Narayanan, Rodnitzky, & Uc, 2013), and this depletion is associated with cognitive dysfunction in PD. Recent findings also show that adenosine A_{2A} receptors play an important role in controlling cognitive function. Furthermore, adenosine A_{2A} receptor antagonists improve the cognitive functioning of PFC-lesioned rats (Kadowaki Horita, Kobayashi, Mori, Jenner, & Kanda, 2013). Therefore, adenosine A_{2A} receptor antagonists may provide a new approach to the treatment of cognitive dysfunction in PD.

In this review, we describe the most recent findings about the clinical features, underlying mechanisms, predictive biomarkers, and treatment of cognitive dysfunction in PD. We also describe the role of adenosine A_{2A} receptors and the effects of the adenosine A_{2A} receptor antagonist on cognitive dysfunction in PD.

2. COGNITIVE DYSFUNCTION IN PD

2.1. Epidemiology

Cognitive dysfunction is a common feature of PD patients, especially of late-stage patients. Several epidemiological studies have shown that the prevalence of cognitive dysfunction in PD patients is roughly 30% and three to six times higher than that in the general population (Aarsland & Kurz, 2010; Svenningsson et al., 2012). In the late-stage disease, cognitive dysfunction and dementia occur in up to 80% of older individuals, and this statistic presumably reflects a pathology involving the loss of cholinergic input to the cortex (Irwin et al., 2013). However, cognitive dysfunction can occur in the early stages of PD, at which time it is characterized by frontal or frontostriatal cognitive deficits such as alterations in executive function, attention, working memory, and episodic memory (McKinlay & Grace, 2011; Pagonabarraga & Kulisevsky, 2012). In addition, 24% of newly diagnosed PD patients show cognitive dysfunction, most frequently on measures of memory and executive functions (Muslimovic, Post, Speelman, & Schmand, 2005). Furthermore, PD patients with cognitive dysfunction experience dementia for an average of 3–4 years during the course of their disease (Kempster, O'Sullivan, Holton, Revesz, & Lees, 2010).

Cognitive dysfunction greatly affects physical functioning, QOL, morbidity, and mortality (Schrag et al., 2000). The health-related costs of PD patients with cognitive dysfunction are also higher than those of non-demented patients. Institutional care is the largest cost factor (Vossius, Larsen, Janvin, & Aarsland, 2011). For example, the average duration of hospitalization is known to be longer in PD patients with cognitive dysfunction than those without (Fletcher, Leake, & Marion, 2011). Even in the early stages of the disease, cognitive dysfunction can interfere with key activities such as driving, and loss of independence is predicted (Narayanan et al., 2013). Therefore, it is important to improve the QOL of these patients and to reduce the financial pressure on society, and understanding the underlying mechanism for cognitive dysfunction in PD is crucial for therapeutic development.

2.2. Risk factors

A number of risk factors have been reported to be associated with cognitive dysfunction in PD. In particular, age is the most important risk factor (Kempster et al., 2010). Cognitive dysfunction in PD has an age of onset similar to that of dementia, regardless of when either condition is diagnosed. Some studies of PD patients with cognitive dysfunction analyzed the mini-mental state examination (MMSE) score, the most commonly used instrument for screening cognitive function. As a result, MMSE scores in PD patients with cognitive dysfunction correlate with the severity of motor disability (Braak, Rüb, Jansen Steur, Del Tredici, & de Vos, 2005). The effects of aging may be added to the severity of motor disturbance. Other reported risk factors include gender, education, bradykinesia, and axial impairment (Levy et al., 2000). Additionally, non-motor symptoms such as visual hallucination, rapid eye movement sleep behavior disorder, and severe olfactory dysfunction are associated with cognitive dysfunction in PD (Baba et al., 2012; Postuma et al., 2012).

Genetic research on cognitive dysfunction in PD has been controversial, although genetic factors may affect the occurrence of cognitive dysfunction in PD. Some studies have focused on α-synuclein and tau protein formed in the neurofibrillary tangle (NFT). Each of the *SNCA* (α-synuclein) multiplication and H1 haplotypes of *MAPT* (microtubule-associated protein tau) may not only be associated with sporadic PD, but they may also affect cognitive function in PD (Goris et al., 2007; Ross et al., 2008). Polymorphisms of *DYRK1A* (dual-specificity tyrosine phosphorylation-regulated kinase), which phosphorylate proteins such as α-synuclein and amyloid precursor protein, have been associated with cognitive dysfunction in PD (Jones, Aarsland, Londos, & Ballard, 2012). In addition, mutations in the *GBA* (glucocerebrosidase) gene, which encodes the lysosomal enzyme, have been shown to increase not only the risk of developing PD, but also the rate of cognitive dysfunction (Sidransky & Lopez, 2012). This finding has been confirmed in a prospective study of newly diagnosed PD cases in which the increased risk of progression to dementia was five times greater in the cases with a *GBA* mutation than those without (Winder-Rhodes et al., 2013). Mutations in the *COMT* (catechol-*O*-methyltransferase) gene are associated with DA-related cognitive dysfunction, but not with dementia risk (Williams-Gray, Evans, et al., 2009). The association between *BDNF* (brain-derived neurotrophic factor) and risk of cognitive dysfunction in PD is inconsistent. The *BDNF* (Met/Met) homozygote genotype is reported to be not only overrepresented in PD patients compared with normal individuals, but also significantly correlated to cognitive dysfunction, age, and PD severity (Guerini et al., 2009). In contrast, another report shows that *BDNF* polymorphism is not associated with cognitive dysfunction in PD (Białecka et al., 2014). The ε4 allele of the *APOE* (apolipoprotein E) genotype is an established risk factor for Alzheimer's disease (AD). In addition to this finding, pathological overlap between AD and cognitive dysfunction in PD has stimulated research to see whether the *APOE* genotype may play a role in the risk of cognitive dysfunction. However, a large meta-analysis has failed to show clinical support for this finding (Williams-Gray, Goris, et al., 2009). Research on genetic susceptibility to cognitive dysfunction in PD has been inconclusive due to discrepancies in the results or differences in sample size. Therefore, large-scale longitudinal studies would be best placed to further evaluate any impact of genetic contribution on cognitive dysfunction in the wider PD population.

2.3. Clinical features

The features of cognitive symptoms in PD are very diverse (Litvan et al., 2012), probably because of the heterogeneity of the underlying neuropathology. Common features of cognitive dysfunctions in PD are impaired attention and concentration, poor executive function and planning, difficulty with multitasking or initiating activity, short-term memory impairment, and poor visuospatial function (Goldman & Holden, 2014). Among these features, attention, working memory, and executive function can be impaired in PD patients with the early stages of cognitive dysfunction. The underlying pathophysiology of cognitive dysfunction in PD overlaps with cholinergic deficiency, and this overlap provides a rationale for treatment with cholinesterase inhibitors, as described later. However, abnormalities in other neurotransmitters, such as DA, serotonin, glutamate, and norepinephrine, also contribute to cognitive dysfunction in PD. Some reports show that executive dysfunction in PD is an early phenomenon of changed dopaminergic tone in the frontal cortex. Executive function includes mental flexibility, planning, and working memory, all of which are mediated by striatal-frontal network (Irwin et al., 2013).

The recognized cognitive symptoms in PD have led to the recent development of clinical criteria. As a result, a task force, organized by the Movement Disorder Society, proposed the first clinical diagnostic criteria for cognitive dysfunction in PD (Emre, Aarsland, Brown, Burn, & Duyckaerts, 2007; Litvan et al., 2012). The proposed diagnostic criteria are based on a literature review and expert consensus. However, the criteria will require validation and possibly refinement as additional research improves our understanding of the epidemiology, neurobiology, assessment, and long-term course of this clinical syndrome. These diagnostic criteria will support future research efforts to identify the increased risk of progressive cognitive dysfunction and dementia at an early stage in PD patients who may benefit from clinical interventions at a predementia stage.

Cognitive dysfunction is a hallmark of PD dementia, as in AD, but other neuropsychiatric symptoms, such as depression, anxiety, apathy, hallucination, and delusions, further complicate management (Irwin et al., 2013). Moreover, some patients may exhibit an impulsive control disorder characterized by pathological gambling, hypersexuality, compulsive shopping, compulsive eating, and/or DA dysregulation syndrome (Weintraub, 2008). The etiology of impulsive control disorder in PD is thought to be due to stimulation of hypersensitive striatal-frontal networks by dopaminergic therapy rather than a direct consensus of the neurodegenerative disease.

2.4. Neuropathology

The neuropathology underlying cognitive dysfunction in PD is heterogeneous and exploring its pathological correlation has proven to be a challenge. Several studies have shown that cortical Lewy body and neuritic pathology is more extensive and severe in PD patients with cognitive dysfunction than those without, and studies using α-synuclein immunochemistry show that cortical α-synuclein pathology is strongly correlated with cognitive dysfunction in PD (Irwin et al., 2013). Furthermore, it has been found that the burden of α-synuclein pathology and diffused amyloid-β (Aβ) plaque is greater in the striatum in PD patients with cognitive dysfunction than in PD without. This striatal pathology has been considered to be possible neuropathologic substrates of PD with cognitive dysfunction. Recent studies have strongly implicated cell-to-cell transmission of pathological species of α-synuclein as a common mechanism for the onset and progression of PD and related disorders (Guo & Lee, 2014). One of these studies found that the intracerebral injection of insoluble α-synuclein fibrils could induce the aggregation of endogenous mouse α-synuclein through a prion-like spreading mechanism (Masuda-Suzukake et al., 2013). Although there are many more questions to be answered, these findings may open up new avenues for therapeutic interventions in neurodegenerative disorders such as PD with or without cognitive dysfunction.

Some studies have also shown that the level of Aβ plaques and tau NFTs, which are the hallmark pathology of AD, correlate with cognitive dysfunction in PD (Compta et al., 2011). Moreover, a combination of Lewy- and AD-type pathologies is a robust pathological correlate of cognitive dysfunction in PD, with quantitative and semiquantitative assessment of Lewy pathology being more informative than Braak α-synuclein stages. In addition, NFTs spread from the limbic areas to neocortex with advancing stages of the disease, whereas Aβ plaques first emerge in the neocortex before arising in the subcortex in AD brains (Guo & Lee, 2014). Therefore, it is important to identify therapeutic targets for the suppression of cell-to-cell transmission of pathological species, such as α-synuclein, Aβ, and tau, in PD with cognitive dysfunction and other neurodegenerative diseases.

PD patients have impaired prefrontal DA signaling, and this impairment can influence cognitive function (Narayanan et al., 2013). Some reports using positron emission tomography (PET) imaging have assessed striatal and cortical DA changes in patients at risk for dementia. As a result, both striatal DA depletion and insular cortex DA receptor changes underlie executive dysfunction (Christopher et al., 2014). In addition, it has been reported that reduced [^{18}F] fluorodopa uptake in the frontal cortex of PD patients is related to impairment in neuropsychological tests measuring verbal fluency, working memory, and attentional functioning, reflecting frontal lobe function (Rinne et al., 2000). In experimental models of PD, DA depletion in the cortex has been reported to produce cognitive dysfunction, partially associated with executive dysfunction and working memory deficits (Clinton, Sucharski, & Finlay, 2006; Decamp & Schneider, 2004). Early cognitive change in younger individuals with PD is characterized by frontal or frontostriatal cognitive dysfunction, such as changes in executive dysfunction, attention, and working memory. These results indicate that DA in the PFC plays an important role in working memory and executive function in the early development of cognitive dysfunction in PD patients. Thus, it may be important to control DA in the PFC, especially for treatment of early cognitive dysfunction in PD.

2.5. Biomarkers

Several possible biomarkers for cognitive dysfunction in PD have been found from data in long-term studies. Lower baseline cerebrospinal fluid (CSF) $A\beta_{1-42}$ is associated with more rapid cognitive dysfunction in PD, but CSF total tau and tau phosphorylated at position threonine 181 (p-tau181p) levels are not significantly associated with cognitive dysfunction (Siderowf et al., 2010). α-Synuclein levels in the CSF have recently begun to be investigated, and despite discrepancies between trial results, some initial results suggest correlations between cognitive status and α-synuclein concentration in the CSF in dementia with Lewy bodies. These findings may be applicable to cognitive dysfunction in PD in the future (Mollenhauer et al., 2011). Serum epidermal growth factor (EGF) levels are related to frontal and temporal cognitive functions in early, drug-naive PD patients and predict performance on frontal and posterior cognitive functions at 2-year follow-up (Pellecchia et al., 2013). EGF is proposed as a potential serum biomarker for early cognitive dysfunction in PD.

Several groups have shown that a quantitative electroencephalogram can serve as a biomarker for the development of cognitive dysfunction in PD. The risk of developing cognitive dysfunction is 13 times higher in patients with a low background rhythm frequency than for those with a normal rhythm (Klassen et al., 2011). Magnetic resonance imaging (MRI) can detect disruptions of white tract matter and atrophy of grey matter of PD patients with cognitive dysfunction (Duncan, Firbank, O'Brien, & Burn, 2013). The results of MRI scans are consistent, however, at present evidence is not enough to adopt such imaging results as biomarkers. Future work should adopt agreed-upon scan protocols, should be adequately powered, and should use carefully diagnosed patients to fully maximize the contribution of MRI.

PET scanning has also been investigated as a biomarker for cognitive dysfunction in PD. A recent study compared three forms of PET scanning: [^{11}C](R)PK11195-PET, which is a marker of microglial activation; [^{11}C]PIB-PET, which is a marker of Aβ; and [^{18}F]FDG-PET, which measures cerebral glucose metabolism (Edison et al., 2013). Results show that PET can detect cortical microglial activation and reduced glucose metabolism in early cognitive dysfunction in PD patients. Microglial activation is inversely correlated with MMSE score, while reduced glucose metabolism correlates directly with MMSE score. However, there is currently no consistent evidence of cortical Aβ changes in PD patients with cognitive dysfunction. When scanned using quantitative single photon emission computed tomography, PD patients with cognitive dysfunction show lower regional cerebral blood flow in the frontal and temporal cortices (Hattori et al., 2013). Yet, despite the apparent promise of these technologies, research on biomarkers of cognitive dysfunction in PD has been inconclusive because of discrepancies in the results or differences in sample size. Therefore, further large-scale longitudinal studies are needed to develop the biomarkers of cognitive dysfunction in PD.

2.6. Treatment

Cognitive dysfunctions in PD are linked to deficits of dopaminergic and cholinergic activities, and possibly to noradrenergic and glutamatergic activities (Klein et al., 2010). Therefore, neurotransmitter-based treatments may improve cognitive dysfunction in PD. Cholinesterase inhibitors have been shown to improve cognitive function and activities of daily living in patients with PD (Rolinski, Fox, Maidment, & McShane, 2012). For example, the Exelon in Parkinson's Disease Dementia Study (EXPRESS), a large, 24-week, randomized, placebo-controlled trial, has shown that rivastigmine moderately alleviate dementia associated with PD (Emre et al., 2004). Currently, rivastigmine is the only therapeutic drug licensed for dementia in PD. Other cholinesterase inhibitors such as donepezil and galantamine are not yet supported by enough evidence to be rated for the treatment of cognitive dysfunction in PD (Seppi et al., 2011). In addition, treatment of early cognitive dysfunction with a cholinesterase inhibitor may worsen motor symptoms in PD (Richard, Justus, Greig, Marshall, & Kurlan, 2002).

It is well known that DA depletion in the frontal cortex produces cognitive deficits, particularly those deficits associated with executive dysfunction (Brozoski, Brown, Rosvold, & Goldman, 1979; Clinton et al., 2006). Thus, treatment with L-DOPA or a DA agonist can improve some aspects of cognitive function such as executive function and working memory, but these drugs do not affect other functions such as attention, associative learning, and spatial recognition memories (Kehagia, Barker, & Robbins, 2010). However, these dopaminergic therapies at high concentrations may worsen cognitive function with deficits in reversal learning, decision-making, and impulse control. For example, pramipexole, a DA D2/D3 receptor agonist, has harmful effects on cognition as compared with pergolide, a DA D1/D2 receptor agonist. In addition, DA agonist-induced dopaminergic overdose is associated with hallucinations in PD (Poewe, 2008). Rasagiline is a potent, selective, irreversible monoamine oxidase type-B inhibitor. A randomized, double-blind, placebo-controlled study of rasagiline has been conducted with PD patients who have cognitive dysfunction, but do not yet fulfill criteria for dementia. This study demonstrated significant improvement in the attention domain with rasagiline, compared with a placebo (Hanagasi et al., 2011). There was no significant benefit on the cognitive domains of executive function, memory, visuospatial function, or language, however.

N-methyl-D-aspartate (NMDA) receptor antagonists may be clinically relevant for the management of cognitive dysfunction in PD. Amantadine may delay the onset of dementia in PD patients and may attenuate its severity, but NMDA can also induce hallucinations and confusion in patients with advanced disease (Inzelberg et al., 2006). There is inconsistent evidence supporting the use of another glutamatergic agent memantine for the treatment of cognitive dysfunction in PD (Seppi et al., 2011). The other approaches to treating cognitive dysfunction in PD focus on the serotonin and norepinephrine systems. Atomoxetine is a serotonin norepinephrine reuptake inhibitor initially approved for the treatment of attention deficit-hyperactivity disorder. A pilot open-label study of atomoxetine in PD showed significant improvements of executive dysfunction based on the Clinical Global Impression-Change Scale (Marsh, Biglan, Gerstenhaber, & Williams, 2009). Atomoxetine is not efficacious for the treatment of clinically significant depressive symptoms in PD, but it is associated with improvements in global cognitive performance as measured by the MMSE (Weintraub et al., 2010). Side effects of atomoxetine include sleep, gastrointestinal disturbances, and hypomania.

Neuroprotective- or disease-modifying drugs may ameliorate cognitive dysfunction in PD. However, no such drugs are yet available. Immune-based therapies targeting α-synuclein may be one potential approach for developing a disease-modifying drug to treat PD. But it is unclear if immunotherapy for α-synuclein would affect cognitive and motor symptoms, especially those in early-onset patients. Novel approaches are needed to treat cognitive dysfunction in PD because cognitive deficits affect the QOL of PD patients (Schrag et al., 2000). In experimental models, adenosine A_{2A} receptor gene disruption has been reported to protect DA neurons from the degeneration caused by the mutant α-synuclein (Kachroo & Schwarzschild, 2012). Moreover, adenosine A_{2A} receptor blockade by SCH58261, an adenosine A_{2A} receptor antagonist, in rats or genetic inactivation in mice prevents synaptotoxicity by Aβ (Canas, Porciúncula, Cunha, Silva, & Machado, 2009). Therefore, utilizing the adenosine A_{2A} receptor antagonist may be one of approaches to control the progression of cognitive dysfunction in PD.

3. THE ROLE OF ADENOSINE A_{2A} RECEPTORS ON COGNITIVE FUNCTION

The overall role of adenosine A_{2A} receptors in the control of cognitive function is not well understood. The pharmacological inactivation of adenosine A_{2A} receptors using selective antagonists has been reported to enhance short-term social memory and enhance long-term memory in passive avoidance (Prediger, Da Cunha, & Takahashi, 2005). Selective adenosine A_{2A} receptor antagonists also improve short-term social memory and spatial learning depressed in spontaneously hypertensive rats and in aged rats (Prediger, Batista, & Takahashi, 2005; Prediger, Fernandes, & Takahashi, 2005; Takahashi, Pamplona, & Prediger, 2008). For working memory, the evidence is conflicting because selective adenosine A_{2A} receptor antagonists reversed MK-801-induced reduction in spontaneous alternation in a mouse Y-maze task, but they did not reverse the scopolamine-induced reduction in spontaneous alternation (Fraser, Fisher, Cooke, Thompson, & Stone, 1997). However, previous studies have demonstrated that overexpression of adenosine A_{2A} receptors in the forebrain caused the impairment of working memory in an object recognition task, radial maze task, and the Morris water maze, whereas the genetic inactivation of adenosine A_{2A} receptors enhanced working memory in the radial maze and Morris water maze, with no effect on reference memory (Costenla et al., 2011; Giménez-Llort et al., 2007; O'Neill & Brown, 2007; Wang, Ma, & van den Buuse, 2006; Wei, Li, & Chen, 2011; Zhou et al., 2009). Overall, these findings suggest that adenosine A_{2A} receptors negatively modulate cognitive function, especially working memory.

4. EFFECTS OF ADENOSINE A_{2A} RECEPTOR ANTAGONISTS ON COGNITIVE DYSFUNCTION IN PD

4.1. Animal studies

Early-stage cognitive dysfunction occurs in younger individuals with PD. Changes in cortical DA content may be responsible for this symptom, but the underlying pathophysiology of the cognitive dysfunction is complex and not fully understood (Kehagia et al., 2010). Dopaminergic afferents to the frontal cortex are degenerated in PD, and leading to a reduction of prefrontal DA content (Scatton, Javoy-Agid, Rouquier, Dubois, & Agid, 1983). Perhaps, importantly, there is also a positive correlation between [^{18}F]fluorodopa uptake in the frontal cortex and the performance of cognitive tasks demanding immediate and working memory and executive strategies (Rinne et al., 2000). In experimental models of PD, DA depletion in the cortex has been reported to produce cognitive deficits, particularly those associated with working memory or executive dysfunction (Brozoski et al., 1979; Clinton et al., 2006). In primates treated with a chronic low dose of the selective dopaminergic toxin, 1-methyl-4-phenyl-1,2,3,6-tetrahydropyridine (MPTP), there are also changes in visuospatial and nonspatial working memory that resemble those seen in PD (Decamp & Schneider, 2004). In addition, caffeine reversed the impairing effect of MPTP injected into the rat substantia nigra, and compact part, on the avoidance scores in the training and test sessions of a two-way active avoidance task (Gevaerd, Takahashi, Silveira, & Da Cunha, 2001). Another animal model widely used to investigate anti-PD treatment is reserpine-treated rodents. This model is known as a pharmacological model of PD, based on the effects of the monoamine-depleting agent on motor activity. More recently, it has been reported that a low dose of reserpine can be used to induce cognitive dysfunction in rats. Interestingly, caffeine or the adenosine A_{2A} receptor antagonist ZM241385 reversed the social recognition disruption by reserpine, while the adenosine A_1 receptor antagonist DPCPX did not (Prediger, Da Cunha et al., 2005).

Recently, it has been reported that the adenosine A_{2A} receptor antagonist istradefylline affects cognitive dysfunction because of deficits in cortical dopaminergic transmission (Kadowaki Horita et al., 2013). The main findings of this report are as follows: (1) PFC DA levels and cognitive performance were significantly reduced by 6-hydroxydopamine (6-OHDA) lesioning in rat PFC; (2) istradefylline, donepezil, and methamphetamine improved the cognitive dysfunction of PFC-lesioned rats; and (3) istradefylline increased DA levels in the PFC in normal and PFC-lesioned rats.

Treatment with istradefylline significantly improved the performance of PFC-lesioned rats in the object recognition task, and its effects reached a plateau at 0.1–0.3 mg/kg (Fig. 8.1). The object recognition task is based on spontaneous exploratory behavior of rats toward a novel object, and it is considered to be a "pure working memory task" that has no reference

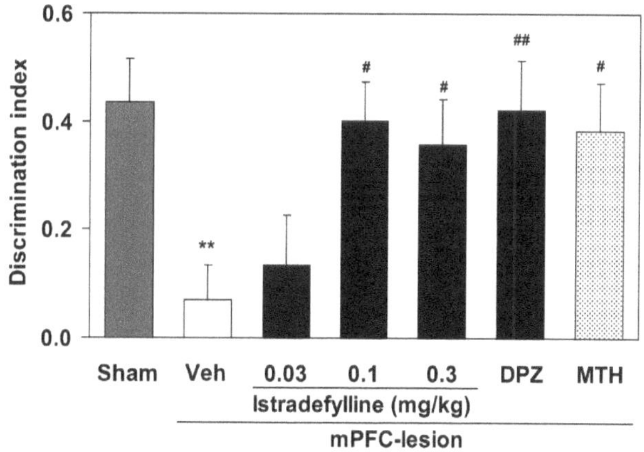

Figure 8.1 Effects of istradefylline, donepezil (DPZ), and methamphetamine (MTH) on cognitive deficits in the object recognition task in mPFC-lesioned rats. Istradefylline (0.03–0.3 mg/kg, po), DPZ (1 mg/kg, po), MTH (3 mg/kg, ip), or a vehicle was administered 60 min before the acquisition trial. Each point represents the mean ± SEM of 7–10 rats. $^{**}p < 0.01$ compared with vehicle-treated sham-operated rats. $^{\#}p < 0.05$, $^{\#\#}p < 0.01$ compared with vehicle-treated mPFC-lesioned rats. *Figure from Kadowaki Horita et al. (2013) with permission.*

memory component, as might result from rule learning and positive or negative reinforcement (Ennaceur & Meliani, 1992). This task has been reported to be sensitive to changes in dopaminergic dysfunction in the PFC but not to hippocampal lesions (Mumby, 2001; Ventura, Pascucci, Catania, Musumeci, & Puglisi-Allegra, 2004). Consistent with previous reports, the dopaminergic lesions of the PFC significantly worsened the performance of the rats in the object recognition task. Performance in this task may have been partially affected by physical and emotional states such as locomotor activity and anxiety. However, the rats' locomotor activity and total exploration time of the novel object were not different between in sham-operated rats and PFC-lesioned rats, and no other abnormal behaviors were observed. Therefore, the decline of discrimination in PFC-lesioned rats is not due to motor dysfunction, but likely due to cognitive dysfunction.

The decreased number of correct responses by PFC-lesioned rats is also reversed by 0.1 mg/kg istradefylline in the delayed alternation task. The delayed alternation task in a T-maze is a standard method for testing spatial working memory, because it requires location awareness related to visual cues around the maze (spatial recognition) and memory retention of the last arm visited (working memory). The former is mainly associated with hippocampal functions and the latter with PFC function (Sánchez-Santed, de

Bruin, Heinsbroek, & Verwer, 1997). Dopaminergic lesions in the PFC, induced by 6-OHDA injection, decreased the correct responses in the delayed alternation task but not in the non-delayed task, as previously reported (Izaki, Maruki, Hori, & Nomura, 2001), indicating a working memory-related cognitive impairment.

Istradefylline has effects on motor function in experimental models of PD with an ED_{50} of 0.05 and 0.26 mg/kg in adenosine A_{2A} receptor agonist CGS21680 and reserpine-induced catalepsy in mice, respectively (Shiozaki et al., 1999). Istradefylline also ameliorates the hypolocomotion (minimum effective dose; 0.16 mg/kg) induced by nigral dopaminergic dysfunction with MPTP or reserpine treatment. In addition, istradefylline also ameliorates motor deficits induced by the nigral injection of 6-OHDA in rats, at doses of more than 0.32 mg/kg (Mihara et al., 2007). Therefore, these results show that istradefylline is effective in reversing working memory impairment related to a loss of PFC dopaminergic input at the same dose levels effective on motor function. Furthermore, this effect of istradefylline would be consistent with that on cognition of PD in humans. These findings suggest that adenosine A_{2A} receptor antagonists, such as istradefylline, may not only be useful therapeutic agents for the treatment of the motor symptoms of PD, but these substances may also be effective in the treatment of both early- and late-stage declines in cognitive function that affect most patients with PD.

4.2. Mechanism of action

The question becomes how the adenosine A_{2A} receptor antagonist improves cognition related to PFC deficits. The mesocortical dopaminergic pathway is critically involved in cognitive function, including working memory, and reduction of DA transmission in the PFC underlies age-related working memory dysfunction (Mizoguchi, Shoji, Tanaka, Maruyama, & Tabira, 2009). This age-related working memory dysfunction was improved by direct stimulation of the prelimbic cortical region of the PFC using the DA D1 receptor agonist SKF 81297. Importantly, increased levels of DA were found in the PFC during delayed alternation tasks in both primates and rodents (Floresco & Magyar, 2006; Watanabe, Kodama, & Hikosaka, 1997). These findings suggest that improved DA transmission in the PFC leads to ameliorated cognitive dysfunction. Indeed, methamphetamine, a DA reuptake inhibitor and releasing agent, improved cognitive dysfunction in PFC-lesioned rats. The blockade of A_{2A} receptors, due to caffeine or the adenosine A_{2A} receptor antagonist SCH58261, facilitated DA transmission in the PFC (Acquas, Tanda, & Di Chiara, 2002). In PFC-lesioned rats, the prefrontal DA levels decline to about 19% of normal content and 42% of normal extracellular concentrations (Kadowaki Horita et al., 2013). Extracellular DA levels in the PFC in both normal and PFC-lesioned rats were significantly increased by 0.1 mg/kg istradefylline (Fig. 8.2). These findings

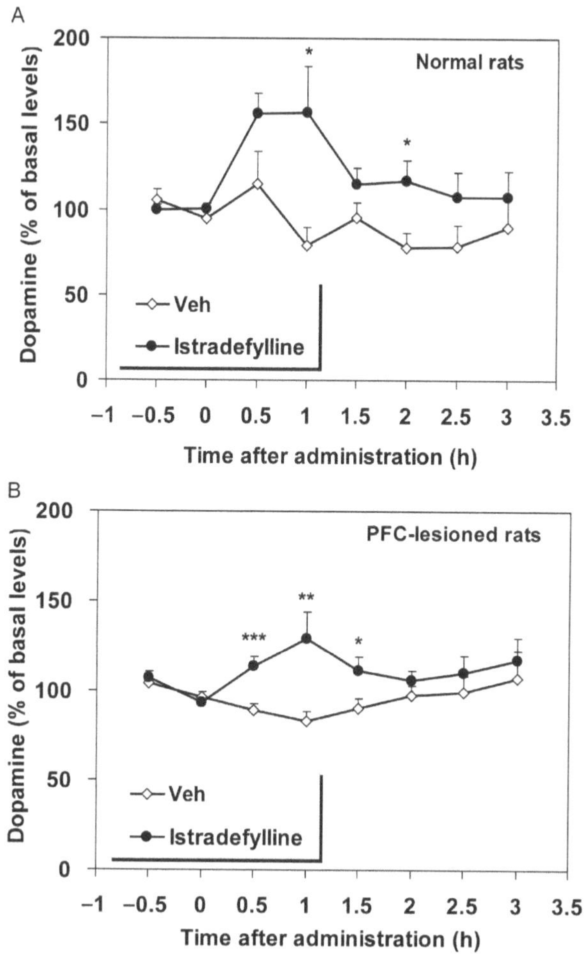

Figure 8.2 Effects of istradefylline on extracellular dopamine levels in the PFC. (A) Effects of istradefylline in normal rats. (B) Effects of istradefylline in PFC-lesioned rats. Istradefylline (0.1 mg/kg) or a vehicle (veh) was administered orally at the time point 0. Dopamine levels are expressed as percentage changes from the basal levels. Each point represents the mean $\pm$ SEM of 7–9 normal rats and 14 lesioned rats. $^{*}p<0.05$; $^{**}p<0.01$; $^{***}p<0.001$ compared with the vehicle-treated group. *Figure from Kadowaki Horita et al. (2013) with permission.*

suggest that a reduction of DA levels in PFC-lesioned rats is responsible for their reduced cognitive performance. This result also strongly supports the concept that cognitive improvement produced by the adenosine A_{2A} receptor antagonist is mediated by the enhancement of PFC DA levels. This conclusion is, in turn, supported by the effects of donepezil, which also ameliorates cognitive dysfunction in PFC-lesioned rats and induces a significant improvement in the object recognition task. Cholinesterase inhibitors have been shown to increase cortical DA levels through both cholinergic-dependent and cholinergic-independent pathways (Liang & Tang, 2006; Shearman et al., 2006). The mechanisms of the adenosine A_{2A} receptor antagonist that contribute to the enhancement of PFC DA release are not fully understood. The local injection of the adenosine A_{2A} receptor agonist CGS21680 into the posterior cingulate cortex impaired social recognition behavior in rats (Pereira et al., 2005), and this result supports the importance of the cortical adenosine A_{2A} receptors in cognitive functions. These findings suggest that an adenosine A_{2A} receptor antagonist such as istradefylline can control DA release by blocking cortical adenosine A_{2A} receptors. In addition, adenosine A_{2A} receptors in the striatum may exert their effect on cortical functions, including working memory, indirectly through a basal ganglia–cortical loop. The neural activity of the striatum markedly affects neuronal activity in the cerebral cortex and alters working memory.

5. CONCLUSION

In this review, we have outlined cognitive dysfunction in PD and the role of adenosine A_{2A} receptors in cognitive dysfunction in PD. It has been shown that DA in the PFC is related to deficits in working memory and executive function in early-stage cognitive dysfunction in PD patients. Moreover, a loss of PFC dopaminergic input impairs working memory performance in animals. This cognitive performance dysfunction in animals can be reversed through the systemic administration of adenosine A_{2A} receptor antagonists, and the reversal appears to result from the facilitation of dopaminergic transmission in the PFC. These findings suggest that an adenosine A_{2A} receptor antagonist, such as istradefylline, may not only be a useful new therapeutic agent for the treatment of the motor symptoms of PD, but it may also be effective in the treatment of both the early- and late-stage declines in cognitive function that affect most patients with PD.

REFERENCES

Aarsland, D., & Kurz, M. W. (2010). The epidemiology of dementia associated with Parkinson disease. *Journal of the Neurological Sciences*, *289*, 18–22.

Acquas, E., Tanda, G., & Di Chiara, G. (2002). Differential effects of caffeine on dopamine and acetylcholine transmission in brain areas of drug-naive and caffeine-pretreated rats. *Neuropsychopharmacology*, *27*, 182–193.

Baba, T., Kikuchi, A., Hirayama, K., Nishio, Y., Hosokai, Y., Kanno, S., et al. (2012). Severe olfactory dysfunction is a prodromal symptom of dementia associated with Parkinson's disease: A 3 year longitudinal study. *Brain*, *135*, 161–169.

Białecka, M., Kurzawski, M., Roszmann, A., Robowski, P., Sitek, E. J., Honczarenko, K., et al. (2014). BDNF G196A (Val66Met) polymorphism associated with cognitive impairment in Parkinson's disease. *Neuroscience Letters*, *561*, 86–90.

Braak, H., Rüb, U., Jansen Steur, E. N., Del Tredici, K., & de Vos, R. A. (2005). Cognitive status correlates with neuropathologic stage in Parkinson disease. *Neurology*, *64*, 1404–1410.

Brozoski, T. J., Brown, R. M., Rosvold, H. E., & Goldman, P. S. (1979). Cognitive deficit caused by regional depletion of dopamine in prefrontal cortex of rhesus monkey. *Science*, *205*, 929–932.

Canas, P. M., Porciúncula, L. O., Cunha, G. M., Silva, C. G., & Machado, N. J. (2009). Adenosine A_{2A} receptor blockade prevents synaptotoxicity and memory dysfunction caused by β-amyloid peptides via p38 mitogen-activated protein kinase pathway. *Journal of Neuroscience*, *29*, 14741–14751.

Christopher, L., Marras, C., Duff-Canning, S., Koshimori, Y., Chen, R., Boileau, I., et al. (2014). Combined insular and striatal dopamine dysfunction are associated with executive deficits in Parkinson's disease with mild cognitive impairment. *Brain*, *137*, 565–575.

Clinton, S. M., Sucharski, I. L., & Finlay, J. M. (2006). Desipramine attenuates working memory impairments induced by partial loss of catecholamines in the rat medial prefrontal cortex. *Psychopharmacology*, *183*, 404–412.

Compta, Y., Parkkinen, L., O'Sullivan, S. S., Vandrovcova, J., Holton, J. L., Collins, C., et al. (2011). Lewy- and Alzheimer-type pathologies in Parkinson's disease dementia: Which is more important? *Brain*, *134*, 1493–1505.

Costenla, A. R., Diógenes, M. J., Canas, P. M., Rodrigues, R. J., Nogueira, C., Maroco, J., et al. (2011). Enhanced role of adenosine A_{2A} receptors in the modulation of LTP in the rat hippocampus upon ageing. *European Journal of Neuroscience*, *34*, 12–21.

Decamp, E., & Schneider, J. S. (2004). Attention and executive function deficits in chronic low dose MPTP-treated non-human primates. *European Journal of Neuroscience*, *20*, 1371–1378.

Duncan, G. W., Firbank, M. J., O'Brien, J. T., & Burn, D. J. (2013). Magnetic resonance imaging: A biomarker for cognitive impairment in Parkinson's disease? *Movement Disorders*, *28*, 425–438.

Edison, P., Ahmed, I., Fan, Z., Hinz, R., Gelosa, G., Ray Chaudhuri, K., et al. (2013). Microglia, amyloid, and glucose metabolism in Parkinson's disease with and without dementia. *Neuropsychopharmacology*, *38*, 938–949.

Emre, M., Aarsland, D., Albanese, A., Byrne, E. J., Deuschl, G., De Deyn, P. P., et al. (2004). Rivastigmine for dementia associated with Parkinson's disease. *New England Journal of Medicine*, *351*, 2509–2518.

Emre, M., Aarsland, D., Brown, R., Burn, D. J., & Duyckaerts, C. (2007). Clinical diagnostic criteria for dementia associated with Parkinson's disease. *Movement Disorders*, *22*, 1689–1707.

Ennaceur, A., & Meliani, K. (1992). A new one-trial test for neurobiological studies of memory in rats. III. Spatial vs. non-spatial working memory. *Behavioural Brain Research*, *51*, 83–92.

Fletcher, P., Leake, A., & Marion, M. H. (2011). Patients with Parkinson's disease dementia stay in the hospital twice as long as those without dementia. *Movement Disorders, 26*, 919.

Floresco, S. B., & Magyar, O. (2006). Mesocortical dopamine modulation of executive functions: Beyond working memory. *Psychopharmacology, 188*, 567–585.

Fraser, C. M., Fisher, A., Cooke, M. J., Thompson, I. D., & Stone, T. W. (1997). Purine modulation of dizocilpine effects on spontaneous alternation. *Psychopharmacology, 130*, 334–342.

Gevaerd, M. S., Takahashi, R. N., Silveira, R., & Da Cunha, C. (2001). Caffeine reverses the memory disruption induced by intra-nigral MPTP-injection in rats. *Brain Research Bulletin, 55*, 101–106.

Giménez-Llort, L., Schiffmann, S. N., Shmidt, T., Canela, L., Camón, L., Wassholm, M., et al. (2007). Working memory deficits in transgenic rats overexpressing human adenosine A_{2A} receptors in the brain. *Neurobiology of Learning and Memory, 87*, 42–56.

Goldman, J. G., & Holden, S. (2014). Treatment of psychosis and dementia in Parkinson's disease. *Current Treatment Options in Neurology, 16*, 281.

Goris, A., Williams-Gray, C. H., Clark, G. R., Foltynie, T., Lewis, S. J., Brown, J., et al. (2007). Tau and α-synuclein in susceptibility to, and dementia in, Parkinson's disease. *Annals of Neurology, 62*, 145–153.

Guerini, F. R., Beghi, E., Riboldazzi, G., Zangaglia, R., Pianezzola, C., Bono, G., et al. (2009). BDNF Val66Met polymorphism is associated with cognitive impairment in Italian patients with Parkinson's disease. *European Journal of Neurology, 16*, 1240–1245.

Guo, J. L., & Lee, V. M. (2014). Cell-to-cell transmission of pathogenic proteins in neurodegenerative diseases. *Nature Medicine, 20*, 130–138.

Hanagasi, H. A., Gurvit, H., Unsalan, P., Horozoglu, H., Tuncer, N., Feyzioglu, A., et al. (2011). The effects of rasagiline on cognitive deficits in Parkinson's disease patients without dementia: A randomized, double-blind, placebo-controlled, multicenter study. *Movement Disorders, 26*, 1851–1858.

Hattori, N., Yabe, I., Hirata, K., Shiga, T., Sakushima, K., Tsuji-Akimoto, S., et al. (2013). Brain regions associated with cognitive impairment in patients with Parkinson disease: Quantitative analysis of cerebral blood flow using ^{123}I iodoamphetamine SPECT. *Clinical Nuclear Medicine, 38*, 315–320.

Hu, M., Cooper, J., Beamish, R., Jones, E., Butterworth, R., Catterall, L., et al. (2011). How well do we recognise non-motor symptoms in a British Parkinson's disease population? *Journal of Neurology, 258*, 1513–1517.

Inzelberg, R., Bonuccelli, U., Schechtman, E., Miniowich, A., Strugatsky, R., Ceravolo, R., et al. (2006). Association between amantadine and the onset of dementia in Parkinson's disease. *Movement Disorders, 21*, 1375–1379.

Irwin, D. J., Lee, V. M., & Trojanowski, J. Q. (2013). Parkinson's disease dementia: Convergence of α-synuclein, tau and amyloid-β pathologies. *Nature Reviews. Neuroscience, 14*, 626–636.

Izaki, Y., Maruki, K., Hori, K., & Nomura, M. (2001). Effects of rat medial prefrontal cortex temporal inactivation on a delayed alternation task. *Neuroscience Letters, 315*, 129–132.

Jones, E. L., Aarsland, D., Londos, E., & Ballard, C. (2012). A pilot study examining associations between DYRK1A and α-synuclein dementias. *Neurodegenerative Diseases, 10*, 229–231.

Kachroo, A., & Schwarzschild, M. A. (2012). Adenosine A_{2A} receptor gene disruption protects in an α-synuclein model of Parkinson's disease. *Annals of Neurology, 71*, 278–282.

Kadowaki Horita, T., Kobayashi, M., Mori, A., Jenner, P., & Kanda, T. (2013). Effects of the adenosine A_{2A} antagonist istradefylline on cognitive performance in rats with a 6-OHDA lesion in prefrontal cortex. *Psychopharmacology, 230*, 345–352.

Kehagia, A. A., Barker, R. A., & Robbins, T. W. (2010). Neuropsychological and clinical heterogeneity of cognitive impairment and dementia in patients with Parkinson's disease. *Lancet Neurology, 9*, 1200–1213.

Kempster, P. A., O'Sullivan, S. S., Holton, J. L., Revesz, T., & Lees, A. J. (2010). Relationships between age and late progression of Parkinson's disease: A clinico-pathological study. *Brain, 133*, 1755–1762.

Klassen, B. T., Hentz, J. G., Shill, H. A., Driver-Dunckley, E., Evidente, V. G., Sabbagh, M. N., et al. (2011). Quantitative EEG as a predictive biomarker for Parkinson disease dementia. *Neurology, 77*, 118–124.

Klein, J. C., Eggers, C., Kalbe, E., Weisenbach, S., Hohmann, C., Vollmar, S., et al. (2010). Neurotransmitter changes in dementia with Lewy bodies and Parkinson disease dementia in vivo. *Neurology, 74*, 885–892.

Levy, G., Tang, M. X., Cote, L. J., Louis, E. D., Alfaro, B., Mejia, H., et al. (2000). Motor impairment in PD: Relationship to incident dementia and age. *Neurology, 55*, 539–544.

Liang, Y. Q., & Tang, X. C. (2006). Comparative studies of huperzine A, donepezil, and rivastigmine on brain acetylcholine, dopamine, norepinephrine, and 5-hydroxytryptamine levels in freely-moving rats. *Acta Pharmacologica Sinica, 27*, 1127–1136.

Litvan, I., Goldman, J. G., Tröster, A. I., Schmand, B. A., Weintraub, D., Petersen, R. C., et al. (2012). Diagnostic criteria for mild cognitive impairment in Parkinson's disease: Movement Disorder Society Task Force guidelines. *Movement Disorders, 27*, 349–356.

Marsh, L., Biglan, K., Gerstenhaber, M., & Williams, J. R. (2009). Atomoxetine for the treatment of executive dysfunction in Parkinson's disease: A pilot open-label study. *Movement Disorders, 24*, 277–282.

Masuda-Suzukake, M., Nonaka, T., Hosokawa, M., Oikawa, T., Arai, T., Akiyama, H., et al. (2013). Prion-like spreading of pathological α-synuclein in brain. *Brain, 136*, 1128–1138.

McKinlay, A., & Grace, R. C. (2011). Characteristic of cognitive decline in Parkinson's disease: A 1-year follow-up. *Applied Neuropsychology, 18*, 269–277.

Mihara, T., Mihara, K., Yarimizu, J., Mitani, Y., Matsuda, R., Yamamoto, H., et al. (2007). Pharmacological characterization of a novel, potent adenosine A_1 and A_{2A} receptor dual antagonist, 5-[5-amino-3-(4-fluorophenyl)pyrazin-2-yl]-1-isopropylpyridine- 2(1H)-one (ASP5854), in models of parkinson's disease and cognition. *Journal of Pharmacology and Experimental Therapeutics, 323*, 708–719.

Mizoguchi, K., Shoji, H., Tanaka, Y., Maruyama, W., & Tabira, T. (2009). Age-related spatial working memory impairment is caused by prefrontal cortical dopaminergic dysfunction in rats. *Neuroscience, 162*, 1192–1201.

Mollenhauer, B., Locascio, J. J., Schulz-Schaeffer, W., Sixel-Döring, F., Trenkwalder, C., & Schlossmacher, M. G. (2011). α-Synuclein and tau concentrations in cerebrospinal fluid of patients presenting with parkinsonism: A cohort study. *Lancet Neurology, 10*, 230–240.

Mumby, D. G. (2001). Perspectives on object-recognition memory following hippocampal damage: Lessons from studies in rats. *Behavioural Brain Research, 127*, 159–181.

Muslimovic, D., Post, B., Speelman, J. D., & Schmand, B. (2005). Cognitive profile of patients with newly diagnosed Parkinson disease. *Neurology, 65*, 1239–1245.

Narayanan, N. S., Rodnitzky, R. L., & Uc, E. Y. (2013). Prefrontal dopamine signaling and cognitive symptoms of Parkinson's disease. *Reviews in the Neurosciences, 24*, 267–278.

O'Neill, M., & Brown, V. J. (2007). The effect of striatal dopamine depletion and the adenosine A_{2A} antagonist KW-6002 on reversal learning in rats. *Neurobiology of Learning and Memory, 88*, 75–81.

Pagonabarraga, J., & Kulisevsky, J. (2012). Cognitive impairment and dementia in Parkinson's disease. *Neurobiology of Disease, 46*, 590–596.

Pellecchia, M. T., Santangelo, G., Picillo, M., Pivonello, R., Longo, K., Pivonello, C., et al. (2013). Serum epidermal growth factor predicts cognitive functions in early, drug-naive Parkinson's disease patients. *Journal of Neurology, 260*, 438–444.

Pereira, G. S., Rossato, J. I., Sarkis, J. J., Cammarota, M., Bonan, C. D., & Izquierdo, I. (2005). Activation of adenosine receptors in the posterior cingulate cortex impairs memory retrieval in the rat. *Neurobiology of Learning and Memory, 83*, 217–223.

Poewe, W. (2008). When a Parkinson's disease patient starts to hallucinate. *Practical Neurology, 8*, 238–241.

Postuma, R. B., Bertrand, J. A., Montplaisir, J., Desjardins, C., Vendette, M., Rios Romenets, S., et al. (2012). Rapid eye movement sleep behavior disorder and risk of dementia in Parkinson's disease: A prospective study. *Movement Disorders, 27*, 720–726.

Prediger, R. D. S., Batista, L. C., & Takahashi, R. N. (2005). Caffeine reverses age-related deficits in olfactory discrimination and social recognition memory in rats: Involvement of adenosine A_1 and A_{2A} receptors. *Neurobiology of Aging, 26*, 957–964.

Prediger, R. D., Da Cunha, C., & Takahashi, R. N. (2005). Antagonistic interaction between adenosine A_{2A} and dopamine D_2 receptors modulates the social recognition memory in reserpine-treated rats. *Behavioral Pharmacology, 16*, 209–218.

Prediger, R. D., Fernandes, D., & Takahashi, R. N. (2005). Blockade of adenosine A_{2A} receptors reverses short-term social memory impairments in spontaneously hypertensive rats. *Behavioural Brain Research, 159*, 197–205.

Richard, I. H., Justus, A. W., Greig, N. H., Marshall, F., & Kurlan, R. (2002). Worsening of motor function and mood in a patient with Parkinson's disease after pharmacologic challenge with oral rivastigmine. *Clinical Neuropharmacology, 25*, 296–299.

Rinne, J. O., Portin, R., Ruottinen, H., Nurmi, E., Bergman, J., Haaparanta, M., et al. (2000). Cognitive impairment and the brain dopaminergic system in Parkinson disease: [^{18}F]fluorodopa positron emission tomographic study. *Archives of Neurology, 57*, 470–475.

Rolinski, M., Fox, C., Maidment, I., & McShane, R. (2012). Cholinesterase inhibitors for dementia with Lewy bodies, Parkinson's disease dementia and cognitive impairment in Parkinson's disease. *Cochrane Database of Systematic Reviews, 3*, CD006504.

Ross, O. A., Braithwaite, A. T., Skipper, L. M., Kachergus, J., Hulihan, M. M., Middleton, F. A., et al. (2008). Genomic investigation of α-synuclein multiplication and parkinsonism. *Annals of Neurology, 63*, 743–750.

Sánchez-Santed, F., de Bruin, J. P., Heinsbroek, R. P., & Verwer, R. W. (1997). Spatial delayed alternation of rats in a T-maze: Effects of neurotoxic lesions of the medial prefrontal cortex and of T-maze rotations. *Behavioural Brain Research, 84*, 73–79.

Scatton, B., Javoy-Agid, F., Rouquier, L., Dubois, B., & Agid, Y. (1983). Reduction of cortical dopamine, noradrenaline, serotonin and their metabolites in Parkinson's disease. *Brain Research, 275*, 321–328.

Schrag, A., Jahanshahi, M., & Quinn, N. (2000). What contributes to quality of life in patients with Parkinson's disease? *Journal of Neurology, Neurosurgery, and Psychiatry, 69*, 308–312.

Seppi, K., Weintraub, D., Coelho, M., Perez-Lloret, S., Fox, S. H., Katzenschlager, R., et al. (2011). The movement disorder society evidence-based medicine review update: Treatments for the non-motor symptoms of Parkinson's disease. *Movement Disorders, 26*, S42–S80.

Shearman, E., Rossi, S., Szasz, B., Juranyi, Z., Fallon, S., Pomara, N., et al. (2006). Changes in cerebral neurotransmitters and metabolites induced by acute donepezil and memantine administrations: A microdialysis study. *Brain Research Bulletin, 69*, 204–213.

Shiozaki, S., Ichikawa, S., Nakamura, J., Kitamura, S., Yamada, K., & Kuwana, Y. (1999). Actions of adenosine A_{2A} receptor antagonist KW-6002 on drug-induced catalepsy and hypokinesia caused by reserpine or MPTP. *Psychopharmacology, 147*, 90–95.

Siderowf, A., Xie, S. X., Hurtig, H., Weintraub, D., Duda, J., Chen-Plotkin, A., et al. (2010). CSF amyloid β 1-42 predicts cognitive decline in Parkinson disease. *Neurology*, *75*, 1055–1061.

Sidransky, E., & Lopez, G. (2012). The link between the GBA gene and parkinsonism. *Lancet Neurology*, *11*, 986–998.

Svenningsson, P., Westman, E., Ballard, C., & Aarsland, D. (2012). Cognitive impairment in patients with Parkinson's disease: Diagnosis, biomarkers, and treatment. *Lancet Neurology*, *11*, 697–707.

Takahashi, R. N., Pamplona, F. A., & Prediger, R. D. S. (2008). Adenosine receptor antagonists for cognitive dysfunction: A review of animal studies. *Frontiers in Bioscience*, *13*, 2614–2632.

Ventura, R., Pascucci, T., Catania, M. V., Musumeci, S. A., & Puglisi-Allegra, S. (2004). Object recognition impairment in Fmr1 knockout mice is reversed by amphetamine: Involvement of dopamine in the medial prefrontal cortex. *Behavioral Pharmacology*, *15*, 433–442.

Vossius, C., Larsen, J. P., Janvin, C., & Aarsland, D. (2011). The economic impact of cognitive impairment in Parkinson's disease. *Movement Disorders*, *26*, 1541–1544.

Wang, J. H., Ma, Y. Y., & van den Buuse, M. (2006). Improved spatial recognition memory in mice lacking adenosine A_{2A} receptors. *Experimental Neurology*, *199*, 438–445.

Watanabe, M., Kodama, T., & Hikosaka, K. (1997). Increase of extracellular dopamine in primate prefrontal cortex during a working memory task. *Journal of Neurophysiology*, *78*, 2795–2798.

Wei, C. J., Li, W., & Chen, J. F. (2011). Normal and abnormal functions of adenosine receptors in the central nervous system revealed by genetic knockout studies. *Biochimica et Biophysica Acta*, *1808*, 1358–1379.

Weintraub, D. (2008). Dopamine and impulse control disorders in Parkinson's disease. *Annals of Neurology*, *64*, S93–S100.

Weintraub, D., Mavandadi, S., Mamikonyan, E., Siderowf, A. D., Duda, J. E., Hurtig, H. I., et al. (2010). Atomoxetine for depression and other neuropsychiatric symptoms in Parkinson disease. *Neurology*, *75*, 448–455.

Williams-Gray, C. H., Evans, J. R., Goris, A., Foltynie, T., Ban, M., Robbins, T. W., et al. (2009). The distinct cognitive syndromes of Parkinson's disease: 5 year follow-up of the CamPaIGN cohort. *Brain*, *132*, 2958–2969.

Williams-Gray, C. H., Goris, A., Saiki, M., Foltynie, T., Compston, D. A., Sawcer, S. J., et al. (2009). Apolipoprotein E genotype as a risk factor for susceptibility to and dementia in Parkinson's disease. *Journal of Neurology*, *256*, 493–498.

Winder-Rhodes, S. E., Evans, J. R., Ban, M., Mason, S. L., Williams-Gray, C. H., Foltynie, T., et al. (2013). Glucocerebrosidase mutations influence the natural history of Parkinson's disease in a community-based incident cohort. *Brain*, *136*, 392–399.

Zhou, S. J., Zhu, M. E., Shu, D., Du, X. P., Song, X. H., Wang, X. T., et al. (2009). Preferential enhancement of working memory in mice lacking adenosine A_{2A} receptors. *Brain Research*, *1303*, 74–83.

CHAPTER NINE

Clinical Nonmotor Aspect of A_{2A} Antagonist in PD Treatment

Masahiro Nomoto[1], Masahiro Nagai, Noriko Nishikawa
Department of Neurology and Clinical Pharmacology, Ehime University Graduate School of Medicine, Shitsukawa Toon, Ehime, Japan
[1]Corresponding author: e-mail address: nomoto@m.ehime-u.ac.jp

Contents

Abstract

The chapter discusses with clinical case study if adenosine A_{2A} receptor antagonist, as a novel therapeutic class for symptomatic improvement of Parkinson's disease (PD), can also be expected to have an improvement of nonmotor symptoms related with PD.

1. INTRODUCTION

Adenosine receptor antagonist, istradefylline is a selective antagonist at the A_{2A} receptor (Kase et al., 2003; Nomoto et al., 2000). It has been found to be useful in the treatment of Parkinson's disease (PD) and has been approved for the treatment of PD (Mizuno, Kondo, & Japanese Istradefylline Study Group, 2013). On clinical trials, the effect of istradefylline was proved to decrease the OFF time in patients with PD, and to improve the akinesia at the ON time. On the other hand, istradefylline improved nonmotor signs of PD. The detail of a few cases showing improvement was described here.

International Review of Neurobiology, Volume 119
ISSN 0074-7742
http://dx.doi.org/10.1016/B978-0-12-801022-8.00009-X

2. CASE REPORT

2.1. Case 1

A 66-year-old lady visited us complaining left leg tremor. The tremor appeared on her sitting or standing, and did not appear on walking. Neurological examination showed rigidity at the neck, arm, and leg. She was diagnosed to have PD and had been treated with anti-parkinsonian agents. At the age of 75 years, she was treated with 650 mg of levodopa with decarboxylase inhibitor (DCI) and could keep her daily life alone at the ON time at home. She, however, needed help of caregivers on coming to the clinic. She was evaluated as grade 3 of Hoehn–Yahr scale at the ON time and grade 4 at the OFF time. She often complained hallucination of image of small kids, which did not cause excitement on her and she could have managed the hallucination. However, she always complained getting tired and easily exhausted in daily life. Physical and blood examinations found nothing particular to cause the fatigue, which caused her depressed. The score of MMSE was 28/30 and she was not demented. She was treated with 20 mg of istradefylline for 12 weeks. Her OFF time was reduced by 1.07 h a day, and her fatigability was improved.

2.2. Case 2

A 60-year-old man was referred to us, because of marked pain at the left leg. He had been treated with anti-parkinsonian agents for the last 3 years. Neurological examination revealed rigidity at the neck and arms, and bradykinesia which were dominant at the left side. Nonsteroidal anti-inflammatory drugs such as loxoprofen or diclofenac were not effective to his pain. X-ray and MRI examination revealed mild deformity of the lumbar vertebrae which did not seem to cause neuralgia. He showed wearing-off in his Parkinsonism and the pain at the left legs was prominent at the OFF time. He was depressive because of the severe pain at the OFF time. The score of MMSE was 30/30 and he was evaluated as grade 2 of Hoehn–Yahr scale at the ON time and grade 2.5 at the OFF time. He was treated with levodopa at the dose of 400 mg with DCI, 5 mg of selegiline, and 1 mg of pramipexole. Twenty milligrams of istradefylline was prescribed, and his OFF time of Parkinsonism decreased markedly by 3.57 h in 12 weeks. His pain of the left legs at the OFF time was also relieved moderately.

2.3. Case 3

A 71-year-old lady referred to us because of chin tremor and bradykinesia. She had been treated with anti-parkinsonian agents for the last 7 years. She was treated with 450 mg of levodopa with DCI and zonisamide at the dose of 25 mg a day. She showed wearing-off in her Parkinsonism and she was evaluated as grade 3 of Hoehn–Yahr scale at the ON time and grade 4 at the OFF time. She could manage her daily life at home; however, she always complained fatigue and cheerless in her life. She was treated with 20 mg of istradefylline at the age of 75 years for 12 weeks. She responded well, and the OFF time reduced by 1.24 h a day and she become cheerful.

3. DISCUSSION

Istradefylline was found to be an adenosine receptor antagonist and especially selective to adenosine A_{2A} receptor. Adenosine receptor antagonists such as caffeine, aminophylline showed stimulation on the central nervous system, and may improve the mood in addition to the motor symptoms. In this case report, all of the three cases, which responded markedly to the treatment of istradefylline, showed marked improvement in the nonmotor signs including depression. In animal model, istradefylline was shown to show antidepressive effects (Yamada, Kobayashi, Mori, Jenner, & Kanda, 2013). Istradefylline may improve mood of patients with PD especially at the time of OFF state.

4. CONCLUSION

Adenosine A_{2A} antagonist, istradefylline is a new anti-parkinsonian drug, which was shown to be effective to improve the wearing-off. Additional more effects such as to improve nonmotor signs should be studied on patients with PD showing wearing-off in their Parkinsonism.

REFERENCES

Kase, H., Aoyama, S., Ichimura, M., Ikeda, K., Ishii, A., Kanda, T., et al. (2003). Progress in pursuit of therapeutic A_{2A} antagonists: The adenosine A_{2A} receptor selective antagonist KW6002: Research and development toward a novel nondopaminergic therapy for Parkinson's disease. *Neurology*, *61*(11 Suppl. 6), S97–S100.

Mizuno, Y., Kondo, T., & Japanese Istradefylline Study Group (2013). Adenosine A_{2A} receptor antagonist istrdefilline reduced daily OFF time in Parkinson's disease. *Movement Disorders*, *28*, 1138–1141.

Nomoto, M., Kaseda, S., Iwata, S., Shimizu, T., Fukuda, T., & Nakagawa, S. (2000). The metabolic rate and vulnerability of dopaminergic neurons, and adenosine dynamics in the cerebral cortex, nucleus accumbens, caudate nucleus, and putamen of the common marmoset. *Journal of Neurology*, *247*(Suppl. 5), V16–V22.

Yamada, K., Kobayashi, M., Mori, A., Jenner, P., & Kanda, T. (2013). Antidepressant like activity of adenosine A_{2A} antagonist istradefylline, in the forced swim test and the tail suspension test in rodents. *Pharmacology, Biochemistry, and Behavior*, *114–115*, 23–30.

CHAPTER TEN

Adenosine Receptors and Huntington's Disease

Chien-fei Lee, Yijuang Chern[1]
Division of Neuroscience, Institute of Biomedical Sciences, Academia Sinica, Taipei, Taiwan
[1]Corresponding author: e-mail address: bmychern@ibms.sinica.edu.tw

Contents

Abstract

Adenosine regulates important pathophysiological functions via four distinct adenosine receptor subtypes (A_1, A_{2A}, A_{2B}, and A_3). The A_1 and A_{2A} adenosine receptors (A_1R and $A_{2A}R$) are major targets of caffeine and have been extensively investigated. Huntington's disease (HD) is a dominant neurodegenerative disease caused by an abnormal CAG expansion in the *Huntingtin* gene. Since the first genetic HD model was created almost two decades ago, tremendous progress regarding the function of the adenosine receptors in HD has been made. Chronic intake of caffeine was recently shown to be positively associated with the disease onset of HD. Moreover, genetic polymorphism of $A_{2A}R$ is believed to impact the age of onset. Given the importance of adenosine receptors as

International Review of Neurobiology, Volume 119
ISSN 0074-7742
http://dx.doi.org/10.1016/B978-0-12-801022-8.00010-6

drug targets for human diseases, this review highlights the recent findings that delineate the roles of adenosine receptors in HD and discusses their potential for serving as drug targets and/or biomarkers for HD.

Adenosine is a purine nucleoside that regulates important physiological functions via four different adenosine receptors (A_1, A_{2A}, A_{2B}, and A_3). These adenosine receptors have seven transmembrane domains and belong to the G protein-coupled receptor family. The functions of the A_1 adenosine receptor (A_1R) and A_{2A} adenosine receptor ($A_{2A}R$) have been investigated relative to HD. In this review, we summarize the recent findings regarding the role of adenosine receptors in HD and discuss the potential application of adenosine receptors as drug targets and biomarkers for HD.

1. PATHOGENETIC MECHANISMS OF HUNTINGTON'S DISEASE

Huntington's disease (HD) is a dominant autosomal degenerative disease caused by the expansion of the CAG repeat in a gene named *Huntingtin* (*HTT*). It is one of the best-studied neurodegenerative diseases, and it exhibits multiple neurological symptoms, including the impairment of motor and cognitive functions, brain atrophy, body weight loss, and shortened life span (Georgiou-Karistianis, Scahill, Tabrizi, Squitieri, & Aylward, 2013; Labbadia & Morimoto, 2013; Reilmann, 2013). A wide variety of animal models of HD, ranging from *C. elegans* to monkeys, are available for exploring the pathogenic mechanisms and potential treatments. We focused on those HD pathogenic pathways that could be modulated by adenosine receptors.

1.1. Mitochondria and metabolic defects

Elevated oxidative damage and reduced cerebral oxidative phosphorylation are common features in mice and humans with HD (Browne, 2008; Powers et al., 2007). Mutant HTT (mHTT) is found on mitochondrial membranes and causes mitochondrial calcium defects and fragmentation, which are hypothesized to contribute to HD pathogenesis (Costa et al., 2010; Panov et al., 2002). Prevention of mitochondrial fission and cristae remodeling has been shown to delay HD progression (Costa et al., 2010; Guo et al., 2013; Song et al., 2011). In addition, mHTT aggregates hinder the mitochondrial transport and movement in neurons, thus further impairing the neuronal energy homeostasis (Chang, Rintoul, Pandipati, & Reynolds, 2006). In line with the above hypothesis, rescue of impaired mitochondria and poor energy homeostasis might lead to the discovery of novel drugs for HD (Guo et al., 2013; Ju et al., 2011; Lin, Cheng, Chang, Chen, & Chern, 2013b; Song et al., 2011).

1.2. Deficiency of the brain-derived neurotrophic factor

Brain-derived neurotrophic factor (BDNF) is a member of the neurotrophin family that is critical for neuronal function and survival. One of the most important functions of normal HTT is to promote the transcription of BDNF by retaining the RE1-silencing transcription factor (REST) in the cytoplasm (Zuccato et al., 2007, 2001, 2003). Cattaneo's group reported that mHTT losses the ability to bind and retain REST in the cytoplasm, and thus disrupts BDNF transcription (Zuccato et al., 2003). In addition, full-length wild-type HTT promotes vesicle transport of BDNF via a molecular motor (Gauthier et al., 2004). The polyQ-expanded mHTT suppresses the transport of BDNF (Gauthier et al., 2004; Wang et al., 2012), impairs neurotrophic support, and causes eventual neuronal atrophy in HD. The reduction of BDNF levels was found in the brains of mice and patients with HD (Zuccato et al., 2008). Several laboratories reported that rescue of the BDNF loss and activation of the BDNF receptor (TrkB) elicit beneficial effects in HD (Gharami, Xie, An, Tonegawa, & Xu, 2008; Jiang et al., 2013; Todd et al., 2014; Xie, Hayden, & Xu, 2010; Zuccato et al., 2008).

1.3. Two major protein degradation systems: Proteasome and autophagy

Suppression of the ubiquitin proteasome system (UPS) by mHTT has been demonstrated in the cells and brains of mice and humans with HD (Seo, Sonntag, & Isacson, 2004; Wang et al., 2008). Global changes in the ubiquitin system, an indicator of the UPS function, were found in HD mouse and patient brains (Bennett et al., 2007). Enhancement of UPS activity, which facilitates the degradation of soluble mHTT at its pathological stage, by expression of PA28γ or several small molecules, have been shown to improve HD symptoms (Jia, Kast, Steffan, & Thomas, 2012; Kim & Seo, 2014; Lin, Chang, et al., 2013a; Lin, Cheng, et al., 2013b; Liu et al., 2014; Seo, Sonntag, Kim, Cattaneo, & Isacson, 2007; Wong et al., 2008).

Alteration in macroautophagy because of an impaired ability of autophagic vacuoles to recognize cytosolic cargo has been demonstrated (Martinez-Vicente et al., 2010). The resultant inferior activity of autophagy causes slower turnover and accumulation of mHTT in HD. In support of the hypothesis that the clearance of mHTT is important, upregulation of autophagy produces beneficial effects (Jia et al., 2012; Koga et al., 2011; Sarkar et al., 2007; Williams et al., 2008).

1.4. Excitotoxicity

Early studies showed that intrastriatal injection of glutamate or quinolinic acid (QA) (an *N*-methyl-D-aspartate (NMDA) glutamate receptor agonist) recapitulates some pathological features of HD (DiFiglia, 1990; Huang et al., 1995). Analyses of HD patients with proton magnetic resonance spectroscopy (^{1}H MRS) revealed an elevation of the striatal glutamate level (Taylor-Robinson et al., 1996, 1994). Alteration in the corticostriatal pathway, the activity of NMDA receptors, and the composition of NMDA receptors were found in HD mouse models (Cepeda, Wu, Andre, Cummings, & Levine, 2007; Fan & Raymond, 2007; Marco et al., 2013). In addition, neurons expressing NMDA receptors are prone to degeneration in the brain of HD patients (Fan & Raymond, 2007; Marco et al., 2013). Excitotoxicity thus is hypothesized to be an important pathogenic mechanism of HD. Studies from genetic mouse models suggest that susceptibility to excitotoxic stress is dynamically regulated during HD progression and dominates at the early HD stage. During HD progression, the NMDA receptor-mediated membrane current is gradually reduced. Symptomatic stage HD mice subsequently become more resistant to excitotoxic stress. Treatments aimed at the reduction of excitotoxicity might be most beneficial for presymptomatic HD carriers (Graham et al., 2009).

1.5. Nonneuronal and peripheral cells

In the brain, mHTT is expressed in neurons and glia (Hsiao & Chern, 2010). Expression of mHTT in astrocytes clearly contributes to HD pathogenesis (Bradford et al., 2009, 2010; Chou et al., 2005; Hsiao, Chen, Chen, Tu, & Chern, 2013). In astrocytes, mHTT alters several major astrocytic functions as follows: impaired glycolysis (Powers et al., 2007), lower expression of GLT-1, which causes lower glutamate uptake (Chen et al., 2012; Shin et al., 2005), greater glutamate synthesis (Lee, Reyes, et al., 2013), inferior GABA release (Wojtowicz, Dvorzhak, Semtner, & Grantyn, 2013), insufficient production and release of trophic factors (Chou et al., 2008; Wang et al., 2012), and higher inflammatory responses (Hsiao et al., 2013, 2014). Similar to the mechanism of other neurodegenerative diseases, microglia play a critical role in HD pathogenesis. Abnormal functions of microglia have been implicated in enhanced inflammatory response (Crotti et al., 2014; Hsiao et al., 2013).

Mutant HTT is also expressed in peripheral cells and altered normal physiology. Specifically, mHTT is expressed in hepatocytes, suppresses the urea cycle activity, caused high blood ammonia (Chiang et al., 2009; Chiu, Mackay, & Bhathal, 1975). The immune system is another important peripheral organ that expresses mHTT. It has been noted that enhanced immune activation in HD mice and patients could be detected in the early stage of HD (Bjorkqvist et al., 2008). It has been proposed that the expression level of mHTT in immune cells might be used as a noninvasive disease biomarker (Weiss et al., 2012).

2. ALTERATION IN ADENOSINE HOMEOSTASIS IN HD

Adenosine is a regulator of the homeostatic bioenergetic network. Endogenous adenosine has been implicated in neuronal plasticity (Fontinha, Diogenes, Ribeiro, & Sebastiao, 2008). The brain adenosine homeostasis is tightly controlled by multiple proteins (Chen, Eltzschig, & Fredholm, 2013; Fredholm, 2013). The major pathway that is responsible for the production of extracellular adenosine is mediated by the membrane associated ecto-5′-nucleotidases (i.e., CD39 and CD73; Fig. 10.1) that transform the extracellular nucleotides into adenosine (Atkinson, Dwyer, Enjyoji, & Robson, 2006; Bulavina et al., 2013; Kulesskaya et al., 2013; Yegutkin, 2008). The sources of ATP include trauma-induced membrane leakage, stretched membranes, exocytosis, and connexin hemichannels (Chen et al., 2013; Eltzschig et al., 2006; Fredholm, 2013; Wall & Dale, 2013; Yegutkin, 2008). In the brain, glia cells play a major role in controlling the brain adenosine tone by producing adenosine from ATP via extracellular ectonucleotidases (Boison, Chen, & Fredholm, 2010). The dominant forms of CD39, which convert ATP to ADP/adenosine monophosphate (AMP) in astrocytes and microglia, are CD39-L1 and CD39, respectively (Bulavina et al., 2013; Farber et al., 2008; Wink et al., 2006). CD73, which hydrolyzes AMP to adenosine, is predominantly located in the basal ganglia core comprised of striatum and globus pallidus (Kulesskaya et al., 2013), the brain areas that are most severely affected in HD. The contribution of these ecto-5′-nucleotidases to HD pathogenesis is currently unknown.

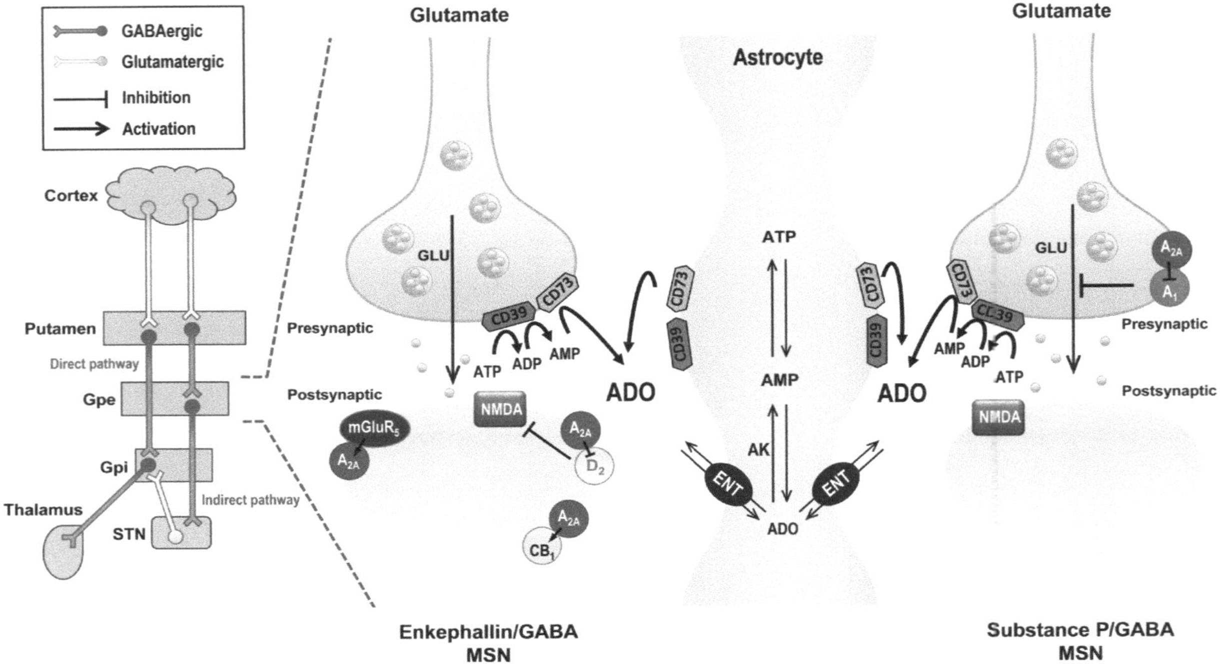

Figure 10.1 A schematic representation showing the localization of A_1 adenosine receptor (A_1R) and A_{2A} adenosine receptor ($A_{2A}R$) in the striatum, the most affected brain area in Huntington's disease (HD). ADO, adenosine; AK, adenosine kinase; CB1, endocannabinoid-CB1 receptor; D2 dopamine receptor; ENT, equilibrative nucleoside transporters; GPe, external globus pallidus; GPi, internal globus pallidus; GLU, glutamate; MSN, medium spiny neuron; NMDA, the *N*-methyl-D-aspartate receptor; STN, subthalamic nucleus.

The major route for the removal of extracellular adenosine is mediated by the uptake of adenosine into cells via equilibrative nucleoside transporters (ENTs) and the concentrative nucleotide transporter (CNT) (Li, Gu, Hertz, & Peng, 2013; Young, Yao, Baldwin, Cass, & Baldwin, 2013). In the brain, these adenosine transporters are distributed ubiquitously in neurons and glia (Li et al., 2013; Li, Quan Lan, Fredholm, Simon, & Boison, 2007). The efficiency of adenosine uptake is controlled at the intracellular adenosine level. The cytosolic adenosine kinase (AK), which converts adenosine into AMP and subsequently reduces the intracellular adenosine level, is critical in maintaining the intracellular and extracellular adenosine levels (Boison, 2006). Thus, AK is hypothesized to attribute to the endogenous adenosine-mediated homeostatic control of synaptic activity (Diogenes et al., 2014).

The adenosine homeostasis is impaired in brains affected by HD. In a transgenic mouse model (R6/1) of HD, the adenosine level in the cerebrospinal fluid of the HD mice is higher than that of the control littermates in the middle stage of HD (Gianfriddo, Melani, Turchi, Giovannini, & Pedata, 2004). It is unknown whether this elevated adenosine tone is associated with disease progression in HD mice. The underlying mechanism is largely uncharacterized. There is no information regarding the level of adenosine tone in human brains affected by HD during disease progression. A microarray analysis of the prefrontal cortex of 157 HD patients showed a significant reduction in the transcript of CD73, suggesting that the synaptic adenosine level converted from nucleotides might be low. The transcript levels of ENTs and AKs in HD patients are higher by one- to threefold and two- to sixfold, respectively, than the levels in non-HD subjects ($n=155$). If shown to be true, the latter observation suggests that the extracellular adenosine concentration in human brains affected by HD would be further reduced by an increased adenosine uptake and lower intracellular adenosine level. An understanding of the regulation of the adenosine modulating enzymes in HD is in its infancy and requires further investigation. This is crucial because a study in a transgenic mouse model (R6/2) showed that an enhancement of the brain adenosine tone by an ENT1 inhibitor is associated with significant improvements of HD symptoms (Huang, Lin, et al., 2011). An adenosine augmentation approach has been used to treat epileptic diseases (Boison, 2013; Williams-Karnesky et al., 2013), and seizures are commonly observed in juvenile HD (Cloud et al., 2012); thus, the brain adenosine tone might emerge as a new therapeutic target of HD. Future studies to explore whether adenosine homeostasis is disrupted during HD progression would determine whether the adenosine augmentation therapy might be applied to HD, as is recommended for epileptic diseases.

3. REGULATION OF ADENOSINE RECEPTORS IN HD

3.1. Biphasic regulation of A_1R during HD progression

Adenosine regulates various physiological functions via four distinct adenosine receptor subtypes (A_1, A_{2A}, A_{2B}, and A_3) (Chen et al., 2013; Jacobson, Balasubramanian, Deflorian, & Gao, 2012). As detailed in the Chapter 1 entitled "Adenosine Receptor Neurobiology: Overview" of this book, adenosine receptors belong to the G protein-coupled receptor (GPCR) family and function through G protein-dependent and/or –independent pathways. The two major adenosine receptors in the brain are A_1R and $A_{2A}R$. A_1R is widely distributed in multiple brain areas including the hippocampus, caudatus, globus pallidus, and cortex. $A_{2A}R$ is enriched in the putamen, nucleus caudatus, nucleus accumbens, and globus pallidus pars lateralis (Svenningsson, Hall, Sedvall, & Fredholm, 1997). A_1R and $A_{2A}R$ have been considered as drug targets for HD. Their modes of action are discussed in the following section. To evaluate the potential of A_1R and $A_{2A}R$ as drug targets and biomarkers for HD, tremendous efforts have been devoted to characterizing the alteration in the expression and the underlying mechanism(s) during HD progression (Table 10.1).

Using a noninvasive PET imaging method, Matusch et al. recently demonstrated a biphasic regulation of the A_1AR protein in the caudate of HD patients. In gene carriers who are at least 10 years before the calculated onset (preHD-A), a slight increase of the A_1R protein was observed. As the disease progresses, the level of A_1R was gradually reduced and became ~20% lower than that of the non-HD subjects after the disease onset (Matusch et al., 2014). A similar level of A_1R reduction was previously demonstrated in the brains of HD patients by a receptor-binding assay (Whitehouse et al., 1985). This dynamic regulation of cerebral A_1R in HD patients occurs before the disease onset and might be used as a noninvasive early predictor for HD. The underlying mechanism and the functional consequence of the changes in A_1R during HD progression are largely unknown. In a transgenic rat model of HD, no change in A_1R was detected (Bauer et al., 2005).

Table 10.1 Regulation of adenosine receptors in Huntington's disease

Disease model	Age	Disease stage	Brain area(s)	Adenosine receptor	Expression level (% of control) (*, protein; #, transcript)	Mechanism	References
Patients	~60 yr	Symptomatic[a]	Corpus striatum	A_1	Decrease* (40%)	n.d.	Whitehouse et al. (1985)
Patients and gene carriers	49.6 ± 16 yr	Presymptomatic and symptomatic	Caudate	A_1	PreHD-A gene carriers: Increase* (120%) HD patients*: Decrease* (75 %)	n.d.	Matusch et al. (2014)
HD rat	24 mo	Symptomatic	Caudate-putamen, nucleus accumbens	A_1	n.c.*	n.d.	Bauer et al. (2005)
Patients	63.2 ± 2.8 yr	Symptomatic (Grade 0–3)	Basal ganglion	A_{2A}	Decrease* (0–35%)	n.d.	Glass, Dragunow, and Faull (2000)
Patients	65.5 ± 5.7 yr	Symptomatic (Chorea grade 3)	Basal ganglion	A_{2A}	Decrease* (26–76%)	n.d.	Martinez-Mir, Probst, and Palacios (1991)

Continued

Table 10.1 Regulation of adenosine receptors in Huntington's disease—cont'd

Disease model	Age	Disease stage	Brain area(s)	Adenosine receptor	Expression level (% of control) (*, protein; #, transcript)	Mechanism	References
Patients	58.1 ± 2.7 yr	Symptomatic (Chorea grade 1–4)	Putamen	A_{2A}	Decrease*	Altered DNA methylation pattern in the 5′UTF of *ADORA2A*	Villar-Menendez et al. (2013)
HD mice (R6/2)	4–12 wk	Presymptomatic and symptomatic	Striatum	A_{2A}	Presymptomatic: Decrease*,# (~75%) Symptomatic: Decrease* (~20%)	n.d.	Cha et al. (1999)
HD mice (R6/1)	15–21 wk	Symptomatic	Striatum	A_{2A}	Decrease* (~10%)	n.d.	Cha et al. (1999)
HD mice (R6/5)	9 mo	Symptomatic	Striatum	A_{2A}	Decrease* (~20%)	n.d.	Cha et al. (1999)
HD mice (R6/2)	9 wk	Symptomatic	Striatum	A_{2A}	Decrease* (~20%)	n.d.	Chou et al. (2005)
HD mice (R6/2)	1–2 wk	Presymptomatic	Striatum	A_{2A}	n.c.*	n.d.	Tarditi et al. (2006)
HD mice (R6/2)	3 wk	Presymptomatic	Striatum	A_{2A}	Decrease* (~50%)	n.d.	Tarditi et al. (2006)

HD mice (R6/1)	12–20 wk	Symptomatic	Striatum	A_{2A}	Decrease* (~50% to ~20%)	Altered DNA methylation pattern in the 5′UTF of *ADORA2A*	Villar-Menendez et al. (2013)
HD mice (R6/2)	12 wk	Symptomatic	Striatum	A_{2A}	Decrease* (~25% to ~20%)	5 hmC level was decreased in 5′UTR of *Adora2a* gene in R6/2	Villar-Menendez et al. (2013)
HD mice (YAC72)	12 mo	Symptomatic	Striatum	A_{2A}	n.c.#	Transcriptional dysregulation mediated by the short N-terminal fragment of mHTT	Chan et al. (2002)
HD mice (HD46)	10–13 mo	Symptomatic	Striatum	A_{2A}	n.c.#	Transcriptional dysregulation mediated by the short N-terminal fragment of mHTT	Chan et al. (2002)
HD mice (HD100)	10–13 mo	Symptomatic	Striatum	A_{2A}	Decrease# (84.8%)	Transcriptional dysregulation mediated by the short N-terminal fragment of mHTT	Chan et al. (2002)
HD mice (N171-82Q)	6 wk	Presymptomatic	Striatum	A_{2A}	Decrease# (~65%)	n.d.	Mievis et al. (2011)

Continued

Table 10.1 Regulation of adenosine receptors in Huntington's disease—cont'd

Disease model	Age	Disease stage	Brain area(s)	Adenosine receptor	Expression level (% of control) (*, protein; #, transcript)	Mechanism	References
HD rat	24 mo	Symptomatic	Caudate-putamen, nucleus accumbens	A_{2A}	Decrease* (40%)	n.d.	Bauer et al. (2005)
HD cells	n.a.	n.d.	n.d.	A_{2A}	Decrease* (20–55%)	mHTT suppress the A_{2A}AR promoter via the PKA/CREB pathway	Chiang, Lee, Huang and Chern (2005)

[a]Vonsattel's classification.

HD, Huntington's disease; n.a., not applicable; n.d., not determined; n.c., no change; hr, hour; wk, week; mo, month; yr, year; CN, caudate nucleus; PU, putamen; Gpe, globus pallidus pars externa; preHD-A gene carriers, >10 years to the calculated onset; preHD-A gene carriers, >10 years to the calculated onset; Presymptomatic: before the onset of motor dysfunction. Symptomatic: after the onset of motor dysfunction.

Because of the lack of information from other genetic HD models, it is unknown whether such discrepancy resulted from the species differences or distinct disease stages.

3.2. Downregulation of $A_{2A}R$ during HD progression

$A_{2A}R$ is highly enriched in striatal medium spiny neurons (MSNs; Fig. 10.1). At least one SNP (rs5751876) in the $A_{2A}R$ gene, which is associated with habitual caffeine consumption (Cornelis, El-Sohemy, & Campos, 2007; Rogers et al., 2010), is tightly associated with HD disease onset (Dhaenens et al., 2009; Taherzadeh-Fard, Saft, Wieczorek, Epplen, & Arning, 2010). Downregulation of $A_{2A}R$ at the protein and transcript levels could be detected even before the onset of motor functions in HD patients and most HD mice that were tested (Table 10.1). Because the expression of polyQ-expanded mHTT in mouse and cell models results in lower $A_{2A}R$ transcript levels, the suppression of $A_{2A}R$ by mHTT appears to occur at the transcript level (Chiang, Lee, Huang, & Chern, 2005; Villar-Menendez et al., 2013). A promoter analysis suggests that mHTT suppresses the $A_{2A}R$ core promoter by interfering with the CREB binding (Chiang et al., 2005). By comparing the expression patterns of HD patients and several different genetic mouse models, it was proposed that the short N-terminal fragment (171 amino acids or shorter) of mHTT is the major contributor of mHTT-mediated transcriptional dysfunction and that $A_{2A}R$ is an early readout of this transcriptional abnormality. The length of the mHTT protein might determine the extent of transcriptional dysregulation, which explains the observation that the levels of $A_{2A}R$ and several other mHTT-sensitive genes (e.g., DARPP-32 and preproenkephalin) were not significantly altered in HD mouse models that express an extended N-terminal fragment (946 amino acids) or a full-length (3144 amino acids) polyQ-expanded mHTT (Chan et al., 2002).

In addition to CREB, expression of $A_{2A}R$ gene could be regulated by two transcriptional factors (NF1 and YY1) and DNA methylation (Buira, Albasanz, et al., 2010a; Buira, Dentesano, et al., 2010b; Lee, Lai, Sun, Chien, & Chern, 2003). The importance of the latter was further demonstrated by a recent study which showed that an increase in 5 mC levels and a reduction in 5 hmC levels occur in the 5′ untranslated region (5′UTR) of $A_{2A}R$ gene and are closely associated with the downregulation of the $A_{2A}R$ transcript. The altered DNA methylation pattern in the $A_{2A}R$ gene is likely to contribute to the dysregulation of $A_{2A}R$ by mHTT

(Villar-Menendez et al., 2013). Given that the $A_{2A}R$ gene is a drug target for HD (Chen et al., 2013; Popoli, Blum, Domenici, Burnouf, & Chern, 2008), further investigation on the possible DNA methylation-dependent suppression of the $A_{2A}R$ gene might lead to the development of novel treatments (such as inhibitors of DNA methyltransferase) that modulate $A_{2A}R$ expression and facilitate the therapeutic efficacy of $A_{2A}R$ drugs.

3.3. Amplification of $A_{2A}R$ signaling during HD progression

Although the expression of $A_{2A}R$ is downregulated in the brain in HD (Table 10.1), the affinity and signaling of $A_{2A}R$ are altered in the cells expressing mHTT. Cattaneo's group reported many years ago that polyQ-expanded mHTT abnormally enhanced the $A_{2A}R$-stimulated adenylyl cyclase activity in striatal-derived cells (Varani et al., 2001). Consistent with this hypothesis, a transient increase in the $A_{2A}R$ density and cAMP signaling in the striatum of HD (R6/2) mice before the onset of the motor degeneration and suppression of the $A_{2A}R$ gene were observed, suggesting that a posttranscriptional compensation of the $A_{2A}R$ signal might occur at the presymptomatic stage of HD (Tarditi et al., 2006). Although the level of striatal $A_{2A}R$ protein was greatly reduced in HD (R6/2) mice at the symptomatic stage, its ability to increase the striatal cAMP signaling is similar to that of the WT mice (Chou et al., 2005). $A_{2A}R$ thus remains functional during HD progression, at least in HD mice. Increased $A_{2A}R$ receptor density in the blood platelets of HD patients has been reported, which positively correlates with the disease onset (Maglione et al., 2006, 2005).

Because the activation of Gαs by $A_{2A}R$ is sensitive to membrane cholesterol content (Charalambous et al., 2008; Varani et al., 2007), the altered cholesterol biosynthetic pathway and the resultant abnormal cholesterol accumulation in HD might be a possible cause for the amplification of $A_{2A}R$ signaling (del Toro et al., 2010; Valenza, Carroll, et al., 2007a; Valenza, Leoni, et al., 2007b; Valenza et al., 2005). Another fascinating possibility is that HTT binds to SAP102, a scaffold protein that physically interacts with the C-terminus of $A_{2A}R$ and enhances the existence of $A_{2A}R$ in the membrane area with low mobility upon stimulation (Fan, Cowan, Zhang, Hayden, & Raymond, 2009; Thurner et al., 2014). Whether the diffusion property of $A_{2A}R$ in HD MSN is altered and accounts for the amplification of $A_{2A}R$ signaling is worth of further investigation.

4. THERAPEUTIC ACTIONS OF ADENOSINE RECEPTOR AGONISTS AND ANTAGONISTS IN HD

4.1. A_1R activation and $A_{2A}R$ blockage are beneficial in chemical- and lesion-induced HD models

Compelling evidence suggests that the inactivation of $A_{2A}R$ by genetic or pharmacological approaches ameliorates the striatal damage evoked by a mitochondrial toxin 3-nitropropionic acid (3-NP), an NMDA receptor agonist (QA), or a mitochondrial complex II inhibitor (malonate) (Table 10.2). The QA- and 3-NP lesion models recapitulate the direct and indirect excitotoxicity of HD, respectively, and trigger the loss of MSN neurons (Alston, Mela, & Bright, 1977; Brouillet, Jacquard, Bizat, & Blum, 2005; Jacobson et al., 2012). Similar to 3-NP, malonate is an inhibitor of succinate dehydrogenase. Intrastriatal injection of malonate has also being used to create a lesion model of HD (Messam, Greene, Greenamyre, & Robinson, 1995). Several $A_{2A}R$ antagonists have been shown to elicit multiple beneficial effects in these chemical- and lesion-induced HD models by reducing the striatal atrophy/degeneration, lowering the glutamate outflow, improving the loss of the GABA content, and increasing the life span. Studies from several laboratories reveal that $A_{2A}R$ functions in a relatively complex manner in the brain. Treatment with 3-NP of different dosages and protocols could lead to opposite effects on the striatum of global $A_{2A}R$ knock-out mice (Blum et al., 2003; Fink et al., 2004; Huang et al., 2006), suggesting the potential involvement of multiple cell types. Using tissue-specific knock-out mice, Huang et al. (2006) showed that $A_{2A}R$ on forebrain neurons does not affect 3-NP-evoked striatal damage. This finding argues against the importance of $A_{2A}R$-mediated glutamate release in 3-NP-induced striatal damage (Huang et al., 2006). Instead, the selective removal of $A_{2A}R$ on bone marrow-derived cells recapitulates the enhanced 3-NP-evoked striatal damage in global $A_{2A}R$ knock-out mice. The possible role of $A_{2A}R$ in controlling nonneuronal cells (e.g., glia) might contribute to the function of $A_{2A}R$ in the brain, which requires further evaluation. An $A_{2A}R$ agonist was shown to increase the 3-NP-induced striatal lesion size (Blum et al., 2003). Inactivation of $A_{2A}R$ appears to be beneficial in the chemical- and lesion-induced HD models.

Few studies on the role of A_1R in HD have been conducted. These studies were conducted in chemical HD models. The results from independent laboratories suggest that activation of A_1R provided beneficial effects,

Table 10.2 The roles of adenosine receptors in chemical- and lesion-induced Huntington's disease (HD) models

Chemical/ injection route	Dosage (# administration and treatment period)	Species	Adenosin receptor	Agonism or antagonism	Compound	Pathophysiological consequence	References
3-NP/i.p.	210 mg/kg (one injection/acute)	Mice	A_1	Agonist	R-PIA 2-CADO	↓ The latency of seizure ↓ Mortality ↑ Life span	Zuchora et al. (2001)
3-NP/mini-pump	56 mg/kg/day (5 d)	Rat	A_1	Agonist	AD	↑ Behavior score ↓ Body weight loss ↓ Lesion size ↓ Neuronal cell death	Blum et al. (2002)
Malonate/ intrastriatal	3–6 μmol (one injection/acute)	Mice Rat	A_1	Antagonist	CPX	↑ Striatal GABA content loss	Alfinito et al. (2003)
QA/intrastriatal	300 nmol (one injection/acute)	Rat	A_{2A}	Antagonist	DMPX	↓ The QA-induced EEG abnormality	Reggio, Pezzola, and Popoli (1999)
QA/intrastriatal	300 μmol (one injection/acute)	Rat	A_{2A}	Antagonist	SCH58261	↓ The QA-induced motor hyperactivity and EEG voltage reduction	Popoli et al. (2002)
QA/intrastriatal	5 mM (probe perfusion)	Rat	A_{2A}	Antagonist	ZM241385	↓ QA-induced glutamate outflow ↑ NMDA-induced toxicity	Tebano et al. (2004)
QA/intrastriatal	300 μmol/1 μl (one injection/acute)	Rat	A_{2A}	Antagonist	SCH58261	↓ Alteration of wall-rearing behavior and anxiety behavior	Scattoni et al. (2007)

QA/intrastriatal	5 mM (probe perfusion)	Rat	A_{2A}	Antagonist	ST1535	↓ The QA-induced glutamate outflow	Galluzzo et al. (2008)
3-NP/mini-pump	56 mg/kg/day (5 d)	Rat	A_{2A}	Antagonist	MSX-3	Low dose of MSX-3 (1 mg/kg) ↓ Lesion size High dose of MSX-3 (5 mg/kg) ↑ Spontaneous locomotion ↑ Lesion size	Blum et al. (2003)
3-NP/i.p.	50 mg/kg (twice a day for 5 d)	Mice	A_{2A}	Antagonist	CSC	↓ Striatal damage	Fink et al. (2004)
Malonate/intrastriatal	3–6 μmol (one injection/acute)	Mice Rat	A_{2A}	Antagonist	DMPX	↓ Loss of striatal GABA content	Alfinito et al. (2003)
3-NP/i.p.	70 mg/kg (twice a day); mild: 10 h interval, analysis at day 4.5 severe: 2 h interval, analysis at day 3	Mice A_{2A}R KO (global)	A_{2A}	Antagonism (knockout)	None	Mild protocol/A_{2A}R KO (global)[a] ↑ Weight loss ↑ Striatal degeneration Severe protocol/A_{2A}R KO[a] ↓ Striatal lesion	Blum et al. (2003)
3-NP/i.p.	50 mg/kg (twice a day for 5 d)	Mice A_{2A}R KO (global)	A_{2A}	Antagonism (knockout)	None	A_{2A}R KO (global)[a] ↓ Striatal degeneration	Fink et al. (2004)

Continued

Table 10.2 The roles of adenosine receptors in chemical- and lesion-induced Huntington's disease (HD) models—cont'd

Chemical/ injection route	Dosage (# administration and treatment period)	Species	Adenosin receptor	Agonism or antagonism	Compound	Pathophysiological consequence	References
3-NP/i.p.	60–80 mg/kg (twice a day for 2d)	Mice $A_{2A}R$ KO (global, forebrain)	A_{2A}	Antagonism (knockout)	n.a.	$A_{2A}R$ KO (global)[a] ↑ Behavior deficiency ↑ Striatal degeneration $A_{2A}R$ KO (forebrain)[a] ◆ Behavior deficiency ◆ Striatal degeneration	Huang et al. (2006)
Malonate/ intrastriatal	3–6 μmol (one injection/acute)	Mice Rat	A_{2A}	Antagonist	DMPX	↓ Loss of striatal GABA content	Alfinito et al. (2003)
QA/i.c.v.	30 nmol (one injection/acute)	Rat	A_{2A}	Agonist	CGS21680	↓ The D-amphetamine-induced locomotor hyperactivity	Popoli, Pezzola, Reggio, Caporali, and Scotti de Carolis (1994)
3-NP/mini-pump	56 mg/kg/day (5 d)	Rat	A_{2A}	Agonist	CGS21680	↑ Striatal lesion size	Blum et al. (2003)

[a]Comparison to WT mice.
n.a., not applicable; QA, quinolinic acid; 3-NP, 3-nitropropionic acid; i.p., intraperitoneal injection; i.c.v., intracerebroventricular administration; KO, knockout; ↑, increase; ↓, decrease; ◆, no change.

whereas blockage of A_1R worsens striatal degeneration (Table 10.2; Alfinito et al., 2003; Blum et al., 2002; Zuchora, Turski, Wielosz, & Urbanska, 2001). Further experiments are required to assess the role of A1R in genetic HD models.

4.2. $A_{2A}R$ activation elicits beneficial effects in genetic HD mouse models

The first genetic mouse model of HD was developed and characterized almost two decades ago (Mangiarini et al., 1996). As described above, mHTT exists in neurons, glia, and many peripheral cell types. The major pathological actions of mHTT include transcriptional dysfunction and transport impairment. Multiple cellular pathways in the cells expressing mHTT have been altered during HD progression. The pathological role of excitotoxicity is regulated in a dynamic pattern. At the presymptomatic stage, HD mice (YAC128) showed enhanced sensitivity to NMDA-mediated apoptosis. As the HD mice aged and the disease progressed, these animals became more resistant to excitotoxic stress (Graham et al., 2009). Activation of $A_{2A}R$ might result in different, if not entirely opposite, effects on WT and HD mice (Martire et al., 2007, 2013). More than 30 genetic mouse models are available from various sources (Lee, Cantle, & Yang, 2013a; Pouladi, Morton, & Hayden, 2013). Some of the findings regarding the functional role of $A_{2A}R$ in genetic HD models are summarized in Table 10.3. The earlier findings regarding the role and regulation of adenosine receptors in HD using the chemical- and lesion-induced HD models might need to be revisited using genetic HD models. This issue is timely because the adenosine receptors are important drug targets for HD as well as other brain disorders (Chen et al., 2013).

The role of $A_{2A}R$ in HD pathogenesis has been extensively investigated using R6/2 mice (Mangiarini et al., 1996), a transgenic mouse model that expresses an ~150 polyQ repeat in the N-terminal fragment and recapitulates a significant portion of the mRNA changes found in HD patients (Cha et al., 1999, 1998; Luthi-Carter et al., 2000). R6/2 mice showed multiple major HD symptoms (e.g., motor impairment, aggregate formation, body weight loss). Similar to the observations in previous studies, intrastriatal injection of an $A_{2A}R$ antagonist (SCH58261) reduced the glutamate out flow (Gianfriddo et al., 2004). A 1-week treatment normalizes the alteration in the emotional response of R6/2 mice and reduces the NMDA-evoked toxicity; however, it exhibited no effect on motor capability (Cipriani et al., 2008; Domenici et al., 2007).

Table 10.3 The roles of adenosine receptors in genetic models of Huntington's disease (HD)

Disease model	Disease stage	Adenosine receptor	Agonism or antagonism (knockout)	Compound (administration route/length)	Pathophysiological consequence	References
HD mice (R6/2)	Presymptomatic	A_{2A}	Antagonist	SCH58261 (daily i.p., 1 wk)	↓ NMDA-induced toxicity ↓ Alteration in emotional response ◆ Rotarod performance	Domenici et al. (2007)
HD mice (R6/2)	Presymptomatic	A_{2A}	Antagonist	SCH58261 (daily i.p., 1 wk)	↓ Reduces BDNF	Potenza et al. (2007)
HD mice (R6/2)	Symptomatic	A_{2A}	Antagonist	SCH58261 (one intrastriatal injection for microdialysis)	↓ Glutamate and adenosine outflow	Gianfriddo et al. (2004)
HD mice (R6/2)	Symptomatic	A_{2A}	Antagonist	SCH58261 (twice a day, 2 wk)	↑ Walking ability (incline plate test) ◆ Rotarod performance	Cipriani et al. (2008)
HD mice (R6/2)	Symptomatic	A_{2A}	Antagonist	SCH58261 (daily i.p., 3 wk)	Remodeling of striatal NMDA receptors	Martire et al. (2010)
HD mice (N171-82Q)/ A_{2A}R-KO	Presymptomatic to symptomatic	A_{2A}	Knockout	n.a.	↓ Survival, ↓ Rotarod performance	Mievis et al. (2011)

HD cells (ST14A/ Q120)	n.a.	A_{2A}	Antagonist	ZM241385	↓ BDNF's protective effect against NMDA toxicity	Martire et al. (2013)
HD mice (R6/2)	Presymptomatic to symptomatic	A_{2A}	Agonist	CGS21680 (daily i.p., 5 wk)	↑ Rotarod performance, ↓ Brain atrophy, ↓ AMPK, ↓ mHTT aggregates.	Chou et al. (2005)
HD mice (R6/2)	Presymptomatic to symptomatic	A_{2A}	Agonist	CGS21680 (daily i.p., 5 wk)	↑ Urea cycle activity, ↓ mHTT aggregates	Chiang et al. (2009)
HD mice (R6/2)	Presymptomatic to symptomatic	A_{2A}	Agonist	CGS (i.p., 5 wk)	↑ BCL2, ↓ AMPK activity	Ju et al. (2011)
HD mice (R6/2)	Presymptomatic to symptomatic	A_{2A}	Agonist	CGS21680 (daily i.p., 5 wk)	↑ Proteasome activity ↓ mHTT aggregates	Lin, Chang, et al. (2013) and Lin, Cheng, et al. (2013)
HD mice (R6/2)	Presymptomatic to symptomatic	A_{2A}	Agonist	T1-11 (drinking water, 5 wk)	↑ Rotarod performance, ↑ Enhanced proteasome activity	Huang, Lin, et al. (2011) and Huang, Yang, et al. (2011)
HD mice (R6/2)	Symptomatic	A_{2A}	Agonist	CGS21680	↑ NMDA toxicity in WT slices, ↓ NMDA toxicity in HD slices	Martire et al. (2007)
HD mice (R6/2)	Symptomatic	A_{2A}	Agonist	CGS21680 (daily i.p., 3 wk)	Remodeling of striatal NMDA receptors	Ferrante et al. (2010)

Continued

Table 10.3 The roles of adenosine receptors in genetic models of Huntington's disease (HD)—cont'd

Disease model	Disease stage	Adenosine receptor	Agonism or antagonism (knockout)	Compound (administration route/length)	Pathophysiological consequence	References
HD mice (R6/2[a])	Symptomatic	A_{2A}	Agonist	CGS21680	↓ EPSCs frequency in WT slices, ↑ EPSCs frequency in HD slices	Cepeda et al. (2010)
HD mice (R6/2)	Symptomatic	A_{2A}	Agonist	CGS21680 (slice perfusion)	↓ NMDA toxicity	Martire et al. (2013)
HD cells (ST14A/Q120)	n.a.	A_{2A}	Agonist	CGS21680	↓ NMDA toxicity	Martire et al. (2013)
HD cells ($STHdh^{Q109}$)	n.a.	A_{2A}	Agonist	CGS21680	↑ BCL2 ↓ AMPK activity ↓ HTT aggregates	Ju et al. (2011)
HD cells (ST14A/mHTTexo1-109Q)	n.a.	A_{2A}	Agonist	CGS21680	↑ Proteasome activity	Lin, Chang, et al. (2013) and Lin, Cheng, et al. (2013)

[a]The number of CAG expansion in this colony is 200–220.

n.a., not applicable; n.d., not determined; n.c., no change; hr, hour; wk, week; mo, month; yr, year; CN, caudate nucleus; PU, putamen; Gpe, globus pallidus pars externa; preHD-A gene carriers, >10 years to the calculated onset; Presymptomatic: before the onset of motor dysfunction. Symptomatic: after the onset of motor dysfunction. ↑, increase; ↓, decrease.

The effect of $A_{2A}R$ blockage in HD mice appears to be complex and could be detrimental because it compromises the function of BDNF (Mievis, Blum, & Ledent, 2011; Fig. 10.2). As discussed above, dysregulated production and transport of BDNF significantly contribute to HD pathogenesis because BDNF protects neurons from mHTT-mediated toxicity. Recent studies have shown that the function of BDNF requires functional $A_{2A}R$. Activation of BDNF receptor (TrkB) can be achieved by agonists of $A_{2A}R$ due to a transactivation of TrkB by $A_{2A}R$ (Lee & Chao, 2001; Wiese et al., 2007). Activation of $A_{2A}R$ triggers the excitatory function of BDNF in a cAMP-dependent pathway (Diogenes, Fernandes, Sebastiao, & Ribeiro, 2004). Tonic activation of $A_{2A}R$ by endogenous adenosine is critical for BDNF to modulate corticostriatal synaptic transmission and sustain a normal BDNF tone (Martire et al., 2013; Potenza et al., 2007). Specifically, the systemic administration of an $A_{2A}R$ antagonist (SCH58261) was shown to reduce the striatal level of BDNF (Potenza et al., 2007). SCH58261

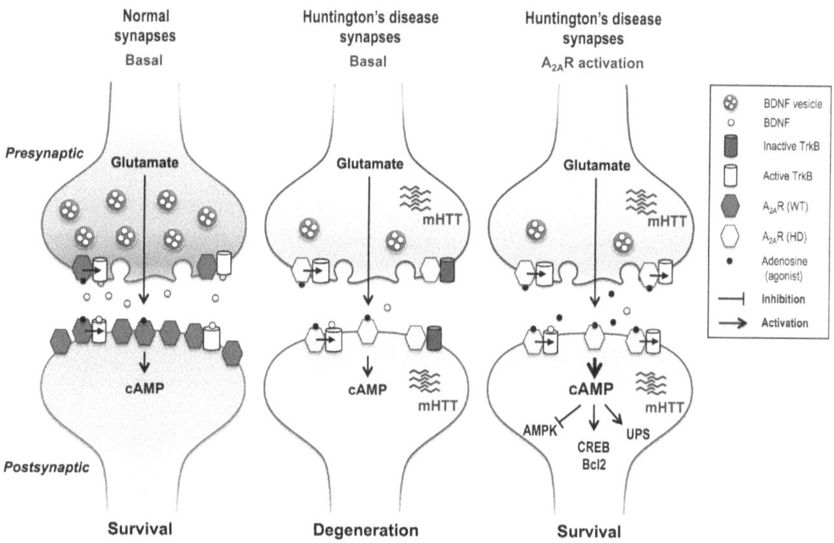

Figure 10.2 A schematic representation showing that the function of brain-derived neurotrophic factor (BDNF) requires functional $A_{2A}R$. Expression of mutant Huntingtin (mHTT) disrupts the production and transport of BDNF. Activation of BDNF receptor (TrkB) can be achieved by agonists of $A_{2A}R$ due to a transactivation of TrkB by $A_{2A}R$. Stimulation of $A_{2A}R$ by endogenous adenosine or exogenous $A_{2A}R$-specific ligands also enhances the cAMP/PKA pathway and protects striatal neurons from the mHTT-evoked toxicity. AMPK, the AMP-activated kinase; UPS, ubiquitin proteasome system.

prevented BDNF-mediated protection against NMDA toxicity in corticostriatal slices prepared from R6/2 mice (Martire et al., 2013). These findings collectively argue against a beneficial role of $A_{2A}R$ antagonists in HD. To consolidate the role of $A_{2A}R$ blockage in HD, Mievis et al. genetically removed $A_{2A}R$ in an HD mouse model (N171-82Q) and found that the absence of $A_{2A}R$ markedly worsens the survival and motor function of HD mice (Mievis et al., 2011), supporting that $A_{2A}R$ blockage might not be an appropriate therapeutic approach for HD. These findings raise an important concern regarding the intake of coffee and other caffeine-containing drinks because $A_{2A}R$ is a major target of caffeine, a nonselective adenosine receptor antagonist. Based on a retrospective analysis of habitual caffeine consumption by 80 HD patients, a recent study suggests that caffeine consumption greater than 190 mg/day is positively associated with an earlier disease onset (Simonin et al., 2013).

It is of great interest to assess whether activation of $A_{2A}R$ might lead to beneficial effects on HD progression. In corticostriatal slices prepared from R6/2 mice, activation of $A_{2A}R$ by CGS21680 prevented NMDA toxicity (Martire et al., 2013). This protective effect of CGS21680 does not require activation of TrkB (a BDNF receptor), whereas that of BDNF depends on $A_{2A}R$ activation (Martire et al., 2013). Activation of $A_{2A}R$ by chronic adenosine tone in the brain is critical for maintaining the proper function of BDNF in HD. Chronic administration of two different $A_{2A}R$ agonists (CGS21680 and T1-11) in R6/2 mice from the presymptomatic to the symptomatic stage leads to improvement in several major HD symptoms including impaired motor function and striatal atrophy. As summarized in Fig. 10.2, activation of $A_{2A}R$ triggers at least three cAMP-dependent protective mechanisms as follows: (i) enhancing the expression of prosurvival genes (e.g., Bcl2), (ii) reducing mHTT by enhancing proteasome activity, and (iii) suppressing the abnormal activation of an energy sensor (AMPK-α1) in the striatal neurons (Chiang et al., 2009; Chou et al., 2005; Huang, Lin, et al., 2011a; Huang, Yang, et al., 2011b; Ju et al., 2011; Lin, Chang, et al., 2013a). In line with the importance of the cAMP-mediated pathway, blockage of a striatum-enriched PDE (PDE10) which enhances the striatal cAMP level also improves several HD symptoms including the striatal atrophy and the formation of mHTT aggregate (Giampa et al., 2010). Given that the electrophysiological properties of MSN in HD mice are dynamically regulated and that excitotoxicity might contribute predominantly in the presymptomatic stage (Graham et al., 2009), a therapeutic window for $A_{2A}R$ agonists might exist in the symptomatic stage. Further investigations are required to evaluate and optimize the potential clinical application of $A_{2A}R$ agonists in HD treatment.

4.3. $A_{2A}R$ receptor dimers as novel drug targets for HD

Accumulating evidence suggests that GPCRs form dimers and, by doing do, create additional levels of function and regulation (for a review, see Ferre et al., 2014). $A_{2A}R$ is enriched in the striatal postsynaptic terminal and forms dimers with several GPCRs including the dopamine D2 receptor (D2R; Canals et al., 2003; Ferre et al., 2014; Ferre, Goldberg, Lluis, & Franco, 2009; Trincavelli et al., 2012; Fig. 10.1). Additionally, $A_{2A}R$ exists in the presynaptic glutamatergic terminals in the striatum and forms a dimer with A_1R to regulate glutamate release (Ciruela et al., 2006). Such dimer formation is important for the treatment development for HD; previous studies suggest that stimulation of the postsynaptic $A_{2A}R$ elicits important neuroprotective effects, whereas stimulation of the presynaptic $A_{2A}R$ complex might enhance excitotoxicity via increasing glutamate release (Blum et al., 2003; Fink et al., 2004; Popoli et al., 2002). Suppression of $A_{2A}R$ during HD progression occurs predominantly at the postsynaptic site, whereas no change in the presynaptic $A_{2A}R$ was reported (Orru et al., 2011). Agonists and antagonists that recognize an $A_{2A}R$ heterodimer (e.g., $A_{2A}R$-D2R vs. $A_{2A}R$-A_1R) with high selectivity are in high demand. Further evaluation of the composition and pathophysiological function of adenosine receptor dimers in the brain in HD might facilitate the development of new therapeutic agents for HD.

5. POSITRON EMISSION TOMOGRAPHY IMAGING FOR ADENOSINE RECEPTOR OCCUPANCY IN HD

Noninvasive brain analyses (such as positron emission tomography (PET) and magnetic resonance technologies) are commonly used to estimate the year remaining to disease onset (YTO) and to evaluate the disease progression of HD patients (Ciarmiello et al., 2012; Feigin et al., 2007; Matusch et al., 2014; Tabrizi et al., 2013). This is of particular interest because lower glucose utilization using ^{18}F-FDG in the brains of premanifest HD carriers has been reported (Ciarmiello et al., 2012; Feigin et al., 2007). A small number of adenosine analogue radiotracers have been developed and successfully employed with PET in the noninvasive imaging of A_1R and $A_{2A}R$. Ubiquitously expressed A_1R provided protective effects against excitotoxicity in a 3-NP-induced HD rat model (Blum et al., 2002;

Zuchora et al., 2001); and biphasic regulation of A_1R in specific brain areas was observed in HD patients during disease progression based on PET analysis using ^{18}F-CPEPX (Matusch et al., 2014). Expression of A_1R is moderately higher in preHD-A gene carriers (>10 years to the calculated onset) and lower in preHD-B gene carriers (<10 years to the calculated onset) and HD patients. This shift of the A_1R profile in the early phase of HD might enable the A_1R to serve as a potential biomarker for early HD progression.

Because $A_{2A}R$ is enriched in the striatum, the brain area that is most susceptible to the mHTT-evoked toxicity, noninvasive mapping of the level of $A_{2A}R$ in the brains affected by HD might provide a method of early detection of striatal atrophy. A few A_{2A} antagonists have been developed and tested with PET imaging (Khanapur et al., 2013). Among them, at least two (e.g., [^{11}C]TMSX and [^{11}C]KW-6002), have been tested in human subjects to evaluate the effect of age or Parkinson's disease on the level of $A_{2A}R$ in the brain (Brooks et al., 2008; Ishiwata, Kimura, Vries, Erik, & Elsinga, 2007; Mishina et al., 2007, 2012; Naganawa et al., 2007). Very limited PET imaging for $A_{2A}R$ has been conducted. In a QA-induced lesion model of HD in rats, PET with [^{11}C]TMSX revealed a reduction of the $A_{2A}R$ level in the lesion area (K. Ishiwata et al., 2002), supporting that $A_{2A}R$ could be a biomarker for early disease progression. Further development of A_{2A}-specific PET tracers with high tracer uptake and a high signal-to-noise ratio would markedly advance the understanding of the pathophysiological role of the A_{2A} receptor in HD.

6. CONCLUDING REMARKS

Despite the tremendous efforts that have been devoted to HD studies globally, no effective disease-modifying treatment for HD is available. Previous studies have successfully highlighted the importance of adenosine receptors in HD. Further investigation using genetic HD models in combination with tissue-/cell type-specific knock-out technology might further clarify the pathophysiological functions of adenosine receptors in HD and in more common degenerative diseases.

ACKNOWLEDGMENTS

This work was supported by grants from the National Science Council (100-2320-B-001-0110MY3) and Institute of Biomedical Sciences, Academia Sinica (103-Academia Sinica Investigation Award-06). We are grateful to Mr. Tsung-Hung Hung for his assistance in preparing the illustrations.

REFERENCES

Alfinito, P. D., Wang, S. P., Manzino, L., Rijhsinghani, S., Zeevalk, G. D., & Sonsalla, P. K. (2003). Adenosinergic protection of dopaminergic and GABAergic neurons against mitochondrial inhibition through receptors located in the substantia nigra and striatum, respectively. *The Journal of Neuroscience*, *23*(34), 10982–10987.

Alston, T. A., Mela, L., & Bright, H. J. (1977). 3-Nitropropionate, the toxic substance of Indigofera, is a suicide inactivator of succinate dehydrogenase. *Proceedings of the National Academy of Sciences of the United States of America*, *74*(9), 3767–3771.

Atkinson, B., Dwyer, K., Enjyoji, K., & Robson, S. C. (2006). Ecto-nucleotidases of the CD39/NTPDase family modulate platelet activation and thrombus formation: Potential as therapeutic targets. *Blood Cells, Molecules & Diseases*, *36*(2), 217–222. http://dx.doi.org/10.1016/j.bcmd.2005.12.025.

Bauer, A., Zilles, K., Matusch, A., Holzmann, C., Riess, O., & von Horsten, S. (2005). Regional and subtype selective changes of neurotransmitter receptor density in a rat transgenic for the Huntington's disease mutation. *Journal of Neurochemistry*, *94*(3), 639–650. http://dx.doi.org/10.1111/j.1471-4159.2005.03169.x.

Bennett, E. J., Shaler, T. A., Woodman, B., Ryu, K. Y., Zaitseva, T. S., Becker, C. H., et al. (2007). Global changes to the ubiquitin system in Huntington's disease. *Nature*, *448*(7154), 704–708. http://dx.doi.org/10.1038/nature06022.

Bjorkqvist, M., Wild, E. J., Thiele, J., Silvestroni, A., Andre, R., Lahiri, N., et al. (2008). A novel pathogenic pathway of immune activation detectable before clinical onset in Huntington's disease. *The Journal of Experimental Medicine*, *205*(8), 1869–1877. http://dx.doi.org/10.1084/jem.20080178.

Blum, D., Galas, M. C., Pintor, A., Brouillet, E., Ledent, C., Muller, C. E., et al. (2003). A dual role of adenosine A2A receptors in 3-nitropropionic acid-induced striatal lesions: Implications for the neuroprotective potential of A2A antagonists. *The Journal of Neuroscience*, *23*(12), 5361–5369.

Blum, D., Gall, D., Galas, M. C., d'Alcantara, P., Bantubungi, K., & Schiffmann, S. N. (2002). The adenosine A1 receptor agonist adenosine amine congener exerts a neuroprotective effect against the development of striatal lesions and motor impairments in the 3-nitropropionic acid model of neurotoxicity. *The Journal of Neuroscience*, *22*(20), 9122–9133.

Boison, D. (2006). Adenosine kinase, epilepsy and stroke: Mechanisms and therapies. *Trends in Pharmacological Sciences*, *27*(12), 652–658. http://dx.doi.org/10.1016/j.tips.2006.10.008.

Boison, D. (2013). Role of adenosine in status epilepticus: A potential new target? *Epilepsia*, *54*(Suppl. 6), 20–22. http://dx.doi.org/10.1111/epi.12268.

Boison, D., Chen, J. F., & Fredholm, B. B. (2010). Adenosine signaling and function in glial cells. *Cell Death and Differentiation*, *17*(7), 1071–1082. http://dx.doi.org/10.1038/cdd.2009.131.

Bradford, J., Shin, J. Y., Roberts, M., Wang, C. E., Li, X. J., & Li, S. (2009). Expression of mutant huntingtin in mouse brain astrocytes causes age-dependent neurological symptoms. *Proceedings of the National Academy of Sciences of the United States of America*, *106*(52), 22480–22485. http://dx.doi.org/10.1073/pnas.0911503106.

Bradford, J., Shin, J. Y., Roberts, M., Wang, C. E., Sheng, G., Li, S., et al. (2010). Mutant huntingtin in glial cells exacerbates neurological symptoms of Huntington disease mice. *The Journal of Biological Chemistry*, *285*(14), 10653–10661. http://dx.doi.org/10.1074/jbc.M109.083287.

Brooks, D. J., Doder, M., Osman, S., Luthra, S. K., Hirani, E., Hume, S., et al. (2008). Positron emission tomography analysis of [11C]KW-6002 binding to human and rat adenosine A2A receptors in the brain. *Synapse*, *62*(9), 671–681. http://dx.doi.org/10.1002/syn.20539.

Brouillet, E., Jacquard, C., Bizat, N., & Blum, D. (2005). 3-Nitropropionic acid: A mitochondrial toxin to uncover physiopathological mechanisms underlying striatal degeneration in Huntington's disease. *Journal of Neurochemistry, 95*(6), 1521–1540. http://dx.doi.org/10.1111/j.1471-4159.2005.03515.x.

Browne, S. E. (2008). Mitochondria and Huntington's disease pathogenesis: Insight from genetic and chemical models. *Annals of the New York Academy of Sciences, 1147*, 358–382. http://dx.doi.org/10.1196/annals.1427.018.

Buira, S. P., Albasanz, J. L., Dentesano, G., Moreno, J., Martin, M., Ferrer, I., et al. (2010a). DNA methylation regulates adenosine A(2A) receptor cell surface expression levels. *Journal of Neurochemistry, 112*(5), 1273–1285. http://dx.doi.org/10.1111/j.1471-4159.2009.06538.x.

Buira, S. P., Dentesano, G., Albasanz, J. L., Moreno, J., Martin, M., Ferrer, I., et al. (2010b). DNA methylation and Yin Yang-1 repress adenosine A2A receptor levels in human brain. *Journal of Neurochemistry, 115*(1), 283–295. http://dx.doi.org/10.1111/j.1471-4159.2010.06928.x.

Bulavina, L., Szulzewsky, F., Rocha, A., Krabbe, G., Robson, S. C., Matyash, V., et al. (2013). NTPDase1 activity attenuates microglial phagocytosis. *Purinergic Signal, 9*(2), 199–205. http://dx.doi.org/10.1007/s11302-012-9339-y.

Canals, M., Marcellino, D., Fanelli, F., Ciruela, F., de Benedetti, P., Goldberg, S. R., et al. (2003). Adenosine A2A-dopamine D2 receptor-receptor heteromerization: Qualitative and quantitative assessment by fluorescence and bioluminescence energy transfer. *The Journal of Biological Chemistry, 278*(47), 46741–46749. http://dx.doi.org/10.1074/jbc.M306451200.

Cepeda, C., Cummings, D. M., Hickey, M. A., Kleiman-Weiner, M., Chen, J. Y., Watson, J. B., et al. (2010). Rescuing the corticostriatal synaptic disconnection in the R6/2 mouse model of Huntington's disease: Exercise, adenosine receptors and ampakines. *PLoS Currents, 2.* http://dx.doi.org/10.1371/currents.RRN1182.

Cepeda, C., Wu, N., Andre, V. M., Cummings, D. M., & Levine, M. S. (2007). The corticostriatal pathway in Huntington's disease. *Progress in Neurobiology, 81*(5–6), 253–271. http://dx.doi.org/10.1016/j.pneurobio.2006.11.001.

Cha, J. H., Frey, A. S., Alsdorf, S. A., Kerner, J. A., Kosinski, C. M., Mangiarini, L., et al. (1999). Altered neurotransmitter receptor expression in transgenic mouse models of Huntington's disease. *Philosophical Transactions of the Royal Society of London. Series B, Biological Sciences, 354*(1386), 981–989.

Cha, J. H., Kosinski, C. M., Kerner, J. A., Alsdorf, S. A., Mangiarini, L., Davies, S. W., et al. (1998). Altered brain neurotransmitter receptors in transgenic mice expressing a portion of an abnormal human Huntington disease gene. *Proceedings of the National Academy of Sciences of the United States of America, 95*(11), 6480–6485.

Chan, E. Y., Luthi-Carter, R., Strand, A., Solano, S. M., Hanson, S. A., DeJohn, M. M., et al. (2002). Increased huntingtin protein length reduces the number of polyglutamine-induced gene expression changes in mouse models of Huntington's disease. *Human Molecular Genetics, 11*(17), 1939–1951.

Chang, D. T., Rintoul, G. L., Pandipati, S., & Reynolds, I. J. (2006). Mutant huntingtin aggregates impair mitochondrial movement and trafficking in cortical neurons. *Neurobiology of Disease, 22*(2), 388–400. http://dx.doi.org/10.1016/j.nbd.2005.12.007.

Charalambous, C., Gsandtner, I., Keuerleber, S., Milan-Lobo, L., Kudlacek, O., Freissmuth, M., et al. (2008). Restricted collision coupling of the A2A receptor revisited: Evidence for physical separation of two signaling cascades. *The Journal of Biological Chemistry, 283*(14), 9276–9288. http://dx.doi.org/10.1074/jbc.M706275200.

Chen, J. F., Eltzschig, H. K., & Fredholm, B. B. (2013). Adenosine receptors as drug targets—What are the challenges? *Nature Reviews. Drug Discovery, 12*(4), 265–286. http://dx.doi.org/10.1038/nrd3955, nrd3955 [pii].

Chen, L. L., Wu, J. C., Wang, L. H., Wang, J., Qin, Z. H., Difiglia, M., et al. (2012). Rapamycin prevents the mutant huntingtin-suppressed GLT-1 expression in cultured astrocytes. *Acta Pharmacologica Sinica*, *33*(3), 385–392. http://dx.doi.org/10.1038/aps.2011.162.

Chiang, M. C., Chen, H. M., Lai, H. L., Chen, H. W., Chou, S. Y., Chen, C. M., et al. (2009). The A2A adenosine receptor rescues the urea cycle deficiency of Huntington's disease by enhancing the activity of the ubiquitin-proteasome system. *Human Molecular Genetics*, *18*(16), 2929–2942. http://dx.doi.org/10.1093/hmg/ddp230.

Chiang, M. C., Lee, Y. C., Huang, C. L., & Chern, Y. (2005). cAMP-response element-binding protein contributes to suppression of the A2A adenosine receptor promoter by mutant Huntingtin with expanded polyglutamine residues. *The Journal of Biological Chemistry*, *280*(14), 14331–14340. http://dx.doi.org/10.1074/jbc.M413279200.

Chiu, E., Mackay, I. R., & Bhathal, P. B. (1975). Hepatic morphology in Huntington's chorea. *Journal of Neurology, Neurosurgery, and Psychiatry*, *38*(10), 1000–1002.

Chou, S. Y., Lee, Y. C., Chen, H. M., Chiang, M. C., Lai, H. L., Chang, H. H., et al. (2005). CGS21680 attenuates symptoms of Huntington's disease in a transgenic mouse model. *Journal of Neurochemistry*, *93*(2), 310–320. http://dx.doi.org/10.1111/j.1471-4159.2005.03029.x.

Chou, S. Y., Weng, J. Y., Lai, H. L., Liao, F., Sun, S. H., Tu, P. H., et al. (2008). Expanded-polyglutamine huntingtin protein suppresses the secretion and production of a chemokine (CCL5/RANTES) by astrocytes. *The Journal of Neuroscience*, *28*(13), 3277–3290. http://dx.doi.org/10.1523/JNEUROSCI.0116-08.2008.

Ciarmiello, A., Giovacchini, G., Orobello, S., Bruselli, L., Elifani, F., & Squitieri, F. (2012). 18F-FDG PET uptake in the pre-Huntington disease caudate affects the time-to-onset independently of CAG expansion size. *European Journal of Nuclear Medicine and Molecular Imaging*, *39*(6), 1030–1036. http://dx.doi.org/10.1007/s00259-012-2114-z.

Cipriani, S., Bizzoco, E., Gianfriddo, M., Melani, A., Vannucchi, M. G., & Pedata, F. (2008). Adenosine A2A receptor antagonism increases nNOS-immunoreactive neurons in the striatum of Huntington transgenic mice. *Experimental Neurology*, *213*(1), 163–170. http://dx.doi.org/10.1016/j.expneurol.2008.05.015.

Ciruela, F., Ferre, S., Casado, V., Cortes, A., Cunha, R. A., Lluis, C., et al. (2006). Heterodimeric adenosine receptors: A device to regulate neurotransmitter release. *Cellular and Molecular Life Sciences*, *63*(21), 2427–2431. http://dx.doi.org/10.1007/s00018-006-6216-2.

Cloud, L. J., Rosenblatt, A., Margolis, R. L., Ross, C. A., Pillai, J. A., Corey-Bloom, J., et al. (2012). Seizures in juvenile Huntington's disease: Frequency and characterization in a multicenter cohort. *Movement Disorders*, *27*(14), 1797–1800. http://dx.doi.org/10.1002/mds.25237.

Cornelis, M. C., El-Sohemy, A., & Campos, H. (2007). Genetic polymorphism of the adenosine A2A receptor is associated with habitual caffeine consumption. *The American Journal of Clinical Nutrition*, *86*(1), 240–244.

Costa, V., Giacomello, M., Hudec, R., Lopreiato, R., Ermak, G., Lim, D., et al. (2010). Mitochondrial fission and cristae disruption increase the response of cell models of Huntington's disease to apoptotic stimuli. *EMBO Molecular Medicine*, *2*(12), 490–503. http://dx.doi.org/10.1002/emmm.201000102.

Crotti, A., Benner, C., Kerman, B. E., Gosselin, D., Lagier-Tourenne, C., Zuccato, C., et al. (2014). Mutant Huntingtin promotes autonomous microglia activation via myeloid lineage-determining factors. *Nature Neuroscience*. http://dx.doi.org/10.1038/nn.3668.

del Toro, D., Xifro, X., Pol, A., Humbert, S., Saudou, F., Canals, J. M., et al. (2010). Altered cholesterol homeostasis contributes to enhanced excitotoxicity in Huntington's disease. *Journal of Neurochemistry*, *115*(1), 153–167. http://dx.doi.org/10.1111/j.1471-4159.2010.06912.x.

Dhaenens, C. M., Burnouf, S., Simonin, C., Van Brussel, E., Duhamel, A., Defebvre, L., et al. (2009). A genetic variation in the ADORA2A gene modifies age at onset in Huntington's disease. *Neurobiology of Disease*, *35*(3), 474–476. http://dx.doi.org/10.1016/j.nbd.2009.06.009.

DiFiglia, M. (1990). Excitotoxic injury of the neostriatum: A model for Huntington's disease. *Trends in Neurosciences*, *13*(7), 286–289.

Diogenes, M. J., Fernandes, C. C., Sebastiao, A. M., & Ribeiro, J. A. (2004). Activation of adenosine A2A receptor facilitates brain-derived neurotrophic factor modulation of synaptic transmission in hippocampal slices. *The Journal of Neuroscience*, *24*(12), 2905–2913. http://dx.doi.org/10.1523/JNEUROSCI.4454-03.2004.

Diogenes, M. J., Neves-Tome, R., Fucile, S., Martinello, K., Scianni, M., Theofilas, P., et al. (2014). Homeostatic control of synaptic activity by endogenous adenosine is mediated by adenosine kinase. *Cerebral Cortex*, *24*(1), 67–80. http://dx.doi.org/10.1093/cercor/bhs284.

Domenici, M. R., Scattoni, M. L., Martire, A., Lastoria, G., Potenza, R. L., Borioni, A., et al. (2007). Behavioral and electrophysiological effects of the adenosine A2A receptor antagonist SCH 58261 in R6/2 Huntington's disease mice. *Neurobiology of Disease*, *28*(2), 197–205. http://dx.doi.org/10.1016/j.nbd.2007.07.009.

Eltzschig, H. K., Eckle, T., Mager, A., Kuper, N., Karcher, C., Weissmuller, T., et al. (2006). ATP release from activated neutrophils occurs via connexin 43 and modulates adenosine-dependent endothelial cell function. *Circulation Research*, *99*(10), 1100–1108. http://dx.doi.org/10.1161/01.RES.0000250174.31269.70.

Fan, J., Cowan, C. M., Zhang, L. Y., Hayden, M. R., & Raymond, L. A. (2009). Interaction of postsynaptic density protein-95 with NMDA receptors influences excitotoxicity in the yeast artificial chromosome mouse model of Huntington's disease. *The Journal of Neuroscience*, *29*(35), 10928–10938. http://dx.doi.org/10.1523/JNEUROSCI.2491-09.2009.

Fan, M. M., & Raymond, L. A. (2007). N-methyl-D-aspartate (NMDA) receptor function and excitotoxicity in Huntington's disease. *Progress in Neurobiology*, *81*(5–6), 272–293. http://dx.doi.org/10.1016/j.pneurobio.2006.11.003.

Farber, K., Markworth, S., Pannasch, U., Nolte, C., Prinz, V., Kronenberg, G., et al. (2008). The ectonucleotidase cd39/ENTPDase1 modulates purinergic-mediated microglial migration. *Glia*, *56*(3), 331–341. http://dx.doi.org/10.1002/glia.20606.

Feigin, A., Tang, C., Ma, Y., Mattis, P., Zgaljardic, D., Guttman, M., et al. (2007). Thalamic metabolism and symptom onset in preclinical Huntington's disease. *Brain*, *130*(Pt 11), 2858–2867. http://dx.doi.org/10.1093/brain/awm217.

Ferrante, A., Martire, A., Armida, M., Chiodi, V., Pezzola, A., Potenza, R. L., et al. (2010). Influence of CGS 21680, a selective adenosine A(2A) receptor agonist, on NMDA receptor function and expression in the brain of Huntington's disease mice. *Brain Research*, *1323*, 184–191. http://dx.doi.org/10.1016/j.brainres.2010.01.080.

Ferre, S., Casado, V., Devi, L. A., Filizola, M., Jockers, R., Lohse, M. J., et al. (2014). G protein-coupled receptor oligomerization revisited: Functional and pharmacological perspectives. *Pharmacological Reviews*, *66*(2), 413–434. http://dx.doi.org/10.1124/pr.113.008052.

Ferre, S., Goldberg, S. R., Lluis, C., & Franco, R. (2009). Looking for the role of cannabinoid receptor heteromers in striatal function. *Neuropharmacology*, *56*(Suppl 1), 226–234. http://dx.doi.org/10.1016/j.neuropharm.2008.06.076, S0028–3908(08)00265–7 [pii].

Fink, J. S., Kalda, A., Ryu, H., Stack, E. C., Schwarzschild, M. A., Chen, J. F., et al. (2004). Genetic and pharmacological inactivation of the adenosine A2A receptor attenuates 3-nitropropionic acid-induced striatal damage. *Journal of Neurochemistry*, *88*(3), 538–544.

Fontinha, B. M., Diogenes, M. J., Ribeiro, J. A., & Sebastiao, A. M. (2008). Enhancement of long-term potentiation by brain-derived neurotrophic factor requires adenosine A2A receptor activation by endogenous adenosine. *Neuropharmacology, 54*(6), 924–933. http://dx.doi.org/10.1016/j.neuropharm.2008.01.011.

Fredholm, B. B. (2013). Adenosine and metabolism–A brief historical note. In S. Masino, & D. Boison (Eds.), *Adenosine: A key link between metabolism and brain activity* (pp. 3–22). New York: Springer. http://dx.doi.org/10.1007/978-1-4614-3903-5.

Galluzzo, M., Pintor, A., Pezzola, A., Grieco, R., Borsini, F., & Popoli, P. (2008). Behavioural and neurochemical characterization of the adenosine A2A receptor antagonist ST1535. *European Journal of Pharmacology, 579*(1–3), 149–152. http://dx.doi.org/10.1016/j.ejphar.2007.10.057.

Gauthier, L. R., Charrin, B. C., Borrell-Pages, M., Dompierre, J. P., Rangone, H., Cordelieres, F. P., et al. (2004). Huntingtin controls neurotrophic support and survival of neurons by enhancing BDNF vesicular transport along microtubules. *Cell, 118*(1), 127–138. http://dx.doi.org/10.1016/j.cell.2004.06.018.

Georgiou-Karistianis, N., Scahill, R., Tabrizi, S. J., Squitieri, F., & Aylward, E. (2013). Structural MRI in Huntington's disease and recommendations for its potential use in clinical trials. *Neuroscience and Biobehavioral Reviews, 37*(3), 480–490. http://dx.doi.org/10.1016/j.neubiorev.2013.01.022, S0149-7634(13)00029-8 [pii].

Gharami, K., Xie, Y., An, J. J., Tonegawa, S., & Xu, B. (2008). Brain-derived neurotrophic factor over-expression in the forebrain ameliorates Huntington's disease phenotypes in mice. *Journal of Neurochemistry, 105*(2), 369–379. http://dx.doi.org/10.1111/j.1471-4159.2007.05137.x.

Giampa, C., Laurenti, D., Anzilotti, S., Bernardi, G., Menniti, F. S., & Fusco, F. R. (2010). Inhibition of the striatal specific phosphodiesterase PDE10A ameliorates striatal and cortical pathology in R6/2 mouse model of Huntington's disease. *PLoS One, 5*(10), e13417. http://dx.doi.org/10.1371/journal.pone.0013417.

Gianfriddo, M., Melani, A., Turchi, D., Giovannini, M. G., & Pedata, F. (2004). Adenosine and glutamate extracellular concentrations and mitogen-activated protein kinases in the striatum of Huntington transgenic mice. Selective antagonism of adenosine A2A receptors reduces transmitter outflow. *Neurobiology of Disease, 17*(1), 77–88. http://dx.doi.org/10.1016/j.nbd.2004.05.008.

Glass, M., Dragunow, M., & Faull, R. L. (2000). The pattern of neurodegeneration in Huntington's disease: A comparative study of cannabinoid, dopamine, adenosine and GABA(a) receptor alterations in the human basal ganglia in Huntington's disease. *Neuroscience, 97*(3), 505–519.

Graham, R. K., Pouladi, M. A., Joshi, P., Lu, G., Deng, Y., Wu, N. P., et al. (2009). Differential susceptibility to excitotoxic stress in YAC128 mouse models of Huntington disease between initiation and progression of disease. *The Journal of Neuroscience, 29*(7), 2193–2204. http://dx.doi.org/10.1523/JNEUROSCI.5473-08.2009, 29/7/2193 [pii].

Guo, X., Disatnik, M. H., Monbureau, M., Shamloo, M., Mochly-Rosen, D., & Qi, X. (2013). Inhibition of mitochondrial fragmentation diminishes Huntington's disease-associated neurodegeneration. *The Journal of Clinical Investigation, 123*(12), 5371–5388. http://dx.doi.org/10.1172/JCI70911.

Hsiao, H. Y., Chen, Y. C., Chen, H. M., Tu, P. H., & Chern, Y. (2013). A critical role of astrocyte-mediated nuclear factor-kappaB-dependent inflammation in Huntington's disease. *Human Molecular Genetics, 22*(9), 1826–1842. http://dx.doi.org/10.1093/hmg/ddt036.

Hsiao, H. Y., & Chern, Y. (2010). Targeting glial cells to elucidate the pathogenesis of Huntington's disease. *Molecular Neurobiology, 41*(2–3), 248–255. http://dx.doi.org/10.1007/s12035-009-8097-5.

Hsiao, H. Y., Chiu, F. L., Chen, C. M., Wu, Y. R., Chen, H. M., Chen, Y. C., et al. (2014). Inhibition of soluble tumor necrosis factor is therapeutic in Huntington's disease. *Human Molecular Genetics*. http://dx.doi.org/10.1093/hmg/ddu151, ddu151 [pii].

Huang, N. K., Lin, J. H., Lin, J. T., Lin, C. I., Liu, E. M., Lin, C. J., et al. (2011a). A new drug design targeting the adenosinergic system for Huntington's disease. *PLoS One*, *6*(6), e20934. http://dx.doi.org/10.1371/journal.pone.0020934.

Huang, Q. Y., Wei, C., Yu, L., Coelho, J. E., Shen, H. Y., Kalda, A., et al. (2006). Adenosine A2A receptors in bone marrow-derived cells but not in forebrain neurons are important contributors to 3-nitropropionic acid-induced striatal damage as revealed by cell-type-selective inactivation. *The Journal of Neuroscience*, *26*(44), 11371–11378. http://dx.doi.org/10.1523/JNEUROSCI.1907-06.2006.

Huang, C. L., Yang, J. M., Wang, K. C., Lee, Y. C., Lin, Y. L., Yang, Y. C., et al. (2011b). Gastrodia elata prevents huntingtin aggregations through activation of the adenosine A(2) A receptor and ubiquitin proteasome system. *Journal of Ethnopharmacology*, *138*(1), 162–168. http://dx.doi.org/10.1016/j.jep.2011.08.075.

Huang, Q., Zhou, D., Sapp, E., Aizawa, H., Ge, P., Bird, E. D., et al. (1995). Quinolinic acid-induced increases in calbindin D28k immunoreactivity in rat striatal neurons in vivo and in vitro mimic the pattern seen in Huntington's disease. *Neuroscience*, *65*(2), 397–407.

Ishiwata, K., Kimura, Y., Vries, De, Erik, F. J., & Elsinga, P. H. (2007). PET tracers for mapping adenosine receptors as probes for diagnosis of CNS disorders. *Central Nervous System Agents in Medicinal Chemistry*, 7(1), 57–77.

Ishiwata, K., Ogi, N., Hayakawa, N., Oda, K., Nagaoka, T., Toyama, H., et al. (2002). Adenosine A2A receptor imaging with [11C]KF18446 PET in the rat brain after quinolinic acid lesion: Comparison with the dopamine receptor imaging. *Annals of Nuclear Medicine*, *16*(7), 467–475.

Jacobson, K. A., Balasubramanian, R., Deflorian, F., & Gao, Z. G. (2012). G protein-coupled adenosine (P1) and P2Y receptors: Ligand design and receptor interactions. *Purinergic Signal*, *8*(3), 419–436. http://dx.doi.org/10.1007/s11302-012-9294-7.

Jia, H., Kast, R. J., Steffan, J. S., & Thomas, E. A. (2012). Selective histone deacetylase (HDAC) inhibition imparts beneficial effects in Huntington's disease mice: Implications for the ubiquitin-proteasomal and autophagy systems. *Human Molecular Genetics*, *21*(24), 5280–5293. http://dx.doi.org/10.1093/hmg/dds379.

Jiang, M., Peng, Q., Liu, X., Jin, J., Hou, Z., Zhang, J., et al. (2013). Small-molecule TrkB receptor agonists improve motor function and extend survival in a mouse model of Huntington's disease. *Human Molecular Genetics*, *22*(12), 2462–2470. http://dx.doi.org/10.1093/hmg/ddt098.

Ju, T. C., Chen, H. M., Lin, J. T., Chang, C. P., Chang, W. C., Kang, J. J., et al. (2011). Nuclear translocation of AMPK-alpha1 potentiates striatal neurodegeneration in Huntington's disease. *The Journal of Cell Biology*, *194*(2), 209–227. http://dx.doi.org/10.1083/jcb.201105010.

Khanapur, S., Waarde, Av, Ishiwata, K., Leenders, K. L., Dierckx, R. A., & Elsinga, P. H. (2013). Adenosine A(2A) receptor antagonists as positron emission tomography (PET) tracers. *Current Medicinal Chemistry*, *21*(3), 312–328.

Kim, W., & Seo, H. (2014). Baclofen, a GABAB receptor agonist, enhances ubiquitin-proteasome system functioning and neuronal survival in Huntington's disease model mice. *Biochemical and Biophysical Research Communications*, *443*(2), 706–711. http://dx.doi.org/10.1016/j.bbrc.2013.12.034.

Koga, H., Martinez-Vicente, M., Arias, E., Kaushik, S., Sulzer, D., & Cuervo, A. M. (2011). Constitutive upregulation of chaperone-mediated autophagy in Huntington's disease. *The Journal of Neuroscience*, *31*(50), 18492–18505. http://dx.doi.org/10.1523/JNEUROSCI.3219-11.2011.

Kulesskaya, N., Voikar, V., Peltola, M., Yegutkin, G. G., Salmi, M., Jalkanen, S., et al. (2013). CD73 is a major regulator of adenosinergic signalling in mouse brain. *PLoS One*, *8*(6), e66896. http://dx.doi.org/10.1371/journal.pone.0066896.

Labbadia, J., & Morimoto, R. I. (2013). Huntington's disease: Underlying molecular mechanisms and emerging concepts. *Trends in Biochemical Sciences*, *38*(8), 378–385. http://dx.doi.org/10.1016/j.tibs.2013.05.003, S0968-0004(13)00078-9 [pii].

Lee, C. Y., Cantle, J. P., & Yang, X. W. (2013a). Genetic manipulations of mutant huntingtin in mice: New insights into Huntington's disease pathogenesis. *The FEBS Journal*, *280*(18), 4382–4394. http://dx.doi.org/10.1111/febs.12418.

Lee, F. S., & Chao, M. V. (2001). Activation of Trk neurotrophin receptors in the absence of neurotrophins. *Proceedings of the National Academy of Sciences of the United States of America*, *98*(6), 3555–3560. http://dx.doi.org/10.1073/pnas.061020198.

Lee, Y. C., Lai, H. L., Sun, C. N., Chien, C. L., & Chern, Y. (2003). Identification of nuclear factor 1 (NF1) as a transcriptional modulator of rat A(2A) adenosine receptor. *Brain Research. Molecular Brain Research*, *111*(1–2), 61–73.

Lee, W., Reyes, R. C., Gottipati, M. K., Lewis, K., Lesort, M., Parpura, V., et al. (2013b). Enhanced Ca(2+)-dependent glutamate release from astrocytes of the BACHD Huntington's disease mouse model. *Neurobiology of Disease*, *58*, 192–199. http://dx.doi.org/10.1016/j.nbd.2013.06.002.

Li, B., Gu, L., Hertz, L., & Peng, L. (2013). Expression of nucleoside transporter in freshly isolated neurons and astrocytes from mouse brain. *Neurochemical Research*, *38*(11), 2351–2358. http://dx.doi.org/10.1007/s11064-013-1146-5.

Li, T., Quan Lan, J., Fredholm, B. B., Simon, R. P., & Boison, D. (2007). Adenosine dysfunction in astrogliosis: Cause for seizure generation? *Neuron Glia Biology*, *3*(4), 353–366. http://dx.doi.org/10.1017/S1740925X0800015X.

Lin, J. T., Chang, W. C., Chen, H. M., Lai, H. L., Chen, C. Y., Tao, M. H., et al. (2013a). Regulation of feedback between protein kinase A and the proteasome system worsens Huntington's disease. *Molecular and Cellular Biology*, *33*(5), 1073–1084. http://dx.doi.org/10.1128/MCB.01434-12.

Lin, Y. S., Cheng, T. H., Chang, C. P., Chen, H. M., & Chern, Y. (2013b). Enhancement of brain-type creatine kinase activity ameliorates neuronal deficits in Huntington's disease. *Biochimica et Biophysica Acta*, *1832*(6), 742–753. http://dx.doi.org/10.1016/j.bbadis.2013.02.006.

Liu, Y., Hettinger, C. L., Zhang, D., Rezvani, K., Wang, X., & Wang, H. (2014). Sulforaphane enhances proteasomal and autophagic activities in mice and is a potential therapeutic reagent for Huntington's disease. *Journal of Neurochemistry*. http://dx.doi.org/10.1111/jnc.12647.

Luthi-Carter, R., Strand, A., Peters, N. L., Solano, S. M., Hollingsworth, Z. R., Menon, A. S., et al. (2000). Decreased expression of striatal signaling genes in a mouse model of Huntington's disease. *Human Molecular Genetics*, *9*(9), 1259–1271.

Maglione, V., Cannella, M., Martino, T., De Blasi, A., Frati, L., & Squitieri, F. (2006). The platelet maximum number of A2A-receptor binding sites (Bmax) linearly correlates with age at onset and CAG repeat expansion in Huntington's disease patients with predominant chorea. *Neuroscience Letters*, *393*(1), 27–30. http://dx.doi.org/10.1016/j.neulet.2005.09.037.

Maglione, V., Giallonardo, P., Cannella, M., Martino, T., Frati, L., & Squitieri, F. (2005). Adenosine A2A receptor dysfunction correlates with age at onset anticipation in blood platelets of subjects with Huntington's disease. *American Journal of Medical Genetics. Part B, Neuropsychiatric Genetics*, *139B*(1), 101–105. http://dx.doi.org/10.1002/ajmg.b.30223.

Mangiarini, L., Sathasivam, K., Seller, M., Cozens, B., Harper, A., Hetherington, C., et al. (1996). Exon 1 of the HD gene with an expanded CAG repeat is sufficient to cause a progressive neurological phenotype in transgenic mice. *Cell*, *87*(3), 493–506.

Marco, S., Giralt, A., Petrovic, M. M., Pouladi, M. A., Martinez-Turrillas, R., Martinez-Hernandez, J., et al. (2013). Suppressing aberrant GluN3A expression rescues synaptic and behavioral impairments in Huntington's disease models. *Nature Medicine*, *19*(8), 1030–1038. http://dx.doi.org/10.1038/nm.3246.

Martinez-Mir, M. I., Probst, A., & Palacios, J. M. (1991). Adenosine A2 receptors: Selective localization in the human basal ganglia and alterations with disease. *Neuroscience*, *42*(3), 697–706.

Martinez-Vicente, M., Talloczy, Z., Wong, E., Tang, G., Koga, H., Kaushik, S., et al. (2010). Cargo recognition failure is responsible for inefficient autophagy in Huntington's disease. *Nature Neuroscience*, *13*(5), 567–576. http://dx.doi.org/10.1038/nn.2528.

Martire, A., Calamandrei, G., Felici, F., Scattoni, M. L., Lastoria, G., Domenici, M. R., et al. (2007). Opposite effects of the A2A receptor agonist CGS21680 in the striatum of Huntington's disease versus wild-type mice. *Neuroscience Letters*, *417*(1), 78–83. http://dx.doi.org/10.1016/j.neulet.2007.02.034.

Martire, A., Ferrante, A., Potenza, R. L., Armida, M., Ferretti, R., Pezzola, A., et al. (2010). Remodeling of striatal NMDA receptors by chronic A(2A) receptor blockade in Huntington's disease mice. *Neurobiology of Disease*, *37*(1), 99–105. http://dx.doi.org/10.1016/j.nbd.2009.09.012.

Martire, A., Pepponi, R., Domenici, M. R., Ferrante, A., Chiodi, V., & Popoli, P. (2013). BDNF prevents NMDA-induced toxicity in models of Huntington's disease: The effects are genotype specific and adenosine A(2A) receptor is involved. *Journal of Neurochemistry*. http://dx.doi.org/10.1111/jnc.12177.

Matusch, A., Saft, C., Elmenhorst, D., Kraus, P. H., Gold, R., Hartung, H. P., et al. (2014). Cross sectional PET study of cerebral adenosine A receptors in premanifest and manifest Huntington's disease. *European Journal of Nuclear Medicine and Molecular Imaging*. http://dx.doi.org/10.1007/s00259-014-2724-8.

Messam, C. A., Greene, J. G., Greenamyre, J. T., & Robinson, M. B. (1995). Intrastriatal injections of the succinate dehydrogenase inhibitor, malonate, cause a rise in extracellular amino acids that is blocked by MK-801. *Brain Research*, *684*(2), 221–224.

Mievis, S., Blum, D., & Ledent, C. (2011). A2A receptor knockout worsens survival and motor behaviour in a transgenic mouse model of Huntington's disease. *Neurobiology of Disease*, *41*(2), 570–576. http://dx.doi.org/10.1016/j.nbd.2010.09.021.

Mishina, M., Ishiwata, K., Kimura, Y., Naganawa, M., Oda, K., Kobayashi, S., et al. (2007). Evaluation of distribution of adenosine A2A receptors in normal human brain measured with [11C]TMSX PET. *Synapse*, *61*(9), 778–784. http://dx.doi.org/10.1002/syn.20423.

Mishina, M., Kimura, Y., Naganawa, M., Ishii, K., Oda, K., Sakata, M., et al. (2012). Differential effects of age on human striatal adenosine A(1) and A(2A) receptors. *Synapse*, *66*(9), 832–839. http://dx.doi.org/10.1002/syn.21573.

Naganawa, M., Kimura, Y., Mishina, M., Manabe, Y., Chihara, K., Oda, K., et al. (2007). Quantification of adenosine A2A receptors in the human brain using [11C]TMSX and positron emission tomography. *European Journal of Nuclear Medicine and Molecular Imaging*, *34*(5), 679–687. http://dx.doi.org/10.1007/s00259-006-0294-0.

Orru, M., Zanoveli, J. M., Quiroz, C., Nguyen, H. P., Guitart, X., & Ferre, S. (2011). Functional changes in postsynaptic adenosine A(2A) receptors during early stages of a rat model of Huntington disease. *Experimental Neurology*, *232*(1), 76–80. http://dx.doi.org/10.1016/j.expneurol.2011.08.005.

Panov, A. V., Gutekunst, C. A., Leavitt, B. R., Hayden, M. R., Burke, J. R., Strittmatter, W. J., et al. (2002). Early mitochondrial calcium defects in Huntington's disease are a direct effect of polyglutamines. *Nature Neuroscience*, *5*(8), 731–736. http://dx.doi.org/10.1038/nn884.

Popoli, P., Blum, D., Domenici, M. R., Burnouf, S., & Chern, Y. (2008). A critical evaluation of adenosine A2A receptors as potentially "druggable" targets in Huntington's disease. *Current Pharmaceutical Design*, *14*(15), 1500–1511.

Popoli, P., Pezzola, A., Reggio, R., Caporali, M. G., & Scotti de Carolis, A. (1994). CGS 21680 antagonizes motor hyperactivity in a rat model of Huntington's disease. *European Journal of Pharmacology*, *257*(1–2), R5–6.

Popoli, P., Pintor, A., Domenici, M. R., Frank, C., Tebano, M. T., Pezzola, A., et al. (2002). Blockade of striatal adenosine A2A receptor reduces, through a presynaptic mechanism, quinolinic acid-induced excitotoxicity: Possible relevance to neuroprotective interventions in neurodegenerative diseases of the striatum. *The Journal of Neuroscience*, *22*(5), 1967–1975.

Potenza, R. L., Tebano, M. T., Martire, A., Domenici, M. R., Pepponi, R., Armida, M., et al. (2007). Adenosine A(2A) receptors modulate BDNF both in normal conditions and in experimental models of Huntington's disease. *Purinergic Signal*, *3*(4), 333–338. http://dx.doi.org/10.1007/s11302-007-9066-y.

Pouladi, M. A., Morton, A. J., & Hayden, M. R. (2013). Choosing an animal model for the study of Huntington's disease. *Nature Reviews. Neuroscience*, *14*(10), 708–721. http://dx.doi.org/10.1038/nrn3570, nrn3570. [pii].

Powers, W. J., Videen, T. O., Markham, J., McGee-Minnich, L., Antenor-Dorsey, J. V., Hershey, T., et al. (2007). Selective defect of in vivo glycolysis in early Huntington's disease striatum. *Proceedings of the National Academy of Sciences of the United States of America*, *104*(8), 2945–2949. http://dx.doi.org/10.1073/pnas.0609833104.

Reggio, R., Pezzola, A., & Popoli, P. (1999). The intrastriatal injection of an adenosine A(2) receptor antagonist prevents frontal cortex EEG abnormalities in a rat model of Huntington's disease. *Brain Research*, *831*(1–2), 315–318.

Reilmann, R. (2013). Pharmacological treatment of chorea in Huntington's disease-good clinical practice versus evidence-based guideline. *Movement Disorders*, *28*(8), 1030–1033. http://dx.doi.org/10.1002/mds.25500.

Rogers, P. J., Hohoff, C., Heatherley, S. V., Mullings, E. L., Maxfield, P. J., Evershed, R. P., et al. (2010). Association of the anxiogenic and alerting effects of caffeine with ADORA2A and ADORA1 polymorphisms and habitual level of caffeine consumption. *Neuropsychopharmacology*, *35*(9), 1973–1983. http://dx.doi.org/10.1038/npp.2010.71.

Sarkar, S., Perlstein, E. O., Imarisio, S., Pineau, S., Cordenier, A., Maglathlin, R. L., et al. (2007). Small molecules enhance autophagy and reduce toxicity in Huntington's disease models. *Nature Chemical Biology*, *3*(6), 331–338. http://dx.doi.org/10.1038/nchembio883.

Scattoni, M. L., Valanzano, A., Pezzola, A., March, Z. D., Fusco, F. R., Popoli, P., et al. (2007). Adenosine A2A receptor blockade before striatal excitotoxic lesions prevents long term behavioural disturbances in the quinolinic rat model of Huntington's disease. *Behavioural Brain Research*, *176*(2), 216–221. http://dx.doi.org/10.1016/j.bbr.2006.10.004.

Seo, H., Sonntag, K. C., & Isacson, O. (2004). Generalized brain and skin proteasome inhibition in Huntington's disease. *Annals of Neurology*, *56*(3), 319–328. http://dx.doi.org/10.1002/ana.20207.

Seo, H., Sonntag, K. C., Kim, W., Cattaneo, E., & Isacson, O. (2007). Proteasome activator enhances survival of Huntington's disease neuronal model cells. *PLoS One*, *2*(2), e238. http://dx.doi.org/10.1371/journal.pone.0000238.

Shin, J. Y., Fang, Z. H., Yu, Z. X., Wang, C. E., Li, S. H., & Li, X. J. (2005). Expression of mutant huntingtin in glial cells contributes to neuronal excitotoxicity. *The Journal of Cell Biology*, *171*(6), 1001–1012. http://dx.doi.org/10.1083/jcb.200508072.

Simonin, C., Duru, C., Salleron, J., Hincker, P., Charles, P., Delval, A., et al. (2013). Association between caffeine intake and age at onset in Huntington's disease. *Neurobiology of Disease*, *58*, 179–182. http://dx.doi.org/10.1016/j.nbd.2013.05.013.

Song, W., Chen, J., Petrilli, A., Liot, G., Klinglmayr, E., Zhou, Y., et al. (2011). Mutant huntingtin binds the mitochondrial fission GTPase dynamin-related protein-1 and increases its enzymatic activity. *Nature Medicine*, *17*(3), 377–382. http://dx.doi.org/10.1038/nm.2313.

Svenningsson, P., Hall, H., Sedvall, G., & Fredholm, B. B. (1997). Distribution of adenosine receptors in the postmortem human brain: An extended autoradiographic study. *Synapse*, *27*(4), 322–335. http://dx.doi.org/10.1002/(SICI)1098-2396(199712)27:4<322::AID-SYN6>3.0.CO;2-E, 10.1002/(SICI)1098–2396(199712)27:4<322::AID-SYN6>3.0.CO;2-E [pii].

Tabrizi, S. J., Scahill, R. I., Owen, G., Durr, A., Leavitt, B. R., Roos, R. A., et al. (2013). Predictors of phenotypic progression and disease onset in premanifest and early-stage Huntington's disease in the TRACK-HD study: Analysis of 36-month observational data. *Lancet Neurology*, *12*(7), 637–649. http://dx.doi.org/10.1016/S1474-4422(13)70088-7.

Taherzadeh-Fard, E., Saft, C., Wieczorek, S., Epplen, J. T., & Arning, L. (2010). Age at onset in Huntington's disease: Replication study on the associations of ADORA2A, HAP1 and OGG1. *Neurogenetics*, *11*(4), 435–439. http://dx.doi.org/10.1007/s10048-010-0248-3.

Tarditi, A., Camurri, A., Varani, K., Borea, P. A., Woodman, B., Bates, G., et al. (2006). Early and transient alteration of adenosine A2A receptor signaling in a mouse model of Huntington disease. *Neurobiology of Disease*, *23*(1), 44–53. http://dx.doi.org/10.1016/j.nbd.2006.01.014.

Taylor-Robinson, S. D., Weeks, R. A., Bryant, D. J., Sargentoni, J., Marcus, C. D., Harding, A. E., et al. (1996). Proton magnetic resonance spectroscopy in Huntington's disease: Evidence in favour of the glutamate excitotoxic theory. *Movement Disorders*, *11*(2), 167–173. http://dx.doi.org/10.1002/mds.870110209.

Taylor-Robinson, S. D., Weeks, R. A., Sargentoni, J., Marcus, C. D., Bryant, D. J., Harding, A. E., et al. (1994). Evidence for glutamate excitotoxicity in Huntington's disease with proton magnetic resonance spectroscopy. *Lancet*, *343*(8906), 1170.

Tebano, M. T., Pintor, A., Frank, C., Domenici, M. R., Martire, A., Pepponi, R., et al. (2004). Adenosine A2A receptor blockade differentially influences excitotoxic mechanisms at pre- and postsynaptic sites in the rat striatum. *Journal of Neuroscience Research*, 77(1), 100–107. http://dx.doi.org/10.1002/jnr.20138.

Thurner, P., Gsandtner, I., Kudlacek, O., Choquet, D., Nanoff, C., Freissmuth, M., et al. (2014). A Two-state model for the diffusion of the A2A adenosine receptor in hippocampal neurons: Agonist-induced switch to slow mobility is modified by synapse-associated protein 102 (SAP102). *The Journal of Biological Chemistry*, *289*(13), 9263–9274. http://dx.doi.org/10.1074/jbc.M113.505685.

Todd, D., Gowers, I., Dowler, S. J., Wall, M. D., McAllister, G., Fischer, D. F., et al. (2014). A monoclonal antibody TrkB receptor agonist as a potential therapeutic for Huntington's disease. *PLoS One*, *9*(2), e87923. http://dx.doi.org/10.1371/journal.pone.0087923.

Trincavelli, M. L., Daniele, S., Orlandini, E., Navarro, G., Casado, V., Giacomelli, C., et al. (2012). A new D(2) dopamine receptor agonist allosterically modulates A(2A) adenosine receptor signalling by interacting with the A(2A)/D(2) receptor heteromer. *Cellular Signalling*, *24*(4), 951–960. http://dx.doi.org/10.1016/j.cellsig.2011.12.018, S0898–6568(11)00410–4 [pii].

Valenza, M., Carroll, J. B., Leoni, V., Bertram, L. N., Bjorkhem, I., Singaraja, R. R., et al. (2007a). Cholesterol biosynthesis pathway is disturbed in YAC128 mice and is modulated by huntingtin mutation. *Human Molecular Genetics*, *16*(18), 2187–2198. http://dx.doi.org/10.1093/hmg/ddm170.

Valenza, M., Leoni, V., Tarditi, A., Mariotti, C., Bjorkhem, I., Di Donato, S., et al. (2007b). Progressive dysfunction of the cholesterol biosynthesis pathway in the R6/2 mouse model of Huntington's disease. *Neurobiology of Disease*, *28*(1), 133–142. http://dx.doi.org/10.1016/j.nbd.2007.07.004.

Valenza, M., Rigamonti, D., Goffredo, D., Zuccato, C., Fenu, S., Jamot, L., et al. (2005). Dysfunction of the cholesterol biosynthetic pathway in Huntington's disease. *The Journal of Neuroscience, 25*(43), 9932–9939. http://dx.doi.org/10.1523/JNEUROSCI.3355-05.2005.

Varani, K., Bachoud-Levi, A. C., Mariotti, C., Tarditi, A., Abbracchio, M. P., Gasperi, V., et al. (2007). Biological abnormalities of peripheral A(2A) receptors in a large representation of polyglutamine disorders and Huntington's disease stages. *Neurobiology of Disease, 27*(1), 36–43. http://dx.doi.org/10.1016/j.nbd.2007.03.011.

Varani, K., Rigamonti, D., Sipione, S., Camurri, A., Borea, P. A., Cattabeni, F., et al. (2001). Aberrant amplification of A(2A) receptor signaling in striatal cells expressing mutant huntingtin. *The FASEB Journal, 15*(7), 1245–1247.

Villar-Menendez, I., Blanch, M., Tyebji, S., Pereira-Veiga, T., Albasanz, J. L., Martin, M., et al. (2013). Increased 5-methylcytosine and decreased 5-hydroxymethylcytosine levels are associated with reduced striatal A2AR levels in Huntington's disease. *Neuromolecular Medicine, 15*(2), 295–309. http://dx.doi.org/10.1007/s12017-013-8219-0.

Wall, M. J., & Dale, N. (2013). Neuronal transporter and astrocytic ATP exocytosis underlie activity-dependent adenosine release in the hippocampus. *The Journal of Physiology, 591*(Pt 16), 3853–3871. http://dx.doi.org/10.1113/jphysiol.2013.253450.

Wang, L., Lin, F., Wang, J., Wu, J., Han, R., Zhu, L., et al. (2012). Expression of mutant N-terminal huntingtin fragment (htt552-100Q) in astrocytes suppresses the secretion of BDNF. *Brain Research, 1449*, 69–82. http://dx.doi.org/10.1016/j.brainres.2012.01.077.

Wang, J., Wang, C. E., Orr, A., Tydlacka, S., Li, S. H., & Li, X. J. (2008). Impaired ubiquitin-proteasome system activity in the synapses of Huntington's disease mice. *The Journal of Cell Biology, 180*(6), 1177–1189. http://dx.doi.org/10.1083/jcb.200709080.

Weiss, A., Trager, U., Wild, E. J., Grueninger, S., Farmer, R., Landles, C., et al. (2012). Mutant huntingtin fragmentation in immune cells tracks Huntington's disease progression. *The Journal of Clinical Investigation, 122*(10), 3731–3736. http://dx.doi.org/10.1172/JCI64565.

Whitehouse, P. J., Trifiletti, R. R., Jones, B. E., Folstein, S., Price, D. L., Snyder, S. H., et al. (1985). Neurotransmitter receptor alterations in Huntington's disease: Autoradiographic and homogenate studies with special reference to benzodiazepine receptor complexes. *Annals of Neurology, 18*(2), 202–210. http://dx.doi.org/10.1002/ana.410180207.

Wiese, S., Jablonka, S., Holtmann, B., Orel, N., Rajagopal, R., Chao, M. V., et al. (2007). Adenosine receptor A2A-R contributes to motoneuron survival by transactivating the tyrosine kinase receptor TrkB. *Proceedings of the National Academy of Sciences of the United States of America, 104*(43), 17210–17215. http://dx.doi.org/10.1073/pnas.0705267104.

Williams, A., Sarkar, S., Cuddon, P., Ttofi, E. K., Saiki, S., Siddiqi, F. H., et al. (2008). Novel targets for Huntington's disease in an mTOR-independent autophagy pathway. *Nature Chemical Biology, 4*(5), 295–305. http://dx.doi.org/10.1038/nchembio.79.

Williams-Karnesky, R. L., Sandau, U. S., Lusardi, T. A., Lytle, N. K., Farrell, J. M., Pritchard, E. M., et al. (2013). Epigenetic changes induced by adenosine augmentation therapy prevent epileptogenesis. *The Journal of Clinical Investigation, 123*(8), 3552–3563. http://dx.doi.org/10.1172/JCI65636.

Wink, M. R., Braganhol, E., Tamajusuku, A. S., Lenz, G., Zerbini, L. F., Libermann, T. A., et al. (2006). Nucleoside triphosphate diphosphohydrolase-2 (NTPDase2/CD39L1) is the dominant ectonucleotidase expressed by rat astrocytes. *Neuroscience, 138*(2), 421–432. http://dx.doi.org/10.1016/j.neuroscience.2005.11.039.

Wojtowicz, A. M., Dvorzhak, A., Semtner, M., & Grantyn, R. (2013). Reduced tonic inhibition in striatal output neurons from Huntington mice due to loss of astrocytic GABA release through GAT-3. *Frontiers in Neural Circuits*, 7, 188. http://dx.doi.org/10.3389/fncir.2013.00188.

Wong, H. K., Bauer, P. O., Kurosawa, M., Goswami, A., Washizu, C., Machida, Y., et al. (2008). Blocking acid-sensing ion channel 1 alleviates Huntington's disease pathology via an ubiquitin-proteasome system-dependent mechanism. *Human Molecular Genetics*, *17*(20), 3223–3235. http://dx.doi.org/10.1093/hmg/ddn218.

Xie, Y., Hayden, M. R., & Xu, B. (2010). BDNF overexpression in the forebrain rescues Huntington's disease phenotypes in YAC128 mice. *The Journal of Neuroscience*, *30*(44), 14708–14718. http://dx.doi.org/10.1523/JNEUROSCI.1637-10.2010.

Yegutkin, G. G. (2008). Nucleotide- and nucleoside-converting ectoenzymes: Important modulators of purinergic signalling cascade. *Biochimica et Biophysica Acta*, *1783*(5), 673–694. http://dx.doi.org/10.1016/j.bbamcr.2008.01.024.

Young, J. D., Yao, S. Y., Baldwin, J. M., Cass, C. E., & Baldwin, S. A. (2013). The human concentrative and equilibrative nucleoside transporter families, SLC28 and SLC29. *Molecular Aspects of Medicine*, *34*(2–3), 529–547. http://dx.doi.org/10.1016/j.mam.2012.05.007.

Zuccato, C., Belyaev, N., Conforti, P., Ooi, L., Tartari, M., Papadimou, E., et al. (2007). Widespread disruption of repressor element-1 silencing transcription factor/neuron-restrictive silencer factor occupancy at its target genes in Huntington's disease. *The Journal of Neuroscience*, *27*(26), 6972–6983. http://dx.doi.org/10.1523/JNEUROSCI.4278-06.2007.

Zuccato, C., Ciammola, A., Rigamonti, D., Leavitt, B. R., Goffredo, D., Conti, L., et al. (2001). Loss of huntingtin-mediated BDNF gene transcription in Huntington's disease. *Science*, *293*(5529), 493–498. http://dx.doi.org/10.1126/science.1059581.

Zuccato, C., Marullo, M., Conforti, P., MacDonald, M. E., Tartari, M., & Cattaneo, E. (2008). Systematic assessment of BDNF and its receptor levels in human cortices affected by Huntington's disease. *Brain Pathology*, *18*(2), 225–238. http://dx.doi.org/10.1111/j.1750-3639.2007.00111.x.

Zuccato, C., Tartari, M., Crotti, A., Goffredo, D., Valenza, M., Conti, L., et al. (2003). Huntingtin interacts with REST/NRSF to modulate the transcription of NRSE-controlled neuronal genes. *Nature Genetics*, *35*(1), 76–83. http://dx.doi.org/10.1038/ng1219.

Zuchora, B., Turski, W. A., Wielosz, M., & Urbanska, E. M. (2001). Protective effect of adenosine receptor agonists in a new model of epilepsy–seizures evoked by mitochondrial toxin, 3-nitropropionic acid, in mice. *Neuroscience Letters*, *305*(2), 91–94.

CHAPTER ELEVEN

Adenosine Receptors and Epilepsy: Current Evidence and Future Potential

Susan A. Masino*,[1], Masahito Kawamura Jr[†], David N. Ruskin*
*Department of Psychology and Neuroscience Program, Trinity College, Hartford, Connecticut, USA
[†]Department of Pharmacology, Jikei University School of Medicine, Minato-ku, Tokyo, Japan
[1]Corresponding author: e-mail address: susan.masino@trincoll.edu

Contents

Abstract

Adenosine receptors are a powerful therapeutic target for regulating epileptic seizures. As a homeostatic bioenergetic network regulator, adenosine is perfectly suited to establish or restore an ongoing balance between excitation and inhibition, and its anticonvulsant efficacy is well established. There is evidence for the involvement of multiple adenosine receptor subtypes in epilepsy, but in particular the adenosine A_1 receptor subtype can powerfully and bidirectionally regulate seizure activity. Mechanisms that regulate adenosine itself are increasingly appreciated as targets to thus influence receptor activity and seizure propensity. Taken together, established evidence for the powerful potential of adenosine-based epilepsy therapies and new strategies to influence receptor activity can combine to capitalize on this endogenous homeostatic neuromodulator.

1. INTRODUCTION

Adenosine is a ubiquitous and powerful homeostatic bioenergetic network regulator (Boison, Masino, & Geiger, 2011). It exerts these effects

International Review of Neurobiology, Volume 119
ISSN 0074-7742
http://dx.doi.org/10.1016/B978-0-12-801022-8.00011-8

through receptor-dependent and -independent effects, and has the powerful potential to restore homeostasis irrespective of the specific mechanism of upstream disruption of network function. Chronic neurological disorders are characterized by an assortment of metabolic and neuronal changes, and adenosine is in a unique position to address metabolic and neuronal dysfunction broadly. Overall, influencing adenosine receptor function can modulate neuronal activity directly and shift seizure threshold, regardless of whether the seizure's etiology is metabolic, genetic, environmental, or unknown.

Epilepsy is a prevalent neurological disorder that affects about 50 million people worldwide (World Health Organization, 2005). The prevalence of epilepsy is 3–14 per 1000 people in children (Hauser, Annegers, & Kurland, 1991; Medina et al., 2005; Oka et al., 2006; Olafsson & Hauser, 1999; Rwiza et al., 1992) and 5–19 per 1000 people in adults (Hauser et al., 1991; Medina et al., 2005; Olafsson & Hauser, 1999; Rwiza et al., 1992; Tellez-Zenteno, Matijevic, & Wiebe, 2005), indicating that epilepsy is one of the most common disorders of the brain. Epilepsy is defined as "a chronic brain disorder of various etiologies characterized by recurrent seizures due to excessive discharge of cerebral neurones (epileptic seizures), associated with a variety of clinical and laboratory manifestations" in Dictionary of Epilepsy (Gastaut, 1973). It is a (1) chronic disorder of (2) the central nervous system, and involves (3) repetitive seizure events with (4) various causes. It is increasingly appreciated that seizures are comorbid with a primary diagnosis of other common disorders in adults and children, such as Alzheimer's disease and autism (Devinsky, 2003; Gaitatzis, Sisodiya, & Sander, 2012; Gleichmann, Chow, & Mattson, 2011).

One of the hallmarks of epilepsy is its diverse etiology. The underlying causes are categorized as (1) genetic, (2) structural/metabolic (due to injury, environmental triggers, etc.), and (3) unknown. These three terms are intended to replace the traditional terms idiopathic, symptomatic, and cryptogenic (Berg et al., 2010). Genetic epilepsy is caused by known or presumed genetic defects that directly link to seizures (e.g., childhood absence epilepsy, autosomal dominant nocturnal frontal lobe epilepsy, and Dravet syndrome). Structural/metabolic epilepsies include brain injury, stroke, tumor, and infection; some genetic epilepsies cause seizures due to clear structural or metabolic defects. Unfortunately, nearly half of childhood epilepsies have an unknown cause (Wirrell, Grossardt, Wong-Kisiel, & Nickels, 2011).

Epilepsy is frequently associated with a psychiatric disorder and/or cognitive dysfunction. It has been reported that psychiatric comorbidity occurs in 20–50% of epilepsy patients (Akanuma, Hara, Adachi, Hara, & Koutroumanidis, 2008; Cutting, Lauchheimer, Barr, & Devinsky, 2001; Filho, Mazetto, da Silva, Caboclo, & Yacubian, 2011; Trinka et al., 2006). Depression is one of the psychiatric comorbidities with the highest prevalence among patients with epilepsy (Hauser & Hesdorffer, 2001), and anxiety disorder is also a common psychiatric comorbidity (Devinsky, 2003). Autism spectrum disorder, Asperger syndrome, and attention-deficit/hyperactivity disorder are also frequently comorbid with epilepsy (Devinsky, 2003). Interestingly, epilepsy patients with psychiatric comorbidity diagnoses are more likely to be refractory to medication compared to patients without (Akanuma et al., 2008), and psychiatric comorbidities are associated with poor outcomes (Devinsky, 2003). Sleep disruption is also common with an epilepsy diagnosis (Malow, 2005).

Because epilepsy has various underlying causes, and because most often the cause(s) are unknown, it is difficult to treat all epilepsies similarly or even successfully. The most common treatment for epilepsy is therapeutic management of seizures with antiepileptic drugs (AEDs), and about 70% of epileptic patients benefit from AEDs. However ~30% of patients remain medically refractory to AEDs because of inefficacy in controlling seizures or significant adverse events and side effects (St. Louis, 2009; Vajda, 2007). Based on multiple lines of evidence, adenosine remains one of the most promising targets for medically refractory epilepsy as well as epilepsy in general. Here, we describe the involvement of adenosine receptors in control of seizures; where possible, we focus on research within the last decade. We outline adenosine-based antiepileptic and potentially antiepileptogenic effects through receptors, novel adenosine augmentation therapies, and metabolic diet-based therapy.

2. ADENOSINE REGULATES ONGOING NEUROTRANSMISSION

Adenosine is not a classically packaged and released neurotransmitter, yet acts like a transmitter by influencing neuronal activity, primarily via pre- and postsynaptic adenosine A_1 (A_1R) and A_2 (A_2R) receptor subtypes. With this modulatory role on neurotransmission, adenosine is in a unique position to influence the balance of excitation and inhibition without interfering directly with postsynaptic GABA or glutamate receptors. Strong neuronal activity in the nervous system leads directly to elevated extracellular levels of the neuromodulator adenosine. At the same time, A_1Rs inhibit synaptic transmission; thus, adenosine is an inhibitory neuromodulator (Newby, 1984) and so is a target of opportunity in conditions involving neural overactivity such as epilepsy. This activity-induced increase in adenosine serves to increase adenosine "tone" directly in the zone where there is increased neuronal activity. This conceptual framework demonstrates that adenosine is positioned as a homeostatic network modulator: an epilepsy therapy enhances this modulatory role could deliver seizure control and network stability exactly where it is needed most.

Among adenosine receptor subtypes (reviewed by Chen, Lee, & Chern, 2014), direct inhibition of neurotransmission is mediated mostly by the A_1R subtype. Well-established effects on excitatory glutamatergic transmission have been reported for some regions such that overall activity is reduced, but transmission via GABA and other neurotransmitters can also be affected (Sperlágh & Vizi, 2011). Sufficiently high levels of adenosine shut down neurotransmission completely at some synapses (Canals, Larrosa, Pintor, Mena, & Herreras, 2008; Coelho et al., 2006; Kukley, Schwan, Fredholm, & Dietrich, 2005; Martín et al., 2007; Pearson, Damian, Lynas, & Frenguelli, 2006). In addition, adenosine can diffuse away from the site of activation to inhibit neighboring synapses and neurons thus exerting network-level effects (Andersson, Blomstrand, & Hanse, 2007; Chen et al., 2013; Lovatt et al., 2012; Pascual et al., 2005; Serrano, Haddjeri, Lacaille, & Robitaille, 2006).

3. ADENOSINE RECEPTOR SUBTYPES AND EPILEPSY

As noted above, adenosine release is particularly prominent after the excessive neuronal activity of seizures or *in vitro* seizure-like activity (Dulla, Frenguelli, Staley, & Masino, 2009; Etherington et al., 2009; Ilie, Raimondo, & Akerman, 2012; Kaku, Jiang, Hada, Morimoto, & Hayashi, 2001; Lovatt et al., 2012; Van Gompel et al., 2014), and tissue content of adenosine rises rapidly after seizure initiation (Winn, Welsh, Rubio, & Berne, 1980). These seizure-related adenosine elevations are believed to be a feedback mechanism that limits seizure intensity or duration (Ilie et al., 2012; Lovatt et al., 2012; Van Gompel et al., 2014). As with neurotransmission *per se*, much evidence from animal research shows that this effect is A_1R-mediated. Systemic injection of A_1R-selective agonists limits seizures and convulsions in numerous types of experimental seizures and epilepsies (Girardi, Auzmendi, Charo, Gori, & Castro, 2010; Girardi, Canitrot, Antonelli, Gonzalez, & Coirini, 2007; Gouder, Fritschy, &

Boison, 2003; Li, Kang, Shi, Liu, & Zhang, 2013; Li & Zhang, 2011; Luszczki, Kozicka, Swiader, & Czuczwar, 2005; Mares, 2010; Moezi, Akbarian, Niknahad, & Shafaroodi, 2013; Muzzi, Coppi, Pugliese, & Chiarugi, 2013; Pometlova, Kubova, & Mares, 2010; Tosh et al., 2012; Vianna, Ferreira, Doná, Cavalheiro, & da Silva Fernandes, 2005; Zuchora, Wielosz, & Urbańska, 2005). In addition, A_1R activation augments the effects of standard AEDs (Borowicz, Łuszczki, & Czuczwar, 2002; Luszczki et al., 2005). Focal intracranial activation of A_1Rs in various brain regions is also anticonvulsant (Guttinger, Padrun, Pralong, & Boison, 2005; Heidarianpour, Sadeghian, Mirnajafi-Zadeh, Fathollahi, & Mohammad-Zadeh, 2006; Hosseinmardi, Mirnajafi-Zadeh, Fathollahi, & Shahabi, 2007; Rezvani, Mirnajafi-Zadeh, Fathollahi, & Palizvan, 2007a,2007b; Zeraati, Mirnajafi-Zadeh, Fathollahi, Namvar, & Rezvani, 2006). Drug development efforts have long been focused on developing central-acting therapeutics and limiting peripheral side effects, particularly cardiac side effects. Recent research suggests targeting highly specific regions of the A_1R may offer an advance in direct A_1R-based antiseizure therapies (Tosh et al., 2012).

Conversely, and further confirming antiseizure effects of A_1Rs, systemic injection of A_1R-selective antagonists worsens seizures (Fukuda et al., 2010) and blunts the beneficial effects of standard AEDs (Chwalczuk, Rubaj, Swiader, & Czuczwar, 2008; Zuchora et al., 2005) and experimental anticonvulsant compounds (Akula & Kulkarni, 2014; Rezvani, Roohbakhsh, Allahtavakoli, & Shamsizadeh, 2010; Uzbay, Kayir, & Ceyhan, 2007). Strikingly, systemic genetic inactivation of this receptor subtype leads to spontaneous electrographic seizures in cerebral cortex, allows focal seizure activity to generalize, and promotes status epilepticus after traumatic brain injury (Fedele, Li, Lan, Fredholm, & Boison, 2006; Kochanek et al., 2006; Masino, Li, et al., 2011).

In brain slices and cultures, seizure-like or epileptiform electrical patterns (repetitive bursts of action potentials, oscillatory field potentials, etc.) can be induced by elevating excitability or limiting inhibition. As expected, adenosine release is prominent during this type of epileptiform activity, and is autoinhibitory via A_1Rs (demonstrated by A_1R-selective antagonist application and use of tissue from A_1R knock-out (KO) mice) (Avsar & Empson, 2004; Dulla et al., 2005, 2009; Etherington & Frenguelli, 2004; Etherington et al., 2009; Gu, Hazra, Aulakh, & Ziburkus, 2014; Hargus, Jennings, Perez-Reyes, Bertram, & Patel, 2012; Klaft et al., 2012; Muzzi et al., 2013; Salmani, Fathollahi, Mirnajafizadeh, & Semnanian, 2011; Wang et al., 2013; Wu, Wong, Wu, Sheppy, & Zhang, 2009; Zhou et al., 2011).

A_1Rs have long been strongly associated with antiseizure effects due to their clear effects on neuronal activity, high affinity, and widespread distribution. They are concentrated and influence ongoing transmission in seizure-prone areas (neocortex and hippocampus). Yet, other adenosine receptors may also play a role and offer therapeutic targets for epilepsy. For example, $A_{2A}Rs$ are excitatory in most brain regions; therefore, $A_{2A}Rs$ antagonists would be more likely to have antiseizure effects. $A_{2A}Rs$ have a strikingly restricted anatomical distribution compared to A_1Rs (for review see Chen et al., 2014; Mishina & Ishikawa, 2014). Thus, it is perhaps not surprising that a number of paradigms have shown it was not involved in convulsions (Akula & Kulkarni, 2014; Hargus et al., 2012; Li & Zhang, 2011; Rezvani et al., 2007a; Uzbay et al., 2007; Wang et al., 2013; Young & Dragunow, 1994), and one study showed that $A_{2A}R$ modulation of convulsions was selective to the basal ganglia (Rosen & Berman, 1987). When significant effects are found, pharmacological evidence from *in vivo* and *in vitro* studies mostly shows that $A_{2A}Rs$ are proconvulsant and proseizure (D'Alimonte et al., 2009; Etherington & Frenguelli, 2004; Li et al., 2012; Mares, 2010; Vianna et al., 2005; Zeraati et al., 2006), though some studies have evidence for anticonvulsant or anatomically specific effects (Hosseinmardi et al., 2007; Moschovos, Kostopoulos, & Papatheodoropoulos, 2012). Regardless of valence, some of these effects were found in regions with few receptors, which could suggest a nonspecificity of drug dose or concentration, although $A_{2A}Rs$ may have functional actions even if expressed at quite low levels (Cunha et al., 1994). Mice with genetic inactivation of $A_{2A}Rs$ are seizure-resistant in several models, further supporting a proconvulsant receptor action (El Yacoubi, Ledent, Parmentier, Costentin, & Vaugeois, 2008, 2009; El Yacoubi et al., 2001).

Regarding other adenosine receptor, though central expression is low, pharmacological evidence for effects of the adenosine A_3 receptor (A_3R) has been reported in a few studies, but results are mixed. Notably, A_3R activation promotes seizure-like activity in hippocampal slices (Etherington & Frenguelli, 2004; Laudadio & Psarropoulou, 2004), whereas an anticonvulsant effect of this receptor was suggested in studies measuring convulsions and using systemic administration (Borowicz, Świąder, Wielosz, & Czuczwar, 2004; Vianna et al., 2005). Thus, there may be region-specific A_3R effects. There are essentially no data on seizure involvement of $A_{2B}Rs$, which are also expressed only at a low level in the brain.

In sum, pharmacological manipulations of ARs have long been targeted in treating epilepsy, with A_1R activation being the most effective. A_1R

agonists as a whole, however, have a number of peripheral side effects (most dangerously involving the heart) such that they have not been pursued clinically – though some recent compounds lack cardiac effects in animal models (Luongo et al., 2012) – and new strategies may produce central nervous system-dominant compounds. Conversely, existing anticonvulsant drugs do not have substantial affinities for ARs (Macdonald & Kelly, 1995) but there may be AEDs with some A_1R activity. A gentler means of promoting adenosinergic transmission to reduce seizures might be to reduce metabolism of adenosine. Allopurinol is an inhibitor of xanthine oxidase that elevates levels of compounds "upstream" of this enzyme in the degradative pathway from adenosine to uric acid (Schmidt et al., 2009). Allopurinol enhances A_1R activity (Essawy & Elbaz, 2013; Schmidt et al., 2009). This drug is very well tolerated in patients, and two out of three double-blind and placebo-controlled studies concluded that it helped with seizure control (Coppola & Pascotto, 1996; Togha, Akhondzadeh, Motamedi, Ahmadi, & Razeghi, 2007; Zagnoni et al., 1994).

4. CHRONIC CHANGES IN ADENOSINE RECEPTOR ACTIVATION

One of the most widely consumed psychoactive drugs, caffeine, blocks adenosine receptors (except A_3R). Caffeine can, on its own, induce seizures, but only at acute doses far above what humans self-administer. At more reasonable doses, caffeine reduces the anticonvulsant activity of some AEDs in animal models and apparently humans as well (Chrościńska-Krawczyk, Jargiełło-Baszak, Wałek, Tylus, & Czuczwar, 2011), and epileptic patients who are not habitual caffeine users should therefore avoid a large acute intake of caffeine. This is particularly true if there is a circadian pattern to their seizures: a large acute dose of caffeine should be avoided around the time when they are most vulnerable to a seizure.

In contrast, for habitual caffeine users, the situation is more complex due to chronic changes in adenosine receptors. Most metabotropic neurotransmitter receptors respond to chronic antagonism by increased expression, an apparently homeostatic response. ARs are not exceptions to this principle: central A_1R and $A_{2A}R$ levels in experimental animals are augmented after chronic caffeine (Daval, Deckert, Weiss, Post, & Marangos, 1989; Fredholm, 1982; Johansson, Georgiev, Lindström, & Fredholm, 1997; Kaplan, Greenblatt, Kent, & Cotreau-Bibbo, 1993; Li, Dai, et al., 2008; Marangos, Boulenger, & Patel, 1984; Rudolphi, Keil, Fastbom, &

Fredholm, 1989; Tsutsui et al., 2004) and after or chronic theophylline (a closely related drug) treatment (Lupica, Berman, & Jarvis, 1991; Murray, 1982; Szot, Sanders, & Murray, 1987). Adenosinergic transmission should be improved with higher receptor expression, and indeed, in some studies, seizures in experimental animals are moderated by prior chronic (>1 week) caffeine (El Yacoubi et al., 2008, 2001). Unsurprisingly, chronic caffeine is also neuroprotective (Cognato et al., 2010; Li, Dai, et al., 2008; Rudolphi et al., 1989; Tsutsui et al., 2004). The ideal situation for a seizure-inducing event might therefore be in a habitual coffee drinker with no or low circulating caffeine; such events are not predictable, unfortunately. One recent study concluded that seizures in epileptic patients were unrelated to either habitual or acute caffeine intake (Samsonsen, Bråthen, Reimers, Helde, & Brodtkorb, 2013).

5. GLIOSIS, ADENOSINE KINASE, AND EPILEPTOGENESIS

Not only does the adenosinergic system modulate seizures, but the reverse is also true. One particularly relevant target of seizure activity is the astrocytic enzyme adenosine kinase (ADK), the major intracellular enzyme for adenosine regulation and rephosphorylation. This enzyme is acutely downregulated after seizures, which is likely to aid in adenosinergic protection against excessive neuronal activity. However, gliosis is prominently involved in some types of epilepsy (traumatic brain injury-related; tumor-related; temporal lobe) and consequently ADK becomes overexpressed (Aronica et al., 2011; de Groot et al., 2012; Luan et al., 2013; Masino, Li, et al., 2011). Transgenic overexpression of ADK in forebrain causes spontaneous seizures which can be ameliorated by adenosine delivery (Li, Ren, et al., 2008). In fact, gliosis-related seizures are specifically caused by ADK overexpression (Li, Lan, & Boison, 2008; Li, Ren, et al., 2008). Such a nonhomeostatic positive feedback among seizures, gliosis, elevated ADK, and depleted adenosine has been posited to underlie epileptogenesis itself (Boison, 2008). Careful pharmacological inhibition of ADK might therefore be useful in preventing epileptic progression and restoring normal network excitability.

6. HUMAN ADENOSINE RECEPTOR POLYMORPHISMS

In humans, some adenosine receptor gene variants have been associated with seizure disorders. An $A_{2A}R$ polymorphism associated with encephalopathy with febrile status epilepticus is related to increased receptor expression and second messenger output (Shinohara et al., 2013), in accordance with the apparent proconvulsant effect of this receptor subtype. Some polymorphisms in A_1Rs are associated with traumatic brain injury-related seizures, particularly late-onset seizures (Wagner et al., 2010), underscoring the experimental association between this receptor subtype and injury-related seizures. Notably, adenosine-related genes that are not receptors (but would be expected to influence adenosine levels) also have seizure-related polymorphisms. Duplications of a chromosomal region containing the gene for adenosine deaminase (an extracellular enzyme metabolizing adenosine to inosine) are associated with childhood epilepsy in some individuals (Chen et al., 2011; Muhle et al., 2010). A polymorphism in the gene for an equilibrative nucleoside transporter-1 (a major means of removal of adenosine from the extracellular space) is associated with withdrawal-associated seizures in alcoholics (Kim et al., 2011).

7. ADENOSINE AND NEUROINFLAMMATION

Evidence indicates that at least some epileptogenic mechanisms involve an important inflammatory component (Vezzani et al., 2000; Vitaliti, Pavone, Mahmood, Nunnari, & Falsaperla, 2014). Inflammation in the brain is associated with a number of seizure-related conditions (central infection, cancer, physical injury, and ischemia) and the inverse is also true in that seizure activity elevates brain inflammation (Riazi, Galic, & Pittman, 2010). Many types of cells involved in inflammation express adenosine receptors, and so adenosine influences immune cell activity in a receptor subtype- and cell type-dependent manner, and adenosine receptors on non-immune cells (such as neurons) also influence inflammation (Blackburn, Vance, Morschl, & Wilson, 2009; Kumar & Sharma, 2009; Linden, 2001). Furthermore, adenosine receptor subtypes differ in desensitization after prolonged activation (Blackburn et al., 2009), such that subtype involvement might change during the course of events. Thus, general statements about a particular subtype being pro- or anti-inflammatory are essentially impossible. As an illustration of the complexity, brain $A_{2A}Rs$ have been concluded to be either proinflammatory or anti-inflammatory in recent studies (Golembiowska, Wardas, Noworyta-Sokolowska, Kaminska, & Gorska, 2013; Gyoneva et al., 2014; Kermanian, Soleimani, Ebrahimzadeh, Haghir, & Mehdizadeh, 2013; Loram et al., 2013; Mecha et al., 2013; Mohamed, Agha, & Nassar, 2012; Simoes et al., 2012; Yao et al., 2012).

Much of this pattern is explainable by $A_{2A}R$ location: using elegant combinations of transplantation and genetic KO, it has been shown in several models that $A_{2A}Rs$ expressed by bone marrow-derived cells (particularly white blood cells) and by native brain cells have opposing effects on central inflammatory insults (Huang et al., 2006; Mills, Kim, Krenz, Chen, & Bynoe, 2012; Yu et al., 2004).

8. ADENOSINE-BASED MECHANISMS UNDERLYING ANTICONVULSANT DIET THERAPY

Fasting is well known to decrease the incidence of seizures. From these historical observations, a ketogenic diet (KD, a low-carbohydrate high-fat protocol) was designed in the 1920s to mimic the metabolic state of fasting and thus reduce epileptic seizures (Wilder, 1921). During the next two decades, development of AEDs gradually replaced fasting and KD as the first choice for an antiepileptic therapy. However, in recent decades, as it nears its hundredth anniversary, KD has increasingly been noted as a useful therapy for epilepsy because it has been demonstrated repeatedly that patients with medically refractory epilepsy benefit from KD therapy (Hallböök, Köhler, Rosén, & Lundgren, 2007). KD is reported to reduce seizure activity both in humans (Freeman, Kossoff, & Hartman, 2007; Henderson, Filloux, Alder, Lyon, & Caplin, 2006; Neal et al., 2008; Wiznitzer, 2008) and rodents (Bough et al., 2006; Hartman, Lyle, Rogawski, & Gasior, 2008; Rho, Kim, Robbins, Anderson, & Schwartzkroin, 1999; Uhlemann & Neims, 1972). Despite nearly 100 years from the advent of KD therapy, the key mechanisms underlying the anticonvulsant effect of KD is still undetermined and an active research topic (Bough & Rho, 2007; Nylen, Likhodii, & Burnham, 2009).

Some of the proposed mechanisms of a KD's antiseizure effects include increased GABA production (Melø, Nehlig, & Sonnewald, 2006; Yudkoff, Daikhin, Nissim, & Lazarow, 2001) (note that these GABA mechanisms are similar to AED mechanisms and cannot explain KD's success in medically refractory seizures), inhibition of glutamate release through reducing glutamate uptake via vesicular glutamate transporters by direct action of ketone bodies (Juge et al., 2010), direct opening of metabolic ATP-sensitive potassium (K_{ATP}) channels with increasing ketone levels (Ma, Berg, & Yellen, 2007), increase of mitochondrial gene expression (Noh et al., 2004; Nylen, Velazquez, et al., 2009), and decreased IL-1β cytokine expression (Vezzani et al., 2000). Alongside these mechanisms, growing lines of evidence support adenosine and adenosine receptors as one of the key mechanisms underlying the anticonvulsant effect of KDs (Masino & Geiger, 2008). Thus far, the most compelling evidence was reported by using ADK-transgenic (ADK-Tg) mice, heterozygous A_1R-KO mice and homozygous A_1R-KO mice. During control diet feeding, these three lines of transgenic mice demonstrated spontaneous electrographic hippocampal seizures via reduction of extracellular adenosine levels (with ADK overexpression) or lacking activation of A_1Rs (in heterozygous A_1R-KO mice, seizures events were smaller and fewer than in homozygous A_1R-KO mice, but in both models spontaneous seizures occurred significantly). KD reduced seizures in ADK-Tg mice and heterozygotic A_1R-KO mice, but did not decrease spontaneous seizures in homozygotic A_1R-KO mice. The selective A_1R inhibitor 8-cyclopentyl-1,3-dipropylxanthine (DPCPX) reversed this KD effect in ADK-Tg mice and heterozygous A_1R-KO mice. In the presence of DPCPX, spontaneous electric seizures in homozygous A_1R-KO mice were unchanged. This report strongly argues that the reduction of seizures with KD is caused by the activation of A_1R (Masino, Li, et al., 2011).

More detailed mechanisms of a KD's effect through activation of A_1R were investigated by using a brain slice preparation which mimicked a KD *in vitro*. Increased ketone body levels with a KD might accelerate intracellular ATP production and decrease blood glucose levels (Masino & Rho, 2012), and it has been reported that a KD increases brain concentrations of ATP in rodents (DeVivo, Leckie, Ferrendelli, & McDougal, 1978; Nakazawa, Kodama, & Matsuo, 1983; Nylen, Velazquez, et al., 2009) and humans (Pan, Bebin, Chu, & Hetherington, 1999) and decreases blood glucose levels in both (Bough et al., 2006; Huttenlocher, 1976). Interestingly, reduction of blood glucose is also known to decrease the influence of seizures in both rodents (Greene, Todorova, McGowan, & Seyfried, 2001; Mantis, Centeno, Todorova, McGowan, & Seyfried, 2004) and humans (Muzykewicz et al., 2009). Based on this evidence, and using a whole-cell patch clamp technique, neuronal intracellular ATP concentration was increased and extracellular glucose concentration was reduced in order to mimic a KD. This *in vitro* condition which mimicked a KD caused hyperpolarization through autocrine activation of A_1R. Reduced glucose caused opening of pannexin-1 channels which are known to be ATP-releasing sites. Opening of pannexin channels released ATP from the intracellular to the extracellular space, and release of ATP increases extracellular adenosine levels after rapid breakdown with ectonucleotidases

(Cunha, Sebastiao, & Ribeiro, 1998; Dunwiddie, Diao, & Proctor, 1997). The resulting A_1R activation opened K_{ATP} channels and hyperpolarized the neuron (Kawamura, Ruskin, & Masino, 2010).

Hyperpolarization with this adenosinergic autocrine modulation described above causes homeostatic reduction of neuronal activity and could exert a strong antiseizure effect. Taken together, these data show that adenosine receptor-based seizure suppression might be one of the main mechanisms of KD-induced anticonvulsant effects. For example, activation of A_1Rs is known to prevent $GABA_A$ receptor blocker-induced seizure-like activity (Thompson, Haas, & Gähwiler, 1992) because adenosine receptors decrease excitatory synaptic transmission by inhibiting voltage-gated calcium channels and hyperpolarize neurons by opening of potassium channels (Schulte & Fredholm, 2003). Therefore, by activating adenosine/A_1R-based mechanisms, KDs act on different pathways than AEDs and might explain why KDs are highly effective in treating medically refractory epilepsies.

9. CONCLUSIONS

Epilepsy has been documented as an ongoing part of human civilization from 2000 BC to present (Magiorkinis, Sidiropoulou, & Diamantis, 2010). Therefore, epilepsy has always coexisted with us, and it remains a perennial and prevalent problem. Today, AEDs and surgical therapies can successfully manage many epilepsy patients. However, 30% of patients are medically refractory, and we need strategies for reversing epilepsy and for treating the millions of people in this patient population.

Adenosine has been recognized as an anticonvulsant for several decades, and it is a perennial therapeutic target. This seemingly permanent place for adenosine receptors on the landscape of epilepsy treatments acknowledges their capacity to treat all types of seizures—metabolic, genetic, environmental, or unknown origin—and to treat seizures for which no other therapies are effective. Improved adenosine-based options may be better tolerated than some of the current AEDs, and could help some patients avoid surgery. This is particularly true if adenosine can exert a combination of receptor-based antiseizure effects and receptor-independent antiepileptic effects. For example, local adenosine augmentation therapies and adenosine-based anticonvulsant diets might be new strategies for medically refractory epilepsy.

One of the issues of present AEDs is that AEDs often control seizures but may exacerbate psychiatric comorbidities (Devinsky, 2003). The potential of a KD (perhaps through adenosine-based mechanisms) is that it can improve not only epilepsy as well as various types of neurological disorders (Masino, Kawamura, Wasser, Pomeroy, & Ruskin, 2009), including autism which is often comorbidity with epilepsy (Evangeliou et al., 2003; Ruskin et al., 2013). Furthermore adenosine is reported to improve autism spectrum disorders (Masino, Kawamura, Cote, Williams, & Ruskin, 2013; Masino, Kawamura, et al., 2011, Masino, Li, et al., 2011). Also, local adenosine augmentation therapies ameliorate the cognitive symptoms and psychomotor-related symptoms in schizophrenia (Shen et al., 2012). These adenosine-based strategies might shed light on desired therapies that reduce seizures and improve psychiatric comorbidity simultaneously.

REFERENCES

Akanuma, N., Hara, E., Adachi, N., Hara, K., & Koutroumanidis, M. (2008). Psychiatric comorbidity in adult patients with idiopathic generalized epilepsy. *Epilepsy and Behavior, 13*, 248–251.

Akula, K. K., & Kulkarni, S. K. (2014). Effect of curcumin against pentylenetetrazol-induced seizure threshold in mice: Possible involvement of adenosine A1 receptors. *Phytotherapy Research, 28*(5), 714–721.

Andersson, M., Blomstrand, F., & Hanse, E. (2007). Astrocytes play a critical role in transient heterosynaptic depression in the rat hippocampal CA1 region. *Journal of Physiology, 585*, 843–852.

Aronica, E., Zurolo, E., Iyer, A., de Groot, M., Anink, J., Carbonell, C., et al. (2011). Upregulation of adenosine kinase in astrocytes in experimental and human temporal lobe epilepsy. *Epilepsia, 52*, 1645–1655.

Avsar, E., & Empson, R. M. (2004). Adenosine acting via A1 receptors, controls the transition to status epilepticus-like behaviour in an in vitro model of epilepsy. *Neuropharmacology, 47*, 427–437.

Berg, A. T., Berkovic, S. F., Brodie, M. J., Buchhalter, J., Cross, J. H., van Emde Boas, W., et al. (2010). Revised terminology and concepts for organization of seizures and epilepsies: Report of the ILAE Commission on Classification and Terminology, 2005–2009. *Epilepsia, 51*, 676–685.

Blackburn, M. R., Vance, C. O., Morschl, E., & Wilson, C. N. (2009). Adenosine receptors and inflammation. In C. N. Wilson, & S. J. Mustafa (Eds.), *Adenosine receptors in health and disease* (pp. 217–269). Berlin Heidelberg: Springer-Verlag.

Boison, D. (2008). The adenosine kinase hypothesis of epileptogenesis. *Progress in Neurobiology, 84*, 249–262.

Boison, D., Masino, S. A., & Geiger, J. D. (2011). Homeostatic bioenergetic network regulation: A novel concept to avoid pharmacoresistance in epilepsy. *Expert Opinion on Drug Discovery, 6*, 713–724.

Borowicz, K. K., Łuszczki, J., & Czuczwar, S. J. (2002). 2-Chloroadenosine, a preferential agonist of adenosine A_1 receptors, enhances the anticonvulsant activity of carbamazepine and clonazepam in mice. *European Neuropsychopharmacology, 12*, 173–179.

Borowicz, K. K., Świąder, M., Wielosz, M., & Czuczwar, S. J. (2004). Influence of the combined treatment of LY 300164 (an AMPA/kainate receptor antagonist) with adenosine receptor agonists on the electroconvulsive threshold in mice. *European Neuropsychopharmacology*, *14*, 407–412.

Bough, K. J., & Rho, J. M. (2007). Anticonvulsant mechanisms of the ketogenic diet. *Epilepsia*, *48*, 43–58.

Bough, K. J., Wetherington, J., Hassel, B., Pare, J. F., Gawryluk, J. W., Greene, J. G., et al. (2006). Mitochondrial biogenesis in the anticonvulsant mechanism of the ketogenic diet. *Annals of Neurology*, *60*, 223–235.

Canals, S., Larrosa, B., Pintor, J., Mena, M. A., & Herreras, O. (2008). Metabolic challenge to glia activates an adenosine-mediated safety mechanism that promotes neuronal survival by delaying the onset of spreading depression waves. *Journal of Cerebral Blood Flow and Metabolism*, *28*, 1835–1844.

Chen, J.-F., Lee, C.-f., & Chern, Y. (2014). Adenosine receptor neurobiology: Overview. In A. Mori (Ed.), *International review of neurobiology: Vol. 119. Adenosine receptors in neurology and psychiatry* (pp. 1–50). Amsterdam: Elsevier.

Chen, C.-P., Lin, S.-P., Chen, M., Su, Y.-N., Chern, S.-R., Wang, T.-Y., et al. (2011). Mosaic supernumerary r(1)(p13.2q23.3) in a 10-year-old girl with epilepsy, facial asymmetry, psychomotor retardation, kyphoscoliosis, dermatofibrosarcoma and multiple exostoses. *Genetic Counseling*, *22*, 273–280.

Chen, J., Tan, Z., Zeng, L., Zhang, X., He, Y., Gao, W., et al. (2013). Heterosynaptic long-term depression mediated by ATP released from astrocytes. *Glia*, *61*, 178–191.

Chrościńska-Krawczyk, M., Jargiełło-Baszak, M., Wałek, M., Tylus, B., & Czuczwar, S. J. (2011). Caffeine and the anticonvulsant potency of antiepileptic drugs: Experimental and clinical data. *Pharmacological Reports*, *63*, 12–18.

Chwalczuk, K., Rubaj, A., Swiader, M., & Czuczwar, S. J. (2008). Influence of the antagonist of adenosine A1 receptors, 8-cyclopentyl-1,3-dipropylxanthine, upon the anticonvulsant activity of antiepileptic drugs in mice. *Przegląd Lekarski*, *65*, 759–763.

Coelho, J. E., Rebola, N., Fragata, I., Ribeiro, J. A., De Mendonca, A., & Cunha, R. A. (2006). Hypoxia-induced desensitization and internalization of adenosine A_1 receptors in the rat hippocampus. *Neuroscience*, *138*, 1195–1203.

Cognato, G. P., Agostinho, P. M., Hockemeyer, J., Müller, C. E., Souza, D. O., & Cunha, R. A. (2010). Caffeine and an adenosine A_{2A} receptor antagonist prevent memory impairment and synaptotoxicity in adult rats triggered by a convulsive episode in early life. *Journal of Neurochemistry*, *112*, 453–462.

Coppola, G., & Pascotto, A. (1996). Double-blind, placebo-controlled, cross-over trial of allopurinol as add-on therapy in childhood refractory epilepsy. *Brain Dev*, *18*, 50–52.

Cunha, R. A., Johansson, B., van der Ploeg, I., Sebastião, A. M., Ribeiro, J. A., & Fredholm, B. B. (1994). Evidence for functionally important adenosine A_{2a} receptors in the rat hippocampus. *Brain Research*, *649*, 208–216.

Cunha, R. A., Sebastiao, A. M., & Ribeiro, J. A. (1998). Inhibition by ATP of hippocampal synaptic transmission requires localized extracellular catabolism by ecto-nucleotidases into adenosine and channeling to adenosine A_1 receptors. *Journal of Neuroscience*, *18*, 1987–1995.

Cutting, S., Lauchheimer, A., Barr, W., & Devinsky, O. (2001). Adult-onset idiopathic generalized epilepsy: Clinical and behavioral features. *Epilepsia*, *42*, 1395–1398.

D'Alimonte, I., D'Auro, M., Citraro, R., Biagioni, F., Jiang, S., Nargi, E., et al. (2009). Altered distribution and function of A2A adenosine receptors in the brain of WAG/Rij rats with genetic absence epilepsy, before and after appearance of the disease. *European Journal of Neuroscience*, *30*, 1023–1035.

Daval, J.-L., Deckert, J., Weiss, S. R. B., Post, R. M., & Marangos, P. J. (1989). Upregulation of adenosine A1 receptors and forskolin binding sites following chronic treatment with caffeine or carbamazepine: A quantitative autoradiographic study. *Epilepsia*, *30*, 26–33.

de Groot, M., Iyer, A., Zurolo, E., Anink, J., Heimans, J. J., Boison, D., et al. (2012). Overexpression of ADK in human astrocytic tumors and peritumoral tissue is related to tumor-associated epilepsy. *Epilepsia*, *53*, 58–66.

Devinsky, O. (2003). Psychiatric comorbidity in patients with epilepsy: Implications for diagnosis and treatment. *Epilepsy and Behavior*, *4*(Suppl. 4), S2–S10.

DeVivo, D. C., Leckie, M. P., Ferrendelli, J. S., & McDougal, D. B., Jr. (1978). Chronic ketosis and cerebral metabolism. *Annals of Neurology*, *3*, 331–337.

Dulla, C. G., Dobelis, P., Pearson, T., Frenguelli, B. G., Staley, K. J., & Masino, S. A. (2005). Adenosine and ATP link P_{CO2} to cortical excitability via pH. *Neuron*, *48*, 1011–1023.

Dulla, C. G., Frenguelli, B. G., Staley, K. J., & Masino, S. A. (2009). Intracellular acidification causes adenosine release during states of hyperexcitability in the hippocampus. *Journal of Neurophysiology*, *102*, 1984–1993.

Dunwiddie, T. V., Diao, L., & Proctor, W. R. (1997). Adenine nucleotides undergo rapid, quantitative conversion to adenosine in the extracellular space in rat hippocampus. *Journal of Neuroscience*, *17*, 7673–7682.

El Yacoubi, M., Ledent, C., Parmentier, M., Costentin, J., & Vaugeois, J. M. (2008). Evidence for the involvement of the adenosine A(2A) receptor in the lowered susceptibility to pentylenetetrazol-induced seizures produced in mice by long-term treatment with caffeine. *Neuropharmacology*, *55*, 35–40.

El Yacoubi, M., Ledent, C., Parmentier, M., Costentin, J., & Vaugeois, J. M. (2009). Adenosine A2A receptor deficient mice are partially resistant to limbic seizures. *Naunyn-Schmiedeberg's Archives of Pharmacology*, *380*, 223–232.

El Yacoubi, M., Ledent, C., Parmentier, M., Daoust, M., Costentin, J., & Vaugeois, J.-M. (2001). Absence of the adenosine A_{2A} receptor or its chronic blockade decrease ethanol withdrawal-induced seizures in mice. *Neuropharmacology*, *40*, 424–432.

Essawy, S. S., & Elbaz, A. A. (2013). Role of adenosine receptors in the anti-nociceptive effects of allopurinol in mice. *European Review for Medical and Pharmacological Sciences*, *17*, 1857–1863.

Etherington, L. A., & Frenguelli, B. G. (2004). Endogenous adenosine modulates epileptiform activity in rat hippocampus in a receptor subtype-dependent manner. *European Journal of Neuroscience*, *19*, 2539–2550.

Etherington, L.-A. V., Patterson, G. E., Meechan, L., Boison, D., Irving, A. J., Dale, N., et al. (2009). Astrocytic adenosine kinase regulates basal synaptic adenosine levels and seizure activity but not activity-dependent adenosine release in the hippocampus. *Neuropharmacology*, *56*, 429–437.

Evangeliou, A., Vlachonikolis, I., Mihailidou, H., Spilioti, M., Skarpalezou, A., Makaronas, N., et al. (2003). Application of a ketogenic diet in children with autistic behavior: Pilot study. *Journal of Child Neurology*, *18*, 113–118.

Fedele, D. E., Li, T., Lan, J. Q., Fredholm, B. B., & Boison, D. (2006). Adenosine A_1 receptors are crucial in keeping an epileptic focus localized. *Experimental Neurology*, *200*, 184–190.

Filho, G. M., Mazetto, L., da Silva, J. M., Caboclo, L. O., & Yacubian, E. M. (2011). Psychiatric comorbidity in patients with two prototypes of focal versus generalized epilepsy syndromes. *Seizure*, *20*, 383–386.

Fredholm, B. B. (1982). Adenosine actions and adenosine receptors after 1 week treatment with caffeine. *Acta Physiologica Scandinavica*, *115*, 283–286.

Freeman, J. M., Kossoff, E. H., & Hartman, A. L. (2007). The ketogenic diet: One decade later. *Pediatrics*, *119*, 535–543.

Fukuda, M., Suzuki, Y., Hino, H., Kuzume, K., Morimoto, T., & Ishii, E. (2010). Adenosine A1 receptor blockage mediates theophylline-associated seizures. *Epilepsia*, *51*, 483–487.

Gaitatzis, A., Sisodiya, S. M., & Sander, J. W. (2012). The somatic comorbidity of epilepsy: A weighty but often unrecognized burden. *Epilepsia*, *53*, 1282–1293.

Gastaut, H. (1973). *Dictionary of epilepsy. Part 1. Definitions*. Geneva: World Health Organization.

Girardi, E., Auzmendi, J., Charo, N., Gori, M. B., & Castro, M. (2010). 3-mercaptopropionic acid-induced seizures decrease NR2B expression in Purkinje cells: Cyclopentyladenosine effect. *Cellular and Molecular Neurobiology*, *30*, 985–990.

Girardi, E. S., Canitrot, J., Antonelli, M., Gonzalez, N. N., & Coirini, H. (2007). Differential expression of cerebellar metabotropic glutamate receptors mGLUR2/3 and mGLUR4a after the administration of a convulsant drug and the adenosine analogue cyclopentyladenosine. *Neurochemical Research*, *32*, 1120–1128.

Gleichmann, M., Chow, V. W., & Mattson, M. P. (2011). Homeostatic disinhibition in the aging brain and alzheimer's disease. *Journal of Alzheimer's Disease*, *24*, 15–24.

Golembiowska, K., Wardas, J., Noworyta-Sokolowska, K., Kaminska, K., & Gorska, A. (2013). Effects of adenosine receptor antagonists on the in vivo LPS-induced inflammation model of Parkinson's disease. *Neurotoxicity Research*, *24*, 29–40.

Gouder, N., Fritschy, J.-M., & Boison, D. (2003). Seizure suppression by adenosine A_1 receptor activation in a mouse model of pharmacoresistant epilepsy. *Epilepsia*, *44*, 877–885.

Greene, A. E., Todorova, M. T., McGowan, R., & Seyfried, T. N. (2001). Caloric restriction inhibits seizure susceptibility in epileptic EL mice by reducing blood glucose. *Epilepsia*, *42*, 1371–1378.

Gu, F., Hazra, A., Aulakh, A., & Ziburkus, J. (2014). Purinergic control of hippocampal circuit hyperexcitability in Dravet syndrome. *Epilepsia*, *55*, 245–255.

Guttinger, M., Padrun, V., Pralong, W. F., & Boison, D. (2005). Seizure suppression and lack of adenosine A1 receptor desensitization after focal long-term delivery of adenosine by encapsulated myoblasts. *Experimental Neurology*, *193*, 53–64.

Gyoneva, S., Shapiro, L., Lazo, C., Garnier-Amblard, E., Smith, Y., Miller, G. W., et al. (2014). Adenosine A receptor antagonism reverses inflammation-induced impairment of microglial process extension in a model of Parkinson's disease. *Neurobiology of Disease*, *67*, 191–202.

Hallböök, T., Köhler, S., Rosén, I., & Lundgren, J. (2007). Effects of ketogenic diet on epileptiform activity in children with therapy resistant epilepsy. *Epilepsy Research*, *77*, 134–140.

Hargus, N. J., Jennings, C., Perez-Reyes, E., Bertram, E. H., & Patel, M. K. (2012). Enhanced actions of adenosine in medial entorhinal cortex layer II stellate neurons in temporal lobe epilepsy are mediated via A(1)-receptor activation. *Epilepsia*, *53*, 168–176.

Hartman, A. L., Lyle, M., Rogawski, M. A., & Gasior, M. (2008). Efficacy of the ketogenic diet in the 6-Hz seizure test. *Epilepsia*, *49*, 334–339.

Hauser, W. A., Annegers, J. F., & Kurland, L. T. (1991). Prevalence of epilepsy in Rochester, Minnesota: 1940–1980. *Epilepsia*, *32*, 429–445.

Hauser, W. A., & Hesdorffer, D. C. (2001). Psychosis, depression and epilepsy: Epidemiologic considerations. *Psychiatric issues in epilepsy: A practical guide to diagnosis and treatment*. pp. 7–18, Philadelphia, PA: Lippincott Williams & Wilkins.

Heidarianpour, A., Sadeghian, E., Mirnajafi-Zadeh, J., Fathollahi, Y., & Mohammad-Zadeh, M. (2006). Anticonvulsant effects of N6-cyclohexyladenosine microinjected into the CA1 region of the hippocampus on entorhinal cortex-kindled seizures in rats. *Epileptic Disorders*, *8*, 259–266.

Henderson, C. B., Filloux, F. M., Alder, S. C., Lyon, J. L., & Caplin, D. A. (2006). Efficacy of the ketogenic diet as a treatment option for epilepsy: Meta-analysis. *Journal of Child Neurology*, *21*, 193–198.

Hosseinmardi, N., Mirnajafi-Zadeh, J., Fathollahi, Y., & Shahabi, P. (2007). The role of adenosine A1 and A2A receptors of entorhinal cortex on piriform cortex kindled seizures in rats. *Pharmacological Research*, *56*, 110–117.

Huang, Q.-Y., Wei, C., Yu, L., Coelho, J. E., Shen, H.-Y., Kalda, A., et al. (2006). Adenosine A_{2A} receptors in bone marrow-derived cells but not in forebrain neurons are important contributors to 3-nitropropionic acid-induced striatal damage as revealed by cell-type-selective inactivation. *Journal of Neuroscience*, *26*, 11371–11378.

Huttenlocher, P. R. (1976). Ketonemia and seizures: Metabolic and anticonvulsant effects of two ketogenic diets in childhood epilepsy. *Pediatric Research*, *10*, 536–540.

Ilie, A., Raimondo, J. V., & Akerman, C. J. (2012). Adenosine release during seizures attenuates $GABA_A$ receptor-mediated depolarization. *Journal of Neuroscience*, *32*, 5321–5332.

Johansson, B., Georgiev, V., Lindström, K., & Fredholm, B. B. (1997). A_1 and A_{2A} adenosine receptors and A_1 mRNA in mouse brain: Effect of long-term caffeine treatment. *Brain Research*, *762*, 153–164.

Juge, N., Gray, J. A., Omote, H., Miyaji, T., Inoue, T., Hara, C., et al. (2010). Metabolic control of vesicular glutamate transport and release. *Neuron*, *68*, 99–112.

Kaku, T., Jiang, M. H., Hada, J., Morimoto, K., & Hayashi, Y. (2001). Sodium nitroprusside-induced seizures and adenosine release in rat hippocampus. *European Journal of Pharmacology*, *413*, 199–205.

Kaplan, G. B., Greenblatt, D. J., Kent, M. A., & Cotreau-Bibbo, M. M. (1993). Caffeine treatment and withdrawal in mice: Relationships between dosage, concentrations, locomotor activity and A_1 adenosine receptor binding. *Journal of Pharmacology and Experimental Therapeutics*, *266*, 1563–1572.

Kawamura, M., Jr., Ruskin, D. N., & Masino, S. A. (2010). Metabolic autocrine regulation of neurons involves cooperation among pannexin hemichannels, adenosine receptors, and K_{ATP} channels. *Journal of Neuroscience*, *30*, 3886–3895.

Kermanian, F., Soleimani, M., Ebrahimzadeh, A., Haghir, H., & Mehdizadeh, M. (2013). Effects of adenosine A2a receptor agonist and antagonist on hippocampal nuclear factor-kB expression preceded by MDMA toxicity. *Metabolic Brain Disease*, *28*, 45–52.

Kim, J. H., Karpyak, V. M., Biernacka, J. M., Nam, H. W., Lee, M. R., Preuss, U. W., et al. (2011). Functional role of the polymorphic 647 T/C variant of ENT1 (SLC29A1) and its association with alcohol withdrawal seizures. *PLoS One*, *6*, e16331.

Klaft, Z. J., Schulz, S. B., Maslarova, A., Gabriel, S., Heinemann, U., & Gerevich, Z. (2012). Extracellular ATP differentially affects epileptiform activity via purinergic P2X7 and adenosine A1 receptors in naive and chronic epileptic rats. *Epilepsia*, *53*, 1978–1986.

Kochanek, P. M., Vagni, V. A., Janesko, K. L., Washington, C. B., Crumrine, P. K., Garman, R. H., et al. (2006). Adenosine A1 receptor knockout mice develop lethal status epilepticus after experimental traumatic brain injury. *Journal of Cerebral Blood Flow and Metabolism*, *26*, 565–575.

Kukley, M., Schwan, M., Fredholm, B. B., & Dietrich, D. (2005). The role of extracellular adenosine in regulating mossy fiber synaptic plasticity. *Journal of Neuroscience*, *25*, 2832–2837.

Kumar, V., & Sharma, A. (2009). Adenosine: an endogenous modulator of innate immune system with therapeutic potential. *European Journal of Pharmacology*, *616*, 7–15.

Laudadio, M. A., & Psarropoulou, C. (2004). The A3 adenosine receptor agonist 2-Cl-IB-MECA facilitates epileptiform discharges in the CA3 area of immature rat hippocampal slices. *Epilepsy Research*, *59*, 83–94.

Li, W., Dai, S., An, J., Li, P., Chen, X., Xiong, R., et al. (2008). Chronic but not acute treatment with caffeine attenuates traumatic brain injury in the mouse cortical impact model. *Neuroscience*, *151*, 1198–1207.

Li, X., Kang, H., Liu, X., Liu, Z., Shu, K., Chen, X., et al. (2012). Effect of adenosine A2A receptor antagonist ZM241385 on amygdala-kindled seizures and progression of amygdala kindling. *Journal of Huazhong University of Science and Technology. Medical Sciences*, *32*, 257–264.

Li, M., Kang, R., Shi, J., Liu, G., & Zhang, J. (2013). Anticonvulsant activity of B2, an adenosine analog, on chemical convulsant-induced seizures. *PLoS One*, *8*, e67060.

Li, T., Lan, J. Q., & Boison, D. (2008). Uncoupling of astrogliosis from epileptogenesis in adenosine kinase (ADK) transgenic mice. *Neuron Glia Biology*, *4*, 91–99.

Li, T., Ren, G., Lusardi, T., Wilz, A., Lan, J. Q., Iwasato, T., et al. (2008). Adenosine kinase is a target for the prediction and prevention of epileptogenesis in mice. *Journal of Clinical Investigation*, *118*, 571–582.

Li, W., & Zhang, J. J. (2011). Sedative, hypnotic and anticonvulsive effects of an adenosine analogue WS090501. *Yao Xue Xue Bao*, *46*, 742–746.

Linden, J. (2001). Molecular approach to adenosine receptors: Receptor-mediated mechanisms of tissue protection. *Annual Review of Pharmacology and Toxicology*, *41*, 775–787.

Loram, L. C., Taylor, F. R., Strand, K. A., Harrison, J. A., RzasaLynn, R., Sholar, P., et al. (2013). Intrathecal injection of adenosine 2A receptor agonists reversed neuropathic allodynia through protein kinase (PK)A/PKC signaling. *Brain, Behavior, and Immunity*, *33*, 112–122.

Lovatt, D., Xu, Q., Liu, W., Takano, T., Smith, N. A., Schnermann, J., et al. (2012). Neuronal adenosine release, and not astrocytic ATP release, mediates feedback inhibition of excitatory activity. *Proceedings of the National Academy of Sciences of the United States of America*, *109*, 6265–6270.

Luan, G., Gao, Q., Guan, Y., Zhai, F., Zhou, J., Liu, C., et al. (2013). Upregulation of adenosine kinase in Rasmussen encephalitis. *Journal of Neuropathology and Experimental Neurology*, *72*, 1000–1008.

Luongo, L., Petrelli, R., Gatta, L., Giordano, C., Guida, F., Vita, P., et al. (2012). 5′-Chloro-5′-deoxy-(±)-ENBA, a potent and selective adenosine A_1 receptor agonist, alleviates neuropathic pain in mice through functional glial and microglial changes without affecting motor or cardiovascular functions. *Molecules*, *17*, 13712–13726.

Lupica, C. R., Berman, R. F., & Jarvis, M. F. (1991). Chronic theophylline treatment increases adenosine A_1, but not A_2, receptor binding in the rat brain: An autoradiographic study. *Synapse*, *9*, 95–102.

Luszczki, J. J., Kozicka, M., Swiader, M. J., & Czuczwar, S. J. (2005). 2-Chloro-N6-cyclopentyladenosine enhances the anticonvulsant action of carbamazepine in the mouse maximal electroshock-induced seizure model. *Pharmacological Reports*, *57*, 787–794.

Ma, W., Berg, J., & Yellen, G. (2007). Ketogenic diet metabolites reduce firing in central neurons by opening K_{ATP} channels. *Journal of Neuroscience*, *27*, 3618–3625.

Macdonald, R. L., & Kelly, K. M. (1995). Antiepileptic drug mechanisms of action. *Epilepsia*, *36*(Suppl. 2), S2–S12.

Magiorkinis, E., Sidiropoulou, K., & Diamantis, A. (2010). Hallmarks in the history of epilepsy: Epilepsy in antiquity. *Epilepsy and Behavior*, *17*, 103–108.

Malow, B. (2005). Sleep and epilepsy. *Neurologic Clinics*, *23*, 1127–1147.

Mantis, J. G., Centeno, N. A., Todorova, M. T., McGowan, R., & Seyfried, T. N. (2004). Management of multifactorial idiopathic epilepsy in EL mice with caloric restriction and the ketogenic diet: Role of glucose and ketone bodies. *Nutrition & Metabolism (London)*, *1*, 11.

Marangos, P. J., Boulenger, J.-P., & Patel, J. (1984). Effects of chronic caffeine on brain adenosine receptors: Regional and ontogenetic studies. *Life Sciences*, *34*, 899–907.

Mares, P. (2010). Anticonvulsant action of 2-chloroadenosine against pentetrazol-induced seizures in immature rats is due to activation of A1 adenosine receptors. *Journal of Neural Transmission*, *117*, 1269–1277.

Martín, E. D., Fernández, M., Perea, G., Pascual, O., Haydon, P. G., Araque, A., et al. (2007). Adenosine released by astrocytes contributes to hypoxia-induced modulation of synaptic transmission. *Glia*, *55*, 36–45.

Masino, S. A., & Geiger, J. D. (2008). Are purines mediators of the anticonvulsant/neuroprotective effects of ketogenic diets? *Trends in Neurosciences*, *31*, 273–278.

Masino, S. A., Kawamura, M., Jr., Cote, J. L., Williams, R. B., & Ruskin, D. N. (2013). Adenosine and autism: A spectrum of opportunities. *Neuropharmacology*, *68*, 116–121.

Masino, S. A., Kawamura, M., Jr., Plotkin, L. M., Svedova, J., DiMario, F. J., Jr., & Eigsti, I.-M. (2011). The relationship between the neuromodulator adenosine and behavioral symptoms of autism. *Neuroscience Letters*, *500*, 1–5.

Masino, S. A., Kawamura, M., Wasser, C. A., Pomeroy, L. T., & Ruskin, D. N. (2009). Adenosine, ketogenic diet and epilepsy: The emerging therapeutic relationship between metabolism and brain activity. *Current Neuropharmacology*, *7*, 257–268.

Masino, S. A., Li, T., Theofilas, P., Sandau, U., Ruskin, D. N., Fredholm, B. B., et al. (2011). A ketogenic diet suppresses seizures in mice through adenosine A_1 receptors. *Journal of Clinical Investigation*, *121*, 2679–2683.

Masino, S. A., & Rho, J. M. (2012). Mechanisms of ketogenic diet action. In J. L. Noebels, M. Avoli, M. A. Rogawski, R. W. Olsen, & A. V. Delgado-Escueta (Eds.), *Jasper's basic mechanisms of the epilepsies* (pp. 1001–1022) (4th ed.). Bethesda, MD: National Center for Biotechnology Information.

Mecha, M., Feliu, A., Inigo, P. M., Mestre, L., Carrillo-Salinas, F. J., & Guaza, C. (2013). Cannabidiol provides long-lasting protection against the deleterious effects of inflammation in a viral model of multiple sclerosis: A role for A2A receptors. *Neurobiology of Disease*, *59*, 141–150.

Medina, M. T., Duron, R. M., Martinez, L., Osorio, J. R., Estrada, A. L., Zuniga, C., et al. (2005). Prevalence, incidence, and etiology of epilepsies in rural Honduras: The Salama study. *Epilepsia*, *46*, 124–131.

Melø, T. M., Nehlig, A., & Sonnewald, U. (2006). Neuronal-glial interactions in rats fed a ketogenic diet. *Neurochemistry International*, *48*, 498–507.

Mills, J. H., Kim, D. G., Krenz, A., Chen, J. F., & Bynoe, M. S. (2012). A2A adenosine receptor signaling in lymphocytes and the central nervous system regulates inflammation during experimental autoimmune encephalomyelitis. *The Journal of Immunology*, *188*(11), 5713–5722.

Mishina, M., & Ishiwata, K. (2014). Adenosine receptor PET imaging in human brain. In A. Mori (Ed.), *International review of neurobiology: Vol. 119. Adenosine receptors in neurology and psychiatry* (pp. 51–70). Amsterdam: Elsevier.

Moezi, L., Akbarian, R., Niknahad, H., & Shafaroodi, H. (2013). The interaction of adenosine and morphine on pentylenetetrazole-induced seizure threshold in mice. *Neuropharmacology*, *72*, 1–8.

Mohamed, R. A., Agha, A. M., & Nassar, N. N. (2012). SCH58261 the selective adenosine A(2A) receptor blocker modulates ischemia reperfusion injury following bilateral carotid occlusion: Role of inflammatory mediators. *Neurochemical Research*, *37*, 538–547.

Moschovos, C., Kostopoulos, G., & Papatheodoropoulos, C. (2012). Endogenous adenosine induces NMDA receptor-independent persistent epileptiform discharges in dorsal and ventral hippocampus via activation of A2 receptors. *Epilepsy Research*, *100*, 157–167.

Muhle, H., Steinich, I., von Spiczak, S., Franke, A., Weber, Y., Lerche, H., et al. (2010). A duplication in 1q21.3 in a family with early onset and childhood absence epilepsy. *Epilepsia*, *51*, 2453–2456.

Murray, T. F. (1982). Up-regulation of rat cortical adenosine receptors following chronic administration of theophylline. *European Journal of Pharmacology*, *82*, 113–114.

Muzykewicz, D. A., Lyczkowski, D. A., Memon, N., Conant, K. D., Pfeifer, H. H., & Thiele, E. A. (2009). Efficacy, safety, and tolerability of the low glycemic index treatment in pediatric epilepsy. *Epilepsia*, *50*, 1118–1126.

Muzzi, M., Coppi, E., Pugliese, A. M., & Chiarugi, A. (2013). Anticonvulsant effect of AMP by direct activation of adenosine A1 receptor. *Experimental Neurology*, *250*, 189–193.

Nakazawa, M., Kodama, S., & Matsuo, T. (1983). Effects of ketogenic diet on electroconvulsive threshold and brain contents of adenosine nucleotides. *Brain and Development, 5*, 375–380.

Neal, E. G., Chaffe, H., Schwartz, R. H., Lawson, M. S., Edwards, N., Fitzsimmons, G., et al. (2008). The ketogenic diet for the treatment of childhood epilepsy: A randomised controlled trial. *Lancet Neurology*, 7, 500–506.

Newby, A. C. (1984). Adenosine and the concept of "retaliatory metabolites" *Trends in Biochemical Sciences, 9*, 42–44.

Noh, H. S., Lee, H. P., Kim, D. W., Kang, S. S., Cho, G. J., Rho, J. M., et al. (2004). A cDNA microarray analysis of gene expression profiles in rat hippocampus following a ketogenic diet. *Brain Research. Molecular Brain Research, 129*, 80–87.

Nylen, K., Likhodii, S., & Burnham, W. M. (2009). The ketogenic diet: Proposed mechanisms of action. *Neurotherapeutics, 6*, 402–405.

Nylen, K., Velazquez, J. L. P., Sayed, V., Gibson, K. M., Burnham, W. M., & Snead, O. C., III (2009). The effects of a ketogenic diet on ATP concentrations and the number of hippocampal mitochondria in *Aldh5a1*$^{-/-}$ mice. *Biochimica et Biophysica Acta, 1790*, 208–212.

Oka, E., Ohtsuka, Y., Yoshinaga, H., Murakami, N., Kobayashi, K., & Ogino, T. (2006). Prevalence of childhood epilepsy and distribution of epileptic syndromes: A population-based survey in Okayama, Japan. *Epilepsia, 47*, 626–630.

Olafsson, E., & Hauser, W. A. (1999). Prevalence of epilepsy in rural Iceland: A population-based study. *Epilepsia, 40*, 1529–1534.

Pan, J. W., Bebin, E. M., Chu, W. J., & Hetherington, H. P. (1999). Ketosis and epilepsy: ^{31}P spectroscopic imaging at 4.1 T. *Epilepsia, 40*, 703–707.

Pascual, O., Casper, K. B., Kubera, C., Zhang, J., Revilla-Sanchez, R., Sul, J.-Y., et al. (2005). Astrocytic purinergic signaling coordinates synaptic networks. *Science, 310*, 113–116.

Pearson, T., Damian, K., Lynas, R. E., & Frenguelli, B. G. (2006). Sustained elevation of extracellular adenosine and activation of A_1 receptors underlie the post-ischaemic inhibition of neuronal function in rat hippocampus *in vitro*. *Journal of Neurochemistry, 97*, 1357–1368.

Pometlova, M., Kubova, H., & Mares, P. (2010). Effects of 2-chloroadenosine on cortical epileptic afterdischarges in immature rats. *Pharmacological Reports, 62*, 62–67.

Rezvani, M. E., Mirnajafi-Zadeh, J., Fathollahi, Y., & Palizvan, M. R. (2007a). Anticonvulsant effect of A_1 but not A_{2A} adenosine receptors of piriform cortex in amygdala-kindled rats. *Canadian Journal of Physiology and Pharmacology, 85*, 606–612.

Rezvani, M. E., Mirnajafi-Zadeh, J., Fathollahi, Y., & Palizvan, M. R. (2007b). Changes in neuromodulatory effect of adenosine A_1 receptors on piriform cortex field potentials in amygdala kindled rats. *European Journal of Pharmacology, 565*, 60–67.

Rezvani, M. E., Roohbakhsh, A., Allahtavakoli, M., & Shamsizadeh, A. (2010). Anticonvulsant effect of aqueous extract of Valeriana officinalis in amygdala-kindled rats: Possible involvement of adenosine. *Journal of Ethnopharmacology, 127*, 313–318.

Rho, J. M., Kim, D. W., Robbins, C. A., Anderson, G. D., & Schwartzkroin, P. A. (1999). Age-dependent differences in flurothyl seizure sensitivity in mice treated with a ketogenic diet. *Epilepsy Research, 37*, 233–240.

Riazi, K., Galic, M. A., & Pittman, Q. J. (2010). Contributions of peripheral inflammation to seizure susceptibility: Cytokines and brain excitability. *Epilepsy Research, 89*, 34–42.

Rosen, J. B., & Berman, R. F. (1987). Differential effects of adenosine analogs on amygdala, hippocampus, and caudate nucleus kindled seizures. *Epilepsia, 28*, 658–666.

Rudolphi, K. A., Keil, M., Fastbom, J., & Fredholm, B. B. (1989). Ischaemic damage in gerbil hippocampus is reduced following upregulation of adenosine (A_1) receptors by caffeine treatment. *Neuroscience Letters, 103*, 275–280.

Ruskin, D. N., Svedova, J., Cote, J. L., Sandau, U., Rho, J. M., Kawamura, M., Jr., et al. (2013). Ketogenic Diet Improves Core Symptoms of Autism in BTBR Mice. *PLoS One*, *8*, e65021.

Rwiza, H. T., Kilonzo, G. P., Haule, J., Matuja, W. B., Mteza, I., Mbena, P., et al. (1992). Prevalence and incidence of epilepsy in Ulanga, a rural Tanzanian district: A community-based study. *Epilepsia*, *33*, 1051–1056.

Salmani, M. E., Fathollahi, Y., Mirnajafizadeh, J., & Semnanian, S. (2011). Epileptogenic insult alters endogenous adenosine control on long-term changes in synaptic strength by theta pattern stimulation in hippocampus area CA1. *Synapse*, *65*, 189–197.

Samsonsen, C., Bråthen, G., Reimers, A., Helde, G., & Brodtkorb, E. (2013). Is dietary caffeine involved in seizure precipitation? *Epilepsy and Behavior*, *28*, 147–150.

Schmidt, A. P., Böhmer, A. E., Antunes, C., Schallenberger, C., Porciúncula, L. O., Elisabetsky, E., et al. (2009). Anti-nociceptive properties of the xanthine oxidase inhibitor allopurinol in mice: Role of A_1 adenosine receptors. *British Journal of Pharmacology*, *156*, 163–172.

Schulte, G., & Fredholm, B. B. (2003). Signalling from adenosine receptors to mitogen-activated protein kinases. *Cellular Signalling*, *15*, 813–827.

Serrano, A., Haddjeri, N., Lacaille, J. C., & Robitaille, R. (2006). GABAergic network activation of glial cells underlies hippocampal heterosynaptic depression. *Journal of Neuroscience*, *26*, 5370–5382.

Shen, H. Y., Singer, P., Lytle, N., Wei, C. J., Lan, J. Q., Williams-Karnesky, R. L., et al. (2012). Adenosine augmentation ameliorates psychotic and cognitive endophenotypes of schizophrenia. *Journal of Clinical Investigation*, *122*, 2567–2577.

Shinohara, M., Saitoh, M., Nishizawa, D., Ikeda, K., Hirose, S., Takanashi, J., et al. (2013). ADORA2A polymorphism predisposes children to encephalopathy with febrile status epilepticus. *Neurology*, *80*, 1571–1576.

Simoes, A. P., Duarte, J. A., Agasse, F., Canas, P. M., Tome, A. R., Agostinho, P., et al. (2012). Blockade of adenosine A2A receptors prevents interleukin-1beta-induced exacerbation of neuronal toxicity through a p38 mitogen-activated protein kinase pathway. *Journal of Neuroinflammation*, *9*, 204.

Sperlágh, B., & Vizi, E. S. (2011). The role of extracellular adenosine in chemical neurotransmission in the hippocampus and basal ganglia: Pharmacological and clinical aspects. *Current Topics in Medicinal Chemistry*, *11*, 1034–1046.

St Louis, E. K. (2009). Minimizing AED adverse effects: Improving quality of life in the interictal state in epilepsy care. *Current Neuropharmacology*, 7, 106–114.

Szot, P., Sanders, R. C., & Murray, T. F. (1987). Theophylline-induced upregulation of A_1-adenosine receptors associated with reduced sensitivity to convulsants. *Neuropharmacology*, *26*, 1173–1180.

Tellez-Zenteno, J. F., Matijevic, S., & Wiebe, S. (2005). Somatic comorbidity of epilepsy in the general population in Canada. *Epilepsia*, *46*, 1955–1962.

Thompson, S. M., Haas, H. L., & Gähwiler, B. H. (1992). Comparison of the actions of adenosine at pre- and postsynaptic receptors in the rat hippocampus in vitro. *Journal of Physiology*, *451*, 347–363.

Togha, M., Akhondzadeh, S., Motamedi, M., Ahmadi, B., & Razeghi, S. (2007). Allopurinol as adjunctive therapy in intractable epilepsy: A double-blind and placebo-controlled trial. *Archives of Medical Research*, *38*, 313–316.

Tosh, D. K., Paoletta, S., Deflorian, F., Phan, K., Moss, S. M., Gao, Z. G., et al. (2012). Structural sweet spot for A1 adenosine receptor activation by truncated (N)-methanocarba nucleosides: Receptor docking and potent anticonvulsant activity. *Journal of Medicinal Chemistry*, *55*, 8075–8090.

Trinka, E., Kienpointner, G., Unterberger, I., Luef, G., Bauer, G., Doering, L. B., et al. (2006). Psychiatric comorbidity in juvenile myoclonic epilepsy. *Epilepsia*, *47*, 2086–2091.

Tsutsui, S., Schnermann, J., Noorbakhsh, F., Henry, S., Yong, V. W., Winston, B. W., et al. (2004). A1 adenosine receptor upregulation and activation attenuates neuroinflammation and demyelination in a model of multiple sclerosis. *Journal of Neuroscience, 24*, 1521–1529.
Uhlemann, E. R., & Neims, A. H. (1972). Anticonvulsant properties of the ketogenic diet in mice. *Journal of Pharmacology and Experimental Therapeutics, 180*, 231–238.
Uzbay, T. I., Kayir, H., & Ceyhan, M. (2007). Effects of tianeptine on onset time of pentylenetetrazole-induced seizures in mice: Possible role of adenosine A_1 receptors. *Neuropsychopharmacology, 32*, 412–416.
Vajda, F. J. (2007). Pharmacotherapy of epilepsy: New armamentarium, new issues. *Journal of Clinical Neuroscience, 14*, 813–823.
Van Gompel, J. J., Bower, M. R., Worrell, G. A., Stead, M., Chang, S. Y., Goerss, S. J., et al. (2014). Increased cortical extracellular adenosine correlates with seizure termination. *Epilepsia, 55*, 233–244.
Vezzani, A., Moneta, D., Conti, M., Richichi, C., Ravizza, T., De Luigi, A., et al. (2000). Powerful anticonvulsant action of IL-1 receptor antagonist on intracerebral injection and astrocytic overexpression in mice. *Proceedings of the National Academy of Sciences of the United States of America, 97*, 11534–11539.
Vianna, E. P. M., Ferreira, A. T., Doná, F., Cavalheiro, E. A., & da Silva Fernandes, M. J. (2005). Modulation of seizures and synaptic plasticity by adenosinergic receptors in an experimental model of temporal lobe epilepsy induced by pilocarpine in rats. *Epilepsia, 46*(Suppl. 5), 166–173.
Vitaliti, G., Pavone, P., Mahmood, F., Nunnari, G., & Falsaperla, R. (2014). Targeting inflammation as a therapeutic strategy for drug-resistant epilepsies: An update of new immune-modulating approaches. *Human Vaccines & Immunotherapeutics, 10*, 868–875.
Wagner, A. K., Miller, M. A., Scanlon, J., Ren, D., Kochanek, P. M., & Conley, Y. P. (2010). Adenosine A1 receptor gene variants associated with post-traumatic seizures after severe TBI. *Epilepsy Research, 90*, 259–272.
Wang, S., Kurada, L., Cilz, N. I., Chen, X., Xiao, Z., Dong, H., et al. (2013). Adenosinergic depression of glutamatergic transmission in the entorhinal cortex of juvenile rats via reduction of glutamate release probability and the number of releasable vesicles. *PLoS One, 8*, e62185.
Wilder, R. M. (1921). The effects of ketonemia on the course of epilepsy. *Mayo Clinic Bulletin, 2*, 307–308.
Winn, H. R., Welsh, J. E., Rubio, R., & Berne, R. M. (1980). Changes in brain adenosine during bicuculline-induced seizures in rats. Effects of hypoxia and altered systemic blood pressure. *Circulation Research, 47*, 568–577.
Wirrell, E. C., Grossardt, B. R., Wong-Kisiel, L. C., & Nickels, K. C. (2011). Incidence and classification of new-onset epilepsy and epilepsy syndromes in children in Olmsted County, Minnesota from 1980 to 2004: A population-based study. *Epilepsy Research, 95*, 110–118.
Wiznitzer, M. (2008). From observations to trials: The ketogenic diet and epilepsy. *Lancet Neurology, 7*, 471–472.
World Health Organization (2005). *Atlas: Epilepsy care in the world.* Geneva: World Health Organization.
Wu, C., Wong, T., Wu, X., Sheppy, E., & Zhang, L. (2009). Adenosine as an endogenous regulating factor of hippocampal sharp waves. *Hippocampus, 19*, 205–220.
Yao, S.-Q., Li, Z.-Z., Huang, Q.-Y., Li, F., Wang, Z.-W., Augusto, E., et al. (2012). Genetic inactivation of the adenosine A_{2A} receptor exacerbates brain damage in mice with experimental autoimmune encephalomyelitis. *Journal of Neurochemistry, 123*, 100–112.
Young, D., & Dragunow, M. (1994). Status epilepticus may be caused by loss of adenosine anticonvulsant mechanisms. *Neuroscience, 58*, 245–261.

Yu, L., Huang, Z., Mariani, J., Wang, Y., Moskowitz, M., & Chen, J. F. (2004). Selective inactivation or reconstitution of adenosine A_{2A} receptors in bone marrow cells reveals their significant contribution to the development of ischemic brain injury. *Nature Medicine*, *10*, 1081–1087.

Yudkoff, M., Daikhin, Y., Nissim, I., & Lazarow, A. (2001). Brain amino acid metabolism and ketosis. *Journal of Neuroscience Research*, *66*, 272–281.

Zagnoni, P. G., Bianchi, A., Zolo, P., Canger, R., Cornaggia, C., D'Alessandro, P., et al. (1994). Allopurinol as add-on therapy in refractory epilepsy: A double-blind placebo-controlled randomized study. *Epilepsia*, *35*, 107–112.

Zeraati, M., Mirnajafi-Zadeh, J., Fathollahi, Y., Namvar, S., & Rezvani, M. E. (2006). Adenosine A1 and A2A receptors of hippocampal CA1 region have opposite effects on piriform cortex kindled seizures in rats. *Seizure*, *15*, 41–48.

Zhou, S. Y., Xu, S. J., Yan, Y. G., Yu, H. M., Ling, S. C., & Luo, J. H. (2011). Decreased purinergic inhibition of synaptic activity in a mouse model of Niemann-Pick disease type C. *Hippocampus*, *21*, 212–219.

Zuchora, B., Wielosz, M., & Urbańska, E. M. (2005). Adenosine A_1 receptors and the anticonvulsant potential of drugs effective in the model of 3-nitropropionic acid-induced seizures in mice. *European Neuropsychopharmacology*, *15*, 85–93.

CHAPTER TWELVE

Adenosine Receptor Control of Cognition in Normal and Disease

Jiang-Fan Chen[*,†,1]
[*]Department of Neurology, Boston University School of Medicine, Boston, Massachusetts, USA
[†]The Molecular Medicine Institute, Wenzhou Medical University, Wenzhou, Zhejiang, PR China
[1]Corresponding author: e-mail address: chenjf@bu.edu

Contents

Abstract

Adenosine and adenosine receptors (ARs) are increasingly recognized as important therapeutic targets for controlling cognition under normal and disease conditions for its dual roles of neuromodulation as well as of homeostatic function in the brain. This chapter first presents the unique ability of adenosine, by acting on the inhibitory A_1 and facilitating A_{2A} receptor, to integrate dopamine, glutamate, and BNDF signaling and to modulate synaptic plasticity (e.g., long-term potentiation and long-term

International Review of Neurobiology, Volume 119
ISSN 0074-7742
http://dx.doi.org/10.1016/B978-0-12-801022-8.00012-X

depression) in brain regions relevant to learning and memory, providing the molecular and cellular bases for adenosine receptor (AR) control of cognition. This led to the demonstration of AR modulation of social recognition memory, working memory, reference memory, reversal learning, goal-directed behavior/habit formation, Pavlovian fear conditioning, and effort-related behavior. Furthermore, human and animal studies support that AR activity can also, through cognitive enhancement and neuroprotection, reverse cognitive impairments in animal models of Alzheimer's disease (AD), Parkinson's disease (PD), Huntington's disease, and schizophrenia. Lastly, epidemiological evidence indicates that regular human consumption of caffeine, the most widely used psychoactive drug and nonselective AR antagonists, is associated with the reduced cognitive decline in aging and AD patients, and with the reduced risk in developing PD. Thus, there is a convergence of the molecular studies revealing AR as molecular targets for integrating neurotransmitter signaling and controlling synaptic plasticity, with animal studies demonstrating the strong procognitive impact upon AR antagonism in normal and disease brains and with epidemiological and clinical evidences in support of caffeine and AR drugs for therapeutic modulation of cognition. Since some of adenosine A_{2A} receptor antagonists are already in phase III clinical trials for motor benefits in PD patients with remarkable safety profiles, additional animal and human studies to better understand the mechanism underlying the AR-mediated control of cognition under normal and disease conditions will provide the required rationale to stimulate the necessary clinical investigation to rapidly translate adenosine and AR drug as a novel strategy to control memory impairment in neuropsychiatric disorders.

1. ADENOSINE AS AN UPSTREAM REGULATOR OF DOPAMINE, GLUTAMATE, AND BRAIN-DERIVED NEUROTROPHIC FACTOR SIGNALING: A MOLECULAR BASIS FOR AR CONTROL OF COGNITION

Adenosine acts in parallel as a general homeostatic role by its paracrine signaling ability to coordinate metabolic activity in all eukaryotic cells, and as a specific neuromodulator in the brain, modulating neuronal function and information processing by controlling neuronal excitability, releasing various neurotransmitters, modulating synaptic plasticity and neuro-inflammation and cell death (Sebastiao & Ribeiro, 1996). The dual roles of adenosine are often difficult to disentangle as the adenosine control of neuronal function is probably intrinsically linked with its coordinate metabolic activity in neuronal circuits. As such, adenosine is critical in setting the stage for plasticity by balancing inhibition and excitation toward homeostasis and by a mechanism of adenosine-mediated metaplasticity (Dias, Rombo, Ribeiro, Henley, & Sebastiao, 2013). In addition, the idea of a nonsynaptic transmission of adenosine is postulated: astrocyte-derived adenosine acts long-range at the A_1 receptor to produce global (not a single synapse) heterosynaptic inhibition, cooperating with the local synaptic activation of the A_{2A} receptor, to increase signal-to-noise ratio (Gomes, Kaster, Tome, Agostinho, & Cunha, 2011). The homeostatic and neuromodulator controls of neuronal processes underlie the ability of adenosine to control cognition since adenosine kinase (ADK)-mediated adenosine homeostatic function is necessary and permissive to synaptic actions of adenosine acting on multiple pathways (Diogenes et al., 2014).

1.1. Adenosine A_1 receptor

The adenosine A_1 receptor (A_1R) is widely distributed throughout the central nervous system, with the highest levels occurring in the cerebral cortex, hippocampus, cerebellum, thalamus, brain stem, and spinal cord of the rat (Dixon, Gubitz, Sirinathsinghji, Richardson, & Freeman, 1996; Reppert, Weaver, Stehle, & Rivkees, 1991). A_1Rs are coupled to pertussis toxin-sensitive Gi-proteins (Londos, Cooper, & Wolff, 1980; van Calker, Muller, & Hamprecht, 1978) to inhibit adenylate cyclase. The most prominent effect of A_1Rs on the brain is to depress excitatory transmission (Dunwiddie & Fredholm, 1997; Dunwiddie & Masino, 2001; Ribeiro, Sebastiao, & de Mendonca, 2002). At the presynaptic site, A_1R activation inhibits synaptic transmission by the suppression of N-type calcium channels (Gundlfinger et al., 2007; Wu & Saggau, 1994) and by a direct downregulation of the release apparatus (Fredholm, Proctor, Van der Ploeg, & Dunwiddie, 1989; Hasuo, Shoji, Gallagher, & Akasu, 1992). At postsynaptic sites, A_1Rs are located in the postsynaptic density where they can influence the responsiveness to excitatory stimuli by a simultaneous control of N-type calcium channels and *N*-methyl-D-aspartate receptors (NMDARs) (Dunwiddie & Masino, 2001; Scanziani, Capogna, Gahwiler, & Thompson, 1992). In addition, A_1R in neuronal cells is also located nonsynaptically where activation of A_1Rs results in G-protein-dependent activation of inwardly rectifying K^+ channels (GIRKs), leading to hyperpolarization of the resting membrane potential (Kirsch, Codina, Birnbaumer, & Brown, 1990; Wan, Yao, & Wang, 1999). Thus, the A_1R can affect neuronal excitability and control of "basal" synaptic transmission (i.e., under conditions where synaptic plasticity is not engaged) by the activation of A_1R located presynaptically and postsynaptically as well as nonsynaptically.

1.2. Adenosine A_{2A} receptor

Adenosine A_{2A} receptors (A_{2A}Rs) are highly enriched in the striatum where the expression is largely localized to striatopallidal medium spiny neurons (MSNs) of the indirect pathway (Fink et al., 1992; Schiffmann & Vanderhaeghen, 1993). In the striatopallidal neurons, A_{2A}Rs are colocalized and interacted with striatal dopamine D_2 receptors (D_2Rs) (Canals et al., 2003; Fuxe et al., 2003; Hillion et al., 2002) or NMDARs (Gerevich, Wirkner, & Illes, 2002; Wirkner et al., 2000) in an *antagonistic* manner, as well as with metabotropic glutamate 5 receptors (mGlu$_5$Rs) (Coccurello, Breysse, & Amalric, 2004; Ferre et al., 2002; Kachroo et al., 2005) or cannabinoid CB1 receptors (CB$_1$Rs) (Ferre et al., 2010; Lerner, Horne, Stella, & Kreitzer, 2010) in a *synergistic* manner. For example, activation of the A_{2A}R inhibits the D_2R binding and antagonizes the D_2R-mediated inhibition of neurotransmitter release and c-Fos expression as D_2R-mediated behaviors (Ferre, Fredholm, Morelli, Popoli, & Fuxe, 1997; Ongini & Fredholm, 1996). Similarly, activation of postsynaptic A_{2A}R inhibits *N*-methyl-D-aspartate (NMDA) current in the striatum neurons (Gerevich et al., 2002; Wirkner et al., 2000), contributing to A_{2A}R-mediated control of synaptic plasticity. A_{2A}Rs also modulate brain-derived neurotrophic factor (BDNF) function in the striatum (Tebano et al., 2008), another controller of synaptic plasticity. Lastly, striatopallidal A_{2A}Rs also control GABAergic neurotransmission in the striatum and globus pallidus by dual mechanisms and may contribute to not only motor effect but also cognitive modulation (Mori & Shindou, 2003). Thus, striatal A_{2A}Rs are uniquely positioned in brain to modulate neurotransmission, intracellular signaling, and synaptic plasticity that underlie cognitive behavior.

Outside the striatum, A_{2A}Rs are also weakly expressed in the hippocampus and cortex, other brain regions, and in glial cells. A_{2A}Rs in these extrastriatal neuronal populations similarly interact with other receptors such as mGlu$_5$Rs (Rodrigues, Alfaro, Rebola, Oliveira, & Cunha, 2005; Tebano et al., 2005). Hippocampal A_{2A}Rs appear to be essential for LTP at mossy fiber–CA3 (Rebola, Lujan, Cunha, & Mulle, 2008) and CA3–CA1 (Fontinha, Diogenes, Ribeiro, & Sebastiao, 2008) synapses. Cortical A_{2A}Rs located at corticostriatal projections (Rebola et al., 2005; Rosin, Hettinger, Lee, & Linden, 2003) are capable of modulating glutamate release in the striatum (Ciruela et al., 2006; Rebola et al., 2005). In contrast to the prominent inhibitory role of A_1Rs, A_{2A}Rs seem to have limited impact on the control of "basal" synaptic transmission but play a crucial role in controlling synaptic plasticity (reviewed in Gomes et al., 2011). This presynaptic $A_{2A}R$ control of the glutamate release plays a major role in locally shutting down the profound A_1R-mediated inhibition of synaptic transmission (Ciruela et al., 2006; Lopes, Cunha, & Ribeiro, 1999).

A_1 and A_2 receptors are also expressed in glial cells and contribute to modulation of synaptic plasticity by modulating astrocytic glutamate uptake and astrocytic release of ATP and adenosine (Boison, Chen, & Fredholm, 2010). For example, we recently found that $A_{2A}Rs$ control glutamate transmission through its modulation of the expression of astrocytic glutamate transporter-1 (GLT-1), a critical controller of the bulk of glutamate uptake (Matos et al., 2012). Furthermore, astrocytic $A_{2A}R$ activation also inhibits GLT-1 activity by a mechanism involving physical association of $A_{2A}Rs$ with NKA-α2s in astrocytes, as gauged by *in situ* proximity ligation assays (Matos, Augusto, Agostinho, Cunha, & Chen, 2013). This astrocytic $A_{2A}R$:NPA:GLT-1 interaction provides a novel mechanism linking neuronal activity to ion homeostasis controlling glutamatergic activity and synaptic plasticity.

BDNF plays a critical role in modulating synaptic plasticity (Chao, 2003; Poo, 2001) and cognition (Bozdagi et al., 2008; Yee, Zhu, Mohammed, & Feldon, 2007). $A_{2A}R$ activation transactivates the major BDNF receptors, TrkA and TrkB, in the absence of neurotrophins (Lee & Chao, 2001) and facilitates BDNF release and BDNF-induced potentiation of synaptic transmission (Diogenes, Fernandes, Sebastiao, & Ribeiro, 2004; Tebano et al., 2008), whereas genetic and pharmacological $A_{2A}R$ inactivation decreases BDNF levels (Tebano et al., 2008), reduces TrkB activation, decreases the expression of the TrkB receptor in the hippocampus (Jeronimo-Santos et al., 2014), and impairs BDNF-dependent LTP in the hippocampus (Assaife-Lopes et al., 2010; Jeronimo-Santos et al., 2014; Lee & Chao, 2001). $A_{2A}R$ activation also induces a BDNF-independent recruitment of TrkB receptors to lipid rafts, with functional consequences for TrkB- and BDNF-induced modulation of hippocampal plasticity (Assaife-Lopes, Sousa, Pereira, Ribeiro, & Sebastiao, 2014). Thus, $A_{2A}R$ activity may influence cognitive behaviors by modulating BDNF activity in the brain. The potential consequences of chronic blockade of $A_{2A}Rs$ on BNDF signaling and synaptic plasticity have to take this into consideration in drug development, particularly $A_{2A}R$ antagonists.

The role of the $A_{2A}R$ in modulating memory deficits is further strengthened by the signaling transducing systems operated by the AR. For example, $A_{2A}Rs$ are positively coupled to adenylate cyclase to increase cAMP production and CREB phosphorylation (Fredholm, Chen, Masino, & Vaugeois, 2005; Fredholm, Chern, Franco, & Sitkovsky, 2007; Fredholm, IJzerman, Jacobson, Klotz, & Linden, 2001). Accordingly, the $A_{2A}R$ activation has been well established to control the recruitment of CREB (Cabeza de Vaca et al., 2007; Cheng, Shih, & Chern, 2002; Chiang, Lee, Huang, & Chern, 2005; Fredholm et al., 2007), a transcription factor that has been closely associated with memory performance (Benito & Barco, 2010; Hansen & Zhang, 2013; Kim, Kwon, Kim, & Han, 2013; Rouaux, Loeffler, & Boutillier, 2004; Sakamoto, Karelina, & Obrietan, 2011). Using transgenic mice expressing β-galactosidase (lacZ) under the control of seven repeated CRE sites or the dominant negative form of CREB, Nam et al. (2013)found that reduced CREB activity in the striatum was causally associated with decreased $A_{2A}R$ signaling and increased goal-directed ethanol drinking. Activation of $A_{2A}R$ to also control the recruitment of MAP kinases (Charalambous et al., 2008; Fredholm et al., 2007; Melani et al., 2006, 2009; Quiroz et al., 2006; Schulte & Fredholm, 2003), another group of signaling transducing molecules associated with memory performance and neurodegeneration (Adams & Sweatt, 2002; Munoz & Ammit, 2010; Thomas & Huganir, 2004; Xia & Storm, 2012). The ability of $A_{2A}R$ to prevent synaptotoxicity and cognitive behavior in animal models of Alzheimer's disease (AD) was related to the aberrant recruitment of one MAP kinase, p38 (Canas et al., 2009).

2. ADENOSINE RECEPTOR MODULATION OF SYNAPTIC PLASTICITY: A CELLULAR BASIS FOR AR CONTROL OF COGNITION

Hebbian forms of synaptic plasticity such as long-term potentiation (LTP) and long-term depression (LTD) are generally thought to form the cellular correlates of associated learning and memory processes thought. It is suggested that the homeostatic function of adenosine may provide the permissive condition to set the stage for Hebbian forms of plasticity (Dias et al., 2013). As a neuromodulator, adenosine generated from different sources may preferentially act at different ARs to exert different controls of synaptic plasticity: astrocyte-derived adenosine acting at A_1R to exert a global inhibition through astrocytic-driven volume transmission while neuron-derived adenosine via ATP conversion to adenosine by ectonucleotidases acting at $A_{2A}R$ to exert a synapse-restricted facilitation of plasticity (Gomes et al., 2011). Moreover, the global A_1R-mediated inhibition in the circuit (i.e., heterosynaptic depression) cooperates with the local synaptic activation of A_{2A}Rs in the activated synapse, to maximize signal-to-noise ratio (salience between the activated and nontetanized synapses). By control of multiple neurotransmitter release and glutamate and dopamine and BDNF signaling in brain (Ribeiro, 1999), AR plays a critical role in modulation of Hebbian plasticity in various brain regions (de Mendonca & Ribeiro, 1997), including thalamocortical project (Blundon, Bayazitov, & Zakharenko, 2011), somatosensory cortex (Marquez-Ruiz et al., 2012), hippocampus (CA3–CA1 synapse) (Rebola et al., 2008), corticostriatal projections (Shen, Flajolet, Greengard, & Surmeier, 2008), hypothalamus (Xia et al., 2009), and neuronal muscle junction (Todd, Darabid, & Robitaille, 2010) (for review, see Dias et al., 2013).

2.1. A_1 receptor modulation of synaptic plasticity

Despite the consistent inhibitory effect of A_1R on glutamatergic transmission in various brain regions, the A_1R has been shown to differentially affect short-term and long-term synaptic plasticity in different brain regions. In the hippocampus, local activation of A_1Rs is not critical to the basal release probability and plasticity at mossy fiber synapses since the magnitude of frequency facilitation, paired-pulse facilitation, or posttetanic potentiation was not affected by genetic A_1R knockout (KO) or pharmacological A_1R antagonism (Kukley, Schwan, Fredholm, & Dietrich, 2005). In the Shaffer collateral-CA1 pathway of hippocampal slices, the deletion of A_1R did not alter in LTP elicited by either tetanic stimulation or theta burst as well as CA1 LTD but it did impair pair pulse facilitation (Gimenez-Llort et al., 2005). However, other studies with the A_1R antagonist DPCPX, genetic A_1R KO, or removal of adenosine by adenosine deaminase have shown to selectively augment mossy fiber basal transmission in the hippocampus, but attenuated both short-term plasticity (e.g., frequency facilitation and paired pulse facilitation) and LTP at this synapse (Moore, Nicoll, & Schmitz, 2003). The A_1R activation after NMDA stimulation contributes to GIRKcurrent in the hippocampus and participates in depotentiation of the previously potentiated LTP at Shaffer collateral synapses (Chung et al., 2009). A_1R antagonists enhanced the depolarization and increase in the amount of stable LTP induced by five theta bursts (Arai & Lynch, 1992) and reversed synaptic deficits induced by sleep-deprivation (Florian, Vecsey, Halassa, Haydon, & Abel, 2011) and the impairment in CA3–CA1 hippocampal LTP induced by high-frequency stimulation in chronic morphine-treated mice (Lu et al., 2010). Thus, under pathological conditions, A_1R activation is capable of inhibiting CA3–CA1 LTP. Additional studies are required to clarify the exact role of A_1R modulation of hippocampal synaptic plasticity in normal and disease.

In the striatum, genetic A_1R KO abolished the depression of the field excitatory postsynaptic potential/population spike by application of either NMDA or adenosine, suggesting A_1Rs modulation of NMDAR-triggered LTD in the striatum (Schotanus, Fredholm, & Chergui, 2006). However, the A_1R antagonists block and A_1R agonists induced short-term depression but had no effect on LTD at these synapses (Lovinger & Choi, 1995). These discrepancies were attributed in part to some differences in experimental conditions and different experimental approaches (such as genetic KO vs. pharmacological blockade). In developing neocortex, presynaptic A_1R activation by local extracellular adenosine levels at synapses is critical to developmental shift in release probability at synapses of this neuronal population, contributing to synapse strength and long-term synaptic plasticity (Kerr, Wall, & Richardson, 2013). In cerebellar Purkinje cells where A_1R activity and type-1 metabotropic glutamate receptor (mGluR1) are closely colocalized and formed heterodimeric complex, activation of the A_1R blocked mGluR1-mediated LTD (*glu-LTD*) and affects cerebellar motor learning possibly through the A_1R-mGluR1 receptor heterodimerization (Kamikubo et al., 2013). In hypothalamus and neuronal muscle junction, bath application of A_1R antagonists enhances LTP induced by high-frequency stimulation in hypocretin neurons (Xia et al., 2009) and potentiates postsynaptic potentials (Todd et al., 2010), suggesting A_1R-mediated posttetanic depression.

2.2. A_{2A} receptor modulation of synaptic plasticity

The prominent function of $A_{2A}R$ seems to be the control of aberrant synaptic plasticity. Thus, pharmacological or genetic blockade of the $A_{2A}R$ did not affect basal synaptic transmission or short-term plasticity (paired-pulse facilitation) but reduced LTP in corticoaccumbens (d'Alcantara, Ledent, Swillens, & Schiffmann, 2001) or hippocampal synapses (Rebola et al., 2008). In a trace eyeblink conditioning paradigm, recording activity in behaving mice demonstrated the inhibitory impact of $A_{2A}R$ blockade on the field excitatory postsynaptic potential slope at the CA3–CA1 synapse and experimentally evoked LTP in the hippocampus, on the conditioned response behaviors (Fontinha et al., 2009). Interestingly, in contrast to the conventional presynaptic plasticity at hippocampal mossy fiber synapses, hippocampus $A_{2A}Rs$ are localized postsynaptically at synapses between mossy fibers and CA3-pyramidal cells and control a unique form of *postsynaptic NMDA-dependent LTP* induced by short bursts of mossy fiber stimulation (Rebola et al., 2008). This postsynaptic LTP depends on G-protein activation, and Src kinase activity, and on a postsynaptic Ca^{2+} rise with the activation of NMDA and mGluR5 receptors as potential sources. Activation of the hippocampal $A_{2A}R$ also facilitates *AMPA-evoked LTP* in CA1 pyramidal neurons and the potentiation of LTP at the CA3–CA1 synapse by a PKA dependent of GluR1 subunits phosphorylated at the Ser-845 site. With increased extracellular level of adenosine, $A_{2A}R$ activation enhances the endogenous GluR1 Ser-845 phosphorylation and facilitates the synaptic insertion of GluR1-containing AMPA receptor from extrasynaptic pool for reinforcement of synaptic strength (Dias, Ribeiro, & Sebastiao, 2012). At hippocampal mossy fiber synapse, $A_{2A}R$ activation is also required for the kainate receptors (KARs)-mediated LTD (*KAR-LTD*) induced by high-frequency mossy fiber stimulation and natural spike patterns. Thus, hippocampal $A_{2A}R$ may modulate this special form of glutamate receptor plasticity induced by natural spike patterns to regulate cellular excitability (Chamberlain, Sadowski, Teles-Grilo Ruivo, Atherton, & Mellor, 2013). In addition, $A_{2A}R$ stimulation may augment *BDNF-induced LTP* in the hippocampus (Fontinha et al., 2008). Indeed, $A_{2A}R$-KO mice or mice treated with the selective $A_{2A}R$ antagonist ZM241385 displayed a loss of BDNF-induced LTP and reduced BDNF levels in the hippocampus (Tebano et al., 2008). Thus, activation of hippocampal $A_{2A}Rs$ modulates synaptic plasticity through multiple mechanisms involving a postsynaptic NMDAR-, or GluR1-containing AMPAR-, kainate receptor-, and BDNF-mediated LTP.

Striatal $A_{2A}R$ activity did not affect basal synaptic transmission nor the paired-pulse facilitation index of presynaptic function but plays an important role in modulating striatal LTP since pharmacological blockade or gene deletion of $A_{2A}Rs$ impair LTP at corticoaccumbal synapses (d'Alcantara et al., 2001) and spike-timing-dependent LTP at glutamatergic synapses onto striatopallidal neurons (Shen, Flajolet, et al., 2008). Activation of $A_{2A}R$, in the setting of FGFR coactivation, also promotes LTP at the corticostriatopallidal synapses (Flajolet et al. 2008). This phenotype was recapitulated by treatment with a specific PKA inhibitor, suggesting that the $A_{2A}R$-dependent LTP process occurs through a PKA-dependent pathway. A recent study with combined optogenetics, two-photon microscopy, and glutamate uncaging further shows that a protein kinase

A (PKA)-dependent mechanism of Ca^{2+} entry through NMDA-type glutamate receptors at corticostriatal terminals is attenuated by the D_2R but enhanced by $A_{2A}R$ activation (Higley & Sabatini, 2010). Thus, $A_{2A}R$ activation controls the striatal plasticity by countering D_2R-mediated inhibition of NMDA-dependent Ca^{2+} influx in the striatopallidal neurons. LTD is the best-characterized form of long-term plasticity in the striatum. To the best of our knowledge, the impact of postsynaptic striatal $A_{2A}R$ activity on striatal LTD has not yet been directly examined. Endocannabinoid (eCB)-dependent LTD, the best-characterized form of striatal plasticity, is dependent on endocannabinoid release and D_2R activation and is found to be restricted to striatopallidal MSNs, where $A_{2A}Rs$ are highly concentrated (Kreitzer & Malenka, 2007) and $A_{2A}Rs$ interact with both D_2Rs and CB1Rs, which may contribute to the modulation of striatal LTD (Calabresi, Maj, Mercuri, & Bernardi, 1992; Calabresi, Maj, Pisani, Mercuri, & Bernardi, 1992; Pan, Hillard, & Liu, 2008; Tebano et al., 2009). The release of endocannabinoids from striatal MSNs requires activation of $mGlu_5Rs$ and L-type calcium channels that are bidirectionally modulated by D_2Rs (Kreitzer & Malenka, 2005). Two distinct signaling cascades (i.e., PLC-dependent and PLC-independent cascades) mediate eCB production in striatopallidal neurons, and both signaling cascades were modulated by D_2Rs and $A_{2A}Rs$ through their modulation of cAMP/PKA pathways (Lerner & Kreitzer, 2012). Furthermore, concomitant stimulation of $A_{2A}R$ and D_2R shifts the striatopallidal MSN plasticity response from a D_2R-induced LTD response to a LTP response (Shen, Flajolet, et al., 2008). Thus, activation of striatal $A_{2A}R$ in indirect pathway MSNs is required for the induction of LTP at the corticostriatal synapse. Thus, striatal $A_{2A}Rs$ can integrate incoming information (glutamate signals) and sensitivity to this incoming information (dopamine endocannabinoids signals) to control striatal neurotransmission, synaptic plasticity (LTP and LTD).

2.3. A_3 receptor modulation of synaptic plasticity

Several pharmacological studies with the selective A_3 receptor (A_3R) agonists and antagonists suggest a functional role of A_3Rs in the modulation of synaptic plasticity in the hippocampus. In the CA1 region of hippocampal slices, the selective A_3Rs agonist Cl-IBMECA increased the theta-burst-induced LTP and -attenuated LTD, the effect that was blockaded by the selective A_3R antagonist, MRS-1191 (Costenla, Lopes, de Mendonca, & Ribeiro, 2001). However, other study found that activation of A_3Rs had no direct effects on synaptically evoked excitatory responses, LTP, or synaptic facilitation at hippocampal CA3–CA1 synapses (Hagena & Manahan-Vaughan, 2010). Interestingly, LTP impairment at hippocampal CA3–CA1 synapses (Maggi et al., 2009) or EPSC depression (Piccinin et al., 2010) induced by the chemokine CX3CL1 was shown to be prevented in A_3R-KO mice. Thus, under normal physiological conditions, endogenous adenosine acts at A_3Rs to influence synaptic plasticity, but when extracellular adenosine is elevated in response to brain insults, A_3R activation might limit the protective effects of adenosine.

3. ADENOSINE RECEPTOR MODULATES LEARNING AND MEMORY IN NORMAL ANIMALS

Adenosine acts mainly at inhibitory A_1Rs and excitatory $A_{2A}Rs$ to modulate neurotransmitter signaling, neuronal excitability, and synaptic plasticity (e.g., LTP and LTD) in the brain, contributing to adenosine control of learning and memory. The precise contribution of A_1Rs and $A_{2A}Rs$ to adenosine regulation of cognitive functions, however, remains to be established. For its wide and abundant expression patterns of A_1R in various brain regions associated with learning and memory and for its profound effect on neurotransmission, A_1Rs were traditionally thought to execute adenosine's potential modulatory effects on cognition. By contrast, $A_{2A}R$ function is believed to be mainly associated with motor function for its highly concentrated expression pattern of the $A_{2A}R$ in the striatum. However, neurochemical, pharmacological, and genetic KO studies have demonstrated the $A_{2A}R$ function far beyond the striatum, to cerebral cortex and hiappocampus where they are particular effective in the control of cognition. Studies in AR KO mice are beginning to reveal the complexities and vastness of adenosine's functions on cognition. The importance of $A_{2A}Rs$ for some forms of learning and memory is now gradually reaching the forefront.

3.1. Adenosine levels by ENT-1, ADK, and CD73 and learning and memory

Adenosine fulfills a unique role in the brain integrating glutamate and dopamine signaling and affects cognition via activation of A_1 and A_{2A} receptors. Genetic deletion of (ethanol sensitive) equilibrative nucleoside transporter-1 (ENT-1) increased vulnerability to the progression of excessive ethanol drinking during operant conditioning (Nam et al., 2013). Altered adenosine levels by transgenic expression of ADK in forebrain have profound effects on multiple learning impairments in the Morris water maze task and in the Pavlovian conditioning (Yee, Singer, Chen, Feldon, & Boison, 2007). Ectonucleotidases CD73-mediated formation of extracellular adenosine is critical for the striatal $A_{2A}R$ control of cognition (Augusto et al., 2013; Ena, De Backer, Schiffmann, & de Kerchove d'Exaerde, 2013). In the striatum, CD73 is selectively expressed in the indirect pathway by fluorescence-activated cell sorting array profiling (Ena et al., 2013) and is physically associated with the $A_{2A}R$ in the striatopallidal neurons as demonstrated by coimmunoprecipitation and proximity ligation assay (Augusto et al., 2013). Furthermore, global gene deletion or AAV-mediated focal knockdown of CD73 increased working memory performance (Augusto et al., 2013) and affected motor learning (Ena et al., 2013), mimicking the phenotype of global or forebrain-$A_{2A}R$ KO mice, as well as upon pharmacological $A_{2A}R$ blockade. Thus, ENT-1-, ADK-, and CD73-mediated formation of extracellular adenosine are responsible for the AR control of cognition.

3.2. A_1 receptors and learning and memory

In line with the evidence of the A_1R control of synaptic plasticity, earlier pharmacological studies showing that hippocampal A_1Rs influence working memory (Ohno & Watanabe, 1996), prevent scopolamine-induced working memory deficits (Hooper, Fraser, & Stone, 1996), and prevent morphine-induced impairment in the retrieval of a spatial reference memory (Lu et al., 2010). There is strong evidence that astrocyte-released adenosine acting at A_1R is critical to the memory deficit induced by sleep deprivation (Halassa et al., 2009). Consistent with the "effect inversion" feature of A_1R drug (Jacobson, von Lubitz, Daly, & Fredholm, 1996), subchronic blockade of the A_1R agonist and antagonist produced opposite effect of the acute treatment of these agents on cognition performance (Von Lubitz, Paul, Bartus, & Jacobson, 1993). These findings suggest that A_1R activity influence learning and memory, particularly gaining its relevance under certain pathological conditions. On the other hand, studies from A_1R KO mice suggest that A_1Rs may not play as critical a role as once believed in mediating some of the mnemonic effects of adenosine since global deletion of the A_1R failed to produce any performance effect in the water maze from two different KO mouse lines (Gimenez-Llort et al., 2002, 2005; Lang et al., 2003). In these studies, A_1R KO mice showed normal acquisition and retention of a spatial reference memory, normal spatial working memory, and normal ability to learn the new position of a fixed platform during reversal learning.

These findings in A_1R KO mice suggest that the A_1R is not critical for the expression of normal spatial reference memory or working memory, at least under physiological conditions. However, old A_1R KO mice (19 months old) appear to show some spatial working memory deficits in the six-arm radial maze (Gimenez-Llort et al., 2005). It should nonetheless be noted that the A_1R KO mice have an altered emotional status (Gimenez-Llort et al., 2002; Johansson et al., 2001) and display a reduced habituation to known environments (Gimenez-Llort et al., 2005), both of which could potentially influence their cognitive performance. Also a possible confounding developmental effect of A_1R KO mice on A_1R control of cognition cannot be ruled out.

3.3. A_{2A} receptors and learning and memory

Consistent with the ability of the $A_{2A}R$ to integrate dopamine, glutamate, and BDNF signaling and to modulate synaptic plasticity (LTP in the hippocampus and LTP/LTD in the striatum) (d'Alcantara et al., 2001; Rebola et al., 2008), a growing body of evidence from pharmacological and genetic studies supports that brain $A_{2A}R$ activity contribute to modulation of learning and memory (Cunha, 2008; Cunha, Ferre, Vaugeois, & Chen, 2008; Ferre et al., 2008; Schiffmann, Fisone, Moresco, Cunha, & Ferre, 2007; Shen et al., 2008).

(i) *Short-term recognition memory:* Systemic administration of caffeine (a nonselective AR antagonist) and selective $A_{2A}R$ antagonists reverse memory deficits in aging animals and spontaneously in hypertensive rats using olfactory discrimination and social recognition memory (Prediger, Fernandes, & Takahashi, 2005; Prediger, Pamplona, Fernandes, & Takahashi, 2005; Prediger & Takahashi, 2005). Global genetic inactivation of the $A_{2A}R$ enhanced spatial recognition memory and novelty exploration in Y-maze testing in mice (Wang, Ma, & van den Buuse, 2006).

(ii) *Working memory:* Working memory is defined as the short-term storage of information for processing in order to produce an appropriate response (Goldman-Rakic, 1995). Working memory is critical to human reasoning and judgment and is at the core of pathophysiology for many neuropsychiatric disorders such as AD (Baddeley, Bressi, Della Sala, Logie, & Spinnler, 1991; Kensinger, Shearer, Locascio, Growdon, & Corkin, 2003; Lim et al., 2008) and schizophrenia (Goldman-Rakic, Castner, Svensson, Siever, & Williams, 2004).

The working memory depends primarily on the integrity of prefrontal cortical function, although other neural structures such as hippocampus, inferior parietal cortex, and striatum may also be involved (Goldman-Rakic, 1999). Transgenic overexpression of A_{2A}Rs in the cortex impairs spatial working memory in radial maze tests, repeated trials of the Morris water maze, and objective recognition tests (Gimenez-Llort et al., 2007); conversely, the genetic deletion of A_{2A}R improves working memory (Wei et al., 2011; Zhou et al., 2009), as gauged from the analysis of repeated acquisition paradigm in the Morris water maze or the eight-arm radial maze. Furthermore, an improved working memory in mice can be achieved globally either with genetic deletion of A_{2A}R or with selective deletion in the entire forebrain neuron or only striatal neurons (Wei et al., 2011; Zhou et al., 2009). These phenotypes were independent of motor function and anxiety-like behavior.

(iii) *Reversal learning:* Reversal learning is an accepted measure of cognitive flexibility because successful task performance requires the ability to flexibly manipulate and update relevant information. Genetic A_{2A}R inactivation in forebrain or striatal neurons facilitates spatial reversal learning: escape performance during reversal training was facilitated (Wei et al., 2011). Interestingly, facilitated reversal learning and cognitive flexibility may come at the expense of weakened habit formation (Yu, Gupta, Chen, & Yin, 2009), thereby reducing unproductive search persistence and promoting alternative search behavior.

(iv) *Goal-directed behavior* versus *habit formation:* Striatum-dependent procedural learning and habits are studied using the instrumental conditioning paradigm (Balleine & Dickinson, 1998). Any instrumental behavior, such as pressing a lever for food, can be controlled by two distinct central processes: initially, it may be a goal-directed behavior, sensitive to manipulations like outcome devaluation (i.e., the action–outcome contingency), and later, it may be a stimulus–response, habit-based behavior, which is impervious to changes in the value of the outcome (Balleine & Dickinson, 1998; Graybiel, 2008; Yin & Knowlton, 2006). Recently, we have shown that inactivation of striatal A_{2A}Rs results in reduced habit formation or enhanced goal-directed behavior driven (Yu, Gupta, Chen, & Yin, 2009). Specifically, after training that generated habitual lever pressing in wild-type controls, the performance of striatum A_{2A}R KO mice remained goal-directed during the devaluation test. Moreover, striatum $A_{2A}R$ KO mice significantly outperformed WT mice on the omission test, indicating increased behavioral inhibition and reduced compulsivity. In an ethanol-drinking study with operant instrumental learning, decreasing $A_{2A}R$-mediated CREB activity genetically and pharmacologically in the dorsomedial striatum was causally associated with increased goal-directed ethanol drinking in WT mice, but not in mice lacking ethanol-sensitive adenosine transporter ENT-1(−/−) (Nam et al., 2013). Similarly, $A_{2A}R$ (but not A_1R) antagonists enhance operant response rates (Randall et al., 2011). These findings identify a novel role of $A_{2A}Rs$ in habit formation and further localized their effect to striatopallidal pathway neurons. The mechanism behind this phenotype might relate to $A_{2A}R$ interactions with the dopamine or endocannabinoid systems and with the $A_{2A}R$/CREB signaling, which are linked to habit learning (Choi, Balsam, & Horvitz, 2005; Hilario, Clouse, Yin, & Costa, 2007).

(v) *Fear conditioning:* Pavlovian associative learning is a fundamental form of learning that depends on the integrity of dopaminergic and glutamatergic functions (Dayan & Balleine, 2002; Lynch et al., 1990). A classical eyeblink conditioning and the avoidance behavior using an aversive paradigm, with parallel hippocampal LTP, are attenuated by pharmacological blockade of $A_{2A}Rs$ (Kopf, Melani, Pedata, & Pepeu, 1999). Activation of cerebral ARs by local injection of A_1R and $A_{2A}R$ agonists into the posterior cingulated cortex impairs memory retrieval in a one-trial inhibitory avoidance task (Pereira et al., 2005). Using an asymmetric reversible neurotransmission-blocking technique, selective blocking of the indirect pathway by transmission-blocking tetanus toxin impairs aversive learning that was further exacerbated by $A_{2A}R$ blockade (Hikida et al., 2013). Studies with brain region-specific $A_{2A}R$ KO models further showed that the genetic deletion of $A_{2A}R$ in striatal neurons *enhanced* context and tone fear conditioning, while deletion of the $A_{2A}R$ in the entire forebrain or focal deletion of hippocampal $A_{2A}R$ normalized or attenuated context and tone fear conditioning (Wei et al., 2013). Genetic deletion of striatal $A_{2A}Rs$ impaired active avoidance learning but did not affect conditioned taste aversion (Singer, Wei, Chen, Boison, & Yee, 2013). Thus, $A_{2A}R$ activities in different brain regions seem to have different effects on the fear conditioning behavior, and additional studies are warranted to clarify the specific role of the $A_{2A}Rs$ in different brain regions in control of Pavlovian fear conditioning.

3.3.1 *Effort-related behavior*

The $A_{2A}R$ antagonist MSX-3 (but not the A_1R antagonist CPT) or $A_{2A}R$ KO reversed the effects of haloperidol on effort-related choice in the maze and on c-Fos expression in the striatum (O'Neill & Brown, 2007a). Thus, striatal D2 receptor and $A_{2A}R$ antagonistically regulate effort-related decision making and effort expenditure in mice.

3.3.2 *Conditional temporal probability*

In a task to dissociate the effect of elapsing time in the foreperiod and condi tional temporal probability of the imperative stimulus, the $A_{2A}R$ antagonist KW-6002 has been shown to speed reaction times and increase anticipatory responding (O'Neill & Brown, 2007a). The pattern of behavior in interpreted as the suggestion that $A_{2A}R$ antagonism enhanced the motor preparatory effects of conditional probability rather than speeded a timing process.

3.3.3 *Selective inactivation of striatal A_{2A}Rs alone was sufficient to reproduce the procognitive phenotypes*

Although the striato-cortical interaction is mostly conceived as supporting the control of actions and procedural memory, there is an increasing recognition that striatal circuits are also actively involved in the control of declarative and episodic memory (Scimeca & Badre, 2012). Our analysis of striatum-specific $A_{2A}R$ KO study suggests that selective inactivation of striatal A_{2A}Rs alone was sufficient to reproduce the procognitive phenotypes. Furthermore, our finding of procognitive effects seen after global or striatal $A_{2A}R$ deletion in working memory (Kirsch et al., 1990; Wirkner et al., 2000), reversal learning (Kirsch et al., 1990), and goal-directed behavior (Ferre et al., 2002), Pavlovian fear conditioning phenotype (Wei et al., 2013) suggests that striatopallidal A_{2A}Rs activities assume a common inhibitory control over diverse mnemonic processes. A similar mechanism might also underlie the inhibitory control by the activation of A_{2A}Rs in striatopallidal neurons over drug reward behavior (Ferre et al., 2010; Lerner et al., 2010), psychomotor behaviors (Ferre et al., 1997), and incentive devaluation in instrumental behavior. This view is in agreement with the increasing acceptance that the ventral striatum acts as an integrative unit associated with the adaptive encoding of working memory (Hallock, Arreola, Shaw, & Griffin, 2013; Scimeca & Badre, 2012; Simpson, Kellendonk, & Kandel, 2010) and of reinforcement learning (Johnson, van der Meer, & Redish, 2007; Liljeholm & O'Doherty, 2012; Pennartz, Ito, Verschure, Battaglia, & Robbins, 2011; Piray, 2011; van der Meer & Redish, 2011). This raises the exciting possibility that the striatal $A_{2A}R$ may represent a target for improving cognitive function under normal and pathological conditions.

3.4. A_3 and A_{2B} receptors and learning and memory

There is very limited information on the A_{2B} and A_3 receptor modulation of learning and memory largely due to the low expression of these receptors in the brain. Pharmacological treatment with A_3R agonist IB-MECA did not affect spontaneous alternation and passive avoidance performance, but diminished scopolamine- and MK-801-induced impairment of learning abilities in Y-maze and passive avoidance tasks in mice (Rubaj, Zgodzinski, & Sieklucka-Dziuba, 2003). Thus, the activation of A3 receptor may protect against learning and memory impairments induced by cholinergic (scopolamine) and glutamatergic (MK-801) deficits.

3.4.1 Complexity of AR control of cognition

Earlier work, largely using antagonists and agonists, has suggested a role of adenosine and its receptor targets in learning and memory, but these findings are somewhat inconsistent. Studies using complementary pharmacology and AR KOs have provided some clarification of the potential impact of A_1Rs and suggest an important role of $A_{2A}R$ signaling on some types of learning and memory. However, these pharmacological studies are limited by their partial specificity, and these genetic KO studies may, on the other hand, be confounded with potential developmental effects. It is possible that the diverse array of findings from these earlier studies reflects the different contributions of the different AR subtypes in distinct brain regions, all of which are more difficult to precisely target using only pharmacological tools or even a global AR KO strategy. Differences in the behavioral tasks and cognitive domains examined, both of which are generally restricted in number and scope, may contribute to the apparent discrepancies between these earlier studies. Therefore, a comprehensive and systematic assessment of the impact of each AR subtype in each relevant brain region (e.g., hippocampus, cortex, and striatum) on the multiple phases of learning and memory (e.g., encoding, storage, consolidation, and retrieval) using different behavioral tasks to evaluate distinct cognitive domains may be necessary to adequately dissect the nature of AR modulation of learning and memory.

4. ADENOSINE RECEPTOR CONTROL OF COGNITION IN NEUROPSYCHIATRIC DISORDERS

The ability of AR to integrate dopamine, glutamate, and BNDF signaling and to modulate learning and memory performance in various behavioral tasks suggest that AR activity not only confers procognitive effect in normal animals but also prevents or reverses memory impairments under several pathological conditions. Consequently, the AR may represent a novel target for reversing pathological cognitive impairments. Substantial evidence reviewed below support that AR activity can influence pathological cognitive impairments by cognitive enhancement (as in the normal brain) and neuroprotective mechanism (as in pathological condition). For example, normal BDNF signaling is critical to synaptic plasticity (which requires the $A_{2A}R$ activation) and cognition, and abnormal BDNF signaling is associated with cognitive impairments in neurodegenerative disorders such as Huntington's disease (HD) AD. Thus, $A_{2A}R$-mediated reestablishment of BDNF actions may represent a promising strategy for reversing cognitive impairments in AD. On the other hand, $A_{2A}R$ antagonists and caffeine confer neuroprotection in several neurological disease models, including stroke (Chen et al., 1999), Parkinson's disease (PD) (Chen et al., 2001), traumatic brain injury (Dai et al., 2010), and AD (Canas et al., 2009). A candidate mechanism whereby AR confers neuroprotection is the control of neuroinflammation, given the profound effect of AR activity on inflammatory response and the critical contribution of cytokine-mediated neuroinflammation and glutamate-mediated excitotoxicity to the amplification of neurodegeneration. For the dual role of IL-1 in both learning and brain injury, adenosine-dependent activation of caspase-1 and IL-1β generation has been shown to be responsible for amygdala-based anterograde amnesia after hypoxia/reoxygenation (Chiu et al., 2012). As such, the memory formation in mice exposed to acute hypoxia was accelerated in the mice lacking IL-1 receptor 1, in the mice receiving IL-1 receptor antagonist, and in the mice treated with the caspase-1 inhibitor (Chiu et al., 2012). Importantly, nonselective AR antagonists caffeine and selective A_1/A_{2A} AR antagonists prevented reoxygenation-dependent activation of caspase 1 and attenuated anterograde amnesia (Chiu et al., 2012). $A_{2A}R$ blockade confers neuroprotection by controlling IL-1β-induced exacerbation of excitotoxic neuronal damage, through the control of IL-1β-mediated phosphorylation MAP kinase (p38) and calcium deregulation (Simoes et al., 2012). In the context of AD therapy, $A_{2A}R$ control of neuroinflammation (through p38) of synaptopathy (Canas et al., 2009) and β-amyloid processing (Cao et al., 2009) have been suggested.

4.1. Alzheimer's disease

Cognitive impairment is prevalent on aging and is accelerated in a pathognomonic manner in AD, the neurodegenerative disorders with the greatest socioeconomical impact in the Western world (Murray & Lopez, 1997; Olesen, Gustavsson, Svensson, Wittchen, & Jonsson, 2012; Wimo, Jonsson, Bond, Prince, & Winblad, 2013). AD pathology is characterized by (i) the changes in the brain levels of different forms of β-amyloid peptides, amyloid plaques, neurofibrillary tangles (Andreasen et al., 2001; Galasko et al., 1998; Mattsson et al., 2009; Riemenschneider et al., 2002) argued to be major culprits of AD (Hardy & Selkoe, 2002; Walsh & Selkoe, 2004); (ii) the loss of synapses in defined brain cortical regions, most evident in the hippocampus in MCI and early phases of AD, and neuronal loss and glial responses (Coleman, Federoff, & Kurlan, 2004; Scheff, Price, Schmitt, DeKosky, & Mufson, 2007; Selkoe, 2002). Guided by the β-amyloid cascade hypothesis, and based on the large amount of preclinical data, several clinical trials have evaluated the effect of immunotherapy and inhibitors of β- or γ-secretases, to prevent the buildup of β-amyloid into plaques, and have not yet proved their therapeutic efficacy (Coric et al., 2012; Doody et al., 2013; Fox et al., 2005; Gilman et al., 2005; Howard et al., 2012; Lemere & Masliah, 2010; Winblad et al., 2012). Currently, there is no disease-modifying treatment to slow down or hold the disease progression. Since 1993, FDA has approved three acetylcholinesterase inhibitors and an NMDA receptor antagonist memantine (2004) for improving cognition at early–moderate (AChE inhibitors) and moderate–later stages (memantine) of AD (Aisen, Cummings, & Schneider, 2012). However, these treatments do not have disease-modifying properties, and their use is limited by their poor efficacy, loss of effect over time, and frequent side effects (Aisen et al., 2012; Amanzio, Benedetti, & Vase, 2012; Jones, 2010; Whitehead et al., 2004). There is increasing recognition that other systems (such as ARs) need to be investigated as putative targets.

There is a strong negative correlation between increased plasma levels of homocysteine and reduced adenosine in AD (Selley, 2004). This is due to formation of *S*-adenosylhomocysteine, suggesting the possibility that a deficiency of adenosine may contribute to neurological manifestation of increased homocysteine levels. Studies of brain tissue from patients who died with a confirmed diagnosis of AD revealed a loss of A_1R in the hippocampus, a brain region critically involved in learning and memory (Jaarsma, Sebens, & Korf, 1991; Jansen, Faull, Dragunow, & Synek, 1990; Ulas, Brunner, Nguyen, & Cotman, 1993). This reduction of A_1R in the hippocampus was substantiated by a PET study of AD patients using the A_1R antagonist ^{11}C-MPCX as a ligand (Fukumitsu et al., 2008). In contrast to the reduced levels of A_1R, pathological brain conditions associated with memory impairment (such as AD, stress, and inflammation) are characterized by a local increase of the extracellular levels of adenosine and by upregulation and abnormal signaling of brain $A_{2A}R$ (Chen, Eltzschig, & Fredholm, 2013; Cunha & Agostinho, 2010). Indeed, the density of hippocampal $A_{2A}R$ increases in aged animals (Canas et al., 2009; Munoz & Ammit, 2010; Ribeiro, 1999; Xia & Storm, 2012) as well as in transgenic mice displaying memory impairments (de Mendonca & Ribeiro, 1997; Lee & Chao, 2001; Nam et al., 2013). However, it remains to be defined if this increased density and function of $A_{2A}R$ are a maladaptive consequence of aging and AD or if they play a causal role in the emergence of memory deficits. Following experimental evidence supports a causal role of the AR in the development of cognitive impairments in AD.

(i) The first evidence of neuroprotection by $A_{2A}R$ antagonists and caffeine in AD models was in cultured cerebellum granular cells where caffeine and the $A_{2A}R$ antagonist SCH58261 reduced Aβ peptide-induced aggregation (Dall'lgna, Porciuncula, Souza, Cunha, & Lara, 2003). This protective effect was substantiated in an animal study in which caffeine and $A_{2A}R$ antagonists protected against loss of learning and memory induced by intracerebroventricular infusion of Aβ peptide (Cunha et al., 2008; Dall'lgna et al., 2003).

(ii) Nonselective AR antagonist caffeine and selective $A_{2A}R$ antagonists reverse memory loss induced by aging (Prediger, Batista, & Takahashi, 2005) and by spontaneous hypertension (Prediger & Takahashi, 2005) and also attenuate neurochemical modifications in the hippocampus of streptozotocin-induced diabetic encephalopathy in rats (Duarte, Carvalho, Cunha, & Gruetter, 2009) and in disease models including murine AD models (Canas et al., 2009; Cunha, Canas, et al., 2008; Dall'Igna et al., 2007), as well as attention deficit and hyperactivity disorder (Prediger, Fernandes, et al., 2005).

(iii) Recent studies have established the ability of $A_{2A}R$ to also control the recruitment of MAP kinases (Charalambous et al., 2008; Fredholm et al., 2007; Melani et al., 2006, 2009; Quiroz et al., 2006;

Schulte & Fredholm, 2003) and to prevent synaptotoxicity in animal models of AD was related to the aberrant recruitment of one MAP kinase, p38 (Canas et al., 2009).

(iv) Aβ(1–42) disrupts glutamatergic synapses function by decreasing astrocytic glutamate uptake and GLT-I expression in astrocytes. Apparently, this impact of Aβ(1–42) on glutamate transporters and uptake was dependent on $A_{2A}R$ function since genetic inactivation or pharmacological blockade of $A_{2A}R$ abrogated this effect. Thus, astrocytic $A_{2A}R$ may contribute to modulation of Aβ-induced impairment of glutamate uptake and glutamatergic synaptic dysfunction and excitotoxicity in AD.

(v) Treatment with nonselective adenosine antagonist caffeine and the selective $A_{2A}R$ antagonist SCH58261 reversed age-associated memory impairment (in a novel object recognition task) and to partially normalized the increase in oxygen and nitrogen reactive species levels and the decrease in Na(+)K(+) ATPase activity in brains of aged rats (Leite, Wilhelm, Jesse, Brandao, & Nogueira, 2011). Thus, ARs can exert its precognitive effect by controlling oxidative stress and NKA activity of aged rats.

Accordingly, antagonists of $A_{2A}R$ can reverse memory impairments, as concluded from studies both in aging rodents (Leite et al., 2011; Prediger, Batista, et al., 2005) and in animal models of AD (Canas et al., 2009; Cunha, Canas, et al., 2008; Dall'Igna et al., 2007; Espinosa et al., 2013) and in other experimental conditions leading to an impairment of memory performance (Cognato et al., 2010; Duarte, Agostinho, Carvalho, & Cunha, 2012).

4.1.1 Human caffeine consumption and cognitive decline in aging and AD

The notion that AR modulates cognition is also supported by epidemiological evidence that consumption of caffeine, the most widely used psychoactive compound, improves cognitive performance in aging and AD patients. Caffeine is a nonselective AR antagonist and caffeine exerts its action on brain function including cognition by blocking ARs (Cunha & Agostinho, 2010; Ferre, 2008; Fredholm et al., 2005). Recent efforts to identify potential risk factors for developing AD have provided epidemiological evidence that caffeine may be neuroprotective in AD. At least six longitudinal studies support the inverse relationship between caffeine consumption and both decreased memory impairments associated with aging as well as a reduced risk of developing AD.

- *Maastricht Aging Study (MAAS)*: This prospective population study in southern Netherlands found a cross-sectional association between caffeine consumption and improvements in psychomotor speed and verbal memory performance in the "nondemented" elderly population (Hameleers et al., 2000; van Boxtel, Schmitt, Bosma, & Jolles, 2003).

- *The Canadian Study of Health and Aging (CSHA)*: After surveying 4615 subjects at 5-year follow-up, including 194 with AD and 3894 cognitively normal controls, this population-based prospective study found that regular consumption of caffeine is associated with a reduced risk of developing AD (Lindsay et al., 2002).
- *The FINE study*: This 10-year follow-up study involving 667 healthy men born in Finland, Italy, and the Netherlands between 1900 and 1920 suggests that consuming three cups of coffee per day reduces cognitive decline in elderly men (van Gelder et al., 2007).
- *The French Three Cities Study*: This large population-based prospective study involving 4197 women and 2820 men found that women (but not men) drinking three cups or more of coffee exhibited less decline in verbal retrieval and visuospatial memory over 4 years, even adjusting for other multiple factors contributing to cognitive decline (Ritchie et al., 2007).
- *The Cardiovascular Risk Factors, Aging and Dementia (CAIDE) Study*: This study involved 1409 individuals aged 65–79 after an average 21-year follow-up (total of 61 cases identified as demented, 48 with AD) and found that individuals identified as coffee drinkers at midlife were 65% less likely to develop dementia and AD later in life compared with those drinking little or no coffee (Eskelinen, Ngandu, Tuomilehto, Soininen, & Kivipelto, 2009).
- *The Honolulu-Asia Aging Study*: This study involved 3494 men with a mean age of 52 years at cohort entry in 1965–1968 and found that the men in the highest quartile of caffeine intake were less likely than men in the lowest quartile to have any neuropathological lesions at death in the 226 men with dementia and the 347 men with cognitive impairment who underwent brain autopsy (Gelber, Petrovitch, Masaki, Ross, & White, 2011).

In further support of this inverse correlation, caffeine consumption and cognitive decline by these large, longitudinal studies, epidemiological investigation, animal studies further support a causal role of caffeine in neuroprotection by demonstrating that chronic consumption of caffeine can also prevent and counteract memory impairment in animal models of aging and AD: (i) the chronic consumption of caffeine abrogates memory impairment in animal models of AD (Canas et al., 2009; Dall'Igna et al., 2007; Espinosa et al., 2013); (ii) the same occurs in aging rodents (Costa, Botton, Mioranza,

Souza, & Porciuncula, 2008; Leite et al., 2011; Prediger, Batista, et al., 2005; Vila-Luna et al., 2012); (iii) furthermore, studies of aged AD transgenic (APPsw, Swedish mutation) mice found that long-term (between early 4–9 months and aged 18–19 months old) administration of a 1.5-mg daily dose of caffeine (equivalent to 500 mg in human) to APPsw mice reduced brain Aβ levels and protect against certain cognitive impairments (Arendash et al., 2006, 2009; Cao et al., 2009); (iv) this is in accordance with the ability of caffeine to control synaptic plasticity phenomena in the hippocampus (Alhaider, Aleisa, Tran, Alzoubi, & Alkadhi, 2010; Alzoubi, Srivareerat, Aleisa, & Alkadhi, 2013; Costenla, Cunha, & de Mendonca, 2010), an effect that becomes more evident upon aging (Costenla et al., 2011); (v) caffeine was recently shown to reduce disruption of the BBB induced by a high-cholesterol diet (Chen, Gawryluk, Wagener, Ghribi, & Geiger, 2008), suggesting additional mechanisms for the therapeutic effect of caffeine in AD and other neurodegenerative disorders. This convergence of the epidemiological and animal evidence led to the proposal that caffeine might be a novel prophylactic strategy to alleviate the burden of AD.

This led to the recent case-control study providing the first direct evidence that caffeine/coffee intake is associated with a reduced risk of dementia (Cao et al. 2012). The study found that subjects with plasma caffeine levels greater than 1200 ng/ml at study onset was associated with stable MCI→MCI and no conversion to dementia during the 2–4 year follow-up examination.

4.2. Parkinson's disease

PD is primarily characterized by cardinal motor symptoms, but cognitive changes also occur both in the early and later stages of the disease (Barone et al., 2011; Kehagia, Barker, & Robbins, 2010; Watson & Leverenz, 2010). In fact, owing to the success of the treatment for motor symptoms of this disease, clinical significance of cognitive impairments in PD is increasingly recognized as a major unmet medical need in PD. Approximately 30% of PD patients have dementia in cross-section studies with a cumulative prevalence of dementia at ~70% for life time (Aarsland et al., 2010; Furtado et al., 2012). The International Movement Disorder Society Task Force conclude that "PD-MCI appears to be a clinically significant syndrome, as even mild cognitive deficits or self-rated cognitive deficits in early PD are associated with functional impairment and worse quality of life. Thus, identification and intervention at the earliest stage of PD-MCI is a crucial unmet need for the overall care of PD patients" (Litvan et al.,

2011). These early cognitive impairments in PD encompass deficits within the *executive*, *mnemonic*, and *visuospatial* domains (Kehagia et al., 2010; Litvan et al., 2011; Watson & Leverenz, 2010). The neurochemical basis of cognitive impairment in PD is not established but it may be related to the dysfunction of the DA system. Indeed, the DA phasic firing in the striatum provides the "error" prediction signal (a "reinforcement" feedback signal), justifying that DA depletion may give raise not only to motor deficits but also to cognitive impairments in PD (Kehagia et al., 2010; Lewis, Dove, Robbins, Barker, & Owen, 2003; Sawamoto et al., 2008; Shohamy et al., 2004). In animal models of PD, DA depletion in the cortex produces cognitive deficits, particularly associated with executive dysfunction (Brozoski, Brown, Rosvold, & Goldman, 1979; Clinton, Sucharski, & Finlay, 2006), visuospatial and working memory that resemble those seen in PD (Decamp & Schneider, 2004). DA-dependent cognitive impairments in PD patients are further supported by the finding that DA drugs improve not only motor deficits but also cognitive deficits that are associated with frontal lobe dysfunction (Gotham, Brown, & Marsden, 1988; Lange et al., 1992; Taylor, Saint-Cyr, & Lang, 1987). However, DA treatment is not effective in many aspects of cognitive dysfunction in PD (including attention set-shifting, task switching, and associative learning) (Kehagia et al., 2010). In fact, DA treatment may exacerbate cognitive impairments such as concurrent learning, probabilistic reversal learning, weather prediction classification, gambling, and decision making (Kehagia et al., 2010), as a result of DA overdosing of ventral striatal DA signaling.

Currently, the only available therapy (i.e., cholinesterase inhibitors) targets late-stage dementia, only providing modest benefits. Importantly, the use of cholinesterase inhibitors to manage early cognitive impairments may worsen motor deficits in PD (Chaudhuri & Schapira, 2009; Richard, Justus, Greig, Marshall, & Kurlan, 2002; van Laar, De Deyn, Aarsland, Barone, & Galvin, 2011). There is an urgent need for novel approaches to their treatment. The $A_{2A}R$ is postulated to be a novel and promising therapeutic target for reversing cognitive impairments in PD. In a recent study with bilateral PFC depletion of DA as a model, the $A_{2A}R$ antagonist istradefylline, a drug in clinical phase III trials for motor benefits in PD, is shown to improve cognitive performance (working memory) of PFC-lesioned rats (Kadowaki Horita, Kobayashi, Mori, Jenner, & Kanda, 2013). This improvement is associated with the istradefylline-mediated increase in PFC dopamine levels in both normal and PFC-lesioned rats. Other studies suggest that blockade of both A_1R and $A_{2A}R$ such as orally active dual A_1R and $A_{2A}R$ antagonist ASP5854 are more effective than the $A_{2A}R$ antagonist KW6002 in reversing the scopolamine- or MK-801-induced impairment of spontaneous alternation in the mouse Y-maze test (Mihara et al., 2007). These results demonstrate that AR antagonists, particularly $A_{2A}R$ antagonists, can not only improve motor performance in PD but may also lead to improved cognition. Since striatum and striatal $A_{2A}R$ are important for cognitive flexibility, $A_{2A}R$ antagonists have also be explored but not found to be effective in improving dopamine depletion-induced deficits in flexible shifting of response patterns (O'Neill & Brown, 2007b).

4.2.1 Human caffeine consumption and PD

Additional support for prospective role of $A_{2A}R$ antagonists as a novel therapy for PD came from their potential to attenuate dopaminergic neurodegeneration, as suggested by convergent epidemiological and experimental evidences. Ross and colleagues (2000) reported an inverse relationship between consumption of the nonselective adenosine antagonist caffeine and the risk of developing PD over a 30-year follow-up study in a large prospective study of 8004 Japanese–American men in the Honolulu Heart Program. The age- and smoking-adjusted risk of PD was five times higher among men who reported no coffee consumption compared to men who reported a daily consumption of 28 oz. or more of coffee. This finding was substantiated by a similar inverse relationship between the consumption of caffeinated (but not decaffeinated) coffee and the risk of developing PD in two larger, more ethnically diverse cohorts—the Health Professionals' Follow-Up Study and the Nurses' Health Study—involving 47,351 men and 88,565 women (Ascherio et al., 2001) and also more recently in the Finnish Mobile Clinic Health Examination Survey involving 19,518 men and women (Saaksjarvi et al., 2007). These studies firmly establish a relationship between increased caffeine consumption and decreased risk of developing PD in males. This postulated neuroprotective effects of caffeine further substantiated by our studies with animal models of PD provide a compelling clue about the potentially protective effects of caffeine by demonstrating that pharmacological blockade (by caffeine or selective $A_{2A}R$ antagonists) or genetic depletion of the $A_{2A}R$ attenuates dopaminergic neurotoxicity and neurodegeneration (Chen et al., 2001; Ikeda, Kurokawa, Aoyama, & Kuwana, 2002; Xu, Xu, Chen, & Schwarzschild, 2002). Caffeine as well as $A_{2A}R$ antagonists (including SCH58261, KW-6002, DMPX, and CSC) attenuate MPTP-induced dopaminergic neurotoxicity, suggesting that the protective effects of caffeine are due to its action at the $A_{2A}R$ (Alfinito et al., 2003; Chen et al., 2001; Pierri, Vaudano, Sager, & Englund, 2005). Finally, genetic inactivation of $A_{2A}Rs$ also reduces MPTP-induced dopaminergic neurotoxicity (Chen et al., 2001) and α-synuclein-induced DA neurotoxicity (Kachroo & Schwarzschild, 2012). These studies provide a neurobiological basis for the inverse relationship between increased caffeine consumption and reduced risk of developing PD. The convergence of epidemiological evidence and animal studies also raises the exciting possibility that $A_{2A}R$ antagonists, including caffeine, may offer a neuroprotective strategy to slow or halt dopaminergic neuron degeneration. The validation of this working hypothesis will provide the required rationale to stimulate the necessary clinical investigation to rapidly translate $A_{2A}R$ antagonists to be used as a novel strategy to control memory impairment associated with PD and AD, since $A_{2A}R$ antagonists [e.g., KW-6002 (istradefylline) (Chen et al., 2013; Jenner et al., 2009) and Preladenant (Chen et al., 2013; Hauser et al., 2011)] are already in phase III clinical trials for motor benefits with noted safety profiles.

4.3. Huntington's disease

Huntington's disease (HD) is caused by a CAG trinucleotide expansion in exon 1 of the Huntington (Htt) gene (The Huntington's Disease Collaborative Research Group, 1993), which leads to progressive neuronal loss in the striatum and cortex. HD is characterized by a triad of progressive motor (chorea and dystonia), cognitive, and psychiatric symptoms, invariably leading to death within 15–25 years (Martin & Gusella, 1986; Young, 2003; Zuccato, Valenza, & Cattaneo, 2010). Although HD onset is defined as the time of first clinical presentation of motor symptoms, recent studies indicate that the earliest motor symptoms are often preceded by noticeable cognitive impairments (Duff et al., 2010; Ho et al., 2003; Lawrence, Hodges, et al., 1998; Lawrence, Weeks, et al., 1998; Stout et al., 2011) and mood disturbances (depression) (Julien et al., 2007; Rosenblatt, 2007). The largest and most comprehensive study of HD cognitive impairment (*PREDICT-HD study*, including 738 HD prodromes) (Stout et al., 2011) and several others (Duff et al., 2010; Lawrence, Hodges, et al., 1998; Papp, Kaplan, & Snyder, 2011) conclusively demonstrated cognitive impairments in the HD prodrome, 5–15 years before the onset of motor symptoms. Neurodegeneration in HD apparently starts as a selective loss of GABA containing medium spiny projection neurons in the striatum (Ferrante et al., 1997; Reiner et al., 1988; Vonsattel & DiFiglia, 1998). Neurons in the associative neocortex that project to the basal ganglia are also affected, resulting in a frontostriatal dysfunction characteristic of HD pathology (Cepeda, Wu, Andre, Cummings, & Levine, 2007; Miller, Walker, Shah, Barton, & Rebec, 2008; Walker, Miller, Fritsch, Barton, & Rebec, 2008). These earliest cognitive symptoms of HD precede neuronal loss (Vonsattel & DiFiglia, 1998), suggesting that these symptoms reflect an underlying neuronal *dysfunction* rather than resulting from neuronal *death*. This is in agreement with the long-standing view that HD begins with a synaptic dysfunction (synaptopathy) preceding neuronal death, which selectively affects corticostriatal projections (Li, Plomann, & Brundin, 2003). There is no cure or disease-modifying treatment for this devastating disorder, and symptomatic treatment is limited (Munoz-Sanjuan & Bates, 2011).

Several lines of the evidence support the involvement of striatal $A_{2A}R$ in HD pathogenesis. (i) A prevailing feature of HD pathology is NMDA-mediated excitotoxicity (DiFiglia, 1990; Starling et al., 2005; Zeron et al., 2002). By a presynaptic mechanism, the $A_{2A}R$ antagonist SCH 58261 reduces quinolinic acid-induced glutamate outflow (Gianfriddo, Melani, Turchi, Giovannini, & Pedata, 2004; Popoli et al., 2002; Tebano et al., 2004) and NMDA-induced toxicity in R6/2 mice (Domenici et al., 2007). At postsynaptic site, activation of $A_{2A}Rs$ inhibits NMDA receptor-mediated synaptic currents in rat neostriatal neurons (Norenberg, Wirkner, Assmann, Richter, & Illes, 1998; Norenberg, Wirkner, & Illes, 1997; Wirkner et al., 2000). Consequently, blockade of postsynaptic $A_{2A}Rs$ is likely detrimental to striatal neurons. Consistent with this notion, $A_{2A}R$ agonists have been shown to attenuate HD pathology in R6 models (Chou et al., 2005). (ii) Gain-of-toxicity and loss-of-function experiments have mechanistically linked BDNF with the underlying HD defect (Zuccato et al., 2001, 2003). In this context, $A_{2A}R$ activation transactivates TrkA and TrkB receptors in the absence of neurotrophins (Lee & Chao, 2001) and facilitates BDNF release and BDNF-induced potentiation of synaptic transmission (Diogenes et al., 2004; Tebano et al., 2008). Thus, $A_{2A}R$ activity may influence the development of HD pathology by modulating BDNF activity in the striatum. (iii) A dysregulated ubiquitin–proteasome system (UPS) is closely associated with the formation of HTT aggregates and HD pathogenesis (Bennett et al., 2007; Hunter, Lesort, & Johnson, 2007; Seo, Sonntag, & Isacson, 2004; Wang et al., 2008; Zhou et al., 2003). In genetic mouse models of HD, the $A_{2A}R$ agonist CGS21680 can ameliorate several major symptoms (Chiang et al., 2009; Chou et al.,

2005). The rescue effect of CGS21680 is attributed to the activation of survival signals such as CREB via a cAMP/PKA-dependent pathway (Chiang et al., 2005) and to its ability to enhance UPS activity (Chiang et al., 2009). (iv) In support of $A_{2A}R$ involvement in the development of HD, a recent genetic association study of 791 unrelated HD patients found that a single nucleotide polymorphism in the $A_{2A}R$ gene (1976 C/T, rs5751876) is associated with the reduced age at onset of the disease by 3.8 years (Dhaenens et al., 2009). Thus, the unique anatomical localization of the $A_{2A}R$ in the striatopallidal neurons, the ability of $A_{2A}Rs$ to fine tune glutamate activity at presynaptic and postsynaptic sites, and the $A_{2A}R$'s ability to facilitate BDNF function and stimulate UPS activity, all make the $A_{2A}R$ a novel and promising therapeutic target for HD.

On the other hand, to the best of our knowledge, the effect of ARs on cognition in HD models has not examined yet.

4.4. Schizophrenia

Schizophrenia is an incurable mental disorder with a presumed neurodevelopmental origin (Ross, Margolis, Reading, Pletnikov, & Coyle, 2006). The spectrum of symptomatology includes not only positive and negative symptoms but also cognitive symptoms, characterized by working memory and executive deterioration (van Os & Kapur, 2009). Existing pharmacotherapy and conventional models of schizophrenia have long emphasized dopaminergic dopamine dysfunction as the prime neurochemical basis for the disorder, building upon the notion that striatal dopamine hyperfunction seems to be implicated in the positive symptoms of schizophrenia (Snyder, 1976). However, the limitations of the dopamine model to account for negative and cognitive symptoms of schizophrenia lead to the emergence of glutamatergic models, based on the observation that psychotomimetic agents such as phencyclidine and dizocilpine (MK-801) induce psychotic and cognitive disturbances similar to those observed in schizophrenia by blocking NMDA receptor (Moghaddam & Javitt, 2012). In fact, cortical glutamatergic hypofunction is well positioned to explain not only the positive and negative symptoms of schizophrenia but also the cognitive decline that is at the premorbid stage of schizophrenia (Ranganath, Minzenberg, & Ragland, 2008). This seems to result from the ability of NMDAR to control the impaired interactions between multiple brain regions, which are a hallmark of the schizophrenic brain (Field, Walker, & Conn, 2011; Lisman et al., 2008), rather than a particular locus of dysfunction.

Based on the neuromodulatory and homeostatic actions of adenosine, adenosine dysfunction has been postulated to contribute to the neurobiological and clinical features of schizophrenia by Lara, Dall'Igna, Ghisolfi, and Brunstein (2006) and Boison, Singer, Shen, Feldon, and Yee (2012). Adenosine, via activation of inhibitory A_1 and facilitatory A_{2A} receptors, has the unique capability to integrate and fine-tune glutamatergic and dopaminergic outputs. For adenosine modulation of dopaminergic signaling, an adenosine deficit can enhance dopamine release because of the inhibitory effect of A_1R on dopamine release (Golembiowska & Zylewska, 1998; Solinas et al., 2002) and potentiate amphetamine-induced locomotion (Popoli, Pezzola, & de Carolis, 1994). Antagonistic interactions between $A_{2A}R$ and D_2R (Ferre et al., 1991; Ferre, Herrera-Marschitz, Grabowska-Anden, Ungerstedt, et al., 1991; Ferre et al., 2001; Sebastiao & Ribeiro, 2000) predict that increased basal D_2R occupancy in schizophrenia patients (Abi-Dargham et al., 2000) could reduce $A_{2A}R$'s effect on D_2R and thereby increases the affinity of the D_2R for dopamine (Ferre, Herrera-Marschitz, Grabowska-Anden, Casas, et al., 1991; Ferre, Herrera-Marschitz, Grabowska-Anden, Ungerstedt, et al., 1991; Ferre, von Euler, Johansson, Fredholm, & Fuxe, 1991). These mechanisms could provide the rationale for an antipsychotic-like profile of AR agonists (Ferre, 1997; Kafka & Corbett, 1996). Furthermore, the astroglial GLT-1 levels are increased in the cortex of some schizophrenia patients and further modified by psychotomimetics and antipsychotic drugs (Deng et al., 2004; Matute, Melone, Vallejo-Illarramendi, & Conti, 2005; Spangaro et al., 2012). Astroglial $A_{2A}R$ tightly regulates GLT-1 activity (Matos et al., 2013, 2012), suggesting a possible astroglial link of schizophrenia, and might contribute to the pathogenesis of schizophrenia. This regulatory function of adenosine might provide a missing link for the functional integration of the dopamine and glutamate hypotheses of schizophrenia.

AR gene polymorphism and AR gene expression patterns have been examined in human schizophrenia samples to seek the correlative evidence for the AR involvement in schizophrenia. A recent genetic analysis of 29 polymorphic markers in 200 schizophrenic patients and 210 healthy controls of Japanese population found that three statistically different markers of *Adora1* polymorphisms may be involved in the pathophysiological mechanisms of schizophrenia in a Japanese population (Gotoh et al., 2009). An association study of the *Adora2a* gene (1976 T > C) polymorphism in 227 schizophrenia patients of Chinese population, however, did not find the evidence that the *Adora2a* 1976 T > C polymorphism plays a major role in the pathogenesis of schizophrenia (Hong, Liu, Liu, Liao, & Tsai, 2005). Evaluation of the expression pattern of A_1R and $A_{2A}R$ in schizophrenia patients has so far produced mixed results with either decreased (Villar-Menendez et al., 2014), increased (Hwang et al., 2013), or no change (Uriguen et al., 2009) of $A_{2A}R$ expression in postmortem schizophrenia brains. For example, by comparing subjects suffering schizophrenia ($n=31$) who mainly died by suicide, matched with nonschizophrenia suicide victims ($n=13$) and nonsuicide controls ($n=33$), Uriguen et al. (2009) found that the density of receptor proteins and their relative mRNA expression of $A_{2A}R$ were comparable among three groups. However, another study using RNA-Seq data from the hippocampus identified upregulation of $A_{2A}R$ as one of 144 differentially expressed genes in schizophrenia cases as compared with unaffected controls, representing an inflammatory response (Hwang et al., 2013). An association study showed that there was no significant difference in peripheral $A_{2A}R$ gene expression between medication-free schizophrenia ($n=31$) and healthy ($n=21$) groups among groups, and $A_{2A}R$ expression was not correlated with sensory gating (Zhang et al., 2012). However, 6-week antipsychotic treatment caused a significant upregulation of $A_{2A}R$ expression was positively association with baseline P50 amplitudes in the schizophrenia group (Zhang et al., 2012). Interestingly, a recent study found that $A_{2A}R$ (but not A_1R) mRNA levels in schizophrenia postmortem cortex was reduced by 50% with respect to age-matched controls, correlating with motor disturbances. Importantly, the reduction in the $A_{2A}R$ mRNA is apparently associated with an increase in 5-methylcytosine percentage in the 5′ UTR region of *Adora2a* gene (Villar-Menendez et al., 2014). These finding led authors to suggest that the reduced striatal $A_{2A}R$ levels and altered motor phenotype may define a molecular subgroup of SZ patients. This adenosine deficit in schizophrenia is further substantiated by the finding that the activity of ectonucleotidases in the postmortem putamen was reduced in SZ patients compared with aged-matched controls (Aliagas et al., 2013).

In support of the adenosine hypothesis of schizophrenia, striatal $A_{2A}R$ activation and extrastriatal $A_{2A}R$ blockade are proposed to confer antipsychotic action against dopamine hyperfunction or NMDAR hypofunction (Rimondini, Ferre, Ogren, & Fuxe, 1997; Shen, Coelho, et al., 2008). In animal models of schizophrenia, A_1R and $A_{2A}R$ agonists have been repeatedly shown to be effective against the behavioral as well as neurophysiological (EEG and prepulse inhibition) effects induced by NMDAR antagonists (Kafka & Corbett, 1996; Popoli, Reggio, Pezzola, Fuxe, & Ferre, 1998), thus lending support for their potential antipsychotic efficacy in humans. Furthermore, overexpression of ADK, the key enzyme of adenosine clearance, is associated with attention deficit, abnormal psychomotor response to amphetamine and working memory deficits, all of which are reversed by augmentation of extracellular adenosine with pharmacological inhibition of ADK or intrastriatal implants of engineered adenosine releasing (Boison et al., 2012; Shen et al., 2012).

Based on the emerging evidence for adenosine deficit hypothesis of schizophrenia, allopurinol and dipyridamole, which are uptake inhibitors of adenosine to indirectly increase adenosine level, have been explored in a pilot clinical study to determine their possible antischizophrenia effect (Akhondzadeh, Shasavand, Jamilian, Shabestari, & Kamalipour, 2000; Wonodi et al., 2011). Early study to explore dipyridamole as an adjunctive therapy with haloperidol showed that combination of haloperidol and dipyridamole was significantly better than haloperidol alone in improving positive and general psychopathology symptoms (Akhondzadeh et al., 2000). Dipyridamole as a monotherapy of 200 mg/day was found to show a trend of improvement of positive and negative symptoms (Wonodi et al., 2011). These findings support for examining dipyridamole (200 mg/day) as monotherapy or adjunct to symptomatic antipsychotic-treated schizophrenia patients. A meta-analysis of nine clinical studies [including six studies in schizophrenia (total $n=457$) and three studies in bipolar disorder (total $n=289$)] shows that adenosine modulators were superior to placebo in PANSS total scores (SMD $=-1.07$, $p=0.01$) and positive and general but not negative symptom subscale scores in schizophrenia (Hirota & Kishi, 2013). Thus, adenosine modulator adjuvant therapy is more beneficial in overall psychopathology (especially positive symptoms) in schizophrenia, a finding needed to be validated in a large clinical study of schizophrenia patient.

5. CONCLUDING REMARKS

Over the last two decades, the convergence of the molecular, animal, and epidemiological evidences supports the proposal that adenosine and ARs present important therapeutic targets for controlling cognition under normal and disease conditions. This proposal is rooted in the dual roles of adenosine and AR as neuromodulation as well as of homeostatic function in the brain and in the unique ability of adenosine, by acting on the inhibitory A_1 and facilitating A_{2A} receptor, to integrate dopamine, glutamate, and BNDF signaling and to modulate synaptic plasticity (e.g., LTP and LTD) in brain regions relevant to learning and memory, providing the molecular and cellular bases for AR control of cognition. The validity of this target is supported by experimental finding that AR antagonism not only enhances working memory, reference memory, reversal learning, goal-directed behavior, effort-related behavior, and Pavlovian conditioning in normal animals but also reverses working memory and short-term memory impairments in animal models of AD, PD, and HD. This proposal is further substantiated by epidemiological evidences supporting an inverse relationship between human caffeine consumption and the risk of developing cognitive impairments in aging and AD and reduced risk in developing PD. The confirmation of this proposal will constitute a novel strategy for controlling cognitive deficit associated with AD by the selective manipulation of $A_{2A}R$ activity. Since $A_{2A}R$ antagonists [e.g., KW-6002 (istradefylline), and Preladenant] are already in phase III clinical trials for motor benefits in PD patients with noted safety profiles (Chen, Eltzschig, Fredholm, 2013), the validation of this proposal will provide the required rationale to stimulate the necessary clinical investigation to rapidly translate AR antagonists to be used as a novel strategy to control memory impairment associated with neuropsychiatric disorders.

ACKNOWLEDGMENTS

This work was supported by grants from the National Institute of Health Grant NS041083-10 and NS073947 (USA).

REFERENCES

Aarsland, D., Bronnick, K., Williams-Gray, C., Weintraub, D., Marder, K., Kulisevsky, J., et al. (2010). Mild cognitive impairment in Parkinson disease: A multicenter pooled analysis. *Neurology, 75*, 1062–1069.

Abi-Dargham, A., Rodenhiser, J., Printz, D., Zea-Ponce, Y., Gil, R., Kegeles, L. S., et al. (2000). Increased baseline occupancy of D2 receptors by dopamine in schizophrenia. *Proceedings of the National Academy of Sciences of the United States of America, 97*, 8104–8109.

Adams, J. P., & Sweatt, J. D. (2002). Molecular psychology: Roles for the ERK MAP kinase cascade in memory. *Annual Review of Pharmacology and Toxicology, 42*, 135–163.

Aisen, P. S., Cummings, J., & Schneider, L. S. (2012). Symptomatic and nonamyloid/tau based pharmacologic treatment for Alzheimer disease. *Cold Spring Harbor Perspectives in Medicine, 2*, a006395.

Akhondzadeh, S., Shasavand, E., Jamilian, H., Shabestari, O., & Kamalipour, A. (2000). Dipyridamole in the treatment of schizophrenia: Adenosine–dopamine receptor interactions. *Journal of Clinical Pharmacy and Therapeutics, 25*, 131–137.

Alfinito, P. D., Wang, S. P., Manzino, L., Rijhsinghani, S., Zeevalk, G. D., & Sonsalla, P. K. (2003). Adenosinergic protection of dopaminergic and GABAergic neurons against mitochondrial inhibition through receptors located in the substantia nigra and striatum, respectively. *Journal of Neuroscience, 23*, 10982–10987.

Alhaider, I. A., Aleisa, A. M., Tran, T. T., Alzoubi, K. H., & Alkadhi, K. A. (2010). Chronic caffeine treatment prevents sleep deprivation-induced impairment of cognitive function and synaptic plasticity. *Sleep, 33*, 437–444.

Aliagas, E., Villar-Menendez, I., Sevigny, J., Roca, M., Romeu, M., Ferrer, I., et al. (2013). Reduced striatal ecto-nucleotidase activity in schizophrenia patients supports the "adenosine hypothesis". *Purinergic Signal, 9*, 599–608.

Alzoubi, K. H., Srivareerat, M., Aleisa, A. M., & Alkadhi, K. A. (2013). Chronic caffeine treatment prevents stress-induced LTP impairment: The critical role of phosphorylated CaMKII and BDNF. *Journal of Molecular Neuroscience, 49*, 11–20.

Amanzio, M., Benedetti, F., & Vase, L. (2012). A systematic review of adverse events in the placebo arm of donepezil trials: The role of cognitive impairment. *International Psychogeriatrics, 24*, 698–707.

Andreasen, N., Minthon, L., Davidsson, P., Vanmechelen, E., Vanderstichele, H., Winblad, B., et al. (2001). Evaluation of CSF-tau and CSF-Abeta42 as diagnostic markers for Alzheimer disease in clinical practice. *Archives of Neurology, 58*, 373–379.

Arai, A., & Lynch, G. (1992). Factors regulating the magnitude of long-term potentiation induced by theta pattern stimulation. *Brain Research, 598*, 173–184.

Arendash, G. W., Mori, T., Cao, C., Mamcarz, M., Runfeldt, M., Dickson, A., et al. (2009). Caffeine reverses cognitive impairment and decreases brain amyloid-beta levels in aged Alzheimer's disease mice. *Journal of Alzheimer's Disease, 17*, 661–680.

Arendash, G. W., Schleif, W., Rezai-Zadeh, K., Jackson, E. K., Zacharia, L. C., Cracchiolo, J. R., et al. (2006). Caffeine protects Alzheimer's mice against cognitive impairment and reduces brain beta-amyloid production. *Neuroscience, 142*, 941–952.

Ascherio, A., Zhang, S. M., Hernan, M. A., Kawachi, I., Colditz, G. A., Speizer, F. E., et al. (2001). Prospective study of caffeine consumption and risk of Parkinson's disease in men and women. *Annals of Neurology, 50*, 56–63.

Assaife-Lopes, N., Sousa, V. C., Pereira, D. B., Ribeiro, J. A., Chao, M. V., & Sebastiao, A. M. (2010). Activation of adenosine A2A receptors induces TrkB translocation and increases BDNF-mediated phospho-TrkB localization in lipid rafts: Implications for neuromodulation. *Journal of Neuroscience, 30*, 8468–8480.

Assaife-Lopes, N., Sousa, V. C., Pereira, D. B., Ribeiro, J. A., & Sebastiao, A. M. (2014). Regulation of TrkB receptor translocation to lipid rafts by adenosine A receptors and its functional implications for BDNF-induced regulation of synaptic plasticity. *Purinergic Signalling, 10*, 2510267.

Augusto, E., Matos, M., Sevigny, J., El-Tayeb, A., Bynoe, M. S., Muller, C. E., et al. (2013). Ecto-5′-nucleotidase (CD73)-mediated formation of adenosine is critical for the striatal adenosine A2A receptor functions. *Journal of Neuroscience, 33*, 11390–11399.

Baddeley, A. D., Bressi, S., Della Sala, S., Logie, R., & Spinnler, H. (1991). The decline of working memory in Alzheimer's disease. A longitudinal study. *Brain, 114*(Pt. 6), 2521–2542.

Balleine, B. W., & Dickinson, A. (1998). Goal-directed instrumental action: Contingency and incentive learning and their cortical substrates. *Neuropharmacology, 37*, 407–419.

Barone, P., Aarsland, D., Burn, D., Emre, M., Kulisevsky, J., & Weintraub, D. (2011). Cognitive impairment in nondemented Parkinson's disease. *Movement Disorders, 26*, 2483–2495.

Benito, E., & Barco, A. (2010). CREB's control of intrinsic and synaptic plasticity: Implications for CREB-dependent memory models. *Trends in Neurosciences, 33*, 230–240.

Bennett, E. J., Shaler, T. A., Woodman, B., Ryu, K.-Y., Zaitseva, T. S., Becker, C. H., et al. (2007). Global changes to the ubiquitin system in Huntington's disease. *Nature, 448*, 704–708.

Blundon, J. A., Bayazitov, I. T., & Zakharenko, S. S. (2011). Presynaptic gating of postsynaptically expressed plasticity at mature thalamocortical synapses. *Journal of Neuroscience*, *31*, 16012–16025.
Boison, D., Chen, J. F., & Fredholm, B. B. (2010). Adenosine signaling and function in glial cells. *Cell Death and Differentiation*, *17*, 1071–1082.
Boison, D., Singer, P., Shen, H. Y., Feldon, J., & Yee, B. K. (2012). Adenosine hypothesis of schizophrenia—Opportunities for pharmacotherapy. *Neuropharmacology*, *62*, 1527–1543.
Bozdagi, O., Rich, E., Tronel, S., Sadahiro, M., Patterson, K., Shapiro, M. L., et al. (2008). The neurotrophin-inducible gene Vgf regulates hippocampal function and behavior through a brain-derived neurotrophic factor-dependent mechanism. *Journal of Neuroscience*, *28*, 9857–9869.
Brozoski, T. J., Brown, R. M., Rosvold, H. E., & Goldman, P. S. (1979). Cognitive deficit caused by regional depletion of dopamine in prefrontal cortex of rhesus monkey. *Science*, *205*, 929–932.
Cabeza de Vaca, S., Kannan, P., Pan, Y., Jiang, N., Sun, Y., & Carr, K. D. (2007). The adenosine A2A receptor agonist, CGS-21680, blocks excessive rearing, acquisition of wheel running, and increases nucleus accumbens CREB phosphorylation in chronically food-restricted rats. *Brain Research*, *1142*, 100–109.
Calabresi, P., Maj, R., Mercuri, N. B., & Bernardi, G. (1992). Coactivation of D1 and D2 dopamine receptors is required for long-term synaptic depression in the striatum. *Neuroscience Letters*, *142*, 95–99.
Calabresi, P., Maj, R., Pisani, A., Mercuri, N. B., & Bernardi, G. (1992). Long-term synaptic depression in the striatum: Physiological and pharmacological characterization. *Journal of Neuroscience*, *12*, 4224–4233.
Canals, M., Marcellino, D., Fanelli, F., Ciruela, F., de Benedetti, P., Goldberg, S. R., et al. (2003). Adenosine A2A-dopamine D2 receptor–receptor heteromerization: Qualitative and quantitative assessment by fluorescence and bioluminescence energy transfer. *Journal of Biological Chemistry*, *278*, 46741–46749.
Canas, P. M., Porciuncula, L. O., Cunha, G. M., Silva, C. G., Machado, N. J., Oliveira, J. M., et al. (2009). Adenosine A2A receptor blockade prevents synaptotoxicity and memory dysfunction caused by beta-amyloid peptides via p38 mitogen-activated protein kinase pathway. *Journal of Neuroscience*, *29*, 14741–14751.
Cao, C., Cirrito, J. R., Lin, X., Wang, L., Verges, D. K., Dickson, A., et al. (2009). Caffeine suppresses amyloid-beta levels in plasma and brain of Alzheimer's disease transgenic mice. *Journal of Alzheimer's Disease*, *17*, 681–697.
Cao, C., Loewenstein, D. A., Lin, X., Zhang, C., Wang, L., Duara, R., et al. (2012). High blood caffeine levels in MCI linked to lack of progression to dementia. *Journal of Alzheimer's Disease*, *30*(3), 559–572.
Cepeda, C., Wu, N., Andre, V. M., Cummings, D. M., & Levine, M. S. (2007). The corticostriatal pathway in Huntington's disease. *Progress in Neurobiology*, *81*, 253–271.
Chamberlain, S. E., Sadowski, J. H., Teles-Grilo Ruivo, L. M., Atherton, L. A., & Mellor, J. R. (2013). Long-term depression of synaptic kainate receptors reduces excitability by relieving inhibition of the slow afterhyperpolarization. *Journal of Neuroscience*, *33*, 9536–9545.
Chao, M. V. (2003). Neurotrophins and their receptors: A convergence point for many signalling pathways. *Nature Review Neuroscience*, *4*, 299–309.
Charalambous, C., Gsandtner, I., Keuerleber, S., Milan-Lobo, L., Kudlacek, O., Freissmuth, M., et al. (2008). Restricted collision coupling of the A2A receptor revisited: Evidence for physical separation of two signaling cascades. *Journal of Biological Chemistry*, *283*, 9276–9288.
Chaudhuri, K. R., & Schapira, A. H. (2009). Non-motor symptoms of Parkinson's disease: Dopaminergic pathophysiology and treatment. *Lancet Neurology*, *8*, 464–474.

Chen, J. F., Eltzschig, H. K., & Fredholm, B. B. (2013). Adenosine receptors as drug targets—What are the challenges? *Nature Reviews. Drug Discovery*, *12*, 265–286.

Chen, X., Gawryluk, J. W., Wagener, J. F., Ghribi, O., & Geiger, J. D. (2008). Caffeine blocks disruption of blood brain barrier in a rabbit model of Alzheimer's disease. *Journal of Neuroinflammation*, *5*, 12.

Chen, J. F., Huang, Z., Ma, J., Zhu, J., Moratalla, R., Standaert, D., et al. (1999). A(2A) adenosine receptor deficiency attenuates brain injury induced by transient focal ischemia in mice. *Journal of Neuroscience*, *19*, 9192–9200.

Chen, J. F., Xu, K., Petzer, J. P., Staal, R., Xu, Y. H., Beilstein, M., et al. (2001). Neuroprotection by caffeine and A(2A) adenosine receptor inactivation in a model of Parkinson's disease. *Journal of Neuroscience*, *21*, RC143.

Cheng, H. C., Shih, H. M., & Chern, Y. (2002). Essential role of CREB activation by A2A adenosine receptors in rescuing the NGF-induced neurite outgrowth impaired by blockage of the MAPK cascade. *Journal of Biological Chemistry*, *277*, 33930–33942.

Chiang, M. C., Chen, H. M., Lai, H. L., Chen, H. W., Chou, S. Y., Chen, C. M., et al. (2009). The A2A adenosine receptor rescues the urea cycle deficiency of Huntington's disease by enhancing the activity of the ubiquitin–proteasome system. *Human Molecular Genetics*, *18*, 2929–2942.

Chiang, M. C., Lee, Y. C., Huang, C. L., & Chern, Y. (2005). cAMP-response element-binding protein contributes to suppression of the A2A adenosine receptor promoter by mutant Huntingtin with expanded polyglutamine residues. *Journal of Biological Chemistry*, *280*, 14331–14340.

Chiu, G. S., Chatterjee, D., Darmody, P. T., Walsh, J. P., Meling, D. D., Johnson, R. W., et al. (2012). Hypoxia/reoxygenation impairs memory formation via adenosine-dependent activation of caspase 1. *Journal of Neuroscience*, *32*, 13945–13955.

Choi, W. Y., Balsam, P. D., & Horvitz, J. C. (2005). Extended habit training reduces dopamine mediation of appetitive response expression. *Journal of Neuroscience*, *25*, 6729–6733.

Chou, S. Y., Lee, Y. C., Chen, H. M., Chiang, M. C., Lai, H. L., Chang, H. H., et al. (2005). CGS21680 attenuates symptoms of Huntington's disease in a transgenic mouse model. *Journal of Neurochemistry*, *93*, 310–320.

Chung, H. J., Ge, W. P., Qian, X., Wiser, O., Jan, Y. N., & Jan, L. Y. (2009). G protein-activated inwardly rectifying potassium channels mediate depotentiation of long-term potentiation. *Proceedings of the National Academy of Sciences of the United States of America*, *106*, 635–640.

Ciruela, F., Casado, V., Rodrigues, R. J., Lujan, R., Burgueno, J., Canals, M., et al. (2006). Presynaptic control of striatal glutamatergic neurotransmission by adenosine A1–A2A receptor heteromers. *Journal of Neuroscience*, *26*, 2080–2087.

Clinton, S. M., Sucharski, I. L., & Finlay, J. M. (2006). Desipramine attenuates working memory impairments induced by partial loss of catecholamines in the rat medial prefrontal cortex. *Psychopharmacology*, *183*, 404–412.

Coccurello, R., Breysse, N., & Amalric, M. (2004). Simultaneous blockade of adenosine A2A and metabotropic glutamate mGlu5 receptors increase their efficacy in reversing Parkinsonian deficits in rats. *Neuropsychopharmacology*, *29*, 1451–1461.

Cognato, G. P., Agostinho, P. M., Hockemeyer, J., Muller, C. E., Souza, D. O., & Cunha, R. A. (2010). Caffeine and an adenosine A(2A) receptor antagonist prevent memory impairment and synaptotoxicity in adult rats triggered by a convulsive episode in early life. *Journal of Neurochemistry*, *112*, 453–462.

Coleman, P., Federoff, H., & Kurlan, R. (2004). A focus on the synapse for neuroprotection in Alzheimer disease and other dementias. *Neurology*, *63*, 1155–1162.

Coric, V., van Dyck, C. H., Salloway, S., Andreasen, N., Brody, M., Richter, R. W., et al. (2012). Safety and tolerability of the gamma-secretase inhibitor avagacestat in a phase 2 study of mild to moderate Alzheimer disease. *Archives of Neurology*, *69*, 1430–1440.

Costa, M. S., Botton, P. H., Mioranzza, S., Souza, D. O., & Porciuncula, L. O. (2008). Caffeine prevents age-associated recognition memory decline and changes brain-derived neurotrophic factor and tirosine kinase receptor (TrkB) content in mice. *Neuroscience*, *153*, 1071–1078.

Costenla, A. R., Cunha, R. A., & de Mendonca, A. (2010). Caffeine, adenosine receptors, and synaptic plasticity. *Journal of Alzheimer's Disease*, *20*(Suppl. 1), S25–S34.

Costenla, A. R., Diogenes, M. J., Canas, P. M., Rodrigues, R. J., Nogueira, C., Maroco, J., et al. (2011). Enhanced role of adenosine A(2A) receptors in the modulation of LTP in the rat hippocampus upon ageing. *European Journal of Neuroscience*, *34*, 12–21.

Costenla, A. R., Lopes, L. V., de Mendonca, A., & Ribeiro, J. A. (2001). A functional role for adenosine A3 receptors: Modulation of synaptic plasticity in the rat hippocampus. *Neuroscience Letters*, *302*, 53–57.

Cunha, R. A. (2008). Caffeine, adenosine receptors, memory and Alzheimer disease. *Medicina Clínica (Barcelona)*, *131*, 790–795.

Cunha, R. A., & Agostinho, P. M. (2010). Chronic caffeine consumption prevents memory disturbance in different animal models of memory decline. *Journal of Alzheimer's Disease*, *20*(Suppl. 1), S95–S116.

Cunha, G. M., Canas, P. M., Melo, C. S., Hockemeyer, J., Muller, C. E., Oliveira, C. R., et al. (2008). Adenosine A2A receptor blockade prevents memory dysfunction caused by beta-amyloid peptides but not by scopolamine or MK-801. *Experimental Neurology*, *210*, 776–781.

Cunha, R. A., Ferre, S., Vaugeois, J. M., & Chen, J. F. (2008). Potential therapeutic interest of adenosine A2A receptors in psychiatric disorders. *Current Pharmaceutical Design*, *14*, 1512–1524.

Dai, S. S., Li, W., An, J. H., Wang, H., Yang, N., Chen, X. Y., et al. (2010). Adenosine A2A receptors in both bone marrow cells and non-bone marrow cells contribute to traumatic brain injury. *Journal of Neurochemistry*, *113*, 1536–1544.

d'Alcantara, P., Ledent, C., Swillens, S., & Schiffmann, S. N. (2001). Inactivation of adenosine A2A receptor impairs long term potentiation in the accumbens nucleus without altering basal synaptic transmission. *Neuroscience*, *107*, 455–464.

Dall'Igna, O. P., Fett, P., Gomes, M. W., Souza, D. O., Cunha, R. A., & Lara, D. R. (2007). Caffeine and adenosine A(2a) receptor antagonists prevent beta-amyloid (25–35)-induced cognitive deficits in mice. *Experimental Neurology*, *203*, 241–245.

Dall'lgna, O. P., Porciuncula, L. O., Souza, D. O., Cunha, R. A., & Lara, D. R. (2003). Neuroprotection by caffeine and adenosine A(2A) receptor blockade of beta-amyloid neurotoxicity. *British Journal of Pharmacology*, *138*, 1207–1209.

Dayan, P., & Balleine, B. W. (2002). Reward, motivation, and reinforcement learning. *Neuron*, *36*, 285–298.

de Mendonca, A., & Ribeiro, J. A. (1997). Adenosine and neuronal plasticity. *Life Sciences*, *60*, 245–251.

Decamp, E., & Schneider, J. S. (2004). Attention and executive function deficits in chronic low-dose MPTP-treated non-human primates. *European Journal of Neuroscience*, *20*, 1371–1378.

Deng, X., Shibata, H., Ninomiya, H., Tashiro, N., Iwata, N., Ozaki, N., et al. (2004). Association study of polymorphisms in the excitatory amino acid transporter 2 gene (SLC1A2) with schizophrenia. *BMC Psychiatry*, *4*, 21.

Dhaenens, C. M., Burnouf, S., Simonin, C., Van Brussel, E., Duhamel, A., Defebvre, L., et al. (2009). A genetic variation in the ADORA2A gene modifies age at onset in Huntington's disease. *Neurobiology of Disease*, *35*, 474–476.

Dias, R. B., Ribeiro, J. A., & Sebastiao, A. M. (2012). Enhancement of AMPA currents and GluR1 membrane expression through PKA-coupled adenosine A(2A) receptors. *Hippocampus*, *22*, 276–291.

Dias, R. B., Rombo, D. M., Ribeiro, J. A., Henley, J. M., & Sebastiao, A. M. (2013). Adenosine: Setting the stage for plasticity. *Trends in Neurosciences*, *36*, 248–257.
DiFiglia, M. (1990). Excitotoxic injury of the neostriatum: A model for Huntington's disease. *Trends in Neurosciences*, *13*, 286–289.
Diogenes, M. J., Fernandes, C. C., Sebastiao, A. M., & Ribeiro, J. A. (2004). Activation of adenosine A2A receptor facilitates brain-derived neurotrophic factor modulation of synaptic transmission in hippocampal slices. *Journal of Neuroscience*, *24*, 2905–2913.
Diogenes, M. J., Neves-Tome, R., Fucile, S., Martinello, K., Scianni, M., Theofilas, P., et al. (2014). Homeostatic control of synaptic activity by endogenous adenosine is mediated by adenosine kinase. *Cerebral Cortex*, *24*, 67–80.
Dixon, A. K., Gubitz, A. K., Sirinathsinghji, D. J., Richardson, P. J., & Freeman, T. C. (1996). Tissue distribution of adenosine receptor mRNAs in the rat. *British Journal of Pharmacology*, *118*, 1461–1468.
Domenici, M. R., Scattoni, M. L., Martire, A., Lastoria, G., Potenza, R. L., Borioni, A., et al. (2007). Behavioral and electrophysiological effects of the adenosine A2A receptor antagonist SCH 58261 in R6/2 Huntington's disease mice. *Neurobiology of Disease*, *28*, 197–205.
Doody, R. S., Raman, R., Farlow, M., Iwatsubo, T., Vellas, B., Joffe, S., et al. (2013). A phase 3 trial of semagacestat for treatment of Alzheimer's disease. *New England Journal of Medicine*, *369*, 341–350.
Duarte, J. M., Agostinho, P. M., Carvalho, R. A., & Cunha, R. A. (2012). Caffeine consumption prevents diabetes-induced memory impairment and synaptotoxicity in the hippocampus of NONcZNO10/LTJ mice. *PLoS One*, *7*, e21899.
Duarte, J. M., Carvalho, R. A., Cunha, R. A., & Gruetter, R. (2009). Caffeine consumption attenuates neurochemical modifications in the hippocampus of streptozotocin-induced diabetic rats. *Journal of Neurochemistry*, *111*, 368–379.
Duff, K., Paulsen, J., Mills, J., Beglinger, L. J., Moser, D. J., Smith, M. M., et al. (2010). Mild cognitive impairment in prediagnosed Huntington disease. *Neurology*, *75*, 500–507.
Dunwiddie, T. V., & Fredholm, B. B. (1997). Adenosine neuromodulation. In M. F. Jarvis, & K. A. Jacobson (Eds.), *Purinergic approaches in experimental therapeutics* (pp. 359–382). New York: Wiley-Liss.
Dunwiddie, T. V., & Masino, S. A. (2001). The role and regulation of adenosine in the central nervous system. *Annual Review of Neuroscience*, *24*, 31–55.
Ena, S. L., De Backer, J. F., Schiffmann, S. N., & de Kerchove d'Exaerde, A. (2013). FACS array profiling identifies Ecto-5′ nucleotidase as a striatopallidal neuron-specific gene involved in striatal-dependent learning. *Journal of Neuroscience*, *33*, 8794–8809.
Eskelinen, M. H., Ngandu, T., Tuomilehto, J., Soininen, H., & Kivipelto, M. (2009). Midlife coffee and tea drinking and the risk of late-life dementia: A population-based CAIDE study. *Journal of Alzheimer's Disease*, *16*, 85–91.
Espinosa, J., Rocha, A., Nunes, F., Costa, M. S., Schein, V., Kazlauckas, V., et al. (2013). Caffeine consumption prevents memory impairment, neuronal damage, and adenosine A2A receptors upregulation in the hippocampus of a rat model of sporadic dementia. *Journal of Alzheimer's Disease*, *34*, 509–518.
Ferrante, R. J., Gutekunst, C. A., Persichetti, F., McNeil, S. M., Kowall, N. W., Gusella, J. F., et al. (1997). Heterogeneous topographic and cellular distribution of Huntingtin expression in the normal human neostriatum. *J Neurosci*, *17*, 3052–3063.
Ferre, S. (1997). Adenosine–dopamine interactions in the ventral striatum. Implications for the treatment of schizophrenia. *Psychopharmacology*, *133*, 107–120.
Ferre, S. (2008). An update on the mechanisms of the psychostimulant effects of caffeine. *Journal of Neurochemistry*, *105*, 1067–1079.
Ferre, S., Fredholm, B. B., Morelli, M., Popoli, P., & Fuxe, K. (1997). Adenosine–dopamine receptor–receptor interactions as an integrative mechanism in the basal ganglia. *Trends in Neurosciences*, *20*, 482–487.

Ferre, S., Herrera-Marschitz, M., Grabowska-Anden, M., Casas, M., Ungerstedt, U., & Anden, N. E. (1991). Postsynaptic dopamine/adenosine interaction: II. Postsynaptic dopamine agonism and adenosine antagonism of methylxanthines in short-term reserpinized mice. *European Journal of Pharmacology*, *192*, 31–37.

Ferre, S., Herrera-Marschitz, M., Grabowska-Anden, M., Ungerstedt, U., Casas, M., & Anden, N. E. (1991). Postsynaptic dopamine/adenosine interaction: I. Adenosine analogues inhibit dopamine D2-mediated behaviour in short-term reserpinized mice. *European Journal of Pharmacology*, *192*, 25–30.

Ferre, S., Karcz-Kubicha, M., Hope, B. T., Popoli, P., Burgueno, J., Gutierrez, M. A., et al. (2002). Synergistic interaction between adenosine A2A and glutamate mGlu5 receptors: Implications for striatal neuronal function. *Proceedings of the National Academy of Sciences of the United States of America*, *99*, 11940–11945.

Ferre, S., Lluis, C., Justinova, Z., Quiroz, C., Orru, M., Navarro, G., et al. (2010). Adenosine-cannabinoid receptor interactions. Implications for striatal function. *British Journal of Pharmacology*, *160*, 443–453.

Ferre, S., Popoli, P., Gimenez-Llort, L., Rimondini, R., Muller, C. E., Stromberg, I., et al. (2001). Adenosine/dopamine interaction: Implications for the treatment of Parkinson's disease. *Parkinsonism & Related Disorders*, *7*, 235–241.

Ferre, S., Quiroz, C., Woods, A. S., Cunha, R., Popoli, P., Ciruela, F., et al. (2008). An update on adenosine A2A–dopamine D2 receptor interactions: Implications for the function of G protein-coupled receptors. *Current Pharmaceutical Design*, *14*, 1468–1474.

Ferre, S., von Euler, G., Johansson, B., Fredholm, B. B., & Fuxe, K. (1991). Stimulation of high-affinity adenosine A2 receptors decreases the affinity of dopamine D2 receptors in rat striatal membranes. *Proceedings of the National Academy of Sciences of the United States of America*, *88*, 7238–7241.

Field, J. R., Walker, A. G., & Conn, P. J. (2011). Targeting glutamate synapses in schizophrenia. *Trends in Molecular Medicine*, *17*, 689–698.

Fink, J. S., Weaver, D. R., Rivkees, S. A., Peterfreund, R. A., Pollack, A. E., Adler, E. M., et al. (1992). Molecular cloning of the rat A2 adenosine receptor: Selective co-expression with D2 dopamine receptors in rat striatum. *Brain Research. Molecular Brain Research*, *14*, 186–195.

Florian, C., Vecsey, C. G., Halassa, M. M., Haydon, P. G., & Abel, T. (2011). Astrocyte-derived adenosine and A1 receptor activity contribute to sleep loss-induced deficits in hippocampal synaptic plasticity and memory in mice. *Journal of Neuroscience*, *31*, 6956–6962.

Fontinha, B. M., Delgado-Garcia, J. M., Madronal, N., Ribeiro, J. A., Sebastiao, A. M., & Gruart, A. (2009). Adenosine A(2A) receptor modulation of hippocampal CA3–CA1 synapse plasticity during associative learning in behaving mice. *Neuropsychopharmacology*, *34*, 1865–1874.

Fontinha, B. M., Diogenes, M. J., Ribeiro, J. A., & Sebastiao, A. M. (2008). Enhancement of long-term potentiation by brain-derived neurotrophic factor requires adenosine A2A receptor activation by endogenous adenosine. *Neuropharmacology*, *54*, 924–933.

Fox, N. C., Black, R. S., Gilman, S., Rossor, M. N., Griffith, S. G., Jenkins, L., et al. (2005). Effects of Abeta immunization (AN1792) on MRI measures of cerebral volume in Alzheimer disease. *Neurology*, *64*, 1563–1572.

Fredholm, B., Chen, J.-F., Masino, S. A., & Vaugeois, J.-M. (2005). Actions of adenosine at its receptors in the CNS: Insights from knockouts and drugs. *Annual Review of Pharmacology and Toxicology*, *45*, 385–412.

Fredholm, B. B., Chern, Y., Franco, R., & Sitkovsky, M. (2007). Aspects of the general biology of adenosine A2A signaling. *Progress in Neurobiology*, *83*, 263–276.

Fredholm, B. B., IJzerman, A., Jacobson, K. A., Klotz, K. N., & Linden, J. (2001). International Union of Pharmacology. XXV. Nomenclature and classification of adenosine receptors. *Pharmacological Reviews*, *53*, 527–552.

Fredholm, B. B., Proctor, W., Van der Ploeg, I., & Dunwiddie, T. V. (1989). In vivo pertussis toxin treatment attenuates some, but not all, adenosine A1 effects in slices of the rat hippocampus. *European Journal of Pharmacology*, *172*, 249–262.

Fukumitsu, N., Ishii, K., Kimura, Y., Oda, K., Hashimoto, M., Suzuki, M., et al. (2008). *Adenosine A(1) receptors using 8-dicyclopropylmethyl-1-[(11)C]methyl-3-propylxanthine PET in Alzheimer's disease*, *22*, 841–847.

Furtado, J. M., Lansingh, V. C., Carter, M. J., Milanese, M. F., Pena, B. N., Ghersi, H. A., et al. (2012). Causes of blindness and visual impairment in Latin America. *Survey of Ophthalmology*, *57*, 149–177.

Fuxe, K., Agnati, L. F., Jacobsen, K., Hillion, J., Canals, M., Torvinen, M., et al. (2003). Receptor heteromerization in adenosine A2A receptor signaling: Relevance for striatal function and Parkinson's disease. *Neurology*, *61*, S19–S23.

Galasko, D., Chang, L., Motter, R., Clark, C. M., Kaye, J., Knopman, D., et al. (1998). High cerebrospinal fluid tau and low amyloid beta42 levels in the clinical diagnosis of Alzheimer disease and relation to apolipoprotein E genotype. *Archives of Neurology*, *55*, 937–945.

Gelber, R. P., Petrovitch, H., Masaki, K. H., Ross, G. W., & White, L. R. (2011). Coffee intake in midlife and risk of dementia and its neuropathologic correlates. *Journal of Alzheimer's Disease*, *23*, 607–615.

Gerevich, Z., Wirkner, K., & Illes, P. (2002). Adenosine A2A receptors inhibit the N-methyl-D-aspartate component of excitatory synaptic currents in rat striatal neurons. *European Journal of Pharmacology*, *451*, 161–164.

Gianfriddo, M., Melani, A., Turchi, D., Giovannini, M. G., & Pedata, F. (2004). Adenosine and glutamate extracellular concentrations and mitogen-activated protein kinases in the striatum of Huntington transgenic mice. Selective antagonism of adenosine A2A receptors reduces transmitter outflow. *Neurobiology of Disease*, *17*, 77–88.

Gilman, S., Koller, M., Black, R. S., Jenkins, L., Griffith, S. G., Fox, N. C., et al. (2005). Clinical effects of Abeta immunization (AN1792) in patients with AD in an interrupted trial. *Neurology*, *64*, 1553–1562.

Gimenez-Llort, L., Fernandez-Teruel, A., Escorihuela, R. M., Fredholm, B. B., Tobena, A., Pekny, M., et al. (2002). Mice lacking the adenosine A1 receptor are anxious and aggressive, but are normal learners with reduced muscle strength and survival rate. *European Journal of Neuroscience*, *16*, 547–550.

Gimenez-Llort, L., Masino, S. A., Diao, L., Fernandez-Teruel, A., Tobena, A., Halldner, L., et al. (2005). Mice lacking the adenosine A(1) receptor have normal spatial learning and plasticity in the CA1 region of the hippocampus, but they habituate more slowly. *Synapse*, *57*, 8–16.

Gimenez-Llort, L., Schiffmann, S. N., Shmidt, T., Canela, L., Camon, L., Wassholm, M., et al. (2007). Working memory deficits in transgenic rats overexpressing human adenosine A2A receptors in the brain. *Neurobiology of Learning and Memory*, *87*, 42–56.

Goldman-Rakic, P. S. (1995). Cellular basis of working memory. *Neuron*, *14*, 477–485.

Goldman-Rakic, P. S. (1999). The physiological approach: Functional architecture of working memory and disordered cognition in schizophrenia. *Biological Psychiatry*, *46*, 650–661.

Goldman-Rakic, P. S., Castner, S. A., Svensson, T. H., Siever, L. J., & Williams, G. V. (2004). Targeting the dopamine D1 receptor in schizophrenia: Insights for cognitive dysfunction. *Psychopharmacology*, *174*, 3–16.

Golembiowska, K., & Zylewska, A. (1998). Agonists of A1 and A2A adenosine receptors attenuate methamphetamine-induced overflow of dopamine in rat striatum. *Brain Research*, *806*, 202–209.

Gomes, C. V., Kaster, M. P., Tome, A. R., Agostinho, P. M., & Cunha, R. A. (2011). Adenosine receptors and brain diseases: Neuroprotection and neurodegeneration. *Biochimica et Biophysica Acta*, *1808*, 1380–1399.

Gotham, A. M., Brown, R. G., & Marsden, C. D. (1988). 'Frontal' cognitive function in patients with Parkinson's disease 'on' and 'off' levodopa. *Brain, 111*(Pt. 2), 299–321.
Gotoh, L., Mitsuyasu, H., Kobayashi, Y., Oribe, N., Takata, A., Ninomiya, H., et al. (2009). Association analysis of adenosine A1 receptor gene (ADORA1) polymorphisms with schizophrenia in a Japanese population. *Psychiatric Genetics, 19*, 328–335.
Graybiel, A. M. (2008). Habits, rituals, and the evaluative brain. *Annual Review of Neuroscience, 31*, 359–387.
Gundlfinger, A., Bischofberger, J., Johenning, F. W., Torvinen, M., Schmitz, D., & Breustedt, J. (2007). Adenosine modulates transmission at the hippocampal mossy fibre synapse via direct inhibition of presynaptic calcium channels. *Journal of Physiology, 582*, 263–277.
Hagena, H., & Manahan-Vaughan, D. (2010). Frequency facilitation at mossy fiber-CA3 synapses of freely behaving rats contributes to the induction of persistent LTD via an adenosine-A1 receptor-regulated mechanism. *Cerebral Cortex, 20*, 1121–1130.
Halassa, M. M., Florian, C., Fellin, T., Munoz, J. R., Lee, S. Y., Abel, T., et al. (2009). Astrocytic modulation of sleep homeostasis and cognitive consequences of sleep loss. *Neuron, 61*, 213–219.
Hallock, H. L., Arreola, A. C., Shaw, C. L., & Griffin, A. L. (2013). Dissociable roles of the dorsal striatum and dorsal hippocampus in conditional discrimination and spatial alternation T-maze tasks. *Neurobiology of Learning and Memory, 100*, 108–116.
Hameleers, P. A., Van Boxtel, M. P., Hogervorst, E., Riedel, W. J., Houx, P. J., Buntinx, F., et al. (2000). Habitual caffeine consumption and its relation to memory, attention, planning capacity and psychomotor performance across multiple age groups. *Human Psychopharmacology, 15*, 573–581.
Hansen, R. T., III, & Zhang, H. T. (2013). Senescent-induced dysregulation of cAMP/CREB signaling and correlations with cognitive decline. *Brain Research, 1516*, 93–109.
Hardy, J., & Selkoe, D. J. (2002). The amyloid hypothesis of Alzheimer's disease: Progress and problems on the road to therapeutics. *Science, 297*, 353–356.
Hasuo, H., Shoji, S., Gallagher, J. P., & Akasu, T. (1992). Adenosine inhibits the synaptic potentials in rat septal nucleus neurons mediated through pre- and postsynaptic A1-adenosine receptors. *Neuroscience Research, 13*, 281–299.
Hauser, R. A., Cantillon, M., Pourcher, E., Micheli, F., Mok, V., Onofrj, M., et al. (2011). Preladenant in patients with Parkinson's disease and motor fluctuations: A phase 2, double-blind, randomised trial. *Lancet Neurology, 10*, 221–229.
Higley, M. J., & Sabatini, B. L. (2010). Competitive regulation of synaptic Ca2+ influx by D2 dopamine and A2A adenosine receptors. *Nature Neuroscience, 13*, 958–966.
Hikida, T., Yawata, S., Yamaguchi, T., Danjo, T., Sasaoka, T., Wang, Y., et al. (2013). Pathway-specific modulation of nucleus accumbens in reward and aversive behavior via selective transmitter receptors. *Proceedings of the National Academy of Sciences of the United States of America, 110*, 342–347.
Hilario, M. R., Clouse, E., Yin, H. H., & Costa, R. M. (2007). Endocannabinoid signaling is critical for habit formation. *Frontiers in Integrative Neuroscience, 1*, 6.
Hillion, J., Canals, M., Torvinen, M., Casado, V., Scott, R., Terasmaa, A., et al. (2002). Coaggregation, cointernalization, and codesensitization of adenosine A2A receptors and dopamine D2 receptors. *Journal of Biological Chemistry, 277*, 18091–18097.
Hirota, T., & Kishi, T. (2013). Adenosine hypothesis in schizophrenia and bipolar disorder: A systematic review and meta-analysis of randomized controlled trial of adjuvant purinergic modulators. *Schizophrenia Research, 149*, 88–95.
Ho, A. K., Sahakian, B. J., Brown, R. G., Barker, R. A., Hodges, J. R., Ane, M. N., et al. (2003). Profile of cognitive progression in early Huntington's disease. *Neurology, 61*, 1702–1706.

Hong, C. J., Liu, H. C., Liu, T. Y., Liao, D. L., & Tsai, S. J. (2005). Association studies of the adenosine A2a receptor (1976 T > C) genetic polymorphism in Parkinson's disease and schizophrenia. *Journal of Neural Transmission, 112*, 1503–1510.

Hooper, N., Fraser, C., & Stone, T. W. (1996). Effects of purine analogues on spontaneous alternation in mice. *Psychopharmacology, 123*, 250–257.

Howard, R., McShane, R., Lindesay, J., Ritchie, C., Baldwin, A., Barber, R., et al. (2012). Donepezil and memantine for moderate-to-severe Alzheimer's disease. *New England Journal of Medicine, 366*, 893–903.

Hunter, J. M., Lesort, M., & Johnson, G. V. (2007). Ubiquitin–proteasome system alterations in a striatal cell model of Huntington's disease. *Journal of Neuroscience Research, 85*, 1774–1788.

Hwang, Y., Kim, J., Shin, J. Y., Kim, J. I., Seo, J. S., Webster, M. J., et al. (2013). Gene expression profiling by mRNA sequencing reveals increased expression of immune/inflammation-related genes in the hippocampus of individuals with schizophrenia. *Translational Psychiatry, 3*, e321.

Ikeda, K., Kurokawa, M., Aoyama, S., & Kuwana, Y. (2002). Neuroprotection by adenosine A2A receptor blockade in experimental models of Parkinson's disease. *Journal of Neurochemistry, 80*, 262–270.

Jaarsma, D., Sebens, J. B., & Korf, J. (1991). Reduction of adenosine A1-receptors in the perforant pathway terminal zone in Alzheimer hippocampus. *Neuroscience Letters, 121*, 111–114.

Jacobson, K. A., von Lubitz, D. K., Daly, J. W., & Fredholm, B. B. (1996). Adenosine receptor ligands: Differences with acute versus chronic treatment. *Trends in Pharmacological Sciences, 17*, 108–113.

Jansen, K. L., Faull, R. L., Dragunow, M., & Synek, B. L. (1990). Alzheimer's disease: Changes in hippocampal N-methyl-D-aspartate, quisqualate, neurotensin, adenosine, benzodiazepine, serotonin and opioid receptors—An autoradiographic study. *Neuroscience, 39*, 613–627.

Jenner, P., Mori, A., Hauser, R., Morelli, M., Fredholm, B. B., & Chen, J. F. (2009). Adenosine, adenosine A 2A antagonists, and Parkinson's disease. *Parkinsonism & Related Disorders, 15*, 406–413.

Jeronimo-Santos, A., Batalha, V. L., Muller, C. E., Baqi, Y., Sebastiao, A. M., Lopes, L. V., et al. (2014). Impact of in vivo chronic blockade of adenosine A receptors on the BDNF-mediated facilitation of LTP. *Neuropharmacology, 83*, 99–106.

Johansson, B., Halldner, L., Dunwiddie, T. V., Masino, S. A., Poelchen, W., Gimenez-Llort, L., et al. (2001). Hyperalgesia, anxiety, and decreased hypoxic neuroprotection in mice lacking the adenosine A1 receptor. *Proceedings of the National Academy of Sciences of the United States of America, 98*, 9407–9412.

Johnson, A., van der Meer, M. A., & Redish, A. D. (2007). Integrating hippocampus and striatum in decision-making. *Current Opinion in Neurobiology, 17*, 692–697.

Jones, R. W. (2010). A review comparing the safety and tolerability of memantine with the acetylcholinesterase inhibitors. *International Journal of Geriatric Psychiatry, 25*, 547–553.

Julien, C. L., Thompson, J. C., Wild, S., Yardumian, P., Snowden, J. S., Turner, G., et al. (2007). Psychiatric disorders in preclinical Huntington's disease. *Journal of Neurology, Neurosurgery and Psychiatry, 78*, 939–943.

Kachroo, A., Orlando, L. R., Grandy, D. K., Chen, J. F., Young, A. B., & Schwarzschild, M. A. (2005). Interactions between metabotropic glutamate 5 and adenosine A2A receptors in normal and parkinsonian mice. *Journal of Neuroscience, 25*, 10414–10419.

Kachroo, A., & Schwarzschild, M. A. (2012). Adenosine A2A receptor gene disruption protects in an alpha-synuclein model of Parkinson's disease. *Annals of Neurology, 71*, 278–282.

Kadowaki Horita, T., Kobayashi, M., Mori, A., Jenner, P., & Kanda, T. (2013). Effects of the adenosine A2A antagonist istradefylline on cognitive performance in rats with a 6-OHDA lesion in prefrontal cortex. *Psychopharmacology*, *230*, 345–352.

Kafka, S. H., & Corbett, R. (1996). Selective adenosine A2A receptor/dopamine D2 receptor interactions in animal models of schizophrenia. *European Journal of Pharmacology*, *295*, 147–154.

Kamikubo, Y., Shimomura, T., Fujita, Y., Tabata, T., Kashiyama, T., Sakurai, T., et al. (2013). Functional cooperation of metabotropic adenosine and glutamate receptors regulates postsynaptic plasticity in the cerebellum. *Journal of Neuroscience*, *33*, 18661–18671.

Kehagia, A. A., Barker, R. A., & Robbins, T. W. (2010). Neuropsychological and clinical heterogeneity of cognitive impairment and dementia in patients with Parkinson's disease. *Lancet Neurology*, *9*, 1200–1213.

Kensinger, E. A., Shearer, D. K., Locascio, J. J., Growdon, J. H., & Corkin, S. (2003). Working memory in mild Alzheimer's disease and early Parkinson's disease. *Neuropsychology*, *17*, 230–239.

Kerr, M. I., Wall, M. J., & Richardson, M. J. (2013). Adenosine A1 receptor activation mediates the developmental shift at layer 5 pyramidal cell synapses and is a determinant of mature synaptic strength. *Journal of Physiology*, *591*, 3371–3380.

Kim, J., Kwon, J. T., Kim, H. S., & Han, J. H. (2013). CREB and neuronal selection for memory trace. *Frontiers in Neural Circuits*, *7*, 44.

Kirsch, G. E., Codina, J., Birnbaumer, L., & Brown, A. M. (1990). Coupling of ATP-sensitive K + channels to A1 receptors by G proteins in rat ventricular myocytes. *American Journal of Physiology*, *259*, H820–H826.

Kopf, S. R., Melani, A., Pedata, F., & Pepeu, G. (1999). Adenosine and memory storage: Effect of A(1) and A(2) receptor antagonists. *Psychopharmacology*, *146*, 214–219.

Kreitzer, A. C., & Malenka, R. C. (2005). Dopamine modulation of state-dependent endocannabinoid release and long-term depression in the striatum. *Journal of Neuroscience*, *25*, 10537–10545.

Kreitzer, A. C., & Malenka, R. C. (2007). Endocannabinoid-mediated rescue of striatal LTD and motor deficits in Parkinson's disease models. *Nature*, *445*, 643–647.

Kukley, M., Schwan, M., Fredholm, B. B., & Dietrich, D. (2005). The role of extracellular adenosine in regulating mossy fiber synaptic plasticity. *Journal of Neuroscience*, *25*, 2832–2837.

Lang, U. E., Lang, F., Richter, K., Vallon, V., Lipp, H. P., Schnermann, J., et al. (2003). Emotional instability but intact spatial cognition in adenosine receptor 1 knock out mice. *Behavioural Brain Research*, *145*, 179–188.

Lange, K. W., Robbins, T. W., Marsden, C. D., James, M., Owen, A. M., & Paul, G. M. (1992). L-DOPA withdrawal in Parkinson's disease selectively impairs cognitive performance in tests sensitive to frontal lobe dysfunction. *Psychopharmacology*, *107*, 394–404.

Lara, D. R., Dall'Igna, O. P., Ghisolfi, E. S., & Brunstein, M. G. (2006). Involvement of adenosine in the neurobiology of schizophrenia and its therapeutic implications. *Progress in Neuropsychopharmacology and Biological Psychiatry*, *30*, 617–629.

Lawrence, A. D., Hodges, J. R., Rosser, A. E., Kershaw, A., ffrench-Constant, C., Rubinsztein, D. C., et al. (1998). Evidence for specific cognitive deficits in preclinical Huntington's disease. *Brain*, *121*(Pt. 7), 1329–1341.

Lawrence, A. D., Weeks, R. A., Brooks, D. J., Andrews, T. C., Watkins, L. H., Harding, A. E., et al. (1998). The relationship between striatal dopamine receptor binding and cognitive performance in Huntington's disease. *Brain*, *121*(Pt. 7), 1343–1355.

Lee, F. S., & Chao, M. V. (2001). Activation of Trk neurotrophin receptors in the absence of neurotrophins. *Proceedings of the National Academy of Sciences of the United States of America*, *98*, 3555–3560.

Leite, M. R., Wilhelm, E. A., Jesse, C. R., Brandao, R., & Nogueira, C. W. (2011). Protective effect of caffeine and a selective A2A receptor antagonist on impairment of memory and oxidative stress of aged rats. *Experimental Gerontology, 46*, 309–315.

Lemere, C. A., & Masliah, E. (2010). Can Alzheimer disease be prevented by amyloid-beta immunotherapy? *Nature Reviews. Neurology, 6*, 108–119.

Lerner, T. N., Horne, E. A., Stella, N., & Kreitzer, A. C. (2010). Endocannabinoid signaling mediates psychomotor activation by adenosine A2A antagonists. *Journal of Neuroscience, 30*, 2160–2164.

Lerner, T. N., & Kreitzer, A. C. (2012). RGS4 is required for dopaminergic control of striatal LTD and susceptibility to parkinsonian motor deficits. *Neuron, 73*, 347–359.

Lewis, S. J., Dove, A., Robbins, T. W., Barker, R. A., & Owen, A. M. (2003). Cognitive impairments in early Parkinson's disease are accompanied by reductions in activity in frontostriatal neural circuitry. *Journal of Neuroscience, 23*, 6351–6356.

Li, J. Y., Plomann, M., & Brundin, P. (2003). Huntington's disease: A synaptopathy? *Trends in Molecular Medicine, 9*, 414–420.

Liljeholm, M., & O'Doherty, J. P. (2012). Contributions of the striatum to learning, motivation, and performance: An associative account. *Trends in Cognitive Science, 16*, 467–475.

Lim, H. K., Juh, R., Pae, C. U., Lee, B. T., Yoo, S. S., Ryu, S. H., et al. (2008). Altered verbal working memory process in patients with Alzheimer's disease: An fMRI investigation. *Neuropsychobiology, 57*, 181–187.

Lindsay, J., Laurin, D., Verreault, R., Hebert, R., Helliwell, B., Hill, G. B., et al. (2002). Risk factors for Alzheimer's disease: A prospective analysis from the Canadian Study of Health and Aging. *American Journal of Epidemiology, 156*, 445–453.

Lisman, J. E., Coyle, J. T., Green, R. W., Javitt, D. C., Benes, F. M., Heckers, S., et al. (2008). Circuit-based framework for understanding neurotransmitter and risk gene interactions in schizophrenia. *Trends in Neurosciences, 31*, 234–242.

Litvan, I., Aarsland, D., Adler, C. H., Goldman, J. G., Kulisevsky, J., Mollenhauer, B., et al. (2011). MDS Task Force on mild cognitive impairment in Parkinson's disease: Critical review of PD-MCI. *Movement Disorders, 26*, 1814–1824.

Londos, C., Cooper, D. M., & Wolff, J. (1980). Subclasses of external adenosine receptors. *Proceedings of the National Academy of Sciences of the United States of America, 77*, 2551–2554.

Lopes, L. V., Cunha, R. A., & Ribeiro, J. A. (1999). Cross talk between A(1) and A(2A) adenosine receptors in the hippocampus and cortex of young adult and old rats. *Journal of Neurophysiology, 82*, 3196–3203.

Lovinger, D. M., & Choi, S. (1995). Activation of adenosine A1 receptors initiates short-term synaptic depression in rat striatum. *Neuroscience Letters, 199*, 9–12.

Lu, G., Zhou, Q. X., Kang, S., Li, Q. L., Zhao, L. C., Chen, J. D., et al. (2010). Chronic morphine treatment impaired hippocampal long-term potentiation and spatial memory via accumulation of extracellular adenosine acting on adenosine A1 receptors. *Journal of Neuroscience, 30*, 5058–5070.

Lynch, M. A., Errington, M. L., Clements, M. P., Bliss, T. V., Redini-Del Negro, C., & Laroche, S. (1990). Increases in glutamate release and phosphoinositide metabolism associated with long-term potentiation and classical conditioning. *Progress in Brain Research, 83*, 251–256.

Maggi, L., Trettel, F., Scianni, M., Bertollini, C., Eusebi, F., Fredholm, B. B., et al. (2009). LTP impairment by fractalkine/CX3CL1 in mouse hippocampus is mediated through the activity of adenosine receptor type 3 (A3R). *Journal of Neuroimmunology, 215*, 36–42.

Marquez-Ruiz, J., Leal-Campanario, R., Sanchez-Campusano, R., Molaee-Ardekani, B., Wendling, F., Miranda, P. C., et al. (2012). Transcranial direct-current stimulation modulates synaptic mechanisms involved in associative learning in behaving rabbits. *Proceedings of the National Academy of Sciences of the United States of America*, *109*, 6710–6715.

Martin, J. B., & Gusella, J. F. (1986). Huntington's disease. Pathogenesis and management. *New England Journal of Medicine*, *315*, 1267–1276.

Matos, M., Augusto, E., Agostinho, P., Cunha, R. A., & Chen, J. F. (2013). Antagonistic interaction between adenosine A2A receptors and Na+/K+-ATPase-alpha2 controlling glutamate uptake in astrocytes. *Journal of Neuroscience*, *33*, 18492–18502.

Matos, M., Augusto, E., Santos-Rodrigues, A. D., Schwarzschild, M. A., Chen, J. F., Cunha, R. A., et al. (2012). Adenosine A(2A) receptors modulate glutamate uptake in cultured astrocytes and gliosomes. *Glia*, *60*, 702–716.

Mattsson, N., Zetterberg, H., Hansson, O., Andreasen, N., Parnetti, L., Jonsson, M., et al. (2009). CSF biomarkers and incipient Alzheimer disease in patients with mild cognitive impairment. *JAMA*, *302*, 385–393.

Matute, C., Melone, M., Vallejo-Illarramendi, A., & Conti, F. (2005). Increased expression of the astrocytic glutamate transporter GLT-1 in the prefrontal cortex of schizophrenics. *Glia*, *49*, 451–455.

Melani, A., Cipriani, S., Vannucchi, M. G., Nosi, D., Donati, C., Bruni, P., et al. (2009). Selective adenosine A2a receptor antagonism reduces JNK activation in oligodendrocytes after cerebral ischaemia. *Brain*, *132*, 1480–1495.

Melani, A., Gianfriddo, M., Vannucchi, M. G., Cipriani, S., Baraldi, P. G., Giovannini, M. G., et al. (2006). The selective A2A receptor antagonist SCH 58261 protects from neurological deficit, brain damage and activation of p38 MAPK in rat focal cerebral ischemia. *Brain Research*, *1073–1074*, 470–480.

Mihara, T., Mihara, K., Yarimizu, J., Mitani, Y., Matsuda, R., Yamamoto, H., et al. (2007). Pharmacological characterization of a novel, potent adenosine A1 and A2A receptor dual antagonist, 5-[5-amino-3-(4-fluorophenyl)pyrazin-2-yl]-1-isopropylpyridine-2 (1H)-one (ASP5854), in models of Parkinson's disease and cognition. *Journal of Pharmacology and Experimental Therapeutics*, *323*, 708–719.

Miller, B. R., Walker, A. G., Shah, A. S., Barton, S. J., & Rebec, G. V. (2008). Dysregulated information processing by medium spiny neurons in striatum of freely behaving mouse models of Huntington's disease. *Journal of Neurophysiology*, *100*, 2205–2216.

Moghaddam, B., & Javitt, D. (2012). From revolution to evolution: The glutamate hypothesis of schizophrenia and its implication for treatment. *Neuropsychopharmacology*, *37*, 4–15.

Moore, K. A., Nicoll, R. A., & Schmitz, D. (2003). Adenosine gates synaptic plasticity at hippocampal mossy fiber synapses. *Proceedings of the National Academy of Sciences of the United States of America*, *100*, 14397–14402.

Mori, A., & Shindou, T. (2003). Modulation of GABAergic transmission in the striatopallidal system by adenosine A2A receptors: A potential mechanism for the antiparkinsonian effects of A2A antagonists. *Neurology*, *61*, S44–S48.

Munoz, L., & Ammit, A. J. (2010). Targeting p38 MAPK pathway for the treatment of Alzheimer's disease. *Neuropharmacology*, *58*, 561–568.

Munoz-Sanjuan, I., & Bates, G. P. (2011). The importance of integrating basic and clinical research toward the development of new therapies for Huntington disease. *Journal of Clinical Investigation*, *121*, 476–483.

Murray, C. J., & Lopez, A. D. (1997). Alternative projections of mortality and disability by cause 1990–2020: Global burden of disease study. *Lancet*, *349*, 1498–1504.

Nam, H. W., Hinton, D. J., Kang, N. Y., Kim, T., Lee, M. R., Oliveros, A., et al. (2013). Adenosine transporter ENT1 regulates the acquisition of goal-directed behavior and ethanol drinking through A2A receptor in the dorsomedial striatum. *Journal of Neuroscience*, *33*, 4329–4338.

Norenberg, W., Wirkner, K., Assmann, H., Richter, M., & Illes, P. (1998). Adenosine A2A receptors inhibit the conductance of NMDA receptor channels in rat neostriatal neurons. *Amino Acids*, *14*, 33–39.

Norenberg, W., Wirkner, K., & Illes, P. (1997). Effect of adenosine and some of its structural analogues on the conductance of NMDA receptor channels in a subset of rat neostriatal neurones. *British Journal of Pharmacology, 122,* 71–80.

Ohno, M., & Watanabe, S. (1996). Working memory failure by stimulation of hippocampal adenosine A1 receptors in rats. *Neuroreport, 7,* 3013–3016.

Olesen, J., Gustavsson, A., Svensson, M., Wittchen, H. U., & Jonsson, B. (2012). The economic cost of brain disorders in Europe. *European Journal of Neurology, 19,* 155–162.

O'Neill, M., & Brown, V. J. (2007a). Amphetamine and the adenosine A(2A) antagonist KW-6002 enhance the effects of conditional temporal probability of a stimulus in rats. *Behavioral Neuroscience, 121,* 535–542.

O'Neill, M., & Brown, V. J. (2007b). The effect of striatal dopamine depletion and the adenosine A2A antagonist KW-6002 on reversal learning in rats. *Neurobiology of Learning and Memory, 88,* 75–81.

Ongini, E., & Fredholm, B. B. (1996). Pharmacology of adenosine A2A receptors. *Trends in Pharmacological Sciences, 17,* 364–372.

Pan, B., Hillard, C. J., & Liu, Q. S. (2008). D2 dopamine receptor activation facilitates endocannabinoid-mediated long-term synaptic depression of GABAergic synaptic transmission in midbrain dopamine neurons via cAMP-protein kinase A signaling. *Journal of Neuroscience, 28,* 14018–14030.

Papp, K. V., Kaplan, R. F., & Snyder, P. J. (2011). Biological markers of cognition in prodromal Huntington's disease: A review. *Brain and Cognition, 77,* 280–291.

Pennartz, C. M., Ito, R., Verschure, P. F., Battaglia, F. P., & Robbins, T. W. (2011). The hippocampal–striatal axis in learning, prediction and goal-directed behavior. *Trends in Neurosciences, 34,* 548–559.

Pereira, G. S., Rossato, J. I., Sarkis, J. J., Cammarota, M., Bonan, C. D., & Izquierdo, I. (2005). Activation of adenosine receptors in the posterior cingulate cortex impairs memory retrieval in the rat. *Neurobiology of Learning and Memory, 83,* 217–223.

Piccinin, S., Di Angelantonio, S., Piccioni, A., Volpini, R., Cristalli, G., Fredholm, B. B., et al. (2010). CX3CL1-induced modulation at CA1 synapses reveals multiple mechanisms of EPSC modulation involving adenosine receptor subtypes. *Journal of Neuroimmunology, 224,* 85–92.

Pierri, M., Vaudano, E., Sager, T., & Englund, U. (2005). KW-6002 protects from MPTP induced dopaminergic toxicity in the mouse. *Neuropharmacology, 48,* 517–524.

Piray, P. (2011). The role of dorsal striatal D2-like receptors in reversal learning: A reinforcement learning viewpoint. *Journal of Neuroscience, 31,* 14049–14050.

Poo, M. M. (2001). Neurotrophins as synaptic modulators. *Nature Review Neuroscience, 2,* 24–32.

Popoli, P., Pezzola, A., & de Carolis, A. S. (1994). Modulation of striatal adenosine A1 and A2 receptors induces rotational behaviour in response to dopaminergic stimulation in intact rats. *European Journal of Pharmacology, 257,* 21–25.

Popoli, P., Pintor, A., Domenici, M. R., Frank, C., Tebano, M. T., Pezzola, A., et al. (2002). Blockade of striatal adenosine A2A receptor reduces, through a presynaptic mechanism, quinolinic acid-induced excitotoxicity: Possible relevance to neuroprotective interventions in neurodegenerative diseases of the striatum. *Journal of Neuroscience, 22,* 1967–1975.

Popoli, P., Reggio, R., Pezzola, A., Fuxe, K., & Ferre, S. (1998). Adenosine A1 and A2A receptor antagonists stimulate motor activity: Evidence for an increased effectiveness in aged rats. *Neuroscience Letters, 251,* 201–204.

Prediger, R. D., Batista, L. C., & Takahashi, R. N. (2005). Caffeine reverses age-related deficits in olfactory discrimination and social recognition memory in rats. Involvement of adenosine A1 and A2A receptors. *Neurobiology of Aging, 26,* 957–964.

Prediger, R. D., Fernandes, D., & Takahashi, R. N. (2005). Blockade of adenosine A2A receptors reverses short-term social memory impairments in spontaneously hypertensive rats. *Behavioural Brain Research, 159,* 197–205.

Prediger, R. D., Pamplona, F. A., Fernandes, D., & Takahashi, R. N. (2005). Caffeine improves spatial learning deficits in an animal model of attention deficit hyperactivity disorder (ADHD)—The spontaneously hypertensive rat (SHR). *International Journal of Neuropsychopharmacology*, *8*, 583–594.

Prediger, R. D., & Takahashi, R. N. (2005). Modulation of short-term social memory in rats by adenosine A1 and A(2A) receptors. *Neuroscience Letters*, *376*, 160–165.

Quiroz, C., Gomes, C., Pak, A. C., Ribeiro, J. A., Goldberg, S. R., Hope, B. T., et al. (2006). Blockade of adenosine A2A receptors prevents protein phosphorylation in the striatum induced by cortical stimulation. *Journal of Neuroscience*, *26*, 10808–10812.

Randall, P. A., Nunes, E. J., Janniere, S. L., Stopper, C. M., Farrar, A. M., Sager, T. N., et al. (2011). Stimulant effects of adenosine antagonists on operant behavior: Differential actions of selective A2A and A1 antagonists. *Psychopharmacology*, *216*, 173–186.

Ranganath, C., Minzenberg, M. J., & Ragland, J. D. (2008). The cognitive neuroscience of memory function and dysfunction in schizophrenia. *Biological Psychiatry*, *64*, 18–25.

Rebola, N., Lujan, R., Cunha, R. A., & Mulle, C. (2008). Adenosine A2A receptors are essential for long-term potentiation of NMDA-EPSCs at hippocampal mossy fiber synapses. *Neuron*, *57*, 121–134.

Rebola, N., Rodrigues, R. J., Lopes, L. V., Richardson, P. J., Oliveira, C. R., & Cunha, R. A. (2005). Adenosine A1 and A2A receptors are co-expressed in pyramidal neurons and co-localized in glutamatergic nerve terminals of the rat hippocampus. *Neuroscience*, *133*, 79–83.

Reiner, A., Albin, R. L., Anderson, K. D., D'Amato, C. J., Penney, J. B., & Young, A. B. (1988). Differential loss of striatal projection neurons in Huntington disease. *Proceedings of the National Academy of Sciences of the United States of America*, *85*, 5733–5737.

Reppert, S. M., Weaver, D. R., Stehle, J. H., & Rivkees, S. A. (1991). Molecular cloning and characterization of a rat A1-adenosine receptor that is widely expressed in brain and spinal cord. *Molecular Endocrinology*, *5*, 1037–1048.

Ribeiro, J. A. (1999). Adenosine A2A receptor interactions with receptors for other neurotransmitters and neuromodulators. *European Journal of Pharmacology*, *375*, 101–113.

Ribeiro, J. A., Sebastiao, A. M., & de Mendonca, A. (2002). Adenosine receptors in the nervous system: Pathophysiological implications. *Progress in Neurobiology*, *68*, 377–392.

Richard, I. H., Justus, A. W., Greig, N. H., Marshall, F., & Kurlan, R. (2002). Worsening of motor function and mood in a patient with Parkinson's disease after pharmacologic challenge with oral rivastigmine. *Clinical Neuropharmacology*, *25*, 296–299.

Riemenschneider, M., Lautenschlager, N., Wagenpfeil, S., Diehl, J., Drzezga, A., & Kurz, A. (2002). Cerebrospinal fluid tau and beta-amyloid 42 proteins identify Alzheimer disease in subjects with mild cognitive impairment. *Archives of Neurology*, *59*, 1729–1734.

Rimondini, R., Ferre, S., Ogren, S. O., & Fuxe, K. (1997). Adenosine A2A agonists: A potential new type of atypical antipsychotic. *Neuropsychopharmacology*, *17*, 82–91.

Ritchie, K., Carrière, I., Portet, F., de Mendonca, A., Dartigues, J. F., Rouaud, O., et al. (2007). The neuro-protective effects of caffeine: A prospective population study (the Three City Study). *Neurology*, *69*, 536–545.

Rodrigues, R. J., Alfaro, T. M., Rebola, N., Oliveira, C. R., & Cunha, R. A. (2005). Co-localization and functional interaction between adenosine A(2A) and metabotropic group 5 receptors in glutamatergic nerve terminals of the rat striatum. *Journal of Neurochemistry*, *92*, 433–441.

Rosenblatt, A. (2007). Neuropsychiatry of Huntington's disease. *Dialogues in Clinical Neuroscience*, *9*, 191–197.

Rosin, D. L., Hettinger, B. D., Lee, A., & Linden, J. (2003). Anatomy of adenosine A2A receptors in brain: Morphological substrates for integration of striatal function. *Neurology*, *61*, S12–S18.

Ross, G. W., Abbott, R. D., Petrovitch, H., Morens, D. M., Grandinetti, A., Tung, K. H., et al. (2000). Association of coffee and caffeine intake with the risk of Parkinson disease. *JAMA, 283*, 2674–2679.

Ross, C. A., Margolis, R. L., Reading, S. A., Pletnikov, M., & Coyle, J. T. (2006). Neurobiology of schizophrenia. *Neuron, 52*, 139–153.

Rouaux, C., Loeffler, J. P., & Boutillier, A. L. (2004). Targeting CREB-binding protein (CBP) loss of function as a therapeutic strategy in neurological disorders. *Biochemical Pharmacology, 68*, 1157–1164.

Rubaj, A., Zgodzinski, W., & Sieklucka-Dziuba, M. (2003). The influence of adenosine A3 receptor agonist: IB-MECA, on scopolamine- and MK-801-induced memory impairment. *Behavioural Brain Research, 141*, 11–17.

Saaksjarvi, K., Knekt, P., Rissanen, H., Laaksonen, M. A., Reunanen, A., & Mannisto, S. (2007). Prospective study of coffee consumption and risk of Parkinson's disease. *European Journal of Clinical Nutrition, 62*, 908–915.

Sakamoto, K., Karelina, K., & Obrietan, K. (2011). CREB: A multifaceted regulator of neuronal plasticity and protection. *Journal of Neurochemistry, 116*, 1–9.

Sawamoto, N., Piccini, P., Hotton, G., Pavese, N., Thielemans, K., & Brooks, D. J. (2008). Cognitive deficits and striato-frontal dopamine release in Parkinson's disease. *Brain, 131*, 1294–1302.

Scanziani, M., Capogna, M., Gahwiler, B. H., & Thompson, S. M. (1992). Presynaptic inhibition of miniature excitatory synaptic currents by baclofen and adenosine in the hippocampus. *Neuron, 9*, 919–927.

Scheff, S. W., Price, D. A., Schmitt, F. A., DeKosky, S. T., & Mufson, E. J. (2007). Synaptic alterations in CA1 in mild Alzheimer disease and mild cognitive impairment. *Neurology, 68*, 1501–1508.

Schiffmann, S. N., Fisone, G., Moresco, R., Cunha, R. A., & Ferre, S. (2007). Adenosine A2A receptors and basal ganglia physiology. *Progress in Neurobiology, 83*, 277–292.

Schiffmann, S. N., & Vanderhaeghen, J. J. (1993). Adenosine A2 receptors regulate the gene expression of striatopallidal and striatonigral neurons. *Journal of Neuroscience, 13*, 1080–1087.

Schotanus, S. M., Fredholm, B. B., & Chergui, K. (2006). NMDA depresses glutamatergic synaptic transmission in the striatum through the activation of adenosine A1 receptors: Evidence from knockout mice. *Neuropharmacology, 51*, 272–282.

Schulte, G., & Fredholm, B. B. (2003). Signalling from adenosine receptors to mitogen-activated protein kinases. *Cellular Signalling, 15*, 813–827.

Scimeca, J. M., & Badre, D. (2012). Striatal contributions to declarative memory retrieval. *Neuron, 75*, 380–392.

Sebastiao, A. M., & Ribeiro, J. A. (1996). Adenosine A2 receptor-mediated excitatory actions on the nervous system. *Progress in Neurobiology, 48*, 167–189.

Sebastiao, A. M., & Ribeiro, J. A. (2000). Fine-tuning neuromodulation by adenosine. *Trends in Pharmacological Sciences, 21*, 341–346.

Selkoe, D. J. (2002). Alzheimer's disease is a synaptic failure. *Science, 298*, 789–791.

Selley, M. L. (2004). Increased homocysteine and decreased adenosine formation in Alzheimer's disease. *Neurological Research, 26*, 554–557.

Seo, H., Sonntag, K. C., & Isacson, O. (2004). Generalized brain and skin proteasome inhibition in Huntington's disease. *Annals of Neurology, 56*, 319–328.

Shen, H. Y., Coelho, J. E., Ohtsuka, N., Canas, P. M., Day, Y. J., Huang, Q. Y., et al. (2008). A critical role of the adenosine A2A receptor in extrastriatal neurons in modulating psychomotor activity as revealed by opposite phenotypes of striatum and forebrain A2A receptor knock-outs. *Journal of Neuroscience, 28*, 2970–2975.

Shen, W., Flajolet, M., Greengard, P., & Surmeier, D. J. (2008). Dichotomous dopaminergic control of striatal synaptic plasticity. *Science, 321*, 848–851.

Shen, H. Y., Singer, P., Lytle, N., Wei, C. J., Lan, J. Q., Williams-Karnesky, R. L., et al. (2012). Adenosine augmentation ameliorates psychotic and cognitive endophenotypes of schizophrenia. *Journal of Clinical Investigation, 122*, 2567–2577.

Shohamy, D., Myers, C. E., Grossman, S., Sage, J., Gluck, M. A., & Poldrack, R. A. (2004). Cortico-striatal contributions to feedback-based learning: Converging data from neuroimaging and neuropsychology. *Brain, 127*, 851–859.

Simoes, A. P., Duarte, J. A., Agasse, F., Canas, P. M., Tome, A. R., Agostinho, P., et al. (2012). Blockade of adenosine A2A receptors prevents interleukin-1beta-induced exacerbation of neuronal toxicity through a p38 mitogen-activated protein kinase pathway. *Journal of Neuroinflammation, 9*, 204.

Simpson, E. H., Kellendonk, C., & Kandel, E. (2010). A possible role for the striatum in the pathogenesis of the cognitive symptoms of schizophrenia. *Neuron, 65*, 585–596.

Singer, P., Wei, C. J., Chen, J. F., Boison, D., & Yee, B. K. (2013). Deletion of striatal adenosine A(2A) receptor spares latent inhibition and prepulse inhibition but impairs active avoidance learning. *Behavioural Brain Research, 242*, 54–61.

Snyder, S. H. (1976). The dopamine hypothesis of schizophrenia: Focus on the dopamine receptor. *The American Journal of Psychiatry, 133*, 197–202.

Solinas, M., Ferre, S., You, Z. B., Karcz-Kubicha, M., Popoli, P., & Goldberg, S. R. (2002). Caffeine induces dopamine and glutamate release in the shell of the nucleus accumbens. *Journal of Neuroscience, 22*, 6321–6324.

Spangaro, M., Bosia, M., Zanoletti, A., Bechi, M., Cocchi, F., Pirovano, A., et al. (2012). Cognitive dysfunction and glutamate reuptake: Effect of EAAT2 polymorphism in schizophrenia. *Neuroscience Letters, 522*, 151–155.

Starling, A. J., Andre, V. M., Cepeda, C., de Lima, M., Chandler, S. H., & Levine, M. S. (2005). Alterations in N-methyl-D-aspartate receptor sensitivity and magnesium blockade occur early in development in the R6/2 mouse model of Huntington's disease. *Journal of Neuroscience Research, 82*, 377–386.

Stout, J. C., Paulsen, J. S., Queller, S., Solomon, A. C., Whitlock, K. B., Campbell, J. C., et al. (2011). Neurocognitive signs in prodromal Huntington disease. *Neuropsychology, 25*, 1–14.

Taylor, A. E., Saint-Cyr, J. A., & Lang, A. E. (1987). Parkinson's disease. Cognitive changes in relation to treatment response. *Brain, 110*(Pt. 1), 35–51.

Tebano, M. T., Martire, A., Chiodi, V., Pepponi, R., Ferrante, A., Domenici, M. R., et al. (2009). Adenosine A2A receptors enable the synaptic effects of cannabinoid CB1 receptors in the rodent striatum. *Journal of Neurochemistry, 110*, 1921–1930.

Tebano, M. T., Martire, A., Potenza, R. L., Gro, C., Pepponi, R., Armida, M., et al. (2008). Adenosine A(2A) receptors are required for normal BDNF levels and BDNF-induced potentiation of synaptic transmission in the mouse hippocampus. *Journal of Neurochemistry, 104*, 279–286.

Tebano, M. T., Martire, A., Rebola, N., Pepponi, R., Domenici, M. R., Gro, M. C., et al. (2005). Adenosine A2A receptors and metabotropic glutamate 5 receptors are co-localized and functionally interact in the hippocampus: A possible key mechanism in the modulation of N-methyl-D-aspartate effects. *Journal of Neurochemistry, 95*, 1188–1200.

Tebano, M. T., Pintor, A., Frank, C., Domenici, M. R., Martire, A., Pepponi, R., et al. (2004). Adenosine A2A receptor blockade differentially influences excitotoxic mechanisms at pre- and postsynaptic sites in the rat striatum. *Journal of Neuroscience Research, 77*, 100–107.

The Huntington's Disease Collaborative Research Group. (1993). A novel gene containing a trinucleotide repeat that is expanded and unstable on Huntington's disease chromosomes. The Huntington's Disease Collaborative Research Group. *Cell, 72*, 971–983.

Thomas, G. M., & Huganir, R. L. (2004). MAPK cascade signalling and synaptic plasticity. *Nature Review Neuroscience, 5*, 173–183.

Todd, K. J., Darabid, H., & Robitaille, R. (2010). Perisynaptic glia discriminate patterns of motor nerve activity and influence plasticity at the neuromuscular junction. *Journal of Neuroscience*, *30*, 11870–11882.

Ulas, J., Brunner, L. C., Nguyen, L., & Cotman, C. W. (1993). Reduced density of adenosine A1 receptors and preserved coupling of adenosine A1 receptors to G proteins in Alzheimer hippocampus: A quantitative autoradiographic study. *Neuroscience*, *52*, 843–854.

Uriguen, L., Garcia-Fuster, M. J., Callado, L. F., Morentin, B., La Harpe, R., Casado, V., et al. (2009). Immunodensity and mRNA expression of A2A adenosine, D2 dopamine, and CB1 cannabinoid receptors in postmortem frontal cortex of subjects with schizophrenia: Effect of antipsychotic treatment. *Psychopharmacology*, *206*, 313–324.

van Boxtel, M. P., Schmitt, J. A., Bosma, H., & Jolles, J. (2003). The effects of habitual caffeine use on cognitive change: A longitudinal perspective. *Pharmacology, Biochemistry, and Behavior*, *75*, 921–927.

van Calker, D., Muller, M., & Hamprecht, B. (1978). Adenosine inhibits the accumulation of cyclic AMP in cultured brain cells. *Nature*, *276*, 839–841.

van der Meer, M. A., & Redish, A. D. (2011). Ventral striatum: A critical look at models of learning and evaluation. *Current Opinion in Neurobiology*, *21*, 387–392.

van Gelder, B. M., Buijsse, B., Tijhuis, M., Kalmijn, S., Giampaoli, S., Nissinen, A., et al. (2007). Coffee consumption is inversely associated with cognitive decline in elderly European men: The FINE Study. *European Journal of Clinical Nutrition*, *61*, 226–232.

van Laar, T., De Deyn, P. P., Aarsland, D., Barone, P., & Galvin, J. E. (2011). Effects of cholinesterase inhibitors in Parkinson's disease dementia: A review of clinical data. *CNS Neuroscience and Therapeutics*, *17*, 428–441.

van Os, J., & Kapur, S. (2009). Schizophrenia. *Lancet*, *374*, 635–645.

Vila-Luna, S., Cabrera-Isidoro, S., Vila-Luna, L., Juarez-Diaz, I., Bata-Garcia, J. L., Alvarez-Cervera, F. J., et al. (2012). Chronic caffeine consumption prevents cognitive decline from young to middle age in rats, and is associated with increased length, branching, and spine density of basal dendrites in CA1 hippocampal neurons. *Neuroscience*, *202*, 384–395.

Villar-Menendez, I., Diaz-Sanchez, S., Blanch, M., Albasanz, J. L., Pereira-Veiga, T., Monje, A., et al. (2014). Reduced striatal adenosine A2A receptor levels define a molecular subgroup in schizophrenia. *Journal of Psychiatric Research*, *51*, 49–59.

Von Lubitz, D. K., Paul, I. A., Bartus, R. T., & Jacobson, K. A. (1993). Effects of chronic administration of adenosine A1 receptor agonist and antagonist on spatial learning and memory. *European Journal of Pharmacology*, *249*, 271–280.

Vonsattel, J. P., & DiFiglia, M. (1998). Huntington disease. *Journal of Neuropathology and Experimental Neurology*, *57*, 369–384.

Walker, A. G., Miller, B. R., Fritsch, J. N., Barton, S. J., & Rebec, G. V. (2008). Altered information processing in the prefrontal cortex of Huntington's disease mouse models. *Journal of Neuroscience*, *28*, 8973–8982.

Walsh, D. M., & Selkoe, D. J. (2004). Deciphering the molecular basis of memory failure in Alzheimer's disease. *Neuron*, *44*, 181–193.

Wan, Q., Yao, H., & Wang, F. (1999). Involvement of K(+) channels in the inhibitory effects of adenosine on anoxia-induced [Ca(2+)](i) increase in cultured rat hippocampal CA1 neurons. *Biological Signals and Receptors*, *8*, 309–315.

Wang, J. H., Ma, Y. Y., & van den Buuse, M. (2006). Improved spatial recognition memory in mice lacking adenosine A2A receptors. *Experimental Neurology*, *199*, 438–445.

Wang, J., Wang, C.-E., Orr, A., Tydlacka, S., Li, S.-H., & Li, X.-J. (2008). Impaired ubiquitin–proteasome system activity in the synapses of Huntington's disease mice. *Journal of Cell Biology*, *180*, 1177–1189.

Watson, G. S., & Leverenz, J. B. (2010). Profile of cognitive impairment in Parkinson's disease. *Brain Pathology*, *20*, 640–645.

Wei, C. J., Augusto, E., Gomes, C. A., Singer, P., Wang, Y., Boison, D., et al. (2013). Regulation of fear responses by striatal and extrastriatal adenosine A receptors in forebrain. *Biological Psychiatry, 75*, 855–863.

Wei, C. J., Singer, P., Coelho, J., Boison, D., Feldon, J., Yee, B. K., et al. (2011). Selective inactivation of adenosine A(2A) receptors in striatal neurons enhances working memory and reversal learning. *Learning and Memory, 18*, 459–474.

Whitehead, A., Perdomo, C., Pratt, R. D., Birks, J., Wilcock, G. K., & Evans, J. G. (2004). Donepezil for the symptomatic treatment of patients with mild to moderate Alzheimer's disease: A meta-analysis of individual patient data from randomised controlled trials. *International Journal of Geriatric Psychiatry, 19*, 624–633.

Wimo, A., Jonsson, L., Bond, J., Prince, M., & Winblad, B. (2013). The worldwide economic impact of dementia 2010. *Alzheimer's & Dementia, 9*(1–11), e13.

Winblad, B., Andreasen, N., Minthon, L., Floesser, A., Imbert, G., Dumortier, T., et al. (2012). Safety, tolerability, and antibody response of active Abeta immunotherapy with CAD106 in patients with Alzheimer's disease: Randomised, double-blind, placebo-controlled, first-in-human study. *Lancet Neurology, 11*, 597–604.

Wirkner, K., Assmann, H., Koles, L., Gerevich, Z., Franke, H., Norenberg, W., et al. (2000). Inhibition by adenosine A(2A) receptors of NMDA but not AMPA currents in rat neostriatal neurons. *British Journal of Pharmacology, 130*, 259–269.

Wonodi, I., Gopinath, H. V., Liu, J., Adami, H., Hong, L. E., Allen-Emerson, R., et al. (2011). Dipyridamole monotherapy in schizophrenia: Pilot of a novel treatment approach by modulation of purinergic signaling. *Psychopharmacology, 218*, 341–345.

Wu, L. G., & Saggau, P. (1994). Adenosine inhibits evoked synaptic transmission primarily by reducing presynaptic calcium influx in area CA1 of hippocampus. *Neuron, 12*, 1139–1148.

Xia, J., Chen, F., Ye, J., Yan, J., Wang, H., Duan, S., et al. (2009). Activity-dependent release of adenosine inhibits the glutamatergic synaptic transmission and plasticity in the hypothalamic hypocretin/orexin neurons. *Neuroscience, 162*, 980–988.

Xia, Z., & Storm, D. R. (2012). Role of signal transduction crosstalk between adenylyl cyclase and MAP kinase in hippocampus-dependent memory. *Learning and Memory, 19*, 369–374.

Xu, K., Xu, Y. H., Chen, J. F., & Schwarzschild, M. A. (2002). Caffeine's neuroprotection against 1-methyl-4-phenyl-1,2,3,6-tetrahydropyridine toxicity shows no tolerance to chronic caffeine administration in mice. *Neuroscience Letters, 322*, 13–16.

Yee, B. K., Singer, P., Chen, J. F., Feldon, J., & Boison, D. (2007). Transgenic overexpression of adenosine kinase in brain leads to multiple learning impairments and altered sensitivity to psychomimetic drugs. *European Journal of Neuroscience, 26*, 3237–3252.

Yee, B. K., Zhu, S. W., Mohammed, A. H., & Feldon, J. (2007). Levels of neurotrophic factors in the hippocampus and amygdala correlate with anxiety- and fear-related behaviour in C57BL6 mice. *Journal of Neural Transmission, 114*, 431–444.

Yin, H. H., & Knowlton, B. J. (2006). The role of the basal ganglia in habit formation. *Nature Review Neuroscience, 7*, 464–476.

Young, A. B. (2003). Huntingtin in health and disease. *Journal of Clinical Investigation, 111*, 299–302.

Yu, C., Gupta, J., Chen, J. F., & Yin, H. H. (2009). Genetic deletion of A2A adenosine receptors in the striatum selectively impairs habit formation. *Journal of Neuroscience, 29*, 15100–15103.

Zeron, M. M., Hansson, O., Chen, N., Wellington, C. L., Leavitt, B. R., Brundin, P., et al. (2002). Increased sensitivity to N-methyl-D-aspartate receptor-mediated excitotoxicity in a mouse model of Huntington's disease. *Neuron, 33*, 849–860.

Zhang, J., Abdallah, C. G., Wang, J., Wan, X., Liang, C., Jiang, L., et al. (2012). Upregulation of adenosine A2A receptors induced by atypical antipsychotics and its correlation with sensory gating in schizophrenia patients. *Psychiatry Research*, *200*, 126–132.
Zhou, H., Cao, F., Wang, Z., Yu, Z.-X., Nguyen, H.-P., Evans, J., et al. (2003). Huntingtin forms toxic NH2-terminal fragment complexes that are promoted by the age-dependent decrease in proteasome activity. *Journal of Cell Biology*, *163*, 109–118.
Zhou, S. J., Zhu, M. E., Shu, D., Du, X. P., Song, X. H., Wang, X. T., et al. (2009). Preferential enhancement of working memory in mice lacking adenosine A(2A) receptors. *Brain Research*, *1303*, 74–83.
Zuccato, C., Ciammola, A., Rigamonti, D., Leavitt, B. R., Goffredo, D., Conti, L., et al. (2001). Loss of huntingtin-mediated BDNF gene transcription in Huntington's disease. *Science*, *293*, 493–498.
Zuccato, C., Tartari, M., Crotti, A., Goffredo, D., Valenza, M., Conti, L., et al. (2003). Huntingtin interacts with REST/NRSF to modulate the transcription of NRSE-controlled neuronal genes. *Nature Genetics*, *35*, 76–83.
Zuccato, C., Valenza, M., & Cattaneo, E. (2010). Molecular mechanisms and potential therapeutical targets in Huntington's disease. *Physiological Reviews*, *90*, 905–981.

FURTHER READING

Davies, S. W., Turmaine, M., Cozens, B. A., DiFiglia, M., Sharp, A. H., Ross, C. A., et al. (1997). Formation of neuronal intranuclear inclusions underlies the neurological dysfunction in mice transgenic for the HD mutation. *Cell*, *90*, 537–548.
DiFiglia, M., Sapp, E., Chase, K. O., Davies, S. W., Bates, G. P., Vonsattel, J. P., et al. (1997). Aggregation of huntingtin in neuronal intranuclear inclusions and dystrophic neurites in brain. *Science*, *277*, 1990–1993.
Gutekunst, C. A., Levey, A. I., Heilman, C. J., Whaley, W. L., Yi, H., Nash, N. R., et al. (1995). Identification and localization of huntingtin in brain and human lymphoblastoid cell lines with anti-fusion protein antibodies. *Proceedings of the National Academy of Sciences of the United States of America*, *92*, 8710–8714.
Sapp, E., Schwarz, C., Chase, K., Bhide, P. G., Young, A. B., Penney, J., et al. (1997). Huntingtin localization in brains of normal and Huntington's disease patients. *Annals of Neurology*, *42*, 604–612.

CHAPTER THIRTEEN

Adenosine Receptors in Cerebral Ischemia

Alessia Melani*[,1], Anna Maria Pugliese, Felicita Pedata
Department of Neuroscience, Psychology, Drug Research and Child Health (NEUROFARBA), Division of Pharmacology and Toxicology, University of Florence, 50139 Florence, Italy
[1]Corresponding author: e-mail address: alessia.melani@unifi.it

Contents

Abstract

Ischemic stroke is a complex pathology characterized by a sequence of events that evolve over time and space. It is the second leading cause of death and the main cause of adult long-term disability in developed countries. At the moment, there is no promising pharmacotherapy for acute ischemic stroke. Adenosine receptors (A_1, A_{2A}, A_{2B}, A_3) are important targets for therapeutic implementation in the treatment of stroke because extracellular adenosine concentrations increase dramatically soon after ischemia. Adenosine receptors located both on central nervous system cells and on immune blood cells exert important roles during ischemia. The neuroprotective role of adenosine through A_1 receptor subtype during ischemia is accepted, but the use of selective A_1 agonists is hampered by undesirable side effects such as sedation, bradycardia, and hypotension. Recently, the A_{2A} receptor subtype emerged as a potential therapeutic attractive target in ischemia. Evidence suggests that A_{2A} receptor has dual role: in a first phase of ischemia, it potentiates excitotoxicity, while hours and days after ischemia, A_{2A} receptors on immune blood cells potentiate cell adhesion mechanisms and infiltration in the ischemic parenchyma. Consistently, the use of A_{2A} receptor agonists/antagonists (administered at doses that do not modify blood pressure and heart rate)

*Fellow of the Fondazione Umberto Veronesi

International Review of Neurobiology, Volume 119
ISSN 0074-7742
http://dx.doi.org/10.1016/B978-0-12-801022-8.00013-1

should be carefully evaluated in function of time after ischemia. Although much is still to be known about the role of A_{2B} and A_3 receptor subtypes in brain ischemia, most consistent information indicates their role in regulation of immunosuppression and inflammation.

1. INTRODUCTION

Cerebrovascular diseases are the second leading cause of death and the main cause of adult long-term disability in developed countries. The World Health Organization places the global incidence of stroke at around 200 cases per 100,000 inhabitants, although data vary among countries (Soler & Ruiz, 2010). Ischemic stroke commonly accounts for approximately 80% of all stroke cases, and is caused from occlusion of a major cerebral artery by a thrombus or an embolism, which leads to loss of cerebral blood flow, a condition of oxygen and glucose deprivation (OGD), and subsequently tissue damage in the affected region (Gibson, 2013). Ischemic brain injury results from a complex sequence of pathophysiological events that evolve over time and space. The major pathogenic mechanisms of this cascade include: acute mechanisms of excitotoxicity and periinfarct depolarizations and the delayed mechanisms of inflammation and apoptosis (Dirnagl, Iadecola, & Moskowitz, 1999). A complex interplay of biochemical and molecular mechanisms involving practically any cell type of the brain, concept of "the neurovascular unit," partakes in either salvage or demise of the tissue after a stroke (Dirnagl, 2012). Moreover, the peripheral immune cells play an important role in the responses of brain tissue to ischemic injury (Iadecola & Anrather, 2011) because a massive blood cell infiltration caused by blood–brain barrier (BBB) disruption occurs into the ischemic area.

At the moment, there is no promising pharmacotherapy for acute ischemic stroke, aside from intravenous or intraarterial thrombolysis (Chen et al., 2014). Neuroprotective drugs, such as glutamate receptor antagonists, have shown therapeutic efficacy in animal stroke models but have failed during clinical trials (De Keyser, Sulter, & Luiten, 1999; O'Bryant, Vann, & Xiong, 2014). In the last years, numerous authors have proposed adenosine and its receptors as an important target for therapeutic implementation in the treatment of stroke. Adenosine acts through G-protein-coupled receptors that are present on extracellular membrane (adenosine receptor subtypes A_1, A_{2A}, A_{2B}, and A_3) and exert a variety of physiological effects (Fredholm,

IJzerman, Jacobson, Linden, & Müller, 2011). Adenosine receptors are expressed at significant levels in neurons and glial cells and in peripheral inflammatory cells (such as lymphocytes and neutrophils) (Brodie, Blumberg, & Jacobson, 1998; Fiebich et al., 1996; Hettinger, Lee, Linden, & Rosin, 2001; Peterfreund, MacCollin, Gusella, & Fink, 1996; Svenningsson, Le Moine, Fisone, & Fredholm, 1999; Yu et al., 2004) (Fig. 13.1). The wide distribution is consistent with the multifaceted neurochemical and molecular effects of adenosine and suggests that the role of adenosine in ischemia is the consequence of an interplay among different receptor activation in neuronal, glial, and inflammatory cells, which may vary depending on the time-related development of the pathological condition.

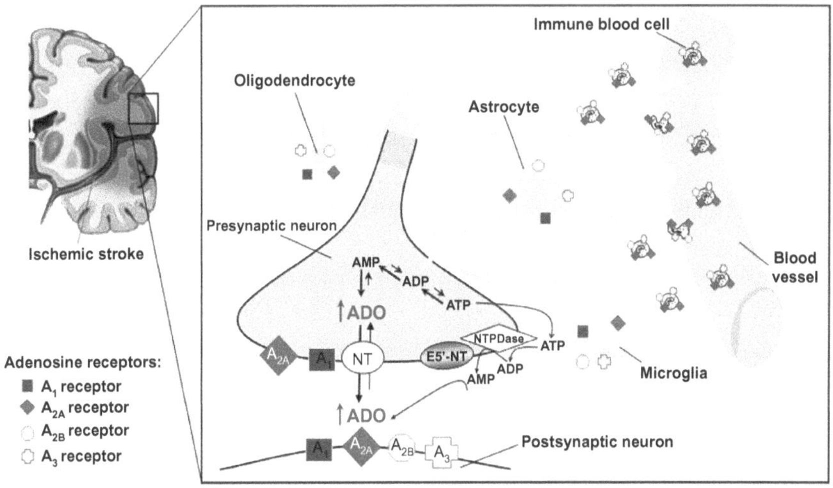

Figure 13.1 Schematic drawing of adenosine receptors on different cell types. All adenosine receptor subtypes are expressed both at the central level on presynaptic and postsynaptic neurons, on astrocytes, on microglia, and on oligodendrocytes and at the peripheral level on leukocytes and vasculature. After cerebral ischemia, leukocytes infiltrate into ischemic tissue due to increased permeability of blood–brain barrier (BBB). During ischemia, extracellular adenosine levels increase mainly due to: (i) extracellular ATP degradation by NTPDase and ecto-5′-nucleotidase enzymes; (ii) release *per se* from cells likely by the equilibrative nucleoside transporter (ENT) (Melani et al., 2014); (iii) inhibition of adenosine uptake processes due to downregulation of concentrative nucleoside transporters (CNT) 2 and 3 and of ENT. AC, adenylate cyclase; ADO, adenosine; ADP, adenosine diphosphate; AMP, adenosine monophosphate; ATP, adenosine triphosphate; E5′-NT, ecto-5′-nucleotidase; NT, nucleoside transporter; NTDPase, ectonucleoside triphosphate diphosphohydrolases. The proportions of the various components of the nervous tissue have not been kept.

Soon after a cerebral ischemic event, the adenosine concentration increases dramatically in the extracellular space (Frenguelli, Wigmore, Llaudet, & Dale, 2007, Matsumoto, Graf, Rosner, Shimada, & Heiss, 1992; Melani, Corti, Stephan, et al., 2012; Melani et al., 1999; Phillis, Smith-Barbour, & O'Regan, 1996) due to different mechanisms: (1) degradation of extracellularly released ATP in the first minutes after stroke (Melani, Corti, Stephan, et al., 2012); (2) adenosine released *per se* from cells likely by the equilibrative nucleoside transporter (ENT)2 (Melani, Corti, Stephan, et al., 2012) due to the enhancement of intracellular adenosine concentration resulting from an imbalance between degradation and resynthesis of intracellular ATP; and (3) inhibition of adenosine uptake processes due to downregulation of concentrative nucleoside transporters 2 and 3 and of the ENT1 (Medina-Pulido et al., 2013). During ischemia, adenosine has long been known to act predominantly as a neuroprotectant endogenous agent (Cunha, 2001; Fredholm, Cunha, & Svenningsson, 2003; Ongini, Adami, Ferri, & Bertorelli, 1997; Pedata et al., 2007; Ribeiro, Sebastiao, & de Mendonca, 2002; Schwarzschild, Chen, & Ascherio, 2002). In fact, adenosine infusion into the ischemic striatum has been shown to significantly ameliorate neurological outcome and reduce infarct volume after transient focal cerebral ischemia (Kitagawa, Mori, Shimada, Mitsumoto, & Kikuchi, 2002).

In this chapter, we summarize the studies that have contributed to current understanding of the mechanisms by which adenosine receptors modulate tissue damage in brain ischemia models.

2. A_1 RECEPTORS IN BRAIN ISCHEMIA

The increase in extracellular adenosine levels found in ischemic brain areas soon after ischemia represents an endogenous neuroprotective response (Latini & Pedata, 2001). One of the prime adaptive mechanisms in response to hypoxia/ischemia is the cellular activation of A_1 receptors, which inhibits excitatory synaptic transmission as demonstrated *in vitro* (Dale, Pearson, & Fringuelli, 2000; Fowler, 1989, 1990; Gribkoff, Bauman, & VanderMaelen, 1990; Latini, Bordoni, Pedata, & Corradetti, 1999; Sebastião, Cunha, de Mendonca, & Ribeiro, 2000) and *in vivo* (Fowler, Gervitz, Hamilton, & Walker, 2003; Gervitz, Davies, Omidvar, & Fowler, 2003; Gervitz, Lutherer, Davies, Pirch, & Fowler, 2001).

Protective effects of adenosine in ischemia are greatly attributed to A_1 receptor activation that reduces Ca^{2+} influx, lowers presynaptic release of neurotransmitters (Andiné, 1993; Corradetti, Lo, Moroni, Passani, & Pepeu, 1984; Dunwiddie, 1984; Kitagawa et al., 2002; Pedata, Latini, Pugliese, & Pepeu, 1993; Zetterstrom & Fillenz, 1990) and in particular of glutamate which exerts an excitotoxic effect during ischemia mainly by overstimulation of NMDA receptors (Choi, 1990). In addition, adenosine stabilizes the neuronal membrane potentials by directly increasing the K^+ and Cl^- ion conductances, thus reducing neuronal excitability (Choi, 1990). Consequently, adenosine by decreasing cellular metabolism, energy consumption (Greene & Haas, 1991), and body/brain temperature (Gourine, Dale, Gourine, & Spyer, 2004) protects against ischemia.

In *in vitro* hypoxia/ischemia models, adenosine and selective A_1 receptor agonists reduce neuronal damage following hypoxia and/or OGD in primary cortical or hippocampal cell cultures (Daval & Nicolas, 1994; Goldberg, Monyer, Weiss, & Choi, 1988; Logan & Sweeney, 1997) and brain slices (Mori, Nishizaki, & Okada, 1992; Newman, Hospod, Trowbridge, Motwani, & Liu, 1998). Marcoli and coworkers (2003) demonstrated that the selective A_1 receptor antagonist, 1,3-dipropyl-8-cyclopentyladenosine (DPCPX), increases OGD-evoked aspartate and glutamate efflux in rat cerebrocortical slices. In accordance, the A_1 receptor-mediated OGD-induced depression of synaptic transmission is reversed by administration of selective A_1 receptor antagonists to rat hippocampal slices (Latini, Bordoni, et al., 1999; Pedata et al., 1993; Tanaka et al., 2001). A_1 receptor antagonists reduce inhibition of synaptic transmission, impair the recovery of synaptic potentials (Sebastião, de Mendonca, Moreira, & Ribeiro, 2001), and shorten the onset of anoxic depolarization (AD) induced by hypoxia in the CA1 region of hippocampal slices (Lee & Lowenkopf, 1993). Activation of A_1 receptor, by endogenous adenosine, transiently enhances inhibitory synaptic transmission in CA1 pyramidal slices from rat exposed to transient forebrain ischemia (Liang, Pang, Deng, & Xu, 2009). Thus, a depression of excitatory synaptic transmission such as that brought about by adenosine A_1 receptors during hypoxia/ischemia is neuroprotective and crucial for the functional recovery of hippocampal circuits upon reoxygenation (Arrigoni, Crocker, Saper, Greene, & Scammell, 2005; Johansson et al., 2001; Sebastião et al., 2001).

However, recent data indicate that synaptic depression by postischemic A_1 receptor activation functionally disrupts the connectivity within the cortical networks. This was suggested by the observation that the A_1 receptor antagonist DPCPX (1.25 mg/kg i.p.), administered at the onset of reperfusion after global cerebral ischemia in rats, accelerated the recovery of the pattern of postischemic burst suppression that is index of impairment of neocortical connectivity (Ilie et al., 2009).

Studies using A_1 receptor knock-out (KO) mice support the neuroprotective role of A_1 adenosine receptor stimulation. Hippocampal slices from A_1 receptor KO mice showed a markedly reduced and delayed protective response to hypoxia compared to slices from wild-type mice (Johansson et al., 2001). In astrocytes prepared from A_1 receptor KO mice, more pronounced hypoxic cytotoxicity was observed (Bjorklund, Shang, Tonazzini, Dare, & Fredholm, 2008). However, Olsson et al. (2004) reported that hippocampal slices from A_1 receptor KO mice subjected to hypoxia show a small, though not statistically significant decrease in damage to neurons in the CA1 region.

The A_1 receptor desensitization after ischemia might be a critical event that can interfere with the maintaining of the neuroprotective efficiency of adenosine. In rat hippocampal slices, hypoxia leads to a rapid (<90 min) desensitization of A_1 receptor-mediated inhibition of synaptic transmission that is likely due to an internalization of A_1 receptors in nerve terminals (Coelho et al., 2006), a process that may result in hyperexcitability and increased brain damage.

In *in vivo* animal models of global cerebral ischemia, local administration of an adenosine analog, 2-chloroadenosine (CADO), and of a nonselective A_1 receptor agonist, N6-(L-2-phenylisopropyl) adenosine (L-PIA), attenuates neuronal loss in the CA1 region of the rat hippocampus (Domenici, de Carolis, & Sagratella, 1996; Evans, Swan, & Meldrum, 1987); acute systemic or intracerebroventricular injection of the A_1 receptor agonists cyclohexyladenosine and *R*-phenylisopropyl-adenosine (*R*-PIA) improves neurological deficits (Heron et al., 1994; von Lubitz & Marangos, 1990; Zhou, Meno, Hsu, & Winn, 1994), protects the CA1 region of the hippocampus (von Lubitz, Dambrosia, Kempski, & Redmond, 1988), and prevents the reduction of adenosine A_1 receptors (Daval, Von Lubitz, Deckert, Redmond, & Marangos, 1989) in rats or gerbils. Similarly, acute administration of the A_1 receptor agonists, N6-cyclopentyladenosine (CPA) and 2-chloro-N(6)-cyclopentyladenosine (CCPA), reduces mortality and the loss of neurons after global forebrain ischemia in the gerbil (von Lubitz, Lin, Melman, et al., 1994). Administration of the A_1 receptor agonist adenosine amine congener (ADAC), which induces fewer undesirable effects, after global ischemia in the gerbil, increased survival, protected neuronal morphology, induced preservation of microtubule-associated protein 2 (MAP-2) and of spatial memory and learning ability (Phillis & Goshgarian, 2001, von Lubitz et al., 1996). In accordance, adenosine A_1 receptor antagonists given acutely exacerbate the changes induced by global ischemia. A nonspecific A_1 receptor antagonist, theophylline, increased mortality (Jarrott & Domer, 1980) and ischemic cell damage (Rudolphi, Keil, Fastbom, & Fredholm, 1989; Rudolphi, Keil, & Hinze, 1987) after global ischemia in gerbils. Similar detrimental effects were also observed after acute administration of the selective A_1 receptor antagonists DPCPX and 8-cyclopentyl-1,3-dimethylxanthine (CPT) (Boissard, Lindner, & Gribkoff, 1992; Phillis, 1995; von Lubitz, Lin, Melman, et al., 1994). Interestingly, a reverse effect appeared after chronic administration of adenosine receptor antagonists. Both caffeine and DPCPX given chronically weeks before an ischemic insult reduced the neuronal injury assessed by magnetic resonance imaging and histopathological examination in rats and gerbils (Rudolphi et al., 1989; Sutherland et al., 1991; von Lubitz, Lin, Melman, et al., 1994). It has been suggested that the beneficial effects seen after chronic administration of adenosine antagonists may be due to upregulation of A_1 receptors (Jacobson, Von Lubitz, Daly, & Fredholm, 1996). In agreement and unlike acute treatment, chronic administration of A_1 receptor agonists worsened survival and increased neuronal loss (Jacobson et al., 1996), a phenomenon thought to depend on A_1 receptor desensitization.

Plastic changes in A_1 receptors are critical to understanding the effect of adenosine A_1 receptor agonists/antagonists but also whether adenosine maintains its neuroprotective efficiency after ischemia. In this regard, several studies have shown that short periods of focal or global ischemia produced a long-lasting decrease in the density of A_1 receptors (Nagasawa, Araki, & Kogure, 1994; Nariai et al., 2003; Onodera, Sato, & Kogure, 1987). On the contrary, an increase in the mRNA level of A_1 receptors was reported in the cerebral cortex after global ischemia in rats and normalization under reperfusion (Lai, Tu, Liu, & Cheng, 2005). Relevantly, it was reported that A_1 receptor KO mice, when exposed to global ischemia, do not show increased neuronal damage in the CA1 region of the hippocampus, in the cortex, or in the striatum compared to wild-type controls (Olsson et al., 2004). These discrepancies may reflect development of compensatory mechanisms after genetic deletion (Wei, Li, & Chen, 2011).

Recently, it has been demonstrated that A_1 adenosine receptors are involved in protection exerted by the chemokine CX3CL1. CX3CL1, intracerebroventricularly injected 20 min before permanent middle cerebral artery occlusion (MCAo), reduces cerebral infarct size and neurological deficits after focal cerebral ischemia in rat and mice (Cipriani et al., 2011). The neuroprotective effect of CX3CL1 requires the activation of A_1 adenosine receptor since the effect is abolished in the presence of the A_1 receptor antagonist, DPCPX, and in A_1 receptor KO mice (Cipriani et al., 2011).

During ischemic preconditioning, brief sublethal anoxic/ischemic insults may "precondition" and protect the brain from a subsequent ischemic event and adenosine, by stimulating A_1 receptors, plays a crucial role in this phenomenon. In fact, an A_1 receptor agonist, CADO, markedly enhanced (Perez-Pinzon, Mumford, Rosenthal, & Sick, 1996) and A_1 receptor antagonists completely prevented (Lange-Asschenfeldt et al., 2004; Perez-Pinzon et al., 1996; Pugliese, Latini, Corradetti, & Pedata, 2003) the protective effect of ischemic preconditioning in rat hippocampal slices and in mouse caudate brain slices (Chauhan, Young, Gibson, & Davidson, 2013). In accordance with *in vitro* data, the selective A_1 receptor antagonist, DPCPX, attenuated the neuroprotective effect of ischemic preconditioning in models of global (Heurteaux, Lauritzen, Widmann, & Lazdunski, 1995) and focal (Cui et al., 2013; Hu et al., 2012; Nakamura, Nakakimura, Matsumoto, & Sakabe, 2002; Yoshida, Nakakimura, Cui, Matsumoto, & Sakabe, 2004) cerebral ischemia.

Unlike in adult rodents, pharmacological A_1 adenosine receptor stimulation in immature brain exacerbates rather than attenuates ischemic injury (Turner, Yan, Schwartz, Othman, & Rivkees, 2002). It should be remembered that in the formation of the central nervous system, A_1 receptor activation potently inhibits the development of axons and can lead to leukomalacia (Rivkees, Zhao, Porter, & Turner, 2001). Leukomalacia is the most common ischemic brain injury in premature infants, and it is characterized by the death or damage and softening of the white matter near the lateral ventricles. Consistent with this result, genetic deletion of A_1 adenosine receptors was neuroprotective in immature mice under hypoxic conditions in which adenosine extracellular levels were increased (Turner et al., 2003).

Although data demonstrate a neuroprotective effect of adenosine through A_1 receptors during ischemia, the use of selective A_1 receptor agonists is hampered by undesirable peripheral effects such as sedation, bradycardia, and hypotension (Fredholm, Chen, Masino, & Vaugeois, 2005; White, Rose'Meyer, & Hope, 1996).

3. A_{2A} RECEPTORS IN BRAIN ISCHEMIA

Adenosine A_{2A} receptor is expressed both at central level on neurons and glial cells (astrocytes, microglia, and oligodendrocytes) and at peripheral level on inflammatory blood cells (lymphocytes and granulocytes) (Varani, Gessi, Dalpiaz, & Borea, 1996, Varani, Gessi, Dalpiaz, Ongini, & Borea, 1997, Varani, Gessi, Dionisotti, Ongini, & Borea, 1998). In recent years, it emerged that both central and peripheral A_{2A} receptors play important roles at different times of brain ischemia.

3.1. A_{2A} receptor antagonism is protective against ischemic brain injury

Gao and Phillis (1994) demonstrated for the first time that the nonselective A_{2A} receptor antagonist, 9-chloro-2-(2-furanyl)-[1,2,4] triazolo[1,5-*c*] quinazolin-5-amine (CGS15943), administered before global forebrain ischemia in the gerbil reduced cerebral ischemic injury. Thereafter, many reports have confirmed the neuroprotective role of A_{2A} receptor antagonists in different models of ischemia. The selective A_{2A} receptor antagonist, 8-(3-chlorostyryl)caffeine (CSC), administered preischemia as well as the less selective antagonists, CGS15943 and 4-amino [1,2,4] triazolo [4,3-*a*] quinoxalines (CP66713), protected against hippocampal cell injury during global forebrain ischemia in gerbils (Phillis, 1995; von Lubitz, Lin, & Jacobson, 1995). The selective A_{2A} receptor antagonist, 4-(2-[7-amino -2-(2-furyl)[1,2,4]triazolo[2,3-*a*][1,3,5]triazin-5-yl-amino]ethyl)phenol (ZM241385), administered preischemia reduced hippocampal injury and improved performance in the Morris water maze in hyperglycemic four-vessel occluded rats (Higashi, Meno, Marwaha, & Winn, 2002). In all the mentioned studies, adenosine A_{2A} receptor antagonists were administered preischemia. Relevant to a possible clinical use of drugs in stroke, in subsequent studies A_{2A} receptor antagonists were administered postischemia. The selective A_{2A} receptor antagonist, 7-(2-phenylethyl)-5-amino-2-(2-furyl)-pyrazolo-[4,3-*e*]-1,2,4,triazolo[1,5-*c*]pyrimidine (SCH58261), acutely administered after hypoxia reduced ischemic brain damage in a rat neonatal model of hypoxia/ischemia (Bona, Aden, Gilland, Fredholm, & Hagberg, 1997). SCH58261, acutely administered soon after ischemia, resulted protective from brain damage evaluated 24 h after focal ischemia in adult rats (Melani et al., 2003; Monopoli, Lozza, Forlani, Mattavelli, & Ongini, 1998). The same antagonist, administered subchronically, was protective against both brain damage and neurological deficit (Melani et al., 2009, 2006; Pedata, Gianfriddo, Turchi, & Melani, 2005) and disorganization of myelin (Melani et al., 2009) 24 h after focal cerebral ischemia in the adult rat. Most recently, in the model of global ischemia (i.e., 7 min Asphyxic Cardiac Arrest) in newborn piglets, posttreatment with SCH58261, infused soon after resuscitation and for 6 hours, improved neurologic recovery and protected striatopallidal neurons after 4 days of recovery (Yang et al., 2013). SCH58261 behaved as a significant protective agent at a dose (0.01 mg/kg) that does not have cardiovascular effects. This low dose did not affect the spontaneous motor behavior in naive animals but decreased contralateral turning behavior that was evident in the first day after monolateral focal ischemia induced by the intraluminal filament technique (Melani et al., 2006, 2003).

A noxious role of A_{2A} receptors during ischemia is supported by the observation that A_{2A} receptor KO mice show a significant decrease of infarct volume after focal cerebral ischemia when compared with their wild-type littermates (Chen et al., 1999; Gui et al., 2009).

Most recently, a question arose: if A_{2A} receptor continuous blockade over an extended time window after ischemia is protective. A recent paper found that CSC continuously administered over 72 h, using subcutaneously implanted osmotic minipumps after permanent MCAo in spontaneosly hypertensive rats, did not decrease brain infarct volume determined by magnetic resonance imaging 1 and 3 days after induction of ischemia (Fronz et al., 2014). In agreement, Melani and coworkers (2012b) observed a lack of protection of the selective A_{2A} antagonist, SCH58261, administered subchronically (three times in the first day) or chronically (twice/day for 7 days) after ischemia, on infarct volume and neurological deficit evaluated 7 days after tMCAo.

3.1.1 A_{2A} receptor antagonism protect from excitotoxicity

Adenosine, by A_{2A} receptor stimulation, promotes glutamate release under normoxic and ischemic conditions *in vivo* and *in vitro* (Corsi, Melani, Bianchi, & Pedata, 2000; Corsi, Melani, Bianchi, Pepeu, & Pedata, 1999; O'Regan, Simpson, Perkins, & Phillis, 1992; Popoli, Betto, Reggio, & Ricciarello, 1995; Simpson, O'Regan, Perkins, & Phillis, 1992). A_{2A} receptors are expressed on neurons at high levels in the striatum (Fink et al., 1992) and at lower levels in all brain regions as detected by real-time PCR (Dixon, Gubitz, Sirinathsinghji, Richardson, & Freeman, 1996). A_{2A} receptors in the striatum are mostly present on GABA–enkephalin neurons (Schiffmann, Jacobs, & Vanderhaeghen, 1991) but are also located presynaptically on glutamatergic terminals (Hettinger et al., 2001; Rosin, Hettinger, Lee, & Linden, 2003), where they can directly regulate glutamate outflow (Marcoli et al., 2004, 2003). Under *in vitro* OGD, A_{2A} receptors play an important modulation of synaptic transmission (Lopes, Sebastião, & Ribeiro, 2011; Sebastião & Ribeiro, 1996) as mostly demonstrated in the hippocampus (Cunha et al., 1994; Lopes, Cunha, & Ribeiro, 1999; O'Kane & Stone, 1998). In the CA1 area of the rat hippocampus, which is a most sensitive region to ischemia, the selective A_{2A} agonist, CGS21680, clearly reduces the depression of synaptic activity brought about by OGD (Latini, Bordoni, Corradetti, Pepeu, & Pedata, 1999). Following A_{2A} receptor stimulation, the increase of extracellular glutamate concentration counteracts depression brought about by adenosine A_1 receptors. In agreement, the selective A_{2A} receptor antagonists, ZM241385 and SCH58261, delay the appearance of AD, a phenomenon strictly related to cell damage and death (Somjen, 2001), protect from the synaptic activity depression brought about by a severe (7 min) OGD period, and protect CA1 neuron and astrocyte from injury (Pugliese et al., 2009). Moreover, ZM241385 also delays the appearance of AD, protects from synaptic failure, and reverses the decreased proliferation of neuroblast induced by a severe (9 min) OGD period in the gyrus dentatus of the hippocampus (Maraula et al., 2013). The time window of the protective effects of the A_{2A} receptor antagonists in the hippocampus overlaps with the delay obtained by treating the slices with glutamate receptor antagonists (Tanaka, Yamamoto, Kudo, Mihara, & Higashi, 1997; Yamamoto et al., 1997), indicating that their effects are attributable to reduced glutamate excitotoxicity.

Several mechanisms contribute to the regulation of extracellular glutamate concentration-mediated by A_{2A} receptor. A_{2A} receptor stimulation might regulate extracellular glutamate not only by reducing release from glutamatergic terminals but also by modulating glutamate uptake transporter. A_{2A} receptors located on astrocytes mediate inhibition of glutamate uptake transporter GLT-1 (Nishizaki et al., 2002; Pinto-Duarte, Coelho, Cunha, Ribeiro, & Sebastião, 2005; Pintor et al., 2004). Most recent data show that acute exposure to the selective A_{2A} receptor agonist, CGS21680, reduces glutamate uptake, and prolonged exposure to the same agonist inhibits glutamate-1 transporter (GLT-1) and glutamate-aspartate transporter mRNA and protein levels from astrocytes (Matos et al., 2012). Such inhibition is exerted through modulation of Na^+/K^+-ATPase (Matos, Augusto, Agostinho, Cunha, & Chen, 2013). An imbalance of A_1/A_{2A} receptor expression might also contribute to the inhibition of excitatory synaptic transmission under ischemia. Short periods of global ischemia decrease A_1 adenosine receptor density in the brain likely due to an internalization of A_1 adenosine receptors in nerve terminals (Coelho et al., 2006). Moreover, a heterodimerization of adenosine A_1 and A_{2A} receptors in striatal glutamatergic nerve terminals might allow adenosine to exert a fine-tuning modulation of glutamatergic neurotransmission. A main biochemical characteristic of the A_1–A_{2A} heteromer is the ability of A_{2A} receptor activation to reduce the affinity of the A_1 receptor for agonists with an ultimate switch mechanism by which low and high concentrations of adenosine inhibit and stimulate, respectively, glutamate release (Ciruela et al., 2006). *In vivo*, a definite overexpression of A_{2A} receptors was found in neurons of the striatum and cortex 24 h after focal ischemia (Trincavelli, Melani, et al., 2008).

The low dose of SCH58261 that protects against tissue damage induced by MCAo or quinolinic acid (QA) excitotoxicity also reduces glutamate outflow (Melani et al., 2003; Popoli et al., 2002). The increased outflow of excitatory amino acids soon after ischemia primes an excitotoxic cascade which ultimately results in cell death (Bruce-Keller, 1999).

On the whole, evidence indicates that protective effects of low doses of A_{2A} receptor antagonists early after brain ischemia are largely due to reduced glutamate excitotoxicity. The neuroprotective properties of A_{2A} receptor antagonists largely reside in a central mechanism of control of excitotoxicity, as demonstrated by the observation that the selective A_{2A} receptor antagonist ZM241385 injected peripherally or directly intra hippocampus is protective against excitotoxicity induced by kainate (Jones, Smith, & Stone, 1998a) and by the combinations of QA and IL-1beta (Stone & Behan, 2007).

These results are consistent with the observation that adenosine A_{2A} receptor KO mice are protected from an excess of striatal glutamate outflow and damage induced by transient MCAo (Chen et al., 1999; Gui et al., 2009).

A further protective effect of A_{2A} receptor antagonism may be attributed to the capability of A_{2A} receptor antagonists to increase GABA outflow during ischemia. The major part of excitatory glutamatergic innervation is modulated by inhibitory GABA-releasing interneurons. Potentiation of GABAergic synaptic transmission has neuroprotective effects in several experimental models of cerebral ischemia (Schwartz-Bloom & Sah, 2001). GABA is strongly increased in the cortex and striatum during ischemia (Melani et al., 1999; O'Regan et al., 1992), and evidence shows that selective A_{2A} receptor stimulation decreases ischemia-evoked GABA outflow (O'Regan et al., 1992; Saransaari & Oja, 2005) and enhances

GABA transport into nerve terminals by restraining PKC inhibition of GAT-1 (Cristóvão-Ferreira, Vaz, Ribeiro, & Sebastião, 2009).

3.1.2 A_{2A} receptor antagonism protects from intracellular messenger activation

Several data indicate that regulation of proteins involved in transcriptional or posttranslational mechanisms plays an important role in the neuroprotective effect of A_{2A} receptor antagonists in ischemia.

All members of the mitogen-activated protein kinases (MAPKs) family are activated up to 24 h after ischemia (Irving, Barone, Reith, Hadingham, & Parsons, 2000; Wu, Ye, Che, & Yang, 2000). While p38 and ERK1/2 MAPKs are activated in neurons and in microglia (Irving et al., 2000; Melani et al., 2009; Piao, Kim, Han, & Lee, 2003; Takagi, Nozaki, Sugino, Hattori, & Hashimoto, 2000), JNK MAPK is activated in neurons and mainly in oligodendrocytes (Melani et al., 2009). Twenty-four hours after MCAo, adenosine A_{2A} receptors are overexpressed on microglia (Trincavelli, Melani, et al., 2008). Subchronic administration of the A_{2A} receptor antagonist SCH58261 reduced p38 activation in microglia (Melani et al., 2006), while it did not affect ERK1/2 activation (Melani et al., 2009). It is known that soon after excitotoxic phenomena, microglial cells initiate a rapid change in their phenotype, referred as microglial activation (Bruce-Keller, 1999). The microglial cells start to proliferate and migrate toward the site of damage (Marks et al., 2001; Stoll, Jander, & Schroeter, 1998) and to produce cytotoxic substances and cytokines. These substances contribute to the inflammatory response that follows an ischemic insult and that causes further exacerbation brain damage (Dirnagl et al., 1999). Since phospho-p38 inhibition has direct neuroprotective effects in hippocampal brain slices after OGD (Barone et al., 2001) and reduces neurological deficit and infarct volume 24 h after permanent MCAo (Barone & Feuerstein, 1999; Barone et al., 2001), a control of p38 activation by A_{2A} receptor antagonism (Melani et al., 2006) might account for protection against brain damage induced by ischemia.

Twenty-four hours after MCAo, subchronic administration of the A_{2A} receptor antagonist SCH58261 also reduces phospho-JNK, expressed mainly in mature oligodendrocytes and oligodendrocyte precursor cells (OPCs) (Melani et al., 2009, 2011). Phospho-JNK is a factor involved in oligodendrocyte death (Howe, Bieber, Warrington, Pease, & Rodriguez, 2004; Jurewicz, Matysiak, Andrzejak, & Selmaj, 2006). Interestingly, activation of JNK has been described in oligodendrocytes in multiple sclerosis lesions where oligodendrocytes are major targets of the disease (Bonetti et al., 1999). A specific peptide inhibitor of JNK protects against cell death induced by OGD *in vitro* (Hirt et al., 2004) and by MCAo *in vivo* (Borsello et al., 2003; Hirt et al., 2004). JNK2/3 KO mice are protected from damage following cerebral ischemia (Gelderblom, Eminel, Herdegen, & Waetzig, 2004; Kuan et al., 2003). Therefore, JNK activation in oligodendrocytes represents a noxious event after ischemia that can damage oligodendrocytes leading to myelin damage and disorganization; a control of JNK activation by A_{2A} receptor antagonism might account for protection against brain damage induced by ischemia (Melani et al., 2009). A_{2A} receptor antagonism also reduces Olig2 that is a transcription factor expressed mostly by OPCs, while mature oligodendrocytes are characterized by lower levels of Olig2 (Kitada & Rowitch, 2006). Data have suggested that the A_{2A} receptor antagonist stimulates OPC differentiation to mature cells after ischemia (Melani et al., 2009). In agreement, we have recently observed in primary OPC cultures that selective stimulation of A_{2A} receptors by CGS21680 inhibits maturation of OPC in the firsts 10 days of differentiation *in vitro* (Coppi, Cellai, Maraula, Pugliese, & Pedata, 2013). The drug also inhibits K^+ "delayed rectifier" channels (K_{DR}) (Coppi et al., 2013) that are known to inhibit proliferation and differentiation of OPCs to mature oligodendrocytes, thus preventing myelin deposition (Attali et al., 1997; Shrager & Novakovic, 1995).

Besides a direct effect of the A_{2A} receptor antagonists on A_{2A} receptors located on oligodendrocytes or microglia, we must consider that the effect of SCH58261 on the reduction of MAPK activation is secondary to overall reduction of the excitotoxic cascade, that in turn primes MAPK activation in the initial hours after ischemia (Melani et al., 2003). Oligodendroglia are extremely sensitive to glutamate receptor overactivation and ensuing oxidative stress (Matute, Alberdi, Ibarretxe, & Sanchez-Gomez, 2002; Matute, Sanchez-Gomez, Martinez-Millan, & Miledi, 1997; McDonald, Althomsons, Hyrc, Choi, & Goldberg, 1998).

It was also reported that the adenosine A_{2A} receptor antagonist, SCH58261, administered 10 min after permanent MCAo, decreases *c-Fos* gene expression in the ischemic hemisphere within 4 h after ischemia (Petroni, Papini, Blasevich, & Galli, 2002). After permanent and temporary focal cerebral ischemia, a rapid (in terms of minutes or hours) increase in different early genes (Lu et al., 2003, 2004) and *c-Fos* expression was demonstrated in glial cells (Yu, Lee, Fu, & Eng, 1995). Products of the Fos family are players in inducing inflammatory gene expression in glial cells (Lewis & Manning, 1999).

3.2. A_{2A} receptor agonism is protective against ischemic brain injury

While many data support that A_{2A} receptor antagonists protect against central excitotoxicity, the protective effect of A_{2A} receptor agonists appears attributable to different mechanisms.

In the traumatic brain injury (TBI) model, it was demonstrated that the A_{2A} receptor antagonist, ZM241385, administered repeatedly (1 mg/kg i.p.) in the 12 h after trauma, was protective 15 min after trauma when extracellular glutamate concentration rose; conversely, the A_{2A} receptor agonist, CGS21680, administered repeatedly (0.1 mg/kg i.p.) in the 12 h after trauma, was protective 3 h after trauma when glutamate concentrations were down (Dai et al., 2010).

A protective role of adenosine A_{2A} receptor was demonstrated in newborn rodents. A_{2A} receptor KO neonatal mice show aggravated hypoxic/ischemic injury in comparison to wild-type mice (Adén et al., 2003), and in immature brain forebrain slices, it was demonstrated that cannabinoids induce robust neuroprotection through both CB(2) and A_{2A} adenosine receptors (Castillo, Tolón, Fernández-Ruiz, Romero, & Martinez-Orgado, 2010). In adult gerbil, a protective effect of adenosine A_{2A} receptor agonists was reported by von Lubitz et al. (1995) who demonstrated that the A_{2A} receptor agonist, APEC, administered systemically before a global ischemia, ameliorated recovery of cerebral blood flow and animal and neuron survival. Moreover Sheardown and Knutsen (1996) demonstrated that a high dose of the selective A_{2A} receptor agonist, CGS21680 (10 mg/kg i.p.), administered 5 min after a global ischemia in gerbil, exhibited a high significant protection against neuronal loss, but was inactive at 3 mg/kg. In these two works in adult gerbils, adenosine agonists were administered before ischemia or at a high dose. A main problem of A_{2A} receptor agonists in considering translation to clinic is their cardiovascular effect: stimulation of adenosine A_{2A} receptors located on blood vessels causes a vasodilatory effect, inducing hypotension and increases heart rate (Schindler et al., 2004, 2005). Schindler et al. (2004, 2005) reported that the decrease of blood pressure induced by CGS21680 (0.5 mg/kg i.p.) in conscious rats is most probably mediated in the periphery, while the increase of heart rate is mediated at the central level. The drug, at dose of 0.1 mg/kg i.p., induced a small increase of heart rate only in the first hour after administration, while at the lower dose of 0.01 mg/kg, it affected neither blood pressure nor heart rate (Melani, Corti, Cellai, Vannucchi, & Pedata, 2014).

Relevantly, recent experiments have demonstrated that the A_{2A} receptor agonist CGS21680, chronically administered (twice/day for 7 days) after transient MCAo, at doses of 0.01 and 0.1 mg/kg, protected from neurological deficit, weight loss, infarct volume, myelin disorganization, and glial activation at 7 days after ischemia (Melani et al., 2014). Moreover, 2 days after transient MCAo, CGS21680 has massively reduced blood cell infiltration into ischemic areas (Melani et al., 2014). These results are in agreement with previous observations that A_{2A} receptor agonists, systemically administered after spinal cord injury (SCI) in mice, protected from neurological and tissue damage, reduced inflammation parameters, and blood cell infiltration (Cassada et al., 2002; Genovese et al., 2009, 2010).

Numerous literature data indicate that the protective effects of A_{2A} agonists are attributable to stimulation of A_{2A} receptors on blood cells (peripheral effects). Jones and coworkers (1998a) showed that peripheral administration of the A_{2A} receptor agonist, CGS21680, protects the hippocampus against kainate-induced excitotoxicity while the direct injection of CGS21680 into the hippocampus failed to afford protection (Jones, Smith, & Stone, 1998b). Similar results were obtained after spinal cord trauma where the A_{2A} receptor agonist CGS21680 protected from damage when injected systemically but not when centrally injected into the injured spinal cord (Paterniti et al., 2011). Li et al. (2006) demonstrated that the protective effect of A_{2A} receptor agonists, systemically administered after spinal trauma, is lost in mice lacking A_{2A} receptors on bone marrow-derived cells (BMDCs), but is restored in A_{2A} receptor KO mice reconstituted with A_{2A} receptors on BMDCs. Results suggested that A_{2A} receptors present on BMDCs are targets responsible for the A_{2A} receptor agonist protective effect from SCI.

Many studies have reported that the selective activation of A_{2A} receptors directly on blood cells, including platelets, monocytes, some mast cells, neutrophils, and T cells inhibits proinflammatory responses (Haskó et al., 2000; Lappas, Day, Marshall, Engelhard, & Linden, 2006; Sitkovsky, 2003), reduces production of adhesion cell factors, reduces platelet aggregation and neutrophil activation, thereby exerting antithrombotic, antioxidant, and anti-inflammatory effects (Sitkovsky et al., 2004). Consistent with its anti-inflammatory and immunosuppressive role, the protective effect of adenosine A_{2A} receptor stimulation has been observed in various models of autoimmune disease, such as rheumatoid arthritis (Odashima et al., 2005), colitis (Choukèr et al., 2008; Genovese et al., 2010), and hepatitis (Howe et al., 2004).

On the whole, data suggest that the protection of A_{2A} receptor agonists, systemically administered after brain ischemia, is exerted at the peripheral level on inflammatory blood cells, resulting ultimately in reduced leukocyte infiltration and reduced inflammatory cascade at the central level during a later phase of ischemia. However, it cannot be excluded that adenosine A_{2A} receptor agonists display protection against brain ischemia also by direct effects on central A_{2A} receptors. Activation of central A_{2A} receptors is known to increase expression and release of neurotrophic factors (Sebastião & Ribeiro, 2009) as NGF in microglia (Heese, Fiebich, Bauer, & Otten, 1997) and BDNF in mice hippocampus (Tebano et al., 2008) and in rat cortical neurons (Jeon et al., 2011). The increase in neurotrophic factor expression by adenosine A_{2A} receptor stimulation may contribute to restore neurological functions and cerebral damage after brain ischemia (Ke, Xiong, & Liu, 2012). Moreover, we must remember that adenosine is also implicated in cerebral blood flow regulation as a vasodilator agent acting on A_{2A} receptors on endothelial cells of brain vessels (Phillis, 2004).

By controlling neuroinflammation, A_{2A} receptors might have a potential for influencing neurogenesis (Ekdahl, Kokaia, & Lindvall, 2009; Gomes, Kaster, Tomé, Agostinho, & Cunha, 2011) and reparative neurogenetic processes after ischemia. Neuroinflammation in fact results in the inhibition of adult neurogenesis (Ekdahl, Claasen, Bonde, Kokaia, & Lindvall, 2003). Adenosine A_{2A} receptor gene is expressed in different areas of the developing rat brain (Weaver, 1993), and the capability of adenosine A_{2A} receptors to control extracellular glutamate concentration and BDNF (Diógenes, Assaife-Lopes, Pinto-Duarte, Ribeiro, & Sebastião, 2007; Diógenes, Fernandes, Sebastião, & Ribeiro, 2004) might be relevant to control on synaptic plasticity and neurogenetic processes after ischemia.

3.3. A_{2A}-receptor-based protection in ischemia?

In conclusion, from literature, it is emerging as a duplex role of A_{2A} receptors in cerebral ischemia. Evidence up to now reported indicates that antagonism and/or stimulation of A_{2A} receptors may represent a protective strategy in relation to the time of different phases of ischemia.

After ischemia, extracellular glutamate concentrations remain elevated at least up to 4 h after permanent MCAo (Melani et al., 2003, 1999) and up to 12 h after brain trauma (Dai et al., 2010). The massive increase of glutamate excitotoxicity triggers acute tissue injury and the start of an inflammatory cascade that is stressed by blood cell infiltration. While central A_{2A} receptors in the first hours after ischemia are critical in increasing glutamate extracellular concentrations, A_{2A} receptors on blood cells are critical hours and days after ischemia in decreasing intercellular adhesion molecule-1 (ICAM-1) (Johnston, Gudjonsson, Sigmundsdottir, Ludviksson, & Valdimarsson, 2005) and blood cell infiltration in brain parenchyma. Altogether, evidence suggests that A_{2A} receptor antagonists provide protection via centrally mediated control of a precocious excitotoxicity (Melani et al., 2003), while A_{2A} receptor agonists provide protection peripherally by controlling immune blood cell infiltration and delayed neuroinflammation (Melani et al., 2014). In agreement, the lack of a protection by A_{2A} receptor antagonist at later time after stroke (Fronz et al., 2014; our unpublished observation) might be attributable to the fact that protection is overwhelmed by subsequent damage brought about by cell infiltration that starts 6 h after ischemia and picks at 2 days thereafter (Gelderblom et al., 2009; Melani et al., 2014; Zhang, Chopp, Chen, & Garcia, 1994).

These observations highlight that a therapeutic strategy with adenosine A_{2A} receptor antagonists/agonists should be carefully evaluated in terms of time from ischemia induction. When considering the use of A_{2A} receptor active drugs to protect against brain ischemia, attention should be given to: (1) the administration time after injury and (2) the drug dose. In fact, a problem is that A_{2A} receptors located on endothelial cells mediate important effect on systemic blood pressure and heart frequency. However, both A_{2A} receptor antagonists (Melani et al., 2009, 2003; Monopoli et al., 1998) and agonists (Melani et al., 2014) were found to be protective in ischemia models at doses that did not modify blood pressure nor heart rate.

The design and development of new adenosine A_{2A} receptor ligands is an area of intense research activity (Cristalli, Müller, & Volpini, 2009; Jacobson, 2013).

4. A_{2B} RECEPTORS IN BRAIN ISCHEMIA

Recent progress in understanding the role of A_{2B} receptors in the regulation of inflammation, immunity, and tissue repair was considerably facilitated by the introduction of new pharmacological and genetic tools (Feoktistov & Biaggioni, 2011). A_{2B} receptors are ubiquitously expressed, and in most cases, A_{2B} receptors are coexpressed with A_{2A} receptors. A_{2B} receptor transcripts are found in neutrophils (Fredholm, Zhang, & van der Ploeg, 1996), lymphocytes (Gessi et al., 2005), and platelets (Amisten, Braun, Bengtsson, & Erlinge, 2008). Moreover, A_{2B} receptors are expressed on the surface of endothelial cells (Feoktistov et al., 2002) and regulate every aspect of endothelial inflammatory processes.

Low levels of A_{2B} receptor expression were found in the rat brain (Dixon et al., 1996; Fredholm et al., 2000; Puffinbarger et al., 1995). The A_{2B} receptor has a low affinity for adenosine; however, it might be activated during conditions of hypoxia or ischemia when the extracellular adenosine levels rise (Koeppen, Eckle, & Eltzschig, 2011). Few studies have investigated the role of A_{2B} receptors in brain ischemia *in vivo*. In the stratum radiatum of CA1 hippocampal slices, the immunoreactive cells for A_{2B} receptor showed distinct morphological characteristics of astrocytes; the number and density of A_{2B} receptors increased after ischemic preconditioning (Zhou et al., 2004). In a recent paper, it was reported that a selective A_{2B} receptor antagonist, MRS-1754, reduced the ceramide production in astrocytes and attenuated inflammatory responses and neuronal damage after global cerebral ischemia (Gu et al., 2013). This effect was related to an early reduction of p38 MAPK activation. In fact, A_{2B} receptor plays a key role in the rapid activation of p38 and in the subsequent inflammatory process (Koscsó et al., 2012; We et al., 2013). *In vitro* experiments have indicated the role of A_{2B} receptors in the inflammatory process. In human astroglial cells, a selective A_{2B} antagonist, *N*-(4-acetylphenyl)-2-[4-(2,3,6,7-tetrahydro-2,6-dioxo-1,3-dipropyl-1*H*-purin-8-yl) henoxy]acetamide (MRS1706), completely prevents elongation of astrocytic processes (a morphological hallmark of *in vivo* reactive astrogliosis) induced by selective stimulation of A_{2B} receptors (Trincavelli et al., 2004). A short-term TNF-α treatment induces A_{2B} adenosine receptor desensitization in human astroglial cells (Trincavelli, Tonazzini, Montali, Abbracchio, & Martini, 2008). These results suggest that in the acute phase of brain ischemia that is characterized by both cytokine and adenosine high release, A_{2B} receptor desensitization on astroglia might represents a cell defense mechanism (Trincavelli, Tonazzini, et al., 2008). On the whole, in cerebral ischemia, A_{2B} receptor antagonism might be protective from ischemic brain damage.

Besides a possible role of central A_{2B} receptor in ischemia, a possible important role is exerted by A_{2B} receptors present on endothelial and blood immune cells. Hypoxia is an important stimulus for the upregulation of A_{2B} receptor expression via the oxygen-regulated transcription factor, HIF-1α (hypoxia-inducible factor), in endothelial cells (Eltzschig et al., 2003). Vascular permeability was significantly increased in vascular organs of A_{2B} receptor KO mice subjected to ambient hypoxia. By contrast, hypoxia-induced vascular leak was not accentuated in A_1, A_{2A}, or A_3 receptor KO mice, suggesting a degree of specificity for the A_{2B} receptor for endothelial cells (Eckle et al., 2007). In agreement, A_{2B} receptor KO mice showed increased basal levels of TNF-α and expression of adhesion molecules such as ICAM-1, P-selectin, and E-selectin in lymphoid cells, resulting in increased leukocyte rolling and adhesion (Yang et al., 2006). A_{2B} receptor KO mice exposed to hypoxia exhibit increased neutrophil infiltration into hypoxic tissues, revealing an inhibitory role for A_{2B} receptors in neutrophil transmigration *in vivo* (Eckle et al., 2008, 2007). Attenuation of hypoxia-associated increases in tissue neutrophil numbers appeared to depend largely on hematopoietic cell A_{2B} receptor signaling (Eckle et al., 2007; Yang et al., 2006). Pharmacologic studies in fact indicate that A_{2B} receptors on neutrophils contribute to their decreased adhesion to endothelial cells and transmigration in tissue parenchyma (Eckle et al., 2007; Eltzschig et al., 2004).

A_{2B} receptors have also a key role in promoting angiogenic response because activation of A_{2B} receptors by adenosine increases endothelial cell proliferation, chemotaxis, and capillary tube formation (Adair, 2005; Grant et al., 2001). Exposure of human umbilical vein endothelial cells to hypoxia increases their expression of A_{2B} receptor, which upon stimulation promotes the release of vascular endothelial growth factor (Feoktistov et al., 2004).

Altogether, these studies indicate that A_{2B} receptor is implicated in dampening hypoxia-induced inflammation via hematopoietic cells and vascular adhesion signals (Koeppen et al., 2011).

5. A_3 RECEPTORS IN BRAIN ISCHEMIA

Despite the low expression of adenosine A_3 receptors in the CNS, several studies have attempted to characterize their effects in the pathophysiology of cerebral ischemia. These studies are rather contradictory and have been a matter of intense debate (Abbracchio & Cattabeni, 1999; Baraldi et al., 2000; Borea, Gessi, Bar-Yehuda, & Fishman, 2009; Jacobson, 1998; Jacobson, Hoffmann, Cattabeni, & Abbracchio, 1999; Pedata, Pugliese, Sebastiao, & Ribeiro, 2010; von Lubitz, 1999; von Lubitz, Lin, Boyd, Bischofberger, & Jacobson, 1999; von Lubitz, Ye, McClellan, & Lin, 1999). The neuroprotective function of A_3 adenosine receptors was first demonstrated using mice with genetic deletion of the A_3 receptors. Mice lacking A_3 receptors showed increased neurodegeneration in response to repeated episodes of moderate hypoxia (induced by carbon monoxide) (Fedorova, Jacobson, Basile, & Jacobson, 2003) or an increase in cerebral infarction after transient ligation of the middle cerebral artery (Chen et al., 2006). Additionally, intracerebroventricular pretreatment with Cl-IB-MECA, a selective A_3 receptor agonist, decreased the size of infarction in the wild-type controls, but not in the A_3 adenosine receptor KO animals, suggesting that Cl-IB-MECA-induced protection was mediated through the A_3 receptors (Chen et al., 2006). These data suggested that A_3 adenosine receptor agonists might afford protection against hypoxia-induced neuronal damage. Accordingly, Hentschel, Lewerenz, and Nieber (2003) demonstrated that under hypoxic conditions, selective activation of A_3 adenosine receptors by a brief (5 min) application of IB-MECA brings about an inhibition of excitatory neurotransmission on cortical neurons, indicating that A_3 receptors may sustain the neuroprotective action of adenosine induced by A_1 receptors. The mechanism by which A_3 adenosine receptor agonist induced neuroprotective effects in hypoxia/ischemia may involve not only the inhibition of excitotoxicity but also nonneuronal targets. It has been in fact demonstrated that glial cells engineered to lack the A_3 receptors were more sensitive to hypoxia than wild-type cells (Bjorklund et al., 2008). In cultured murine astrocytes, it was demonstrated that adenosine, by selective involvement of A_1 and A_3 receptors, decreases HIF-1 expression induced by LPS under both normoxic and hypoxic conditions (Gessi et al., 2013), leading to inhibition of genes involved in inflammation injury (like *iNOS* and A_{2B} *receptor* genes) (Gessi et al., 2013).

Contrary to the above information, an early *in vivo* study in the model of global forebrain ischemia in the gerbil showed that a selective agonist of A_3 adenosine receptor, IB-MECA, acutely administered 15 min prior to global ischemia, impaired postischemic blood flow, increased mortality, and exacerbated the loss of hippocampal neurons (von Lubitz, Lin, Popik, Carter, & Jacobson, 1994). IB-MECA administration 20 min prior to transient cerebral ischemia also resulted in a significant increase in infarct size (von Lubitz, Simpson, & Lin, 2001). In agreement with a possible noxious role of A_3 adenosine receptor in ischemia, it has been shown that several different selective antagonists of A_3 adenosine receptors consistently abolishes or delays the occurrence of AD and significantly protects from the irreversible disruption of excitatory neurotransmission caused by a severe ischemic episode in the CA1 region of rat hippocampal slices (Colotta et al., 2007, 2008, 2009; Pugliese, Coppi, Spalluto, Corradetti, & Pedata, 2006; Pugliese et al., 2007). Moreover, in an ischemic preconditioning model *in vitro*, the selective A_3 adenosine receptor antagonist, 5-propyl-2-ethyl-4-propyl-3-(ethylsulfanylcarbonyl)-6-phenylpyridine-5-carboxylate

(MRS1523), facilitated the full recovery of CA1 hippocampal neurotransmission after a severe, irreversible OGD period (Pugliese et al., 2003). Recently, it was demonstrated that A_3 receptor blockade attenuates ischemic injury and functional loss of the isolated rat optic nerve (González-Fernández, Sánchez-Gómez, Pérez-Samartín, Arellano, & Matute, 2014).

Conflicting results on the deleterious or beneficial role of A_3 adenosine receptors on synaptic activity under hypoxia/ischemia may find an explanation by data reported by Pugliese et al. (2007). Although the selective A_3 adenosine receptor antagonists, MRS1523 and LJ1251, applied before and during a severe (7 min) OGD stimulates a recovery of neurotransmission, application of the antagonist before a brief (2 min) OGD reduces the OGD-induced depression of field excitatory post-synaptic potential (fEPSP) in the CA1 hippocampal area. The latter finding indicates an inhibitory role of A_3 receptors on synaptic transmission during the brief OGD. These results suggest that A_3 receptors may have a synergistic role with A_1 receptors in decreasing synaptic transmission and exert a protective effect in ischemia. A similar protective role of A_3 receptors on neighboring CA3 pyramidal neurons emerged from a study of Dennis et al. (2011). These authors demonstrated that OGD induced a reduction in synaptic AMPA receptors on hippocampal CA3 neurons and that this effect is mediated by mGluR1 and adenosine A_3 receptors that are also able to promote AMPA internalization promoting neuroprotection.

Prolonged ischemic conditions could play a pivotal role in switching the effects of A_3 adenosine receptor stimulation from A_1-like inhibition to potentiation of an excitotoxic glutamate effect. We may speculate that A_3 adenosine receptor-mediated effects would become particularly deleterious during ischemia, when high extracellular levels of adenosine are reached and detrimental effects of A_3 adenosine receptor activation may be due, at least in part, to its sustained stimulation.

Therefore, it appears that the activation of A_3 adenosine receptors can be protective or deleterious after ischemic/hypoxic brain insults, depending on the cell type involved as well as on the duration of activation. It might be considered that this dual effect may be also attributable to desensitization of A_3 receptors. In fact, both human and rat A_3 receptors are desensitized within a few minutes after agonist exposure (Baines, Corrêa, Irving, & Frenguelli, 2011; Palmer, Benovic, & Stiles, 1995; Trincavelli et al., 2000, 2002), and A_3 adenosine receptors are very sensitive to prolonged stress *in vitro* (von Arnim, Timmler, Ludolph, & Riepe, 2000, von Arnim, Verstege, Etrich, & Riepe, 2006). In an *in vitro* OGD model in hippocampal slices, it was found that a long application (before and during

OGD) of 5′-*N*-methylcarboxamidoadenosine derivatives, Cl-IB-MECA, and of news selective A_3 agonists (Volpini et al., 2002, 2007) have effects similar to those of antagonists on the OGD-induced depression of synaptic transmission and on the appearance of AD after the severe OGD period (Pugliese et al., 2007). These effects may be attributed to desensitization of A_3 adenosine receptors. A stimulation as massive as that reached in the presence of endogenous adenosine plus exogenous A_3 adenosine receptor agonists might induce substantial A_3 adenosine receptor plastic adjustments such as desensitization. We might also consider that A_3 receptor desensitization might account for observations that chronic administration (10-day preischemic) of IB-MECA improves postischemic cerebral flow circulation, survival, and neuron preservation in the global forebrain ischemia model in the gerbil (von Lubitz, Lin, et al., 1999; von Lubitz, Lin, Popik, et al., 1994), and that repeated intravenous administration of Cl-IB MECA before MCAo increases locomotor activity and decreases cerebral infarction 2 days thereafter (Chen et al., 2006).

Besides being localized on central cells, adenosine A_3 receptors are also localized on blood cells (Gessi et al., 2004). The state of the art of the role of adenosine A_3 receptors in inflammatory responses appears conflicting also because exposure of blood peripheral cell lines to selective A_3 adenosine receptor agonists results in both anti- and proinflammatory effects (see Borea et al., 2009; Fishman & Bar-Yehuda, 2003; Gessi et al., 2008; Luo et al., 2010). Recently, Choi and colleagues (2011) have demonstrated that postischemic treatment with LJ529, a selective A_3 receptor agonist, markedly reduced cerebral ischemic injury caused by 1.5-h MCAo, followed by 24-h reperfusion in rats. LJ529 prevented the infiltration/migration of microglia and monocytes occurring after MCAo and reperfusion, and also after injection of lipopolysaccharides into the corpus callosum (Choi et al., 2011). Butler et al. (2012) proposed that A_3 receptor agonists can mediate their protective effects via anti-inflammatory signaling (inhibition of proinflammatory cytokines) and/or concomitant inhibition of innate immune cell trafficking because of A_3 receptor desensitization.

Although much is still to be learned about the function of the A_3 receptor, at present it appears that selective A_3 receptor agonists might be helpful in the treatment of ischemic conditions. A_3 receptor agonists, under clinical evaluation for the treatment of inflammatory diseases and cancer (Fredholm et al., 2011; Haskó, Linden, Cronstein, & Pacher, 2008), demonstrated excellent safety and efficacy (Fishman, Bar-Yehuda, Liang, & Jacobson, 2012).

5.1. A_3 receptor-based protection in ischemia?

Together, these data suggest that the outcome of A_3 receptor stimulation on brain function during hypoxic/ischemic phenomena depends on the intensity and duration of stimulation and on cell type localization of these receptors.

Results raise the question of the time-related utility of A_3 receptor antagonists/agonists for treatment of ischemia. It may be speculated that after ischemia, prolonged treatment with A_3 receptor agonists first protects by reducing glutamate-mediated excitotoxicity supporting a depression of neuronal activity and saving of cellular energy. Later along ischemia, the agonist-induced desensitization of central A_3 adenosine receptor would account for the loss of any A_3 adenosine receptor influence on excitotoxicity. Moreover, treatment with A_3 receptor agonists might protect from neuroinflammation by preventing the infiltration/migration of microglia and monocytes into brain ischemic areas.

6. CONCLUSIONS

Ischemic brain injury results from a complex sequence of pathophysiological events that evolve over time and space. The major pathogenic mechanisms of this cascade include acute mechanisms of excitotoxicity and periinfarct depolarizations and delayed mechanisms of neuroinflammation and apoptosis. Days after ischemia, peripheral immune cells play an important role in the responses of brain tissue to ischemic injury because of a massive blood cell infiltration into the ischemic area caused by BBB disruption. All adenosine receptor subtypes, present both on central and on blood cells, partake in either salvage or demise of the tissue during various phases of ischemia. Understanding the time-related roles of the different adenosine receptor subtypes is crucial to find an effective purinergic treatment after stroke.

REFERENCES

Abbracchio, M. P., & Cattabeni, F. (1999). Brain adenosine receptors as targets for therapeutic intervention in neurodegenerative diseases. *Annals of the New York Academy of Sciences*, *890*, 79–92.

Adair, T. H. (2005). Growth regulation of the vascular system: An emerging role for adenosine. *American Journal of Physiology. Regulatory, Integrative and Comparative Physiology*, *289*, R283–R296.

Adén, U., Halldner, L., Lagercrantz, H., Dalmau, I., Ledent, C., & Fredholm, B. B. (2003). Aggravated brain damage after hypoxic ischemia in immature adenosine A_{2A} knockout mice. *Stroke*, *34*, 739–744.

Amisten, S., Braun, O. O., Bengtsson, A., & Erlinge, D. (2008). Gene expression profiling for the identification of G-protein coupled receptors in human platelets. *Thrombosis Research*, *122*, 47–57.

Andiné, P. (1993). Involvement of adenosine in ischemic and postischemic calcium regulation. *Molecular and Chemistry Neuropathology*, *18*, 35–49.

Arrigoni, E., Crocker, A. J., Saper, C. B., Greene, R. W., & Scammell, T. E. (2005). Deletion of presynaptic adenosine A_1 receptors impairs the recovery of synaptic transmission after hypoxia. *Neuroscience*, *132*, 575–580.

Attali, B., Wang, N., Kolot, A., Sobko, A., Cherepanov, V., & Soliven, B. (1997). Characterization of delayed rectifier Kv channels in oligodendrocytes and progenitor cells. *Journal of Neuroscience*, *17*, 8234–8245.

Baines, A. E., Corrêa, S. A., Irving, A. J., & Frenguelli, B. G. (2011). Differential trafficking of adenosine receptors in hippocampal neurons monitored using GFP- and super-ecliptic pHluorin-tagged receptors. *Neuropharmacology*, *61*, 1–11.

Baraldi, P. G., Cacciari, B., Romagnoli, R., Merighi, S., Varani, K., Borea, P. A., et al. (2000). A(3) adenosine receptor ligands: History and perspectives. *Medicinal Research Reviews*, *20*, 103–128.

Barone, F. C., & Feuerstein, G. Z. (1999). Inflammatory mediators and stroke: New opportunities for novel therapeutics. *Journal of Cerebral Blood Flow and Metabolism*, *19*, 819–834.

Barone, F. C., Irving, E. A., Ray, A. M., Lee, J. C., Kassis, S., Kumar, S., et al. (2001). Inhibition of p38 mitogen-activated protein kinase provides neuroprotection in cerebral focal ischemia. *Medicinal Research Reviews*, *21*, 129–145.

Bjorklund, O., Shang, M., Tonazzini, I., Dare, E., & Fredholm, B. B. (2008). Adenosine A_1 and A_3 receptors protect astrocytes from hypoxic damage. *European Journal of Pharmacology*, *596*, 6–13.

Boissard, C. G., Lindner, M. D., & Gribkoff, V. K. (1992). Hypoxia produces cell death in the rat hippocampus in the presence of an A1 adenosine receptor antagonist: An anatomical and behavioral study. *Neuroscience*, *48*, 807–812.

Bona, E., Aden, U., Gilland, E., Fredholm, B. B., & Hagberg, H. (1997). Neonatal cerebral hypoxia-ischemia: The effect of adenosine receptor antagonists. *Neuropharmacology*, *36*, 1327–1338.

Bonetti, B., Stegagno, C., Cannella, B., Rizzato, N., Moretto, G., & Raine, C. S. (1999). Activation of NF-kappaB and c-jun transcription factors in multiple sclerosis lesions. Implications for oligodendrocyte pathology. *American Journal of Pathology*, *155*, 1433–1438.

Borea, P. A., Gessi, S., Bar-Yehuda, S., & Fishman, P. (2009). A3 adenosine receptor: Pharmacology and role in disease. *Handbook of Experimental Pharmacology*, *193*, 297–327.

Borsello, T., Clarke, P. G., Hirt, L., Vercelli, A., Repici, M., Schorderet, D. F., et al. (2003). A peptide inhibitor of c-Jun N-terminal kinase protects against excitotoxicity and cerebral ischemia. *Nature Medicine*, *9*, 1180–1186.

Brodie, C., Blumberg, P. M., & Jacobson, K. A. (1998). Activation of the A_{2A} adenosine receptor inhibits nitric oxide production in glial cells. *FEBS Letters*, *429*, 139–142.

Bruce-Keller, A. J. (1999). Microglial-neuronal interactions in synaptic damage and recovery. *Journal of Neuroscience Research*, *58*, 191–201.

Butler, M., Sanmugalingam, D., Burton, V. J., Wilson, T., Pearson, R., Watson, R. P., et al. (2012). Impairment of adenosine A_3 receptor activity disrupts neutrophil migratory capacity and impacts innate immune function in vivo. *European Journal of Immunology*, *42*, 3358–3368.

Cassada, D. C., Tribble, C. G., Young, J. S., Gangemi, J. J., Gohari, A. R., Butler, P. D., et al. (2002). Adenosine A_{2A} analogue improves neurologic outcome after spinal cord trauma in the rabbit. *Journal of Trauma*, *53*, 225–229.

Castillo, A., Tolón, M. R., Fernández-Ruiz, J., Romero, J., & Martinez-Orgado, J. (2010). The neuroprotective effect of cannabidiol in an in vitro model of newborn hypoxic-ischemic brain damage in mice is mediated by CB(2) and adenosine receptors. *Neurobiology of Disease, 37*, 434–440.

Chauhan, N. K., Young, A. M., Gibson, C. L., & Davidson, C. (2013). Inhibition of pre-ischeamic conditioning in the mouse caudate brain slice by NMDA- or adenosine A1 receptor antagonists. *European Journal of Pharmacology, 698*, 322–329.

Chen, G. J., Harvey, B. K., Shen, H., Chou, J., Victor, A., & Wang, Y. (2006). Activation of adenosine A3 receptors reduces ischemic brain injury in rodents. *Journal of Neuroscience Research, 84*, 1848–1855.

Chen, J. F., Huang, Z., Ma, J., Zhu, J., Moratalla, R., Standaert, D., et al. (1999). A_{2A} adenosine receptor deficiency attenuates brain injury induced by transient focal ischemia in mice. *Journal of Neuroscience, 19*, 9192–9200.

Chen, F., Qi, Z., Luo, Y., Hinchliffe, T., Ding, G., Xia, Y., et al. (2014). Non-pharmaceutical therapies for stroke: Mechanisms and clinical implications. *Progress in Neurobiology, 115C*, 246–269.

Choi, D. W. (1990). Possible mechanisms limiting N-methyl-D-aspartate receptor over-activation and the therapeutic efficacy of N-methyl-D-aspartate antagonists. *Stroke, 21*, III20–III22.

Choi, I. Y., Lee, J. C., Ju, C., Hwang, S., Cho, G. S., Lee, H. W., et al. (2011). A_3 adenosine receptor agonist reduces brain ischemic injury and inhibits inflammatory cell migration in rats. *American Journal of Pathology, 179*, 2042–2052.

Choukèr, A., Thiel, M., Lukashev, D., Ward, J. M., Kaufmann, I., Apasov, S., et al. (2008). Critical role of hypoxia and A_{2A} adenosine receptors in liver tissue-protecting physiological anti-inflammatory pathway. *Molecular Medicine, 14*, 116–123.

Cipriani, R., Villa, P., Chece, G., Lauro, C., Paladini, A., Micotti, E., et al. (2011). CX3CL1 is neuroprotective in permanent focal cerebral ischemia in rodents. *Journal of Neuroscience, 31*, 16327–16335.

Ciruela, F., Casadó, V., Rodrigues, R. J., Luján, R., Burgueño, J., Canals, M., et al. (2006). Presynaptic control of striatal glutamatergic neurotransmission by adenosine A_1-A_{2A} receptor heteromers. *Journal of Neuroscience, 26*, 2080–2087.

Coelho, J. E., Rebola, N., Fragata, I., Ribeiro, J. A., de Mendonça, A., & Cunha, R. A. (2006). Hypoxia-induced desensitization and internalization of adenosine A_1 receptors in the rat hippocampus. *Neuroscience, 138*, 1195–1203.

Colotta, V., Catarzi, D., Varano, F., Capelli, F., Lenzi, O., Filacchioni, G., et al. (2007). New 2-arylpyrazolo[3,4-c]quinoline derivatives as potent and selective human A3 adenosine receptor antagonists. Synthesis, pharmacological evaluation, and ligand-receptor modeling studies. *Journal of Medicinal Chemistry, 50*, 4061–4074.

Colotta, V., Catarzi, D., Varano, F., Lenzi, O., Filacchioni, G., Martini, C., et al. (2008). Synthesis, ligand-receptor modeling studies and pharmacological evaluation of novel 4-modified-2-aryl-1,2,4-triazolo[4,3-a]quinoxalin-1-one derivatives as potent and selective human A3 adenosine receptor antagonists. *Bioorganic & Medicinal Chemistry, 16*, 6086–6102.

Colotta, V., Lenzi, O., Catarzi, D., Varano, F., Filacchioni, G., Martini, C., et al. (2009). Pyrido[2,3-e]-1,2,4-triazolo[4,3-a]pyrazin-1-one as a new scaffold to develop potent and selective human A3 adenosine receptor antagonists. Synthesis, pharmacological evaluation, and ligand-receptor modeling studies. *Journal of Medicinal Chemistry, 52*, 2407–2419.

Coppi, E., Cellai, L., Maraula, G., Pugliese, A. M., & Pedata, F. (2013). Adenosine A_{2A} receptors inhibit delayed rectifier potassium currents and cell differentiation in primary purified oligodendrocyte cultures. *Neuropharmacology, 73*, 301–310.

Corradetti, R., Lo, Conte G., Moroni, F., Passani, M. B., & Pepeu, G. (1984). Adenosine decreases aspartate and glutamate release from rat hippocampal slices. *European Journal of Pharmacology, 104*, 19–26.
Corsi, C., Melani, A., Bianchi, L., & Pedata, F. (2000). Striatal A_{2A} adenosine receptor antagonism differentially modifies striatal glutamate outflow in vivo in young and aged rats. *Neuroreport, 11*, 2591–2595.
Corsi, C., Melani, A., Bianchi, L., Pepeu, G., & Pedata, F. (1999). Striatal A_{2A} adenosine receptors differentially regulate spontaneous and K^+-evoked glutamate release in vivo in young and aged rats. *Neuroreport, 10*, 687–691.
Cristalli, G., Müller, C. E., & Volpini, R. (2009). Recent developments in adenosine A_{2A} receptor ligands. *Handbook of Experimental Pharmacology, 193*, 59–98.
Cristóvão-Ferreira, S., Vaz, S. H., Ribeiro, J. A., & Sebastião, A. M. (2009). Adenosine A_{2A} receptors enhance GABA transport into nerve terminals by restraining PKC inhibition of GAT-1. *Journal of Neurochemistry, 109*, 336–347.
Cui, M., Bai, X., Li, T., Chen, F., Dong, Q., Zhao, Y., et al. (2013). Decreased extracellular adenosine levels lead to loss of hypoxia-induced neuroprotection after repeated episodes of exposure to hypoxia. *PLoS One, 8*, e57065.
Cunha, R. A. (2001). Adenosine as a neuromodulator and as a homeostatic regulator in the nervous system: Different roles, different sources and different receptors. *Neurochemistry International, 38*, 107–125.
Cunha, R. A., Johansson, B., van der Ploeg, I., Sebastião, A. M., Ribeiro, J. A., & Fredholm, B. B. (1994). Evidence for functionally important adenosine A_{2A} receptors in the rat hippocampus. *Brain Research, 649*, 208–216.
Dai, S. S., Zhou, Y. G., Li, W., An, J. H., Li, P., Yang, N., et al. (2010). Local glutamate level dictates adenosine A_{2A} receptor regulation of neuroinflammation and traumatic brain injury. *Journal of Neuroscience, 30*, 5802–5810.
Dale, N., Pearson, T., & Fringuelli, B. G. (2000). Direct measurement of adenosine release during hypoxia in the CA1 region of the rat hippocampal slice. *Journal of Physiology, 526*(Pt1), 143–155.
Daval, J. L., & Nicolas, F. (1994). Opposite effects of cyclohexyladenosine and theophylline on hypoxic damage in cultured neurons. *Neuroscience Letters, 175*, 114–116.
Daval, J. L., Von Lubitz, D. K., Deckert, J., Redmond, D. J., & Marangos, P. J. (1989). Protective effect of cyclohexyladenosine on adenosine A_1-receptors, guanine nucleotide and forskolin binding sites following transient brain ischemia: A quantitative autoradiographic study. *Brain Research, 491*, 212–226.
De Keyser, J., Sulter, G., & Luiten, P. G. (1999). Clinical trials with neuroprotective drugs in acute ischaemic stroke: Are we doing the right thing? *Trends in Neuroscience, 22*, 535–540.
Dennis, S. H., Jaafari, N., Cimarosti, H., Hanley, J. G., Henley, J. M., & Mellor, J. R. (2011). Oxygen/glucose deprivation induces a reduction in synaptic AMPA receptors on hippocampal CA3 neurons mediated by mGluR1 and adenosine A_3 receptors. *The Journal of Neuroscience, 31*, 11941–11952.
Diógenes, M. J., Assaife-Lopes, N., Pinto-Duarte, A., Ribeiro, J. A., & Sebastião, A. M. (2007). Influence of age on BDNF modulation of hippocampal synaptic transmission: Interplay with adenosine A_{2A} receptors. *Hippocampus, 17*, 577–585.
Diógenes, M. J., Fernandes, C. C., Sebastião, A. M., & Ribeiro, J. A. (2004). Activation of adenosine A_{2A} receptor facilitates brain-derived neurotrophic factor modulation of synaptic transmission in hippocampal slices. *Journal of Neuroscience, 24*, 2905–2913.
Dirnagl, U. (2012). Pathobiology of injury after stroke: The neurovascular unit and beyond. *Annals of the New York Academy of Sciences, 1268*, 21–25.
Dirnagl, U., Iadecola, C., & Moskowitz, M. A. (1999). Pathobiology of ischaemic stroke: An integrated view. *Trends in Neuroscience, 22*, 391–397.

Dixon, A. K., Gubitz, A. K., Sirinathsinghji, D. J., Richardson, P. J., & Freeman, T. C. (1996). Tissue distribution of adenosine receptor mRNAs in the rat. *British Journal of Pharmacology, 118*, 1461–1468.

Domenici, M. R., de Carolis, A. S., & Sagratella, S. (1996). Block by N6-L-phenylisopropyl-adenosine of the electrophysiological and morphological correlates of hippocampal ischaemic injury in the gerbil. *British Journal of Pharmacology, 118*, 1551–1557.

Dunwiddie, T. V. (1984). Interactions between the effects of adenosine and calcium on synaptic responses in rat hippocampus in vitro. *Journal of Physiology, 350*, 545–559.

Eckle, T., Faigle, M., Grenz, A., Laucher, S., Thompson, L. F., & Eltzschig, H. K. (2008). A_{2B} adenosine receptor dampens hypoxia-induced vascular leak. *Blood, 111*, 2024–2035.

Eckle, T., Krahn, T., Grenz, A., Köhler, D., Mittelbronn, M., Ledent, C., et al. (2007). Cardioprotection by ecto-5'-nucleotidase (CD73) and A_{2B} adenosine receptors. *Circulation, 115*, 1581–1590.

Ekdahl, C. T., Claasen, J. H., Bonde, S., Kokaia, Z., & Lindvall, O. (2003). Inflammation is detrimental for neurogenesis in adult brain. *Proceedings of the National Academy of Sciences of the United States of America, 100*, 13632–13637.

Ekdahl, C. T., Kokaia, Z., & Lindvall, O. (2009). Brain inflammation and adult neurogenesis: The dual role of microglia. *Neuroscience, 158*, 1021–1029.

Eltzschig, H. K., Ibla, J. C., Furuta, G. T., Leonard, M. O., Jacobson, K. A., Enjyoji, K., et al. (2003). Coordinated adenine nucleotide phosphohydrolysis and nucleoside signaling in posthypoxic endothelium: Role of ectonucleotidases and adenosine A_{2B} receptors. *Journal of Experimental Medicine, 198*, 783–796.

Eltzschig, H. K., Thompson, L. F., Karhausen, J., Cotta, R. J., Ibla, J. C., Robson, S. C., et al. (2004). Endogenous adenosine produced during hypoxia attenuates neutrophil accumulation: Coordination by extracellular nucleotide metabolism. *Blood, 104*, 3986–3992.

Evans, M. C., Swan, J. H., & Meldrum, B. S. (1987). An adenosine analogue, 2-chloroadenosine, protects against long term development of ischaemic cell loss in the rat hippocampus. *Neuroscience Letters, 83*, 287–292.

Fedorova, I. M., Jacobson, M. A., Basile, A., & Jacobson, K. A. (2003). Behavioral characterization of mice lacking the A3 adenosine receptor: Sensitivity to hypoxic neurodegeneration. *Cellular and Molecular Neurobiology, 23*, 431–447.

Feoktistov, I., & Biaggioni, I. (2011). Role of adenosine A_{2B} receptors in inflammation. *Advances in Pharmacology, 61*, 115–144.

Feoktistov, I., Goldstein, A. E., Ryzhov, S., Zeng, D., Belardinelli, L., Voyno-Yasenetskaya, T., et al. (2002). Differential expression of adenosine receptors in human endothelial cells: Role of A_{2B} receptors in angiogenic factor regulation. *Circulation Research, 90*, 531–538.

Feoktistov, I., Ryzhov, S., Zhong, H., Goldstein, A. E., Matafonov, A., Zeng, D., et al. (2004). Hypoxia modulates adenosine receptors in human endothelial and smooth muscle cells toward an A_{2B} angiogenic phenotype. *Hypertension, 44*, 649–654.

Fiebich, B. L., Biber, K., Lieb, K., van Calker, D., Berger, M., Bauer, J., et al. (1996). Cyclooxygenase-2 expression in rat microglia is induced by adenosine A_{2A}-receptors. *Glia, 18*, 152–160.

Fink, J. S., Weaver, D. R., Rivkees, S. A., Peterfreund, R. A., Pollack, A. E., Adler, E. M., et al. (1992). Molecular cloning of the rat A_2 adenosine receptor: Selective co-expression with D_2 dopamine receptors in rat striatum. *Brain Research. Molecular Brain Research, 14*, 186–195.

Fishman, P., & Bar-Yehuda, S. (2003). Pharmacology and therapeutic applications of A_3 receptor subtype. *Current Topics in Medicinal Chemistry, 3*, 463–469.

Fishman, P., Bar-Yehuda, S., Liang, B. T., & Jacobson, K. A. (2012). Pharmacological and therapeutic effects of A_3 adenosine receptor agonists. *Drug Discovery Today, 17*, 359–366.

Fowler, J. C. (1989). Adenosine antagonists delay hypoxia-induced depression of neuronal activity in hippocampal brain slice. *Brain Research*, *490*, 378–384.
Fowler, J. C. (1990). Adenosine antagonists alter the synaptic response to in vitro ischemia in the rat hippocampus. *Brain Research*, *509*, 331–334.
Fowler, J. C., Gervitz, L. M., Hamilton, M. E., & Walker, J. A. (2003). Systemic hypoxia and the depression of synaptic transmission in rat hippocampus after carotid artery occlusion. *Journal of Physiology*, *550*, 961–972.
Fredholm, B. B., Arslan, G., Halldner, L., Kull, B., Schulte, G., & Wasserman, W. (2000). Structure and function of adenosine receptors and their genes. *Naunyn-Schmiedeberg's Archives of Pharmacology*, *362*, 364–374.
Fredholm, B. B., Chen, J. F., Masino, S. A., & Vaugeois, J. M. (2005). Actions of adenosine at its receptors in the CNS: Insights from knockouts and drugs. *Annual Review of Pharmacology and Toxicology*, *45*, 385–412.
Fredholm, B. B., Cunha, R. A., & Svenningsson, P. (2003). Pharmacology of adenosine A_{2A} receptors and therapeutic applications. *Current Topics in Medicinal Chemistry*, *3*, 413–426.
Fredholm, B. B., IJzerman, A. P., Jacobson, K. A., Linden, J., & Müller, C. E. (2011). International Union of Basic and Clinical Pharmacology. LXXXI. Nomenclature and classification of adenosine receptors-an update. *Pharmacological Review*, *63*, 1–34.
Fredholm, B. B., Zhang, Y., & van der Ploeg, I. (1996). Adenosine A_{2A} receptors mediate the inhibitory effect of adenosine on formyl-Met-Leu-Phe-stimulated respiratory burst in neutrophil leucocytes. *Naunyn-Schmiedeberg's Archives of Pharmacology*, *354*, 262–267.
Frenguelli, B. G., Wigmore, G., Llaudet, E., & Dale, N. (2007). Temporal and mechanistic dissociation of ATP and adenosine release during ischaemia in the mammalian hippocampus. *Journal of Neurochemistry*, *101*, 1400–1413.
Fronz, U., Deten, A., Baumann, F., Kranz, A., Weidlich, S., Härtig, W., et al. (2014). Continuous adenosine A_{2A} receptor antagonism after focal cerebral ischemia in spontaneously hypertensive rats. *Naunyn-Schmiedeberg's Archives of Pharmacology*, *387*, 165–173.
Gao, Y., & Phillis, J. W. (1994). CGS 15943, an adenosine A_2 receptor antagonist, reduces cerebral ischemic injury in the Mongolian gerbil. *Life Sciences*, *55*, L61–L65.
Gelderblom, M., Eminel, S., Herdegen, T., & Waetzig, V. (2004). c-Jun N-terminal kinases (JNKs) and the cytoskeleton–functions beyond neurodegeneration. *International Journal of Developmental Neuroscience*, *22*, 559–564.
Gelderblom, M., Leypoldt, F., Steinbach, K., Behrens, D., Choe, C. U., Siler, D. A., et al. (2009). Temporal and spatial dynamics of cerebral immune cell accumulation in stroke. *Stroke*, *40*, 1849–1857.
Genovese, T., Melani, A., Esposito, E., Mazzon, E., Di Paola, R., Bramanti, P., et al. (2009). The selective adenosine A_{2A} receptor agonist CGS21680 reduces JNK MAPK activation in oligodendrocytes in injured spinal cord. *Shock*, *32*, 578–585.
Genovese, T., Melani, A., Esposito, E., Paterniti, I., Mazzon, E., Di Paola, R., et al. (2010). Selective adenosine A_{2A} receptor agonists reduce the apoptosis in an experimental model of spinal cord trauma. *Journal of Biological Regulators & Homeostatic Agents*, *24*, 73–86.
Gervitz, L. M., Davies, D. G., Omidvar, K., & Fowler, J. C. (2003). The effect of acute hypoxemia and hypotension on adenosine-mediated depression of evoked hippocampal synaptic transmission. *Experimental Neurology*, *182*, 507–517.
Gervitz, L. M., Lutherer, L. O., Davies, D. G., Pirch, J. H., & Fowler, J. C. (2001). Adenosine induces initial hypoxic-ischemic depression of synaptic transmission in the rat hippocampus in vivo. *American Journal of Physiology, Regulative Integrative and Comparative Physiology*, *280*, R639–R645.
Gessi, S., Merighi, S., Stefanelli, A., Fazzi, D., Varani, K., & Borea, P. A. (2013). A_1 and A_3 adenosine receptors inhibit LPS-induced hypoxia-inducible factor-1 accumulation in murine astrocytes. *Pharmacological Research*, *76*, 157–170.

Gessi, S., Merighi, S., Varani, K., Leung, E., Mac Lennan, S., & Borea, P. A. (2008). The A_3 adenosine receptor: An enigmatic player in cell biology. *Pharmacology & Therapeutics, 117*, 123–140.

Gessi, S., Varani, K., Merighi, S., Cattabriga, E., Avitabile, A., Gavioli, R., et al. (2004). Expression of A_3 adenosine receptors in human lymphocytes: Up-regulation in T cell activation. *Molecular Pharmacology, 65*, 711–719.

Gessi, S., Varani, K., Merighi, S., Cattabriga, E., Pancaldi, C., Szabadkai, Y., et al. (2005). Expression, pharmacological profile, and functional coupling of A_{2B} receptors in a recombinant system and in peripheral blood cells using a novel selective antagonist radioligand, [3H]MRE 2029-F20. *Molecular Pharmacology, 67*, 2137–2147.

Gibson, C. L. (2013). Cerebral ischemic stroke: Is gender important? *Journal of Cerebral Blood Flow & Metabolism, 33*, 1355–1361.

Goldberg, M. P., Monyer, H., Weiss, J. H., & Choi, D. W. (1988). Adenosine reduces cortical neuronal injury induced by oxygen or glucose deprivation in vitro. *Neuroscience Letters, 89*, 323–327.

Gomes, C. V., Kaster, M. P., Tomé, A. R., Agostinho, P. M., & Cunha, R. A. (2011). Adenosine receptors and brain diseases: Neuroprotection and neurodegeneration. *Biochimica et Biophysica Acta, 1808*, 1380–1399.

González-Fernández, E., Sánchez-Gómez, M. V., Pérez-Samartín, A., Arellano, R. O., & Matute, C. (2014). A3 adenosine receptors mediate oligodendrocyte death and ischemic damage to optic nerve. *Glia, 62*, 199–216.

Gourine, A. V., Dale, N., Gourine, V. N., & Spyer, K. M. (2004). Fever in systemic inflammation: Roles of purines. *Frontiers in Bioscience, 9*, 1011–1022.

Grant, M. B., Davis, M. I., Caballero, S., Feoktistov, I., Biaggioni, I., & Belardinelli, L. (2001). Proliferation, migration, and ERK activation in human retinal endothelial cells through A_{2B} adenosine receptor stimulation. *Investigative Ophthalmology & Visual Science, 42*, 2068–2073.

Greene, R. W., & Haas, H. L. (1991). The electrophysiology of adenosine in the mammalian central nervous system. *Progress in Neurobiology, 36*, 329–341.

Gribkoff, V. K., Bauman, L. A., & VanderMaelen, C. P. (1990). The adenosine antagonist 8-cyclopentyltheophylline reduces the depression of hippocampal neuronal responses during hypoxia. *Brain Research, 512*, 353–357.

Gu, L., Huang, B., Shen, W., Gao, L., Ding, Z., Wu, H., et al. (2013). Early activation of nSMase2/ceramide pathway in astrocytes is involved in ischemia-associated neuronal damage via inflammation in rat hippocampi. *Journal of Neuroinflammation, 10*, 109–124.

Gui, L., Duan, W., Tian, H., Li, C., Zhu, J., Chen, J. F., et al. (2009). Adenosine A_{2A} receptor deficiency reduces striatal glutamate outflow and attenuates brain injury induced by transient focal cerebral ischemia in mice. *Brain Research, 1297*, 185–193.

Haskó, G., Kuhel, D. G., Chen, J. F., Schwarzschild, M. A., Deitch, E. A., Mabley, J. G., et al. (2000). Adenosine inhibits IL-12 and TNF-[alpha] production via adenosine A_{2A} receptor-dependent and independent mechanisms. *FASEB Journal, 14*, 2065–2074.

Haskó, G., Linden, J., Cronstein, B., & Pacher, P. (2008). Adenosine receptors: Therapeutic aspects for inflammatory and immune diseases. *Nature Reviews Drug Discovery*, 7, 759–770.

Heese, K., Fiebich, B. L., Bauer, J., & Otten, U. (1997). Nerve growth factor (NGF) expression in rat microglia is induced by adenosine A_{2A}-receptors. *Neuroscience Letters, 231*, 83–86.

Hentschel, S., Lewerenz, A., & Nieber, K. (2003). Activation of A_3 receptors by endogenous adenosine inhibits synaptic transmission during hypoxia in rat cortical neurons. *Restorative Neurology and Neuroscience, 21*, 55–63.

Heron, A., Lekieffre, D., Le Peillet, E., Lasbennes, F., Seylaz, J., Plotkine, M., et al. (1994). Effects of an A_1 adenosine receptor agonist on the neurochemical, behavioral and histological consequences of ischemia. *Brain Research, 641*, 217–224.

Hettinger, B. D., Lee, A., Linden, J., & Rosin, D. L. (2001). Ultrastructural localization of adenosine A_{2A} receptors suggests multiple cellular sites for modulation of GABAergic neurons in rat striatum. *Journal of Comparative Neurology, 431*, 331–346.
Heurteaux, C., Lauritzen, I., Widmann, C., & Lazdunski, M. (1995). Essential role of adenosine, adenosine A_1 receptors, and ATP-sensitive K^+ channels in cerebral ischemic preconditioning. *Proceedings of the National Academy of Sciences of the United States of America, 92*, 4666–4670.
Higashi, H., Meno, J. R., Marwaha, A. S., & Winn, H. R. (2002). Hippocampal injury and neurobehavioral deficits following hyperglycemic cerebral ischemia: Effect of theophylline and ZM241385. *Journal of Neurosurgery, 96*, 117–126.
Hirt, L., Badaut, J., Thevenet, J., Granziera, C., Regli, L., Maurer, F., et al. (2004). D-JNKI1, a cell-penetrating c-Jun-N-terminal kinase inhibitor, protects against cell death in severe cerebral ischemia. *Stroke, 35*, 1738–1743.
Howe, C. L., Bieber, A. J., Warrington, A. E., Pease, L. R., & Rodriguez, M. (2004). Anti-apoptotic signaling by a remyelination-promoting human antimyelin antibody. *Neurobiology of Disease, 15*, 120–131.
Hu, S., Dong, H., Zhang, H., Wang, S., Hou, L., Chen, S., et al. (2012). Noninvasive limb remote ischemic preconditioning contributes neuroprotective effects via activation of adenosine A_1 receptor and redox status after transient focal cerebral ischemia in rats. *Brain Research, 1459*, 81–90.
Iadecola, C., & Anrather, J. (2011). Stroke research at a crossroad: Asking the brain for directions. *Nature Neuroscience, 14*, 1363–1368.
Ilie, A., Ciocan, D., Constantinescu, A. O., Zagrean, A. M., Nita, D. A., Zagrean, L., et al. (2009). Endogenous activation of adenosine A_1 receptors promotes post-ischemic electrocortical burst suppression. *Neuroscience, 159*, 1070–1078.
Irving, E. A., Barone, F. C., Reith, A. D., Hadingham, S. J., & Parsons, A. A. (2000). Differential activation of MAPK/ERK and p38/SAPK in neurones and glia following focal cerebral ischaemia in the rat. *Brain Research. Molecular Brain Research, 77*, 65–75.
Jacobson, K. A. (1998). Adenosine A3 receptors: Novel ligands and paradoxical effect. *Trends in Pharmacological Sciences, 19*(5), 184–191.
Jacobson, K. A. (2013). Structure-based approaches to ligands for G-protein-coupled adenosine and P2Y receptors, from small molecules to nanoconjugates. *Journal of Medicinal Chemistry, 56*, 3749–3767.
Jacobson, K. A., Hoffmann, C., Cattabeni, F., & Abbracchio, M. P. (1999). Adenosine-induced cell death: Evidence for receptor-mediated signalling. *Apoptosis, 4*, 197–211.
Jacobson, K. A., Von Lubitz, D. K., Daly, J. W., & Fredholm, B. B. (1996). Adenosine receptor ligands: Differences with acute versus chronic treatment. *Trends in Pharmacological Sciences, 17*, 108–113.
Jarrott, D. M., & Domer, F. R. (1980). A gerbil model of cerebral ischemia suitable for drug evaluation. *Stroke, 11*, 203–209.
Jeon, S. J., Rhee, S. Y., Ryu, J. H., Cheong, J. H., Kwon, K., Yang, S. I., et al. (2011). Activation of adenosine A_{2A} receptor up-regulates BDNF expression in rat primary cortical neurons. *Neurochemical Research, 36*, 2259–2269.
Johansson, B., Halldner, L., Dunwiddie, T. V., Masino, S. A., Poelchen, W., Giménez-Llort, L., et al. (2001). Hyperalgesia, anxiety, and decreased hypoxic neuroprotection in mice lacking the adenosine A_1 receptor. *Proceedings of the National Academy of Sciences of the United States of America, 98*, 9407–9412.
Johnston, A., Gudjonsson, J. E., Sigmundsdottir, H., Ludviksson, B. R., & Valdimarsson, H. (2005). The anti-inflammatory action of methotrexate is not mediated by lymphocyte apoptosis, but by the suppression of activation and adhesion molecules. *Clinical Immunology, 114*, 154–163.

Jones, P. A., Smith, R. A., & Stone, T. W. (1998a). Protection against kainate-induced excitotoxicity by adenosine A_{2A} receptor agonists and antagonists. *Neuroscience*, *85*, 229–237.
Jones, P. A., Smith, R. A., & Stone, T. W. (1998b). Protection against hippocampal kainate excitotoxicity by intracerebral administration of an adenosine A_{2A} receptor antagonist. *Brain Research*, *800*, 328–335.
Jurewicz, A., Matysiak, M., Andrzejak, S., & Selmaj, K. (2006). TRAIL-induced death of human adult oligodendrocytes is mediated by JNK pathway. *Glia*, *53*, 158–166.
Ke, R. H., Xiong, J., & Liu, Y. (2012). Adenosine A_{2A} receptor induces GDNF expression by the Stat3 signal in vitro. *Neuroreport*, *23*, 958–962.
Kitada, M., & Rowitch, D. H. (2006). Transcription factor co-expression patterns indicate heterogeneity of oligodendroglial subpopulations in adult spinal cord. *Glia*, *54*, 35–46.
Kitagawa, H., Mori, A., Shimada, J., Mitsumoto, Y., & Kikuchi, T. (2002). Intracerebral adenosine infusion improves neurological outcome after transient focal ischemia in rats. *Neurological Research*, *24*, 317–323.
Koeppen, M., Eckle, T., & Eltzschig, H. K. (2011). Interplay of hypoxia and A_{2B} adenosine receptors in tissue protection. *Advances in Pharmacology*, *61*, 145–186.
Koscsó, B., Csóka, B., Selmeczy, Z., Himer, L., Pacher, P., Virág, L., et al. (2012). Adenosine augments IL-10 production by microglial cells through an A_{2B} adenosine receptor-mediated process. *Journal of Immunology*, *188*, 445–453.
Kuan, C. Y., Whitmarsh, A. J., Yang, D. D., Liao, G., Schloemer, A. J., Dong, C., et al. (2003). A critical role of neural-specific JNK3 for ischemic apoptosis. *Proceedings of the National Academy of Sciences of the United States of America*, *100*, 15184–15189.
Lai, D.-M., Tu, Y.-K., Liu, I.-M., & Cheng, J.-T. (2005). Increase of adenosine A_1 receptor gene expression in cerebral ischemia of Wistar rats. *Neuroscience Letters*, *387*, 59–61.
Lange-Asschenfeldt, C., Raval, A. P., Dave, K. R., Mochly-Rosen, D., Sick, T. J., & Perez-Pinzon, M. A. (2004). Epsilon protein kinase C mediated ischemic tolerance requires activation of the extracellular regulated kinase pathway in the organotypic hippocampal slice. *Journal of Cerebral Blood Flow and Metabolism*, *24*, 636–645.
Lappas, C. M., Day, Y. J., Marshall, M. A., Engelhard, V. H., & Linden, J. (2006). Adenosine A_{2A} receptor activation reduces hepatic ischemia reperfusion injury by inhibiting CD1d-dependent NKT cell activation. *Journal of Experimental Medicine*, *203*, 2639–2648.
Latini, S., Bordoni, F., Corradetti, R., Pepeu, G., & Pedata, F. (1999). Effect of A_{2A} adenosine receptor stimulation and antagonism on synaptic depression induced by in vitro ischaemia in rat hippocampal slices. *British Journal of Pharmacology*, *128*, 1035–1044.
Latini, S., Bordoni, F., Pedata, F., & Corradetti, R. (1999). Extracellular adenosine concentrations during in vitro ischaemia in rat hippocampal slices. *British Journal of Pharmacology*, *127*, 729–739.
Latini, S., & Pedata, F. (2001). Adenosine in the central nervous system: Release mechanisms and extracellular concentrations. *Journal of Neurochemistry*, *79*, 463–484.
Lee, K. S., & Lowenkopf, T. (1993). Endogenous adenosine delays the onset of hypoxic depolarization in the rat hippocampus in vitro via an action at A_1 receptors. *Brain Research*, *609*, 313–315.
Lewis, A. J., & Manning, A. M. (1999). New targets for anti-inflammatory drugs. *Current Opinion in Chemical Biology*, *3*, 489–494.
Li, Y., Oskouian, R. J., Day, Y. J., Rieger, J. M., Liu, L., Kern, J. A., et al. (2006). Mouse spinal cord compression injury is reduced by either activation of the adenosine A_{2A} receptor on bone marrow-derived cells or deletion of the A_{2A} receptor on non-bone marrow-derived cells. *Neuroscience*, *141*, 2029–2039.
Liang, R., Pang, Z. P., Deng, P., & Xu, Z. C. (2009). Transient enhancement of inhibitory synaptic transmission in hippocampal CA1 pyramidal neurons after cerebral ischemia. *Neuroscience*, *160*, 412–418.

Logan, M., & Sweeney, M. I. (1997). Adenosine A_1 receptor activation preferentially protects cultured cerebellar neurons versus astrocytes against hypoxia-induced death. *Molecular and Chemical Neuropathology*, *31*, 119–133.
Lopes, L. V., Cunha, R. A., & Ribeiro, J. A. (1999). ZM241385, an adenosine A_{2A} receptor antagonist, inhibits hippocampal A_1 receptor responses. *European Journal of Pharmacology*, *383*, 395–398.
Lopes, L. V., Sebastião, A. M., & Ribeiro, J. A. (2011). Adenosine and related drugs in brain diseases: Present and future in clinical trials. *Current Topics in Medicinal Chemistry*, *11*, 1087–1101.
Lu, A., Tang, Y., Ran, R., Clark, J. F., Aronow, B. J., & Sharp, F. R. (2003). Genomics of the periinfarction cortex after focal cerebral ischemia. *Journal of Cerebral Blood Flow and Metabolism*, *23*, 786–810.
Lu, X. C., Williams, A. J., Yao, C., Berti, R., Hartings, J. A., Whipple, R., et al. (2004). Microarray analysis of acute and delayed gene expression profile in rats after focal ischemic brain injury and reperfusion. *Journal of Neuroscience Research*, *77*, 843–857.
Luo, C., Yi, B., Tao, G., Li, M., Chen, Z., Tang, W., et al. (2010). Adenosine A_3 receptor agonist reduces early brain injury in subarachnoid haemorrhage. *Neuroreport*, *21*, 892–896.
Maraula, G., Traini, C., Mello, T., Coppi, E., Galli, A., Pedata, F., et al. (2013). Effects of oxygen and glucose deprivation on synaptic transmission in rat dentate gyrus: Role of A_{2A} adenosine receptors. *Neuropharmacology*, *67*, 511–520.
Marcoli, M., Bonfanti, A., Roccatagliata, P., Chiaramonte, G., Ongini, E., Raiteri, M., et al. (2004). Glutamate efflux from human cerebrocortical slices during ischemia: Vesicular-like mode of glutamate release and sensitivity to A_{2A} adenosine receptor blockade. *Neuropharmacology*, *47*, 884–891.
Marcoli, M., Raiteri, L., Bonfanti, A., Monopoli, A., Ongini, E., Raiteri, M., et al. (2003). Sensitivity to selective adenosine A_1 and A_{2A} receptor antagonists of the release of glutamate induced by ischemia in rat cerebrocortical slices. *Neuropharmacology*, *45*, 201–210.
Marks, L., Carswell, H. V., Peters, E. E., Graham, D. I., Patterson, J., Dominiczak, A. F., et al. (2001). Characterization of the microglial response to cerebral ischemia in the stroke-prone spontaneously hypertensive rat. *Hypertension*, *38*, 116–122.
Matos, M., Augusto, E., Agostinho, P., Cunha, R. A., & Chen, J. F. (2013). Antagonistic interaction between adenosine A_{2A} receptors and Na^+/K^+-ATPase-$\alpha2$ controlling glutamate uptake in astrocytes. *Journal of Neuroscience*, *33*, 18492–18502.
Matos, M., Augusto, E., Santos-Rodrigues, A. D., Schwarzschild, M. A., Chen, J. F., Cunha, R. A., et al. (2012). Adenosine A_{2A} receptors modulate glutamate uptake in cultured astrocytes and gliosomes. *Glia*, *60*, 702–716.
Matsumoto, K., Graf, R., Rosner, G., Shimada, N., & Heiss, W. D. (1992). Flow thresholds for extracellular purine catabolite elevation in cat focal ischemia. *Brain Research*, *579*, 309–314.
Matute, C., Alberdi, E., Ibarretxe, G., & Sanchez-Gomez, M. V. (2002). Excitotoxicity in glial cells. *European Journal of Pharmacology*, *447*, 239–246.
Matute, C., Sanchez-Gomez, M. V., Martinez-Millan, L., & Miledi, R. (1997). Glutamate receptor-mediated toxicity in optic nerve oligodendrocytes. *Proceedings of the National Academy of Sciences of the United States of America*, *94*, 8830–8835.
McDonald, J. W., Althomsons, S. P., Hyrc, K. L., Choi, D. W., & Goldberg, M. P. (1998). Oligodendrocytes from forebrain are highly vulnerable to AMPA/kainate receptor-mediated excitotoxicity. *Nature Medicine*, *4*, 291–297.
Medina-Pulido, L., Molina-Arcas, M., Justicia, C., Soriano, E., Burgaya, F., Planas, A. M., et al. (2013). Hypoxia and P1 receptor activation regulate the high-affinity concentrative adenosine transporter CNT2 in differentiated neuronal PC12 cells. *Biochemical Journal*, *454*, 437–445.

Melani, A., Cipriani, S., Vannucchi, M. G., Nosi, D., Donati, C., Bruni, P., et al. (2009). Selective adenosine A_{2A} receptor antagonism reduces JNK activation in oligodendrocytes after cerebral ischaemia. *Brain, 132*, 1480–1495.

Melani, A., Corti, F., Cellai, L., Vannucchi, M. G., & Pedata, F. (2012b). Time-related protection of adenosine A2A antagonists and agonists after focal cerebral ischemia. In *Annual Meeting "Purine Club", Pisa, Italy*.

Melani, A., Corti, F., Cellai, L., Vannucchi, M. G., & Pedata, F. (2014). Low doses of the selective adenosine A_{2A} receptor agonist CGS21680 are protective in a rat model of transient cerebral ischemia. *Brain Research, 1551*, 59–72.

Melani, A., Corti, F., Stephan, H., Müller, C. E., Donati, C., Bruni, P., et al. (2012). Ecto-ATPase inhibition: ATP and adenosine release under physiological and ischemic *in vivo* conditions in the rat striatum. *Experimental Neurology, 233*, 193–204.

Melani, A., Corti, F., Vannucchi, M. G., Nosi, D., Giovannini, M. G., & Pedata, F. (2011). Role of A_{2A} receptors on modulation of oligodendroglia in cerebral ischemia. *Shock, 36*(Suppl. 1), 24.

Melani, A., Gianfriddo, M., Vannucchi, M. G., Cipriani, S., Baraldi, P. G., Giovannini, M. G., et al. (2006). The selective A_{2A} receptor antagonist SCH58261 protects from neurological deficit, brain damage and activation of p38 MAPK in rat focal cerebral ischemia. *Brain Research, 1073–1074*, 470–480.

Melani, A., Pantoni, L., Bordoni, F., Gianfriddo, M., Bianchi, L., Vannucchi, M. G., et al. (2003). The selective A_{2A} receptor antagonist SCH58261 reduces striatal transmitter outflow, turning behavior and ischemic brain damage induced by permanent focal ischemia in the rat. *Brain Research, 959*, 243–250.

Melani, A., Pantoni, L., Corsi, C., Bianchi, L., Monopoli, A., Bertorelli, R., et al. (1999). Striatal outflow of adenosine, excitatory amino acids, γ-aminobutyric acid, and taurine in awake freely moving rats after middle cerebral artery occlusion. Correlations with neurological deficit and histopatological damage. *Stroke, 30*, 2448–2455.

Monopoli, A., Lozza, G., Forlani, A., Mattavelli, A., & Ongini, E. (1998). Blockade of A_{2A} adenosine receptors by SCH58261 results in neuroprotective effects in cerebral ischaemia in rats. *NeuroReport, 9*, 3955–3959.

Mori, M., Nishizaki, T., & Okada, Y. (1992). Protective effect of adenosine on the anoxic damage of hippocampal slice. *Neuroscience, 46*, 301–307.

Nagasawa, H., Araki, T., & Kogure, K. (1994). Alteration of adenosine A_1 receptor binding in the post-ischaemic rat brain. *Neuroreport, 5*, 1453–1456.

Nakamura, M., Nakakimura, K., Matsumoto, M., & Sakabe, T. (2002). Rapid tolerance to focal cerebral ischemia in rats is attenuated by adenosine A_1 receptor antagonist. *Journal of Cerebral Blood Flow and Metabolism, 22*, 161–170.

Nariai, T., Shimada, Y., Ishiwata, K., Nagaoka, T., Shimada, J., Kuroiwa, T., et al. (2003). PET imaging of adenosine A_1 receptors with (11)C-MPDX as an indicator of severe cerebral ischemic insult. *Journal of Nuclear Medicine, 44*, 1839–1844.

Newman, G. C., Hospod, F. E., Trowbridge, S. D., Motwani, S., & Liu, Y. (1998). Restoring adenine nucleotides in a brain slice model of cerebral reperfusion. *Journal of Cerebral Blood Flow and Metabolism, 18*, 675–685.

Nishizaki, T., Nagai, K., Nomura, T., Tada, H., Kanno, T., Tozaki, H., et al. (2002). A new neuromodulatory pathway with a glial contribution mediated via A_{2A} adenosine receptors. *Glia, 39*, 133–147.

O'Bryant, Z., Vann, K. T., & Xiong, Z. G. (2014). Translational strategies for neuroprotection in ischemic stroke-focusing on Acid-sensing ion channel 1a. *Translational Stroke Research, 5*, 59–68.

Odashima, M., Bamias, G., Rivera-Nieves, J., Linden, J., Nast, C. C., Moskaluk, C. A., et al. (2005). Activation of A_{2A} adenosine receptor attenuates intestinal inflammation in animal models of inflammatory bowel disease. *Gastroenterology, 129*, 26–33.

O'Kane, E. M., & Stone, T. W. (1998). Interaction between adenosine A_1 and A_2 receptor-mediated responses in the rat hippocampus in vitro. *European Journal of Pharmacology, 362*, 17–25.

Olsson, T., Cronberg, T., Rytter, A., Asztély, F., Fredholm, B. B., Smith, M. L., et al. (2004). Deletion of the adenosine A_1 receptor gene does not alter neuronal damage following ischaemia in vivo or in vitro. *European Journal of Neuroscience, 20*, 1197–1204.

Ongini, E., Adami, M., Ferri, C., & Bertorelli, R. (1997). Adenosine A_{2A} receptors and neuroprotection. *Annals of the New York Academy of Sciences, 825*, 30–48.

Onodera, H., Sato, G., & Kogure, K. (1987). Quantitative autoradiographic analysis of muscarinic cholinergic and adenosine A_1 binding sites after transient forebrain ischemia in the gerbil. *Brain Research, 415*, 309–322.

O'Regan, M. H., Simpson, R. E., Perkins, L. M., & Phillis, J. W. (1992). The selective A_2 adenosine receptor agonist CGS21680 enhances excitatory transmitter amino acid release from the ischemic rat cerebral cortex. *Neuroscience Letters, 138*, 169–172.

Palmer, T. M., Benovic, J. L., & Stiles, G. L. (1995). Agonist-dependent phosphorylation and desensitization of the rat A_3 adenosine receptor. Evidence for a G-protein-coupled receptor kinase-mediated mechanism. *The Journal of Biological Chemistry, 270*, 29607–29613.

Paterniti, I., Melani, A., Cipriani, S., Corti, F., Mello, T., Mazzon, E., et al. (2011). Selective adenosine A_{2A} receptor agonists and antagonists protect against spinal cord injury through peripheral and central effects. *Journal of Neuroinflammation, 8*, 31–44.

Pedata, F., Gianfriddo, M., Turchi, D., & Melani, A. (2005). The protective effect of adenosine A_{2A} receptor antagonism in cerebral ischemia. *Neurological Research, 27*, 169–174.

Pedata, F., Latini, S., Pugliese, A. M., & Pepeu, G. (1993). Investigations into the adenosine outflow from hippocampal slices evoked by ischemia-like conditions. *Journal of Neurochemistry, 61*, 284–289.

Pedata, F., Pugliese, A. M., Coppi, E., Popoli, P., Morelli, M., Schwarzschild, M. A., et al. (2007). Adenosine in the central nervous system: Effects on neurotransmission and neuroprotection. *Immunology, Endocrine & Metabolic Agents in Medicinal Chemistry, 4*, 304–321.

Pedata, F., Pugliese, A. M., Sebastiao, A. M., & Ribeiro, J. A. (2010). Adenosine A_3 receptor signaling in the central nervous system. In P. A. Borea (Ed.), *A_3 adenosine receptors from cell biology to pharmacology and therapeutics* (pp. 165–188). Netherlands: Springer.

Perez-Pinzon, M. A., Mumford, P. L., Rosenthal, M., & Sick, T. J. (1996). Anoxic preconditioning in hippocampal slices: Role of adenosine. *Neuroscience, 75*, 687–694.

Peterfreund, R. A., MacCollin, M., Gusella, J., & Fink, J. S. (1996). Characterization and expression of the human A_{2A} adenosine receptor gene. *Journal of Neurochemistry, 66*, 362–368.

Petroni, A., Papini, N., Blasevich, M., & Galli, C. (2002). Blockade of A_{2A} adenosine receptors leads to c-fos inhibition in a rat model of brain ischemia. *Pharmacological Research, 45*, 125–128.

Phillis, J. W. (1995). The effects of selective A_1 and A_{2A} adenosine receptor antagonists on cerebral ischemic injury in the gerbil. *Brain Research, 705*, 79–84.

Phillis, J. W. (2004). Adenosine and adenine nucleotides as regulators of cerebral blood flow: Roles of acidosis, cell swelling, and K ATP channels. *Critical Reviews in Neurobiology, 16*, 237–270.

Phillis, J. W., & Goshgarian, H. G. (2001). Adenosine and neurotrauma: Therapeutic perspectives. *Neurological Research, 23*, 183–189.

Phillis, J. W., Smith-Barbour, M., & O'Regan, M. H. (1996). Changes in extracellular amino acid neurotransmitters and purines during and following ischemias of different durations in the rat cerebral cortex. *Neurochemistry International, 29*, 115–120.

Piao, C. S., Kim, J. B., Han, P. L., & Lee, J. K. (2003). Administration of the p38 MAPK inhibitor SB203580 affords brain protection with a wide therapeutic window against focal ischemic insult. *Journal of Neuroscience Research, 73*, 537–544.

Pinto-Duarte, A., Coelho, J. E., Cunha, R. A., Ribeiro, J. A., & Sebastião, A. M. (2005). Adenosine A_{2A} receptors control the extracellular levels of adenosine through modulation of nucleoside transporters activity in the rat hippocampus. *Journal of Neurochemistry, 93*, 595–604.

Pintor, A., Galluzzo, M., Grieco, R., Pezzola, A., Reggio, R., & Popoli, P. (2004). Adenosine A_{2A} receptor antagonists prevent the increase in striatal glutamate levels induced by glutamate uptake inhibitors. *Journal of Neurochemistry, 89*, 152–156.

Popoli, P., Betto, P., Reggio, R., & Ricciarello, G. (1995). Adenosine A_{2A} receptor stimulation enhances striatal extracellular glutamate levels in rats. *European Journal of Pharmacology, 287*, 215–217.

Popoli, P., Pintor, A., Domenici, M. R., Frank, C., Tebano, M. T., Pezzola, A., et al. (2002). Blockade of striatal adenosine A_{2A} receptor reduces, through a presynaptic mechanism, quinolinic acid-induced excitotoxicity: Possible relevance to neuroprotective interventions in neurodegenerative diseases of the striatum. *The Journal of Neuroscience, 22*, 1967–1975.

Puffinbarger, N. K., Hansen, K. R., Resta, R., Laurent, A. B., Knudsen, T. B., Madara, J. L., et al. (1995). Production and characterization of multiple antigenic peptide antibodies to the adenosine A_{2B} receptor. *Molecular Pharmacology, 47*, 1126–1132.

Pugliese, A. M., Coppi, E., Spalluto, G., Corradetti, R., & Pedata, F. (2006). A3 adenosine receptor antagonists delay irreversible synaptic failure caused by oxygen and glucose deprivation in the rat CA1 hippocampus in vitro. *British Journal of Pharmacology, 147*, 524–532.

Pugliese, A. M., Coppi, E., Volpini, R., Cristalli, G., Corradetti, R., Jeong, L. S., et al. (2007). Role of adenosine A3 receptors on CA1 hippocampal neurotransmission during oxygen-glucose deprivation episodes of different duration. *Biochemical Pharmacology, 74*, 768–779.

Pugliese, A. M., Latini, S., Corradetti, R., & Pedata, F. (2003). Brief, repeated, oxygen-glucose deprivation episodes protect neurotransmission from a longer ischemic episode in the in vitro hippocampus: Role of adenosine receptors. *British Journal of Pharmacology, 140*, 305–314.

Pugliese, A. M., Traini, C., Cipriani, S., Gianfriddo, M., Mello, T., Giovannini, M. G., et al. (2009). The adenosine A_{2A} receptor antagonist ZM241385 enhances neuronal survival after oxygen-glucose deprivation in rat CA1 hippocampal slices. *British Journal of Pharmacology, 157*, 818–830.

Ribeiro, J. A., Sebastiao, A. M., & de Mendonca, A. (2002). Adenosine receptors in the nervous system: Pathophysiological implications. *Progress in Neurobiology, 68*, 377–392.

Rivkees, S. A., Zhao, Z., Porter, G., & Turner, C. (2001). Influences of adenosine on the fetus and newborn. *Molecular Genetics and Metabolism, 74*, 160–171.

Rosin, D. L., Hettinger, B. D., Lee, A., & Linden, J. (2003). Anatomy of adenosine A_{2A} receptors in brain: Morphological substrates for integration of striatal function. *Neurology, 61*, S10–S11.

Rudolphi, K. A., Keil, M., Fastbom, J., & Fredholm, B. B. (1989). Ischaemic damage in gerbil hippocampus is reduced following upregulation of adenosine A_1 receptors by caffeine treatment. *Neuroscience Letters, 103*, 275–280.

Rudolphi, K. A., Keil, M., & Hinze, H. J. (1987). Effect of theophylline on ischemically induced hippocampal damage in Mongolian gerbils: A behavioral and histopathological study. *Journal of Cerebral Blood Flow and Metabolism, 7*, 74–81.

Saransaari, P., & Oja, S. S. (2005). GABA release modified by adenosine receptors in mouse hippocampal slices under normal and ischemic conditions. *Neurochemical Research, 30*, 467–473.

Schiffmann, S. N., Jacobs, O., & Vanderhaeghen, J. J. (1991). Striatal restricted adenosine A_2 receptor (RDC8) is expressed by enkephalin but not by substance P neurons: An in situ hybridization histochemistry study. *Journal of Neurochemistry, 57*, 1062–1067.

Schindler, C. W., Karcz-Kubicha, M., Thorndike, E. B., Müller, C. E., Tella, S. R., Ferré, S., et al. (2005). Role of central and peripheral adenosine receptors in the cardiovascular responses to intraperitoneal injections of adenosine A_1 and A_{2A} subtype receptor agonists. *British Journal of Pharmacology, 144*, 642–650.

Schindler, C. W., Karcz-Kubicha, M., Thorndike, E. B., Müller, C. E., Tella, S. R., Goldberg, S. R., et al. (2004). Lack of adenosine A_1 and dopamine D_2 receptor-mediated modulation of the cardiovascular effects of the adenosine A_{2A} receptor agonist CGS21680. *European Journal of Pharmacology, 484*, 269–275.

Schwartz-Bloom, R. D., & Sah, R. (2001). Gamma-aminobutyric acid(A) neurotransmission and cerebral ischemia. *Journal of Neurochemistry, 77*, 353–371.

Schwarzschild, M. A., Chen, J. F., & Ascherio, A. (2002). Caffeinated clues and the promise of adenosine A_{2A} antagonists in PD. *Neurology, 58*, 1154–1160.

Sebastião, A. M., Cunha, R. A., de Mendonca, A., & Ribeiro, J. A. (2000). Modification of adenosine modulation of synaptic transmission in the hippocampus of aged rats. *British Journal of Pharmacology, 131*, 1629–1634.

Sebastião, A. M., de Mendonca, A., Moreira, T., & Ribeiro, J. A. (2001). Activation of synaptic NMDA receptors by action potential-dependent release of transmitter during hypoxia impairs recovery of synaptic transmission on reoxygenation. *The Journal of Neuroscience, 21*, 8564–8571.

Sebastião, A. M., & Ribeiro, J. A. (1996). Adenosine A_2 receptor-mediated excitatory actions on the nervous system. *Progress in Neurobiology, 48*, 167–189.

Sebastião, A. M., & Ribeiro, J. A. (2009). Triggering neurotrophic factor actions through adenosine A_{2A} receptor activation: Implications for neuroprotection. *British Journal of Pharmacology, 158*, 15–22.

Sheardown, M. J., & Knutsen, L. J. S. (1996). Unexpected neuroprotection observed with the adenosine A_{2A} receptor agonist CGS21680. *Drug Development Research, 39*, 108–114.

Shrager, P., & Novakovic, S. D. (1995). Control of myelination, axonal growth, and synapse formation in spinal cord explants by ion channels and electrical activity. *Brain Research. Developmental Brain Research, 88*, 68–78.

Simpson, R. E., O'Regan, M. H., Perkins, L. M., & Phillis, J. W. (1992). Excitatory transmitter amino acid release from the ischemic rat cerebral cortex: Effects of adenosine receptor agonists and antagonists. *Journal of Neurochemistry, 58*, 1683–1690.

Sitkovsky, M. V. (2003). Use of the A_{2A} adenosine receptor as a physiological immunosuppressor and to engineer inflammation in vivo. *Biochemical Pharmacology, 65*, 493–501.

Sitkovsky, M. V., Lukashev, D., Apasov, S., Kojima, H., Koshiba, M., Caldwell, C., et al. (2004). Physiological control of immune response and inflammatory tissue damage by hypoxia-inducible factors and adenosine A_{2A} receptors. *Annual Review of Immunology, 22*, 657–682.

Soler, E. P., & Ruiz, V. C. (2010). Epidemiology and risk factors of cerebral ischemia and ischemic heart diseases: Similarities and differences. *Current Cardiology Reviews, 6*, 138–149.

Somjen, G. G. (2001). Mechanisms of spreading depression and hypoxic spreading depression-like depolarization. *Physiological Reviews, 81*, 1065–1096.

Stoll, G., Jander, S., & Schroeter, M. (1998). Inflammation and glial responses in ischemic brain lesions. *Progress in Neurobiology, 56*, 149–171.

Stone, T. W., & Behan, W. M. (2007). Interleukin-1beta but not tumor necrosis factor-alpha potentiates neuronal damage by quinolinic acid: Protection by an adenosine A_{2A} receptor antagonist. *Journal of Neuroscience Research, 85*, 1077–1085.

Sutherland, G. R., Peeling, J., Lesiuk, H. J., Brownstone, R. M., Rydzy, M., Saunders, J. K., et al. (1991). The effects of caffeine on ischemic neuronal injury as determined by magnetic resonance imaging and histopathology. *Neuroscience*, *42*, 171–182.

Svenningsson, P., Le Moine, C., Fisone, G., & Fredholm, B. B. (1999). Distribution, biochemistry and function of striatal adenosine A_{2A} receptors. *Progress in Neurobiology*, *59*, 355–396.

Takagi, Y., Nozaki, K., Sugino, T., Hattori, I., & Hashimoto, N. (2000). Phosphorylation of c-Jun NH(2)-terminal kinase and p38 mitogen-activated protein kinase after transient forebrain ischemia in mice. *Neuroscience Letters*, *294*, 117–120.

Tanaka, E., Yamamoto, S., Kudo, Y., Mihara, S., & Higashi, H. (1997). Mechanisms underlying the rapid depolarization produced by deprivation of oxygen and glucose in rat hippocampal CA1 neurons in vitro. *Journal of Neurophysiology*, *78*, 891–902.

Tanaka, E., Yasumoto, S., Hattori, G., Niiyama, S., Matsuyama, S., & Higashi, H. (2001). Mechanisms underlying the depression of evoked fast EPSCs following in vitro ischemia in rat hippocampal CA1 neurons. *Journal of Neurophysiology*, *86*, 1095–1103.

Tebano, M. T., Martire, A., Potenza, R. L., Grò, C., Pepponi, R., Armida, M., et al. (2008). Adenosine A_{2A} receptors are required for normal BDNF levels and BDNF-induced potentiation of synaptic transmission in the mouse hippocampus. *Journal of Neurochemistry*, *104*, 279–286.

Trincavelli, M. L., Marroni, M., Tuscano, D., Ceruti, S., Mazzola, A., Mitro, N., et al. (2004). Regulation of A_{2B} adenosine receptor functioning by tumour necrosis factor a in human astroglial cells. *Journal of Neurochemistry*, *91*, 1180–1190.

Trincavelli, M. L., Melani, A., Guidi, S., Cipriani, S., Pedata, F., & Martini, C. (2008). Regulation of A_{2A} adenosine receptor expression and functioning following permanent focal ischemia in rat brain. *Journal of Neurochemistry*, *104*, 479–490.

Trincavelli, M. L., Tonazzini, I., Montali, M., Abbracchio, M. P., & Martini, C. (2008). Short-term TNF-Alpha treatment induced A_{2B} adenosine receptor desensitization in human astroglial cells. *Journal of Cellular Biochemistry*, *104*, 150–161.

Trincavelli, M. L., Tuscano, D., Cecchetti, P., Falleni, A., Benzi, L., Klotz, K. N., et al. (2000). Agonist-induced internalization and recycling of the human A_3 adenosine receptors: Role in receptor desensitization and resensitization. *Journal of Neurochemistry*, *75*, 1493–1501.

Trincavelli, M. L., Tuscano, D., Marroni, M., Falleni, A., Gremigni, V., Ceruti, S., et al. (2002). A_3 adenosine receptors in human astrocytoma cells: Agonist-mediated desensitization, internalization, and down-regulation. *Molecular Pharmacology*, *62*, 1373–1384.

Turner, C. P., Seli, M., Ment, L., Stewart, W., Yan, H., Johansson, B., et al. (2003). A_1 adenosine receptors mediate hypoxia-induced ventriculomegaly. *Proceedings of the National Academy of Sciences of the United States of America*, *100*, 11718–11722.

Turner, C. P., Yan, H., Schwartz, M., Othman, T., & Rivkees, S. A. (2002). A_1 adenosine receptor activation induces ventriculomegaly and white matter loss. *Neuroreport*, *13*, 1199–1204.

Varani, K., Gessi, S., Dalpiaz, A., & Borea, P. A. (1996). Pharmacological and biochemical characterization of purified A_{2A} adenosine receptors in human platelet membranes by [3H]-CGS21680 binding. *British Journal of Pharmacology*, *117*, 1693–1701.

Varani, K., Gessi, S., Dalpiaz, A., Ongini, E., & Borea, P. A. (1997). Characterization of A_{2A} adenosine receptors in human lymphocyte membranes by [3H]-SCH58261 binding. *British Journal of Pharmacology*, *122*, 386–392.

Varani, K., Gessi, S., Dionisotti, S., Ongini, E., & Borea, P. A. (1998). [3H]-SCH 58261 labelling of functional A2A adenosine receptors in human neutrophil membranes. *British Journal of Pharmacology*, *123*, 1723–1731.

Volpini, R., Costanzi, S., Lambertucci, C., Taffi, S., Vittori, S., Klotz, K. N., et al. (2002). N (6)-alkyl-2-alkynyl derivatives of adenosine as potent and selective agonists at the human adenosine A_3 receptor and a starting point for searching A_{2B} ligands. *Journal of Medicinal Chemistry, 45*, 3271–3279.

Volpini, R., Dal Ben, D., Lambertucci, C., Taffi, S., Vittori, S., Klotz, K. N., et al. (2007). N6-methoxy-2-alkynyladenosine derivatives as highly potent and selective ligands at the human A_3 adenosine receptor. *Journal of Medicinal Chemistry, 50*, 1222–1230.

von Arnim, C. A., Timmler, M., Ludolph, A. C., & Riepe, M. W. (2000). Adenosine receptor up-regulation: Initiated upon preconditioning but not upheld. *Neuroreport, 11*, 1223–1226.

von Arnim, C. A., Verstege, E., Etrich, S. M., & Riepe, M. W. (2006). Mechanisms of hypoxic tolerance in presymptomatic APP23 transgenic mice. *Mechanisms of Ageing and Development, 127*, 109–114.

von Lubitz, D. K. (1999). Adenosine and cerebral ischemia: Therapeutic future or death of a brave concept? *European Journal of Pharmacology, 371*, 85–102.

von Lubitz, D. K., Dambrosia, J. M., Kempski, O., & Redmond, D. J. (1988). Cyclohexyl adenosine protects against neuronal death following ischemia in the CA1 region of gerbil hippocampus. *Stroke, 19*, 1133–1139.

von Lubitz, D. K., Lin, R. C., Boyd, M., Bischofberger, N., & Jacobson, K. A. (1999). Chronic administration of adenosine A_3 receptor agonist and cerebral ischemia: Neuronal and glial effects. *European Journal of Pharmacology, 367*, 157–163.

von Lubitz, D. K., Lin, R. C., & Jacobson, K. A. (1995). Cerebral ischemia in gerbils: Effects of acute and chronic treatment with adenosine A_{2A} receptor agonist and antagonist. *European Journal of Pharmacology, 287*, 295–302.

von Lubitz, D. K., Lin, R. C., Melman, N., Ji, X. D., Carter, M. F., & Jacobson, K. A. (1994). Chronic administration of selective adenosine A_1 receptor agonist or antagonist in cerebral ischemia. *European Journal of Pharmacology, 256*, 161–167.

von Lubitz, D. K., Lin, R. C., Paul, I. A., Beenhakker, M., Boyd, M., Bischofberger, N., et al. (1996). Postischemic administration of adenosine amine congener (ADAC): Analysis of recovery in gerbils. *European Journal of Pharmacology, 316*, 171–179.

von Lubitz, D. K., Lin, R. C., Popik, P., Carter, M. F., & Jacobson, K. A. (1994). Adenosine A_3 receptor stimulation and cerebral ischemia. *European Journal of Pharmacology, 263*, 59–67.

von Lubitz, D. K., & Marangos, P. J. (1990). Cerebral ischemia in gerbils: Postischemic administration of cyclohexyl adenosine and 8-sulfophenyl-theophylline. *Journal of Molecular Neuroscience, 2*, 53–59.

von Lubitz, D. K., Simpson, K. L., & Lin, R. C. (2001). Right thing at a wrong time? Adenosine A_3 receptors and cerebroprotection in stroke. *Annals of the New York Academy of Sciences, 939*, 85–96.

von Lubitz, D. K., Ye, W., McClellan, J., & Lin, R. C. (1999). Stimulation of adenosine A_3 receptors in cerebral ischemia. Neuronal death, recovery, or both? *Annals of the New York Academy of Sciences, 890*, 93–106.

We, W., Du, C., Lv, J., Zhao, G., Li, Z., Wu, Z., et al. (2013). Blocking A_{2B} adenosine receptor alleviates pathogenesis of experimental autoimmune encephalomyelitis via inhibition of IL-6 production and Th17 differentiation. *Journal of Immunology, 190*, 138–146.

Weaver, D. R. (1993). A_{2A} adenosine receptor gene expression in developing rat brain. *Brain Research. Molecular Brain Research, 20*, 313–327.

Wei, C. J., Li, W., & Chen, J. F. (2011). Normal and abnormal functions of adenosine receptors in the central nervous system revealed by genetic knockout studies. *Biochimica et Biophysica Acta, 1808*, 1358–1379.

White, P. J., Rose'Meyer, R. B., & Hope, W. (1996). Functional characterization of adenosine receptors in the nucleus tractus solitarius mediating hypotensive responses in the rat. *British Journal of Pharmacology, 117*, 305–308.

Wu, D. C., Ye, W., Che, X. M., & Yang, G. Y. (2000). Activation of mitogen-activated protein kinases after permanent cerebral artery occlusion in mouse brain. *Journal of Cerebral Blood Flow and Metabolism*, *20*, 1320–1330.

Yamamoto, S., Tanaka, E., Shoji, Y., Kudo, Y., Inokuchi, H., & Higashi, H. (1997). Factors that reverse the persistent depolarization produced by deprivation of oxygen and glucose in rat hippocampal CA1 neurons in vitro. *Journal of Neurophysiology*, *78*, 903–911.

Yang, Z. J., Wang, B., Kwansa, H., Heitmiller, K. D., Hong, G., Carter, E. L., et al. (2013). Adenosine A_{2A} receptor contributes to ischemic brain damage in newborn piglet. *Journal of Cerebral Blood Flow and Metabolism*, *33*, 1612–1620.

Yang, D., Zhang, Y., Nguyen, H. G., Koupenova, M., Chauhan, A. K., Makitalo, M., et al. (2006). The A_{2B} adenosine receptor protects against inflammation and excessive vascular adhesion. *The Journal of Clinical Investigation*, *116*, 1913–1923.

Yoshida, M., Nakakimura, K., Cui, Y. J., Matsumoto, M., & Sakabe, T. (2004). Adenosine A_1 receptor antagonist and mitochondrial ATP-sensitive potassium channel blocker attenuate the tolerance to focal cerebral ischemia in rats. *Journal of Cerebral Blood Flow and Metabolism*, *24*, 771–779.

Yu, L., Frith, M. C., Suzuki, Y., Peterfreund, R. A., Gearan, T., Sugano, S., et al. (2004). Characterization of genomic organization of the adenosine A_{2A} receptor gene by molecular and bioinformatics analyses. *Brain Research*, *1000*, 156–173.

Yu, A. C., Lee, Y. L., Fu, W. Y., & Eng, L. F. (1995). Gene expression in astrocytes during and after ischemia. *Progress in Brain Research*, *105*, 245–253.

Zetterstrom, T., & Fillenz, M. (1990). Adenosine agonists can both inhibit and enhance in vivo striatal dopamine release. *European Journal of Pharmacology*, *180*, 137–143.

Zhang, R. L., Chopp, M., Chen, H., & Garcia, J. H. (1994). Temporal profile of ischemic tissue damage, neutrophil response, and vascular plugging following permanent and transient (2H) middle cerebral artery occlusion in the rat. *Journal of the Neurological Sciences*, *125*, 3–10.

Zhou, A. M., Li, W. B., Li, Q. J., Liu, H. Q., Feng, R. F., & Zhao, H. G. (2004). A short cerebral ischemic preconditioning up-regulates adenosine receptors in the hippocampal CA1 region of rats. *Neuroscience Research*, *48*, 397–404.

Zhou, J. G., Meno, J. R., Hsu, S. S., & Winn, H. R. (1994). Effects of theophylline and cyclohexyladenosine on brain injury following normo- and hyperglycemic ischemia: A histopathologic study in the rat. *Journal of Cerebral Blood Flow and Metabolism*, *14*, 166–173.

CHAPTER FOURTEEN

Roles of Adenosine and Its Receptors in Sleep–Wake Regulation

Zhi-Li Huang[1], Ze Zhang, Wei-Min Qu[1]
Department of Pharmacology, State Key Laboratory of Medical Neurobiology, Institute of Brain Science, Shanghai Medical College of Fudan University, Shanghai, China
[1]Corresponding authors: e-mail address: huangzl@fudan.edu.cn; quweimin@fudan.edu.cn

Contents

Abstract

This chapter summarizes the current knowledge about the role of adenosine in the sleep–wake regulation with a focus on adenosine in the brain, regulation of adenosine levels, adenosine receptors, and manipulations of the adenosine system by the use of pharmacological and molecular biological tools. Adenosine is neither stored nor released as a classical neurotransmitter and is thought to be formed inside cells or on their surface, mostly by breakdown of adenine nucleotides. The extracellular level of adenosine increases in the cortex and basal forebrain (BF) during prolonged wakefulness and decreases during the sleep-recovery period. Therefore, adenosine is proposed to act as a homeostatic regulator of sleep. The endogenous somnogen prostaglandin (PG) D_2 increases the extracellular level of adenosine under the subarachnoid space of the BF and promotes physiological sleep. There are four adenosine receptor subtypes: adenosine A_1 receptor (R, A_1R), A_{2A}R, A_{2B}R, and A_3R. Both the A_1R and the A_{2A}R have been reported to be involved in sleep induction. The A_{2A}R plays an important

International Review of Neurobiology, Volume 119
ISSN 0074-7742
http://dx.doi.org/10.1016/B978-0-12-801022-8.00014-3

role in the somnogenic effects of PGD_2. Activation of $A_{2A}R$ by its agonist infused into the brain potently increases sleep and the arousal effect of caffeine, an A_1R and $A_{2A}R$ antagonist, was shown to be dependent on the $A_{2A}R$. On the other hand, inhibition of wake-promoting neurons via the A_1R also mediates the sleep-inducing effects of adenosine, whereas activation of A_1R in the lateral preoptic area induces wakefulness. These findings indicate that $A_{2A}R$ plays a predominant role in sleep induction, whereas A_1R regulates the sleep–wake cycle in a site-dependent manner.

ABBREVIATIONS

AAV adeno-associated virus
ADA adenosine deaminase
AK adenosine kinase
ATP adenosine triphosphate
BF basal forebrain
cAMP cyclic adenosine 3′,5′-monophosphate
cN cytosolic nucleotidase
CPA N^6-cyclopentyladenosine
CSF cerebrospinal fluid
dnSNARE the SNARE domain of the protein synaptobrevin II
DP_1R PGD_2 receptor
GPCR G-protein-coupled receptor
H-PGDS hematopoietic prostaglandin D synthase
KO knockout
LC locus coeruleus
L-PGDS lipocalin-type PGDS
mPFC medial prefrontal cortex
NAc nucleus accumbens
NBTI *S*-(4-nitrobenzyl)-6-thioinosine
NREM non-rapid eye movement
PB parabrachial nucleus
PG prostaglandin
PGDS PGD synthase
R receptor
REM rapid eye movement
SAHH *S*-adenosyl-homocysteine hydrolase
Se^{4+} tetravalent selenium
$SeCl_4$ selenium tetrachloride
SWA slow-wave activity
TMN tuberomammillary nucleus
VLPO ventrolateral preoptic
WT wild type

1. INTRODUCTION

The concepts of sleep–wake regulation emphasize homeostatic (sleep pressure), circadian (daily rhythms), and allostatic (food availability or behavioral stress) factors that control sleep. The homeostatic process is regulated by the sleep propensity, which builds up during wakefulness and declines during sleep (Lazarus, Chen, Urade, & Huang, 2013; Lazarus, Huang, Lu, Urade, & Chen, 2012). Two such endogenous and potent sleep factors are adenosine, a naturally occurring purine nucleoside present in all cells (Fredholm, 2007), and prostaglandin (PG) D_2, an eicosanoid acting as a tissue or local hormone (Huang, Urade, & Hayaishi, 2007, 2011), both of which are released as neuromodulators in the brain. In this chapter, we are discussing four areas of high general complexity: (a) the metabolism of adenosine, which is ubiquitously present in all cells; (b) adenosine is a key signaling molecule for PGD_2-induced sleep; (c) key roles of the A_{2A} receptor ($A_{2A}R$) in sleep induction and neuron circuits; and (d) activation of adenosine A_1R regulates the sleep–wake cycle in a region or site-dependent manner. This chapter briefly reviews these points, with particular emphasis on the recent progress made by the use of biochemical and pharmacological tools, optogenetical and pharmacogenetic approaches, and gene-manipulated mice.

2. FORMATION, METABOLISM, AND TRANSPORT OF ADENOSINE IN THE CENTRAL NERVOUS SYSTEM

Adenosine is a nucleoside composed of a molecule of adenine attached to a ribose sugar molecule (ribofuranose) via a β-N_9-glycosidic bond for its chemical structure (see Fig. 14.1). It is formed inside the cells or on their surface and is mainly formed by the breakdown of intra or extracellular adenine nucleotides (Zimmermann, 2000). Inside the cell, adenosine can be formed from adenosine triphosphate (ATP) by intracellular 5′-nucleotidase and then transported outside the cell via bidirectional nucleoside transporters, or it can be generated outside the cell by the metabolism of released nucleotides by ectonucleotidases. Two intracellular 5′-nucleotidase enzymes have been cloned, with cytosolic nucleotidase (cN)-I breaking down AMP to adenosine and cN-II breaking down inosine 5′-monophosphate and guanosine monophosphate to inosine and

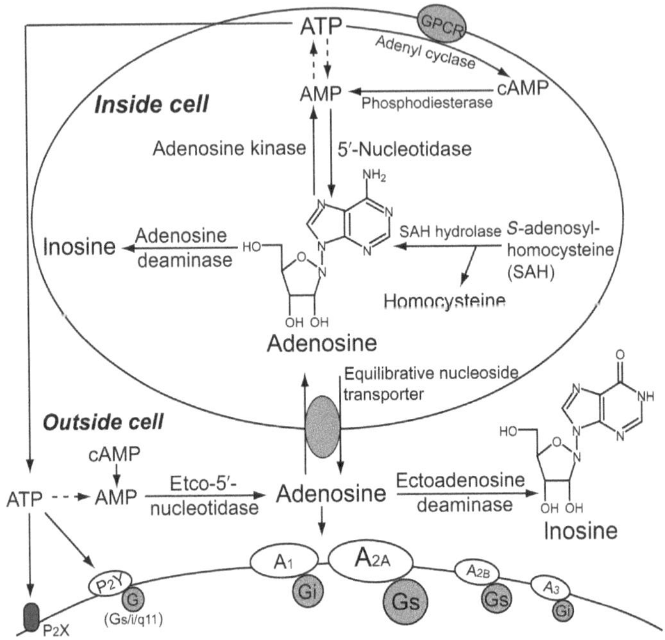

Figure 14.1 Intracellular and extracellular pathways for the formation and metabolism of adenosine. Inside the cell, adenosine is formed from ATP, cAMP, or SAH, while outside the cell it arises from equilibrating nucleoside transporter-mediated release or metabolism from ATP or cAMP. *Adapted from Sawynok and Liu (2003) with modification and permission.*

guanosine, respectively (Sala-Newby, Skladanowski, & Newby, 1999). There is a family of ectonucleotidases that can generate adenosine from ATP, but ecto-5′-nucleotidase is the major enzyme responsible for this under physiological conditions (Zimmermann, 2000). Adenosine can also arise from cyclic adenosine 3′,5′-monophosphate (cAMP), either generated inside the cell by a Gprotein-coupled receptor (GPCR) with subsequent conversion to AMP by phosphodiesterase or following the efflux of cAMP by a probenecid-sensitive transporter and metabolism by ectoenzyme (Brundege, Diao, Proctor, & Dunwiddie, 1997; Rosenberg & Li, 1995). However, the conversion of cAMP to AMP by phosphodiesterase is slow, and activation of adenosine receptors following increased availability of cAMP is limited (Dunwiddie & Masino, 2001). A further source of adenosine inside the cell is hydrolysis of *S*-adenosyl-homocysteine (SAH) (Deussen, Lloyd, & Schrader, 1989), but this pathway is less important in the brain (Latini & Pedata, 2001).

While ATP may function as a neurotransmitter in some brain areas (Burnstock, 2007), adenosine is neither stored nor released as a classical neurotransmitter since it does not accumulate in synaptic vesicles, but is released from the cytoplasm into the extracellular space through a nucleoside transporter. Intra- and extracellular adenosine concentrations are kept in equilibrium by means of equilibrating transporters, which mediate adenosine reuptake, the direction of the transport being dependent upon the concentration gradient on both sides of the membrane (Gu, Foga, Parkinson, & Geiger, 1995). These equilibrating transporters are abundant in most cells, including astrocytes (Peng et al., 2005).

Extracellular adenosine appears through two major mechanisms: (1) the release of adenosine as such through nucleoside transporters (Geiger & Fyda, 1991) after an increase in the intracellular levels of adenosine and (2) the extracellular formation of adenosine through the ectonucleotidase pathway on the release of adenine nucleotides, especially ATP.

The clearance of adenosine depends on nucleoside transporters and three kinds of enzymes, i.e., adenosine kinase (AK), adenosine deaminase (ADA), and *S*-adenosyl-homocysteine hydrolase (SAHH). The clearance of extracellular adenosine mostly occurs through the nonconcentrating nucleoside transporters, and ecto-ADA deaminates adenosine to inosine (Fredholm, Chen, Cunha, Svenningsson, & Vaugeois, 2005). The main intracellular metabolic pathways of adenosine are the formation of AMP by AK and the irreversible breakdown to inosine by ADA. Neurons appear to be enriched in AK, whereas ADA appears to be more abundant in astrocytes (Fredholm et al., 2005) and histaminergic neurons in the tuberomammillary nuclei (TMNs) (Nagy, LaBella, Buss, & Daddona, 1984; Oishi, Huang, Fredholm, Urade, & Hayaishi, 2008). A third enzyme, SAHH, converts adenosine to SAH in cardiomyocytes, yet may be not important in the CNS (Fredholm et al., 2005).

3. ADENOSINE IS A KEY SIGNALING MOLECULE FOR PGD_2-INDUCED SLEEP

Identification of the molecular mechanism by which endogenous PGD_2 promotes sleep led to the discovery that adenosine acts as a mediator of PGD_2-induced sleep (Satoh, Matsumura, Suzuki, & Hayaishi, 1996). PGD_2 was found to be one of the most potent somnogens involved in physiological sleep (Huang et al., 2007, 2011). It is the most abundant PG produced in the brains of rats (Narumiya, Ogorochi, Nakao, &

Hayaishi, 1982) and other mammals including humans (Ogorochi et al., 1984). The PGD_2 concentration in rat cerebrospinal fluid (CSF) shows a circadian fluctuation that parallels the sleep–wake cycle (Pandey, Ram, Matsumura, & Hayaishi, 1995) and becomes elevated with an increase in sleep propensity during sleep deprivation (Ram et al., 1997), suggesting that PGD_2 may play some important function in the CNS. During the search to uncover the neural function of PGD_2, its sleep-inducing activity was discovered (Ueno, Ishikawa, Nakayama, & Hayaishi, 1982), in which microinjection of nanomolar quantities of PGD_2 in the rat brain increased both non-rapid eye movement (non-REM, NREM) and rapid eye movement (REM) sleep (Ueno, Honda, Inoué, & Hayaishi, 1983). Essentially, the same sleep induction was demonstrated in the rhesus monkey (*Macaca mulatta*) during the i.c.v. infusion of PGD_2 (Onoe et al., 1988). The PGD_2-induced sleep was indistinguishable from physiological sleep as judged by several electrophysiological and behavioral criteria.

There are two distinct types of PGD synthase (PGDS) for the production of PGD_2 in the CNS: one is lipocalin-type PGDS (L-PGDS), dominantly expressed in the arachnoid membrane, choroid plexus, and oligodendrocytes (Mizoguchi et al., 2001; Urade et al., 1993); and the other, hematopoietic PGDS (H-PGDS), localized in microglia (Mohri, Eguchi, Suzuki, Urade, & Taniike, 2003). PGD_2 produced by either of these enzymes is secreted into the CSF, and its level there exhibits a circadian alteration in freely moving rats (Pandey et al., 1995). L-PGDS is identical to beta-trace, a major protein in human CSF. Also, the serum L-PGDS/β-trace concentration shows a circadian change with a nocturnal increase, which is suppressed during total sleep deprivation but not affected by deprivation of REM sleep in humans (Jordan et al., 2004). Furthermore, when rats are intraperitoneally given inorganic tetravalent selenium (Se^{4+}) compounds such as selenium tetrachloride ($SeCl_4$), which are specific and reversible inhibitors of PGDS (Islam, Watanabe, Morii, & Hayaishi, 1991), sleep is inhibited time- and dose-dependently and reversibly (Matsumura, Takahata, & Hayaishi, 1991; Takahata et al., 1993). These findings indicate that PGD_2/PGDS plays an important role in sleep induction.

To clarify if PGD_2 is involved in regulating physiological sleep, we examined the effect of inhibition of PGD_2 synthesis by $SeCl_4$ on the sleep of wild-type (WT) mice and knockout (KO) mice for PGDS and PGD_2 receptor (DP_1R), as well as the effect of a DP_1R antagonist, ONO-4127Na, on the sleep of rats. The i.p. injection of $SeCl_4$ into WT mice decreased the PGD_2 content in the brain. It inhibited sleep dose-dependently to induce almost complete insomnia after the administration during the light period when mice normally sleep. The $SeCl_4$-induced insomnia was observed in H-PGDS KO mice but not at all in L-PGDS KO, H- and L-PGDS double KO, or DP_1R KO mice. Furthermore, the DP_1R antagonist ONO-4127Na reduced the sleep of rats during infusion into the subarachnoid space under the rostral basal forebrain (BF) (Qu et al., 2006). These results clearly show that the L-PGDS/PGD_2/DP_1R system plays a pivotal role in the regulation of physiological sleep (Huang et al., 2011).

When PGD_2 was infused to the subarachnoid space of the BF of WT mice, in which DP_1Rs are remarkably abundant, the extracellular adenosine concentration was increased dose-dependently, although the mechanism linking PGD_2 and adenosine accumulation remains unknown. The PGD_2-induced increase in extracellular adenosine was observed in WT mice, but not in KO mice of DP_1Rs (Mizoguchi et al., 2001). Furthermore, the somnogenic effect of PGD_2 was blocked by the intraperitoneal injection of the $A_{2A}R$-specific antagonist KF 17837, indicating that the adenosine increase was dependent on the activation of DP_1R and that endogenous adenosine acting at $A_{2A}Rs$ may be a mediator of the PGD_2-induced sleep.

4. AN INCREASE IN THE EXTRACELLULAR ADENOSINE LEVEL PROMOTES SLEEP

Feldberg and Sherwood (1954) showed that intraventricular injection of adenosine into cats caused a state resembling natural sleep of 30-min duration in 1954. Subsequent pharmacological studies from several laboratories demonstrated that adenosine and its receptor agonists promoted, but antagonists, such as caffeine, inhibited both NREM and REM sleep (for reviews, see Basheer, Strecker, Thakkar, & McCarley, 2004; Huang et al., 2007, 2011; McCarley, 2007).

Under physiological conditions, the actions of highly active ADA, adenosine transport, or AK are important particularly when large amounts of adenosine have to be cleared. Thus, inhibition of adenosine metabolism by blocking AK with ABT-702 (Radek, Decker, & Jarvis, 2004) and ADA with (deoxy)coformycin (Oishi et al., 2008; Okada et al., 2003; Radulovacki, Virus, Djuricic-Nedelson, & Green, 1983), or microdialysis perfusion with an adenosine transport inhibitor *S*-(4-nitrobenzyl)-6-thioinosine (NBTI) in the cholinergic BF (Porkka-Heiskanen et al., 1997), increases extracellular adenosine levels, prolongs sleep, and increases slow-wave activity (SWA) in the EEG.

These pharmacological data in rats are consistent with genetic findings in mice, showing that a genomic region encoding genes which contribute to the regulation of extracellular adenosine levels modifies the rate at which the need for NREM sleep accumulates during wakefulness (Franken, Chollet, & Tafti, 2001). In humans, a genetic variant of ADA, which is associated with the reduced metabolism of adenosine to inosine, has been reported to specifically enhance deep sleep and SWA during sleep, suggesting that genetic variability in the adenosine system contributes to the interindividual variability in brain electrical activity during sleep and wakefulness (Retey et al., 2005).

Sleep deprivation is a useful way to provide a physiological challenge to promote sleep homeostasis. Elevated sleep propensity after prolonged wakefulness is counteracted not only by prolonged sleep duration but also by enhanced sleep intensity (Borb & Achermann, 1999), such as EEG SWA and spindle frequency activity (power within 11–15 Hz). Sleep deprivation increases local adenosine levels in the BF, cortex, and hippocampus in rats and cats during prolonged wakefulness. Enhanced sleep intensity and adenosine levels decline during recovery sleep (Huston et al., 1996; Murillo-Rodriguez, Blanco-Centurion, Gerashchenko, Salin-Pascual, & Shiromani, 2004; Porkka-Heiskanen et al., 1997). Because changes in the adenosine level appear to be more pronounced in the BF than in other cerebral regions (Strecker et al., 2000), the local increase in the extracellular level of adenosine in the BF was proposed to provide a signal for the homeostatic regulation of NREM sleep (see reviews Basheer et al., 2004; McCarley, 2007). Furthermore, nitric oxide production in the BF during sleep deprivation acts to induce sleep through adenosine production (Kalinchuk, McCarley, Porkka-Heiskanen, & Basheer, 2010).

The source of extracellular adenosine from neurons and glia for sleep regulation in the CNS still remains in debate. Sleep is regulated by a homeostatic process that has long been thought to involve adenosine originating from neurons (Jones, 2009). Astrocytes release ATP and glutamate via many pathways including exocytosis (Jourdain et al., 2007; Zhang et al., 2007) and regulate extracellular adenosine by releasing ATP. Conditional astrocyte-selective expression of the SNARE domain of the protein synaptobrevin II (dnSNARE) prevents both tonic and activity-dependent extracellular accumulation of adenosine (Pascual et al., 2005). The dnSNARE transgenic mice lost their compensatory response to sleep deprivation, such as enhancement of SWA and an increase in sleep amount that occurs in WT mice, suggesting the importance of the glia-dependent accumulation of adenosine for sleep drive and homeostasis (Halassa et al., 2009).

Ethanol intake has significant impact on sleep and adenosinergic mechanisms play a role in sleep-promoting effects of ethanol. Studies performed in cell cultures suggest that ethanol inhibits equilibrative nucleoside transporter 1 to block the reuptake of adenosine resulting in increased extracellular adenosine (Williams, Vranizan, Austin, & Krauss, 1993). Local ethanol perfusion in the BF of freely moving rats produced a significant increase in extracellular adenosine with the highest dose of 300 mM ethanol producing a fourfold increase (Sharma, Engemann, Sahota, & Thakkar, 2010). Significant increase in NREM sleep with a concomitant decrease in wakefulness was observed during the first 12 h postethanol, whereas REM sleep remained unaffected. Ethanol administration caused a significant decrease in the number of the BF wake-promoting neurons with c-Fos immunoreactivity. Bilateral microinjections of a selective A_1R antagonist 8-cyclopentyl-1,3-dipropylxanthine into the BF significantly attenuated sleep-promoting effects of ethanol (Thakkar, Engemann, Sharma, & Sahota, 2010). On the other hand, it was reported that the activation of $A_{2A}Rs$ that follows an increase in extracellular adenosine levels caused by the administration of high doses of ethanol plays a role in its hypnotic effects (El Yacoubi, Ledent, Parmentier, Costentin, & Vaugeois, 2003).

5. PREDOMINANT ROLES OF A_{2A} RECEPTOR IN SLEEP REGULATION

There are four adenosine receptor subtypes, all of which are GPCRs: A_1 and A_3 are primarily coupled to the Gi family of G proteins, whereas A_{2A} and A_{2B} are mostly coupled to Gs, like G proteins (Fredholm et al., 2005). Stimulation of A_1Rs inhibits adenylate cyclase through activation of Gi proteins, activates phospholipase C (PLC), opens several types of K^+ channels, and inactivates Q-, P-, and N-type Ca^{2+} channels (Fredholm et al., 2005; Jacobson & Gao, 2006). On the other hand, activation of the $A_{2A}R$ subtype increases adenylate cyclase activity through activation of Gs or Golf (in the striatum) proteins, induces the formation of inositol phosphates, and activates protein kinase C (Fredholm et al., 2005; Jacobson & Gao, 2006). Accumulated findings indicate that both A_1 and $A_{2A}R$ are involved in sleep induction, with $A_{2A}R$ being more important in sleep induction by adenosine. The $A_{2B}R$ is expressed widely, but generally at very low levels, whereas the A_3R is expressed at intermediate levels in the hippocampus and cerebellum (Yaar, Jones, Chen, & Ravid, 2005). Little is known about the functional significance of $A_{2B}R$ and A_3R in sleep. The major properties of adenosine receptors are shown in Table 14.1.

Evidence accumulated in recent years indicates that the $A_{2A}R$ plays a key role in the effects of adenosine on sleep. The $A_{2A}Rs$ are present at high concentration in the CNS, mainly in the striatum, nucleus accumbens (NAc), and olfactory bulb (Fredholm, IJzerman, Jacobson, Klotz, & Linden, 2001). In rats, selective $A_{2A}R$ agonists such as CGS21680 administered to the subarachnoid space adjacent to the BF and lateral preoptic area reliably induce NREM sleep, whereas infusion of A_1R agonists produces weak and variable effects (Satoh, Matsumura, & Hayaishi, 1998; Urade et al., 2003). When infused into the medial pontine reticular formation in rats, CGS 21680 was 10-fold more potent than the A_1R agonist N^6-cyclohexyladenosine in inducing REM sleep (Marks, Shaffery, Speciale, & Birabil, 2003). Sleep induced by the $A_{2A}R$ agonist CGS21680 is followed by a strong rebound of wakefulness after the cessation of CGS21680 infusion (Gerashchenko, Okano, Urade, Inoué, & Hayaishi, 2000). The major site responsible for $A_{2A}R$-mediated sleep is apparently located in or near the rostral BF, as shown by the sleep-promoting effect and c-Fos expression after local infusion of CGS21680 there (Satoh et al., 1999). Furthermore, CGS21680-induced sleep is almost completely abolished in $A_{2A}R$ KO mice, confirming the specificity of CGS21680 for $A_{2A}Rs$ (Urade et al., 2003).

A possible anatomical pathway for $A_{2A}R$-mediated PGD_2-induced sleep was mapped by detecting Fos-positive neurons (Satoh et al., 1999; Scammell et al., 2001). When PGD_2 or the $A_{2A}R$ agonist CGS21680 was infused for 2 h into the PGD_2-sensitive zone of the subarachnoid space of the BF, the number of Fos-positive cells was remarkably increased in the leptomeningeal membrane as well as in the ventrolateral preoptic (VLPO) area, which was concomitant with the induction of NREM sleep (Satoh et al., 1999; Scammell et al., 2001). In contrast, the number of Fos-positive neurons decreased markedly in the TMN of the posterior hypothalamus. The VLPO is known to send specific inhibitory GABAergic and galaninergic efferents to the TMN, the neurons of which contain the ascending histaminergic arousal system (Sherin, Elmquist, Torrealba, & Saper, 1998).

Inhibition of the histaminergic system promotes sleep. *In vivo* microdialysis experiments revealed that infusion of an adenosine $A_{2A}R$ agonist, CGS21680, into the subarachnoid space of the BF inhibited the release of histamine in both the frontal cortex and the medial preoptic area in a dose-dependent manner and increased the GABA release specifically in

Table 14.1 Roles of adenosine receptors in sleep–wake regulation

Receptor	A_1R	$A_{2A}R$	$A_{2B}R$	A_3R
Expression high	Cortex, cerebellum, hippocampus	Caudate putamen, nucleus accumbens, tuberculum olfactorium, olfactory bulb		
Intermediate	Other brain regions		Median eminence	Cerebellum (human?), hippocampus (human?)
Low		Rest of brain		Most of brain (rat, mouse)
G protein	Gi, Go	Gs, Golf	Gs, Gq	Gi
Chromosomal location	chr 1q32.1	chr 22g11.2	chr 17p11.2–12	chr 1p21-13
Roles in animal sleep	Activation of A_1R in BF (Blanco-Centurion et al., 2006), TMN (Oishi et al., 2008), and lateral hypothalamus (Rai et al., 2010; Thakkar, Winston, & McCarley, 2002) induces sleep; whereas activation of A_1R in the lateral preoptic area promotes wakefulness (Methippara, Kumar, Alam, Szymusiak, & McGinty, 2005)	$A_{2A}R$ agonists administered to the brain induce a dramatic increase in sleep (Hong et al., 2005; Satoh et al., 1999, 1996; Satoh, Matsumura, & Hayaishi, 1998); arousal effect of caffeine is seen in A_1R KO, but not in $A_{2A}R$ KO mice (Huang et al., 2005)	None reported	None reported
Roles in human sleep	Prolonged wakefulness induces A_1R upregulation (Elmenhorst et al., 2007)	Variations in $A_{2A}R$ gene contribute to individual sensitivity to caffeine effects on sleep (Retey et al., 2007)	None reported	None reported

From Fredholm et al. (2005) and Fredholm (2007) with modification. ?, not clear.

the TMN but not in the frontal cortex (Hong et al., 2005). The CGS21680-induced inhibition of histamine release was antagonized by perfusion of the TMN with a $GABA_A$ antagonist, picrotoxin, suggesting that the $A_{2A}R$ agonist induced sleep by inhibiting the histaminergic system through an increase in GABA release in the TMN. These results support the original idea of the flip-flop mechanism, whereby sleep is promoted by upregulation of the sleep neurons in the VLPO and at the same time downregulation of the histaminergic wake neurons in the TMN, as mentioned by Saper, Scammell, and Lu (2005).

The local application of CGS21680 increases c-fos expression in the VLPO, but $A_{2A}R$ seems to be undetectable or very weakly expressed in this brain region. Direct activation of sleep-promoting VLPO neurons via postsynaptic stimulation of $A_{2A}R$ was demonstrated in VLPO slices (Gallopin et al., 2005). The intracellular recording of VLPO neurons in rat brain slices demonstrated the existence of two distinct types of VLPO neurons in terms of their responses to serotonin and adenosine. VLPO neurons were inhibited uniformly by two arousal neurotransmitters, noradrenaline and acetylcholine, and mostly by an adenosine A_1R agonist. Serotonin inhibited the type-1 neurons but excited the type-2 neurons.

An $A_{2A}R$ agonist excited postsynaptically the type-2, but not the type-1, neurons. These results suggest that the type-2 neurons are involved in the initiation of sleep and that the type-1 neurons contribute to sleep consolidation, since type-1 neurons are activated only when released from inhibition by arousal systems (Gallopin et al., 2005). Furthermore, the administration of CGS21680 to the rostral BF produced predominant expression of FOS within the shell of the NAc and the medial portion of the olfactory tubercle. The microdialysis perfusion of CGS21680 into the shell of the NAc also exhibited a sleep-promoting effect (Satoh et al., 1999). Scammell et al. (2001) suggested that activation of adenosine $A_{2A}Rs$ in cells of the leptomeninges or NAc would increase the activity of the VLPO. These VLPO neurons may then coordinate the inhibition of multiple wake-promoting regions, resulting in sleep.

Caffeine is the world's most widely used psychoactive drug, and it enhances wakefulness, acting opposite to adenosine. Caffeine binds to A_1R and $A_{2A}R$ with similar high affinities and acts as an antagonist for both receptor subtypes (Fredholm et al., 2001). We used gene-manipulated mice to demonstrate that caffeine promoted wakefulness in WT and A_1R KO mice but not at all in $A_{2A}R$ KO mice (Fig. 14.2), indicating that the arousal

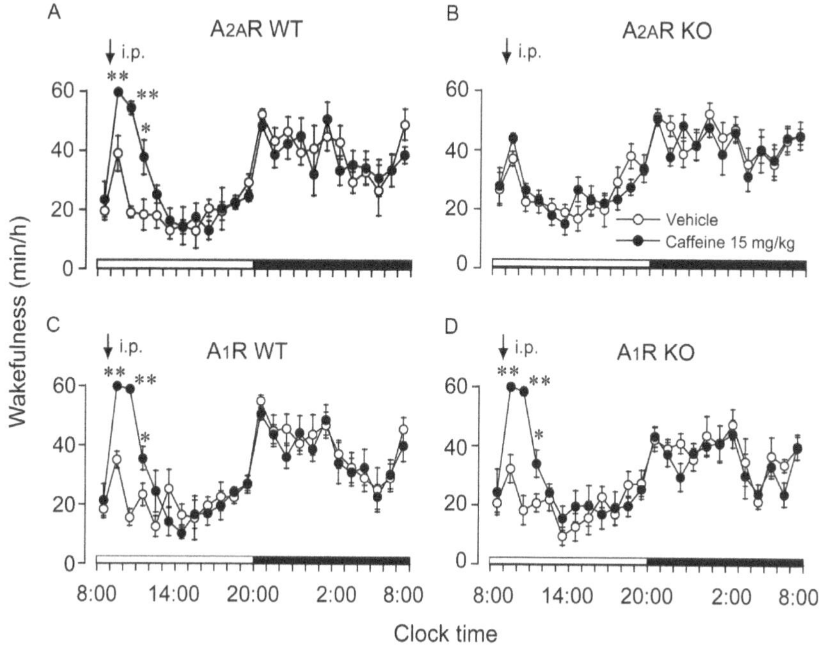

Figure 14.2 Time-course of changes in wakefulness after caffeine 15 mg/kg treatment of A_{2A}R WT (A), A_{2A}R KO (B), A_1R WT (C), and A_1R KO (D) mice. Each circle represents the hourly mean ± SEM ($n = 5$–7). The arrows indicate the injection time (9 AM). $*P < 0.05$; $**P < 0.01$, significantly different from the vehicle, by the paired *t*-test. *Adapted from Huang et al. (2005) with modification and permission.*

effect of caffeine was due to blockade of the A_{2A}R, not the A_1R (Huang et al., 2005). Caffeine may also reduce the hypnotic effects of alcohol via A_{2A}R (El Yacoubi et al., 2003). These results strongly suggest a predominant role of A_{2A}R in sleep regulation.

6. THE INVOLVEMENT OF A_{2A} RECEPTOR IN THE NAc IN SLEEP–WAKE REGULATION

Although caffeine acts as an antagonist for both receptor subtypes with very similar affinities, studies using global genetic KOs of A_1Rs and A_{2A}Rs, in which the receptor is deleted from the entire animal, have found, however, that the A_{2A}R, but not the A_1R, mediates the arousal effect of caffeine (Huang et al., 2005). Moreover, using powerful tools for site-specific gene manipulations, including A_{2A}R KO mice based on the Cre/lox technology and focal $A_{2A}R$ knockdown in rats through local infection with adeno-associated virus (AAV) carrying short-hairpin RNA specific for the mRNA of $A_{2A}Rs$ (Lazarus et al., 2011), we showed that the deletion of the $A_{2A}Rs$ selectively in the NAc shell results in abrogation of caffeine-induced wakefulness. For caffeine to be effective as an antagonist, excitatory $A_{2A}Rs$ must be tonically activated by adenosine. In the NAc shell, this tonic activation is possible because adenosine is available under the most basal conditions and $A_{2A}Rs$ are highly expressed throughout the striatum. Thus, activation of the GABAergic output neurons on the indirect pathway by $A_{2A}Rs$ results in reduced activity of arousal systems in the thalamus, hypothalamus, brainstem, and ultimately, the cerebral cortex.

To understand the projections of $A_{2A}R$-positive neurons in the NAc, Zhang et al. (2013) mapped the projections of $A_{2A}R$ neurons in the NAc by utilizing AAV encoding humanized Renilla green fluorescent protein (hrGFP) as a tracer for long-axonal pathways. The Cre-dependent AAV was injected into the core and shell of the NAc in $A_{2A}R$-Cre mice. Immunohistochemistry was then used to visualize hrGFP, highlighting the perikarya of the $A_{2A}R$ neurons in the injection sites, and their axons in projection regions. The data showed the projections from the NAc distributed to nuclei in the forebrain, diencephalon, and brainstem. In the forebrain, $A_{2A}R$ neurons from all NAc subregions jointly projected to the ventral pallidum, the nucleus of the diagonal band, and the substantia innominata. Heavy projections from the core and the ventral shell, and weaker projections from the medial shell were observed in the lateral hypothalamus and lateral preoptic area. In the brainstem, the NAc projections were found in the ventral tegmental area (VTA), the median raphe nucleus, the dorsal raphe nucleus, and the ventrolateral periaqueductal gray. The results supply a solid base for understanding the roles of the $A_{2A}R$ and $A_{2A}R$ neurons in the NAc, in the regulation of sleep.

In fact, after microinjections of Cre recombinase-dependent AAV carrying light-activated channels (channelrhodopsin) or designer receptors exclusively activated by a designer drug (DREADD) into the NAc of transgenic mice, in which Cre-recombinase is expressed under the $A_{2A}R$ promoter (Durieux et al., 2009), robust induction of NREM sleep can be observed during selective activation of striatopallidal neurons by light or the small molecule clozapine-*N*-oxide (Q. Xu et al., unpublished data).

Thus far identified efferents make it likely that the NAc is able to regulate sleep, probably through inhibition of neuronal populations in the VP, the LHA, the parabrachial nucleus (PB), and the VTA. The NAc can modulate the medial prefrontal cortex (mPFC) via a pathway through the VP and thalamus; in turn, the mPFC projects to arousal-promoting neurons in the hypothalamic TMN, the LHA, and the locus coeruleus (LC) and can provide top-down modulation through its descending projections to these arousal-promoting neurons. The orexinergic and glutamatergic neurons in the LHA not only send major projections to the BF and cerebral cortex but are also reciprocally connected to the VLPO, the TMN, and the LC (Saper, Fuller, Pedersen, Lu, & Scammell, 2010). The NAc shell, but not the NAc core, sends projections to the PB (Li, Chung, Lu, & Cho, 2012), which is an important component of the ascending arousal system and is known to be strongly connected to the BF and LHA. Adenosine acting on excitatory A_{2A}Rs can modulate the activity of GABAergic output neurons in the NAc to inhibit arousal and promote sleep.

A_{2A}Rs are abundantly expressed in the caudate–putamen, NAc, and the tuberculum olfactorium, in which dopamine D_2Rs are colocalized (Missale, Nash, Robinson, Jaber, & Caron, 1998). A_{2A}R and D_2R are involved in sleep–wake regulation in a different and coordinate manner. We found that the D_2R is important for the maintenance of wakefulness (Qu, Huang, Xu, Matsumoto, & Urade, 2008; Qiu et al., 2009; Qu et al., 2010), whereas activation of A_{2A}R induces potent sleep. Opposing effects of A_{2A}R and D_2R have also been shown at the levels of neurotransmitter release, receptor binding, and gene expression (Ferre et al., 2007).

7. A_1 RECEPTORS CONTRIBUTE TO SLEEP INDUCTION IN A REGION-DEPENDENT MANNER

Because of the widespread distribution of cerebral A_1Rs and the inhibition of excitatory neurotransmission following presynaptic A_1R activation, it has been generally assumed that adenosine affects sleep primarily via the A_1R. Several findings from pharmacological studies are compatible with this assumption. For example, i.p. or i.c.v. administration of the A_1R-selective agonist N^6-cyclopentyladenosine (CPA) to rats increases NREM sleep, suppresses REM sleep, and induces changes in the NREM sleep EEG similar to those brought about by prolonged wakefulness (Benington, Kodali, & Heller, 1995; Schwierin, Borbely, & Tobler, 1996). Moreover, microdialysis perfusion of the rat BF with A_1R antisense oligonucleotides reduces NREM sleep and increases wakefulness (Thakkar, Winston, & McCarley, 2003). *In vitro* electrophysiological studies demonstrated that adenosine had a postsynaptic inhibitory effect via A_1R on BF neurons, as well as neurons in the cholinergic laterodorsal tegmental nuclei (Arrigoni, Chamberlin, Saper, & McCarley, 2006; Rainnie, Grunze, McCarley, & Greene, 1994). Christie et al. (2008) suggested that sleep loss induces elevations of BF adenosine acting via A_1R, leading to increased sleepiness and impaired vigilance. During the rat Psychomotor Vigilance Task performance, response latencies and performance lapses of rats increased significantly after adenosine was dialyzed in the BF of rats when compared with baseline (no dialysis) or vehicle dialysis sessions. The codialysis of A_1R antagonist 8-cyclopentyltheophylline with adenosine completely blocked the effects produced by adenosine alone, resulting in performance equivalent to that of the vehicle sessions. These reports suggest that the somnogenic effect of adenosine is mediated by A_1R in the cholinergic region of the BF. In addition, in both humans (Elmenhorst et al., 2007) and rats (Elmenhorst, Basheer, McCarley, & Bauer, 2009), A_1R binding was increased after prolonged wakefulness/sleep deprivation, presumptively due to increases in adenosine, as demonstrated for the BF (Basheer, Bauer, Elmenhorst, Ramesh, & McCarley, 2007).

Although local administration of adenosine or an A_1R agonist into the BF results in an increase in NREM sleep, infusion of the A_1R agonist CPA into the lateral ventricle of mice does not alter the amounts of NREM and REM sleep (Urade et al., 2003), suggesting that activation of A_1R in other brain regions may cause wakefulness. Methippara et al. (2005) studied the effects on sleep of an adenosine transport inhibitor, NBTI, and A_1R agonists/antagonists by microdialyzing them into the lateral preoptic area. The results showed that A_1R stimulation or inhibition of adenosine transport by NBTI induced wakefulness. These observations suggest that adenosine-mediated effects on sleep–wake cycles are site and receptor dependent and that A_1Rs may not be absolutely necessary for sleep homeostasis.

Blanco-Centurion et al. (2006) reported that adenosine levels in the BF did not increase after 6 h of prolonged wakefulness in rats with 95% of their BF cholinergic neurons lesioned. The lesioned rats had an intact sleep drive after 6 and 12 h of prolonged wakefulness. In the absence of cholinergic neurons in the BF, another selective A_1R agonist, N^6-cyclohexyladenosine, effectively induced sleep after administration to the BF. Thus, neither the activity of cholinergic neurons nor the accumulation of adenosine in the BF during wakefulness is necessary for the sleep drive. However, Kalinchuk, McCarley, Stenberg, Porkka-Heiskanen, and Basheer (2008) have found that lesions of the cholinergic BF neurons abolish both the increase of adenosine levels and homeostatic sleep drive. These findings leave open the possibility raised by both studies that A_1R on noncholinergic neurons could affect sleep and that the hypnogenic response to adenosine may not be specific to the BF (Noor Alam, Szymusiak, & McGinty, 2006). To understand the roles of noncholinergic and cholinergic BF neurons in spontaneous sleep–waking and EEG and in homeostatic sleep regulation, Kaur, Junek, Black, and Semba (2008) lesioned noncholinergic BF neurons with ibotenate and cholinergic neurons with 192-IgG saporin and found that the noncholinergic BF neurons promoted cortical activation by inhibiting delta waves, that the cholinergic BF neurons played a nonexclusive role in promoting wakefulness, and that both types of BF neurons played important roles in the increases in NREM sleep and EEG delta power after sleep loss.

Inhibitions of the adenosine system via A_1R in the orexin neuronal regions and histaminergic TMNs are also reported to promote NREM sleep. Thakkar et al. (2002) reported that 30% of the orexin-containing neurons were immunoreactive with A_1R antibody. Perfusion with the A_1R agonist CPA significantly suppressed the sleep–wake discharge activity of perifornical–lateral hypothalamic neurons and suppressed arousal (Rai et al., 2010). We found that adenosine A_1R is also abundant in the TMN. Bilateral injection into the rat TMN of the A_1R agonist CPA significantly increased the amount of NREM sleep. The bilateral injection of adenosine or an inhibitor of ADA, coformycin, into the rat TMN also increased NREM sleep, which was completely abolished by coadministration of a selective A_1R antagonist (Oishi et al., 2008). These results indicate that endogenous adenosine in the TMN suppresses the histaminergic system via A_1Rs to promote NREM sleep.

8. CONCLUSIONS

The roles of adenosine receptors in sleep–wake regulation are summarized in Table 14.1. Adenosine, one of the potent endogenous sleep-promoting factors, accumulates in the brain during wakefulness and induces physiological sleep. Among adenosine receptors responsible for sleep induction, the role of $A_{2A}R$ is predominant in sleep regulation, whereas A_1Rs contribute to sleep induction in a region-dependent manner but may not be absolutely necessary for sleep homeostasis. At the present time, little is known about how and where adenosine originates under physiological conditions or after sleep deprivation.

ACKNOWLEDGMENTS

We thank Drs. Michael Lazarus, Jiang-Fan Chen and Yoshihiro Urade for helpful comments. This study was supported in part by grants-in-aid for scientific research from the National Basic Research Program of China (2011CB711000, 2009ZX09303-006), the National Natural Science Foundation of China (J1210041, 31171010, 31171049, 31121061, 31271164), a key laboratory program of the Education Commission of Shanghai Municipality (ZDSYS14005), the Shanghai Committee of Science and Technology (14JC1400900, 13dz2260700, 13140903100), and the Shanghai Leading Academic Discipline Project (B119).

REFERENCES

Arrigoni, E., Chamberlin, N., Saper, C., & McCarley, R. (2006). Adenosine inhibits basal forebrain cholinergic and noncholinergic neurons in vitro. *Neuroscience*, *140*(2), 403–413.

Basheer, R., Bauer, A., Elmenhorst, D., Ramesh, V., & McCarley, R. W. (2007). Sleep deprivation upregulates A1 adenosine receptors in the rat basal forebrain. *Neuroreport*, *18*(18), 1895–1899.

Basheer, R., Strecker, R. E., Thakkar, M. M., & McCarley, R. W. (2004). Adenosine and sleep-wake regulation. *Progress in Neurobiology*, *73*(6), 379–396.

Benington, J. H., Kodali, S. K., & Heller, H. C. (1995). Stimulation of A1 adenosine receptors mimics the electroencephalographic effects of sleep deprivation. *Brain Research*, *692*(1–2), 79–85.

Blanco-Centurion, C., Xu, M., Murillo-Rodriguez, E., Gerashchenko, D., Shiromani, A. M., Salin-Pascual, R. J., et al. (2006). Adenosine and sleep homeostasis in the basal forebrain. *The Journal of Neuroscience*, *26*(31), 8092–8100.

Borb, A. A., & Achermann, P. (1999). Sleep homeostasis and models of sleep regulation. *Journal of Biological Rhythms*, *14*(6), 559–570.

Brundege, J. M., Diao, L., Proctor, W. R., & Dunwiddie, T. V. (1997). The role of cyclic AMP as a precursor of extracellular adenosine in the rat hippocampus. *Neuropharmacology*, *36*(9), 1201–1210.

Burnstock, G. (2007). Physiology and pathophysiology of purinergic neurotransmission. *Physiological Reviews*, *87*(2), 659–797.

Christie, M. A., Bolortuya, Y., Chen, L. C., McKenna, J. T., McCarley, R. W., & Strecker, R. E. (2008). Microdialysis elevation of adenosine in the basal forebrain produces vigilance impairments in the rat psychomotor vigilance task. *Sleep*, *31*(10), 1393.

Deussen, A., Lloyd, H. G., & Schrader, J. (1989). Contribution of S-adenosylhomocysteine to cardiac adenosine formation. *Journal of Molecular and Cellular Cardiology*, *21*(8), 773–782.

Dunwiddie, T. V., & Masino, S. A. (2001). The role and regulation of adenosine in the central nervous system. *Annual Review of Neuroscience*, *24*(1), 31–55.

Durieux, P. F., Bearzatto, B., Guiducci, S., Buch, T., Waisman, A., Zoli, M., et al. (2009). D2R striatopallidal neurons inhibit both locomotor and drug reward processes. *Nature Neuroscience*, *12*(4), 393–395.

Elmenhorst, D., Basheer, R., McCarley, R. W., & Bauer, A. (2009). Sleep deprivation increases A(1) adenosine receptor density in the rat brain. *Brain Research*, *1258*, 53–58.

Elmenhorst, D., Meyer, P. T., Winz, O. H., Matusch, A., Ermert, J., Coenen, H. H., et al. (2007). Sleep deprivation increases A1 adenosine receptor binding in the human brain: A positron emission tomography study. *The Journal of Neuroscience*, *27*(9), 2410–2415.

El Yacoubi, M., Ledent, C., Parmentier, M., Costentin, J., & Vaugeois, J. M. (2003). Caffeine reduces hypnotic effects of alcohol through adenosine A2A receptor blockade. *Neuropharmacology*, *45*(7), 977–985.

Feldberg, W., & Sherwood, S. (1954). Injections of drugs into the lateral ventricle of the cat. *The Journal of Physiology*, *123*(1), 148–167.

Ferre, S., Diamond, I., Goldberg, S. R., Yao, L., Hourani, S. M., Huang, Z. L., et al. (2007). Adenosine A2A receptors in ventral striatum, hypothalamus and nociceptive circuitry implications for drug addiction, sleep and pain. *Progress in Neurobiology*, *83*(5), 332–347.

Franken, P., Chollet, D., & Tafti, M. (2001). The homeostatic regulation of sleep need is under genetic control. *The Journal of Neuroscience*, *21*(8), 2610–2621.

Fredholm, B. B. (2007). Adenosine, an endogenous distress signal, modulates tissue damage and repair. *Cell Death and Differentiation*, *14*(7), 1315–1323.

Fredholm, B. B., Chen, J. F., Cunha, R. A., Svenningsson, P., & Vaugeois, J. M. (2005). Adenosine and brain function. *International Review of Neurobiology*, *63*, 191–270.

Fredholm, B. B., IJzerman, A. P., Jacobson, K. A., Klotz, K.-N., & Linden, J. (2001). International Union of Pharmacology. XXV. Nomenclature and classification of adenosine receptors. *Pharmacological Reviews*, *53*(4), 527–552.

Gallopin, T., Luppi, P., Cauli, B., Urade, Y., Rossier, J., Hayaishi, O., et al. (2005). The endogenous somnogen adenosine excites a subset of sleep-promoting neurons via A2A receptors in the ventrolateral preoptic nucleus. *Neuroscience*, *134*(4), 1377.

Geiger, J. D., & Fyda, D. M. (1991). Adenosine transport in nervous system tissues. In *Adenosine in the nervous system* (pp. 1–23) London: Academic Press.

Gerashchenko, D., Okano, Y., Urade, Y., Inoué, S., & Hayaishi, O. (2000). Strong rebound of wakefulness follows prostaglandin D2-or adenosine A2a receptor agonist-induced sleep. *Journal of Sleep Research*, *9*(1), 81–87.

Gu, J., Foga, I., Parkinson, F., & Geiger, J. (1995). Involvement of bidirectional adenosine transporters in the release of l-[3H] adenosine from rat brain synaptosomal preparations. *Journal of Neurochemistry*, *64*(5), 2105–2110.

Halassa, M. M., Florian, C., Fellin, T., Munoz, J. R., Lee, S. Y., Abel, T., et al. (2009). Astrocytic modulation of sleep homeostasis and cognitive consequences of sleep loss. *Neuron*, *61*(2), 213–219.

Hong, Z. Y., Huang, Z. L., Qu, W. M., Eguchi, N., Urade, Y., & Hayaishi, O. (2005). An adenosine A2A receptor agonist induces sleep by increasing GABA release in the tuberomammillary nucleus to inhibit histaminergic systems in rats. *Journal of Neurochemistry*, *92*(6), 1542–1549.

Huang, Z. L., Qu, W. M., Eguchi, N., Chen, J. F., Schwarzschild, M. A., Fredholm, B. B., et al. (2005). Adenosine A2A, but not A1, receptors mediate the arousal effect of caffeine. *Nature Neuroscience*, *8*(7), 858–859.

Huang, Z. L., Urade, Y., & Hayaishi, O. (2007). Prostaglandins and adenosine in the regulation of sleep and wakefulness. *Current Opinion in Pharmacology*, 7(1), 33–38.

Huang, Z. L., Urade, Y., & Hayaishi, O. (2011). The role of adenosine in the regulation of sleep. *Current Topics in Medicinal Chemistry*, *11*(8), 1047–1057.

Huston, J., Haas, H., Boix, F., Pfister, M., Decking, U., Schrader, J., et al. (1996). Extracellular adenosine levels in neostriatum and hippocampus during rest and activity periods of rats. *Neuroscience*, *73*(1), 99–107.

Islam, F., Watanabe, Y., Morii, H., & Hayaishi, O. (1991). Inhibition of rat brain prostaglandin D synthase by inorganic selenocompounds. *Archives of Biochemistry and Biophysics*, *289*(1), 161–166.

Jacobson, K. A., & Gao, Z. G. (2006). Adenosine receptors as therapeutic targets. *Nature Reviews Drug Discovery*, *5*(3), 247–264.

Jones, B. E. (2009). Glia, adenosine, and sleep. *Neuron*, *61*(2), 156–157.

Jordan, W., Tumani, H., Cohrs, S., Eggert, S., Rodenbeck, A., Brunner, E., et al. (2004). Prostaglandin D synthase (beta-trace) in healthy human sleep. *Sleep, 27*(5), 867–874.

Jourdain, P., Bergersen, L. H., Bhaukaurally, K., Bezzi, P., Santello, M., Domercq, M., et al. (2007). Glutamate exocytosis from astrocytes controls synaptic strength. *Nature Neuroscience, 10*(3), 331–339.

Kalinchuk, A., McCarley, R., Porkka-Heiskanen, T., & Basheer, R. (2010). Sleep deprivation triggers inducible nitric oxide-dependent nitric oxide production in wake-active basal forebrain neurons. *The Journal of Neuroscience, 30*(40), 13254.

Kalinchuk, A. V., McCarley, R. W., Stenberg, D., Porkka-Heiskanen, T., & Basheer, R. (2008). The role of cholinergic basal forebrain neurons in adenosine-mediated homeostatic control of sleep: Lessons from 192 IgG-saporin lesions. *Neuroscience, 157*(1), 238–253.

Kaur, S., Junek, A., Black, M. A., & Semba, K. (2008). Effects of ibotenate and 192IgG-saporin lesions of the nucleus basalis magnocellularis/substantia innominata on spontaneous sleep and wake states and on recovery sleep after sleep deprivation in rats. *The Journal of Neuroscience, 28*(2), 491–504.

Latini, S., & Pedata, F. (2001). Adenosine in the central nervous system: Release mechanisms and extracellular concentrations. *Journal of Neurochemistry, 79*(3), 463–484.

Lazarus, M., Chen, J. F., Urade, Y., & Huang, Z. L. (2013). Role of the basal ganglia in the control of sleep and wakefulness. *Current Opinion in Neurobiology, 23*(5), 780–785.

Lazarus, M., Huang, Z. L., Lu, J., Urade, Y., & Chen, J. F. (2012). How do the basal ganglia regulate sleep-wake behavior? *Trends in Neurosciences, 35*(12), 723–732.

Lazarus, M., Shen, H. Y., Cherasse, Y., Qu, W. M., Huang, Z. L., Bass, C. E., et al. (2011). Arousal effect of caffeine depends on adenosine A2A receptors in the shell of the nucleus accumbens. *The Journal of Neuroscience, 31*(27), 10067–10075.

Li, C. S., Chung, S., Lu, D. P., & Cho, Y. K. (2012). Descending projections from the nucleus accumbens shell suppress activity of taste-responsive neurons in the hamster parabrachial nuclei. *Journal of Neurophysiology, 108*(5), 1288–1298.

Marks, G., Shaffery, J., Speciale, S., & Birabil, C. (2003). Enhancement of rapid eye movement sleep in the rat by actions at A1 and A2a adenosine receptor subtypes with a differential sensitivity to atropine. *Neuroscience, 116*(3), 913–920.

Matsumura, H., Takahata, R., & Hayaishi, O. (1991). Inhibition of sleep in rats by inorganic selenium compounds, inhibitors of prostaglandin D synthase. *Proceedings of the National Academy of Sciences of the United States of America, 88*(20), 9046–9050.

McCarley, R. W. (2007). Neurobiology of REM and NREM sleep. *Sleep Medicine, 8*(4), 302–330.

Methippara, M. M., Kumar, S., Alam, M. N., Szymusiak, R., & McGinty, D. (2005). Effects on sleep of microdialysis of adenosine A1 and A2a receptor analogs into the lateral preoptic area of rats. *American Journal of Physiology. Regulatory, Integrative and Comparative Physiology, 289*(6), R1715–R1723.

Missale, C., Nash, S. R., Robinson, S. W., Jaber, M., & Caron, M. G. (1998). Dopamine receptors: From structure to function. *Physiological Reviews, 78*(1), 189–225.

Mizoguchi, A., Eguchi, N., Kimura, K., Kiyohara, Y., Qu, W. M., Huang, Z. L., et al. (2001). Dominant localization of prostaglandin D receptors on arachnoid trabecular cells in mouse basal forebrain and their involvement in the regulation of non-rapid eye movement sleep. *Proceedings of the National Academy of Sciences of the United States of America, 98*(20), 11674–11679.

Mohri, I., Eguchi, N., Suzuki, K., Urade, Y., & Taniike, M. (2003). Hematopoietic prostaglandin D synthase is expressed in microglia in the developing postnatal mouse brain. *Glia, 42*(3), 263–274.

Murillo-Rodriguez, E., Blanco-Centurion, C., Gerashchenko, D., Salin-Pascual, R., & Shiromani, P. (2004). The diurnal rhythm of adenosine levels in the basal forebrain of young and old rats. *Neuroscience, 123*(2), 361–370.

Nagy, J., LaBella, L., Buss, M., & Daddona, P. (1984). Immunohistochemistry of adenosine deaminase: Implications for adenosine neurotransmission. *Science*, *224*(4645), 166–168.
Narumiya, S., Ogorochi, T., Nakao, K., & Hayaishi, O. (1982). Prostaglandin D2 in rat brain, spinal cord and pituitary: Basal level and regional distribution. *Life Sciences*, *31*(19), 2093.
Noor Alam, M. D., Szymusiak, R., & McGinty, D. (2006). Adenosinergic regulation of sleep: Multiple sites of action in the brain. *Sleep*, *29*(11), 1384–1385, discussion 1387–1389.
Ogorochi, T., Narumiya, S., Mizuno, N., Yamashita, K., Miyazaki, H., & Hayaishi, O. (1984). Regional distribution of prostaglandins D2, E2, and F2α and related enzymes in postmortem human brain. *Journal of Neurochemistry*, *43*(1), 71–82.
Oishi, Y., Huang, Z. L., Fredholm, B. B., Urade, Y., & Hayaishi, O. (2008). Adenosine in the tuberomammillary nucleus inhibits the histaminergic system via A1 receptors and promotes non-rapid eye movement sleep. *Proceedings of the National Academy of Sciences of the United States of America*, *105*(50), 19992–19997.
Okada, T., Mochizuki, T., Huang, Z. L., Eguchi, N., Sugita, Y., Urade, Y., et al. (2003). Dominant localization of adenosine deaminase in leptomeninges and involvement of the enzyme in sleep. *Biochemical and Biophysical Research Communications*, *312*(1), 29–34.
Onoe, H., Ueno, R., Fujita, I., Nishino, H., Oomura, Y., & Hayaishi, O. (1988). Prostaglandin D2, a cerebral sleep-inducing substance in monkeys. *Proceedings of the National Academy of Sciences of the United States of America*, *85*(11), 4082–4086.
Pandey, H. P., Ram, A., Matsumura, H., & Hayaishi, O. (1995). Concentration of prostaglandin D2 in cerebrospinal fluid exhibits a circadian alteration in conscious rats. *Biochemistry and Molecular Biology International*, *37*(3), 431–437.
Pascual, O., Casper, K. B., Kubera, C., Zhang, J., Revilla-Sanchez, R., Sul, J.-Y., et al. (2005). Astrocytic purinergic signaling coordinates synaptic networks. *Science*, *310*(5745), 113–116.
Peng, L., Huang, R., Yu, A. C., Fung, K. Y., Rathbone, M. P., & Hertz, L. (2005). Nucleoside transporter expression and function in cultured mouse astrocytes. *Glia*, *52*(1), 25–35.
Porkka-Heiskanen, T., Strecker, R. E., Thakkar, M., Bjørkum, A. A., Greene, R. W., & McCarley, R. W. (1997). Adenosine: A mediator of the sleep-inducing effects of prolonged wakefulness. *Science*, *276*(5316), 1265–1268.
Qiu, M., Qu, W., Xu, X., Yan, M., Urade, Y., & Huang, Z. (2009). D (1)/D (2) receptor-targeting L-stepholidine, an active ingredient of the Chinese herb Stephonia, induces non-rapid eye movement sleep in mice. *Pharmacology, Biochemistry, and Behavior*, *94*(1), 16.
Qu, W. M., Huang, Z. L., Xu, X. H., Aritake, K., Eguchi, N., Nambu, F., et al. (2006). Lipocalin-type prostaglandin D synthase produces prostaglandin D2 involved in regulation of physiological sleep. *Proceedings of the National Academy of Sciences of the United States of America*, *103*(47), 17949–17954.
Qu, W. M., Huang, Z. L., Xu, X. H., Matsumoto, N., & Urade, Y. (2008). Dopaminergic D1 and D2 receptors are essential for the arousal effect of modafinil. *The Journal of Neuroscience*, *28*(34), 8462–8469.
Qu, W. M., Xu, X. H., Yan, M. M., Wang, Y. Q., Urade, Y., & Huang, Z. L. (2010). Essential role of dopamine D2 receptor in the maintenance of wakefulness, but not in homeostatic regulation of sleep, in mice. *The Journal of Neuroscience*, *30*(12), 4382–4389.
Radek, R. J., Decker, M. W., & Jarvis, M. F. (2004). The adenosine kinase inhibitor ABT-702 augments EEG slow waves in rats. *Brain Research*, *1026*(1), 74–83.
Radulovacki, M., Virus, R., Djuricic-Nedelson, M., & Green, R. (1983). Hypnotic effects of deoxycorformycin in rats. *Brain Research*, *271*(2), 392–395.
Rai, S., Kumar, S., Alam, M. A., Szymusiak, R., McGinty, D., & Alam, M. N. (2010). A1 receptor mediated adenosinergic regulation of perifornical-lateral hypothalamic area neurons in freely behaving rats. *Neuroscience*, *167*(1), 40.

Rainnie, D. G., Grunze, H. C., McCarley, R. W., & Greene, R. W. (1994). Adenosine inhibition of mesopontine cholinergic neurons: Implications for EEG arousal. *Science, 263*(5147), 689.

Ram, A., Pandey, H., Matsumura, H., Kasahara-Orita, K., Nakajima, T., Takahata, R., et al. (1997). CSF levels of prostaglandins, especially the level of prostaglandin D2, are correlated with increasing propensity towards sleep in rats. *Brain Research, 751*(1), 81.

Retey, J., Adam, M., Honegger, E., Khatami, R., Luhmann, U., Jung, H., et al. (2005). A functional genetic variation of adenosine deaminase affects the duration and intensity of deep sleep in humans. *Proceedings of the National Academy of Sciences of the United States of America, 102*(43), 15676–15681.

Retey, J., Adam, M., Khatami, R., Luhmann, U., Jung, H., Berger, W., et al. (2007). A genetic variation in the adenosine A2A receptor gene (ADORA2A) contributes to individual sensitivity to caffeine effects on sleep. *Clinical Pharmacology & Therapeutics, 81*(5), 692–698.

Rosenberg, P. A., & Li, Y. (1995). Adenylyl cyclase activation underlies intracellular cyclic AMP accumulation, cyclic AMP transport, and extracellular adenosine accumulation evoked by β-adrenergic receptor stimulation in mixed cultures of neurons and astrocytes derived from rat cerebral cortex. *Brain Research, 692*(1), 227–232.

Sala-Newby, G. B., Skladanowski, A. C., & Newby, A. C. (1999). The mechanism of adenosine formation in cells. Cloning of cytosolic 5′-nucleotidase-I. *Journal of Biological Chemistry, 274*(25), 17789–17793.

Saper, C. B., Fuller, P. M., Pedersen, N. P., Lu, J., & Scammell, T. E. (2010). Sleep state switching. *Neuron, 68*(6), 1023–1042.

Saper, C. B., Scammell, T. E., & Lu, J. (2005). Hypothalamic regulation of sleep and circadian rhythms. *Nature, 437*(7063), 1257–1263.

Satoh, S., Matsumura, H., & Hayaishi, O. (1998). Involvement of adenosine A2A receptor in sleep promotion. *The European Journal of Neuroscience, 351*(2), 155–162.

Satoh, S., Matsumura, H., Koike, N., Tokunaga, Y., Maeda, T., & Hayaishi, O. (1999). Region-dependent difference in the sleep-promoting potency of an adenosine A2A receptor agonist. *European Journal of Neuroscience, 11*(5), 1587–1597.

Satoh, S., Matsumura, H., Suzuki, F., & Hayaishi, O. (1996). Promotion of sleep mediated by the A2a-adenosine receptor and possible involvement of this receptor in the sleep induced by prostaglandin D2 in rats. *Proceedings of the National Academy of Sciences of the United States of America, 93*(12), 5980–5984.

Sawynok, J., & Liu, X. J. (2003). Adenosine in the spinal cord and periphery: Release and regulation of pain. *Progress in Neurobiology, 69*(5), 313–340.

Scammell, T., Gerashchenko, D., Mochizuki, T., McCarthy, M., Estabrooke, I., Sears, C., et al. (2001). An adenosine A2a agonist increases sleep and induces Fos in ventrolateral preoptic neurons. *Neuroscience, 107*(4), 653–663.

Schwierin, B., Borbely, A. A., & Tobler, I. (1996). Effects of N6-cyclopentyladenosine and caffeine on sleep regulation in the rat. *European Journal of Pharmacology, 300*(3), 163–171.

Sharma, R., Engemann, S. C., Sahota, P., & Thakkar, M. M. (2010). Effects of ethanol on extracellular levels of adenosine in the basal forebrain: An in vivo microdialysis study in freely behaving rats. *Alcoholism, Clinical and Experimental Research, 34*(5), 813–818.

Sherin, J. E., Elmquist, J. K., Torrealba, F., & Saper, C. B. (1998). Innervation of histaminergic tuberomammillary neurons by GABAergic and galaninergic neurons in the ventrolateral preoptic nucleus of the rat. *The Journal of Neuroscience, 18*(12), 4705–4721.

Strecker, R. E., Morairty, S., Thakkar, M. M., Porkka-Heiskanen, T., Basheer, R., Dauphin, L. J., et al. (2000). Adenosinergic modulation of basal forebrain and preoptic/anterior hypothalamic neuronal activity in the control of behavioral state. *Behavioural Brain Research, 115*(2), 183–204.

Takahata, R., Matsumura, H., Kantha, S. S., Kubo, E., Kawase, K., Sakai, T., et al. (1993). Intravenous administration of inorganic selenium compounds, inhibitors of prostaglandin D synthase, inhibits sleep in freely moving rats. *Brain Research, 623*(1), 65–71.

Thakkar, M. M., Engemann, S. C., Sharma, R., & Sahota, P. (2010). Role of wake-promoting basal forebrain and adenosinergic mechanisms in sleep-promoting effects of ethanol. *Alcoholism, Clinical and Experimental Research, 34*(6), 997–1005.

Thakkar, M. M., Winston, S., & McCarley, R. W. (2002). Orexin neurons of the hypothalamus express adenosine A1 receptors. *Brain Research, 944*(1), 190–194.

Thakkar, M. M., Winston, S., & McCarley, R. W. (2003). A1 receptor and adenosinergic homeostatic regulation of sleep-wakefulness: Effects of antisense to the A1 receptor in the cholinergic basal forebrain. *The Journal of Neuroscience, 23*(10), 4278–4287.

Ueno, R., Honda, K., Inoué, S., & Hayaishi, O. (1983). Prostaglandin D2, a cerebral sleep-inducing substance in rats. *Proceedings of the National Academy of Sciences of the United States of America, 80*(6), 1735–1737.

Ueno, R., Ishikawa, Y., Nakayama, T., & Hayaishi, O. (1982). Prostaglandin D2 induces sleep when microinjected into the preoptic area of conscious rats. *Biochemical and Biophysical Research Communications, 109*(2), 576.

Urade, Y., Eguchi, N., Qu, W. M., Sakata, M., Huang, Z. L., Chen, J. F., et al. (2003). Sleep regulation in adenosine A2A receptor-deficient mice. *Neurology, 61*(11 Suppl. 6), S94–S96.

Urade, Y., Kitahama, K., Ohishi, H., Kaneko, T., Mizuno, N., & Hayaishi, O. (1993). Dominant expression of mRNA for prostaglandin D synthase in leptomeninges, choroid plexus, and oligodendrocytes of the adult rat brain. *Proceedings of the National Academy of Sciences of the United States of America, 90*(19), 9070–9074.

Williams, P. T., Vranizan, K. M., Austin, M. A., & Krauss, R. M. (1993). Associations of age, adiposity, alcohol intake, menstrual status, and estrogen therapy with high-density lipoprotein subclasses. *Arteriosclerosis and Thrombosis, 13*(11), 1654–1661.

Yaar, R., Jones, M., Chen, J. F., & Ravid, K. (2005). Animal models for the study of adenosine receptor function. *Journal of Cellular Physiology, 202*(1), 9–20.

Zhang, Z., Chen, G., Zhou, W., Song, A., Xu, T., Luo, Q., et al. (2007). Regulated ATP release from astrocytes through lysosome exocytosis. *Nature Cell Biology, 9*(8), 945–953.

Zhang, J. P., Xu, Q., Yuan, X. S., Cherasse, Y., Schiffmann, S. N., de Kerchove d'Exaerde, A., et al. (2013). Projections of nucleus accumbens adenosine A2A receptor neurons in the mouse brain and their implications in mediating sleep-wake regulation. *Frontiers in Neuroanatomy, 7*, 43.

Zimmermann, H. (2000). Extracellular metabolism of ATP and other nucleotides. *Naunyn-Schmiedeberg's Archives of Pharmacology, 362*(4), 299–309.

CHAPTER FIFTEEN

Involvement of Adenosine A_{2A} Receptors in Depression and Anxiety

Koji Yamada[1], Minoru Kobayashi, Tomoyuki Kanda

Central Nervous System Research Laboratories, CNS R&D Unit, R&D Division, Kyowa Hakko Kirin Co., Ltd., Shizuoka, Japan

[1]Corresponding author: e-mail address: kohji.yamada@kyowa-kirin.co.jp

Contents

Abstract

When administered to normal healthy patients, a nonselective adenosine A_1/A_{2A} antagonist, caffeine, tended to improve anxiety and depression at low doses and to exacerbate anxiety at high doses. Caffeine also appears to enhance anxiety-related symptoms in patients with panic disorder, and A_{2A} receptor-deficient mice have been reported to exhibit higher anxiety-like behaviors, as well as a lower incidence of depression-like behaviors. Some selective A_{2A} antagonists were reported to ameliorate anxiety-like behaviors in rodents, while others did not affect these behaviors. In addition, most A_{2A} antagonists showed inhibitory effects on depression-like behaviors. The mechanisms underlying the relationship between A_{2A} receptor antagonists and anxiety and depression remain unclear at the present time, although many studies have produced

International Review of Neurobiology, Volume 119
ISSN 0074-7742
http://dx.doi.org/10.1016/B978-0-12-801022-8.00015-5

hypotheses. Given that a selective A_{2A} receptor antagonist has recently become available for use in humans, research on the role of A_{2A} receptors in the treatment of mental illness should progress in the near future.

1. INTRODUCTION

In this stressful contemporary society, the number of patients suffering from psychiatric disorders grows with each passing year. As a result, mental illness has become one of the biggest problems in advanced nations because of the misery of the patients and the large economic burden associated with it (Olesen et al., 2012; Wittchen et al., 2011). Although there are some therapeutic drugs for mental illnesses, it is difficult to say that we can manage these conditions adequately. Therefore, alternative therapeutic options must be developed. Meanwhile, the literature on neuropsychopharmacology increasingly suggests a relationship between mental illness and the modulation of adenosinergic signal transductions in the central nervous system. In particular, depression and anxiety seem to be connected to this modulation, but, up until recently, there have been no agonists or antagonists of adenosine receptors in the brain to serve as potential therapeutic agents.

In 2013, istradefylline was approved as the antiparkinsonian agent in Japan. This is the first therapeutic agent targeting adenosine A_{2A} receptors in the central nervous system, and other candidates for A_{2A} antagonists have emerged in clinical studies. Therefore, in this review, the adenosine receptor-mediated effects on anxiety and depression are summarized, with a focus on the A_{2A} subtype, in order to evaluate whether adenosinergic agents might have therapeutic risks and benefits in the treatment of these disorders.

2. ADENOSINE RECEPTORS AND ANXIETY

2.1. Adenosine receptors and anxiety in humans

2.1.1 Effects of coffee

The molecular targets of caffeine (1,3,7-trimethylxanthine) at physiological concentrations (i.e., by daily coffee drinking) are considered to be the adenosine A_1 and A_{2A} receptors, where caffeine acts as a competitive antagonist, although the agent is also known as a nonselective phosphodiesterase inhibitor at concentrations higher than the physiological conditions (Ferré, 2008). Consequently, caffeine use can provide valuable insights into the effects of adenosine and its possible modulation of human mental conditions, given that there are few adequate compounds available for use in humans at the present time.

The effects of caffeine on anxiety in normal subjects are yet to be defined. In general, higher and lower doses of caffeine appear to be anxiogenic (Bruce, 1990; Green & Suls, 1996; Loke, 1988; Stern, Chait, & Johanson, 1989) and anxiolytic (Haskell, Kennedy, Wesnes, & Scholey, 2005; Lieberman, Tharion, Shukitt-Hale, Speckman, & Tulley, 2002; Lieberman, Wurtman, Emde, & Coviella, 1987), respectively. In addition to dosage level, the effects of caffeine may depend on the habit of drinking coffee or withdrawal time. For patients with a susceptibility to anxiety-related disorders, it has been suggested that caffeine intake can increase anxiety-like symptoms. For example, in patients with panic disorder, the oral administration of caffeine can induce an increase in anxiety-related physical signs more frequent and more intense than in normal control subjects (Boulenger, Uhde, Wolff, & Post, 1984; Charney, Heninger, & Jatlow, 1985). Furthermore, some hormonal parameters that correlate with anxiety may be modified by caffeine. Namely, intravenous administrations of caffeine into healthy subjects have been shown to increase the activity of the hypothalamus–pituitary–adrenal (HPA) axis, as shown by elevations in the adrenocorticotropic hormone (ACTH) and cortisol. In contrast, some reports indicate that caffeine can have beneficial effects in anxiety-prone patients. A recent randomized controlled trial showed that caffeine use led to beneficial clinical responses in 7 out of 12 patients with treatment-resistant obsessive compulsive disorder, noting a 55% reduction of symptom scores in these responders (Koran, Aboujaoude, & Gamel, 2009).

2.1.2 Genetic polymorphism

We may also begin to understand the possible role of A_{2A} receptors in anxiety-related conditions by conducting a polymorphism analysis of the A_{2A} receptor gene. In healthy infrequent caffeine users, research has suggested a significant association between self-reported anxiety level after high-dose caffeine administration and two linked polymorphisms on the A_{2A} receptor gene, namely the 1976C > T and 2592C > T polymorphisms (Alsene, Deckert, Sand, & de Wit, 2003). Moreover, some groups showed the association of this same 1976C > T polymorphism with the incidence of panic disorder in Caucasian populations (Deckert et al., 1998; Hamilton et al., 2004), while the others found no associations in Mongoloid populations (Lam, Hong, & Tsai, 2005; Yamada et al., 2001). Regardless,

A_{2A} receptors seem to have some bearing on the modulation of anxiety-related symptoms in some populations, although the effects of the mentioned polymorphisms on the function of A_{2A} receptors remain unclear. The association between the A_{2A} adenosine receptor gene polymorphism and the physiological response to caffeine in patients with panic disorder is now being targeted in a study conducted by the National Institute of Mental Health in the United States (Lopes, Sebastiao, & Ribeiro, 2011).

2.2. Adenosine receptors and anxiety-like behaviors in animals

Determining physiological effects of adenosine on anxiety in animal models can be difficult because the anxiety state must be indirectly inferred from the animals' behaviors, rather than from spoken or written language, as in the case of human studies. However, more and more useful information has been obtained by evaluating various bioactive substances on adenosine and using animals with modified adenosine-related genes.

2.2.1 Effects of nonselective modulation of adenosine receptors

Broad dose ranges of adenosine have significantly produced anxiolytic effects on wild-type animals in the elevated zero maze and the elevated plus maze (EPM) paradigms (Kulkarni, Singh, & Bishnoi, 2007), which was a popular model of phobic anxiety at the time of the study. In addition, the systemic administration of the neuronal adenosine-uptake inhibitor papaverine, which causes an increase in the extracellular level of adenosine, resulted in a low anxiety state in rats (Zangrossi, Leite, & Graeff, 1992). The micro-injection of an equilibrative nucleoside transporter (ENT) 1-specific antagonist, nitrobenzylthioinosine (nitrobenzylmercaptopurine riboside), into the amygdala of normal mice also reduced anxiety-like behaviors relative to wild-type mice, both in the EPM and light/dark box tests. This effect of the ENT1 antagonist is probably due to the blockade of adenosine influx into cells through ENT1 and then the increased extracellular adenosine under normal physiological conditions. Mice lacking the ENT1 also exhibited the phenotypes with low anxiety (Chen et al., 2007).

In contrast, the acute administration of a nonselective adenosine A_1/A_{2A} receptor antagonist, such as caffeine, has been demonstrated to produce anxiogenic effects in various animal anxiety models, including the EPM, dark/light box, social interaction, and Vogel conflict tests (Baldwin & File, 1989; Bhattacharya, Satyan, & Chakrabarti, 1997; El Yacoubi, Ledent, Parmentier, Costentin, & Vaugeois, 2000; Imaizumi, Miyazaki, & Onodera, 1994; Prediger, Batista, & Takahashi, 2004). Some of the effects depend on dose levels, however, as is seen in humans (Kulkarni et al., 2007). In addition to caffeine, some xanthine-derivative adenosine receptor antagonists with nonspecific affinity for A_1 and A_{2A} receptors exhibited anxiogenic activity. For example, theophylline (1,3-dimethylxanthine), a natural alkaloid found in tea leaves and prescribed as a bronchodilator, showed anxiogenic effects in the EPM, light/dark box, and zero maze tests (Imaizumi et al., 1994; Kulkarni et al., 2007). Other xanthine-derivative nonspecific adenosine receptor antagonists, such as phenyltheophylline (8-phenyl-1,3-dimethylxanthine), aminophylline (theophylline mixed with ethylenediamine in 2:1 ratio), and IBMX (3-isobutyl-1-methylxanthine), also induced anxiety-like behaviors in some of those tests, as did the nonxanthine nonspecific adenosine receptor antagonist CGS15948 (9-chloro-2-(furan-2-yl)-[1,2,4]triazolo[1,5-c]quinazolin-5-amine) (Florio, Prezioso, Papaioannou, & Vertua, 1998; Griebel et al., 1991; Imaizumi et al., 1994; Kulkarni et al., 2007; Zangrossi et al., 1992). As a result, the nonspecific adenosinergic activation is thought to induce the low anxiety state on the whole and *vice versa*.

2.2.2 Effects of genetical manipulations of adenosine A_{2A} receptors

One strain of A_1 receptor deficit (KO) mice, but not another A_1 KO strain (Lang et al., 2003), showed more anxiety-like behaviors than wild-type littermates in several studies, using models such as the dark/light box, hole-board, the plus maze, and open field tests (Gimenez-Llort et al., 2002; Johansson et al., 2001). This difference in behaviors may be due to characteristics of each strain or experimental procedures used in the anxiety tests.

A_{2A} receptor KO mice, already generated in 1990s, breed normally. Their exploratory activity in a novel environment was reduced in an open-field, possibly indicating an increased level of anxiety (Ledent et al., 1997). In addition, the mice exhibited more anxiety-like behaviors than wild-type animals in two models of anxiety. Specifically, they tended to stay in the less threatening dark area of the light/dark box, and they spent more time in the closed, protected arms of an EPM experiment, as opposed to their wild-type littermates (El Yacoubi et al., 2000; Ledent et al., 1997). For these reasons, it appears that complete, lifelong deletion of A_{2A} receptors in the whole brain induces anxiety-like symptoms in animals.

A_{2A} receptors are enriched in the striatum but are also present at lower levels in the extrastriatal forebrain, namely in the hippocampus and cortex (Moreau & Huber, 1999; Schiffmann, Fisone, Moresco, Cunha, & Ferré, 2007; Svenningsson, Le Moine, Fisone, & Fredholm, 1999), where they are thought to contribute to hippocampal (Costenla et al., 2011; Rebola, Lujan, Cunha, & Mulle, 2008) and corticostriatal synaptic plasticity (d'Alcantara, Ledent, Swillens, & Schiffmann, 2001). Selectively deleting A_{2A} receptors in the striatum might then have increased the tendency toward Pavlovian fear conditioning in the mice (both context and tone), and extending the deletion to the rest of the forebrain, including hippocampus and cortex, apparently spared the context fear conditioning and attenuated the tone fear conditioning (Wei et al., 2014). Moreover, focal deletion of hippocampal A_{2A} receptors attenuated the context (but not tone) fear conditioning. Deletion of A_{2A} receptors in the entire forebrain also produced an anxiolytic phenotype in both the EPM and open field tests (Wei et al., 2014). These results suggest that the effects of A_{2A} receptor depletion on anxiety-like behaviors depend on which regions of the brain are manipulated.

2.2.3 Effects of pharmacological modulation of adenosine A_{2A} receptors

The pharmacological modulation of adenosine A_1 receptors in rodents appears to affect the animals' anxiety-like behaviors. The result is similar to the phenotype of A_1 receptor-deficient mice, although there are a few exceptions (Maemoto et al., 2004). Namely, the selective A_1 receptor agonist CPA (N6-cyclopentyladenosine) had anxiolytic effects (Jain, Kemp, Adeyemo, Buchanan, & Stone, 1995) or no effects (El Yacoubi et al., 2000) in the EPM test. The CPA close analogue CCPA (2-chloro-N6-cyclopentyladenosine), another selective A_1 receptor agonist, was reported to be anxiolytic both in the EPM and light/dark box tests, without the significant effects on motor activity (Florio et al., 1998; Prediger et al., 2004). Conversely, the selective A_1 receptor antagonist DPCPX (1,3-dipropyl-8-cyclopentylxanthine) had anxiogenic effects (Imaizumi et al., 1994; Prediger et al., 2004) or no effects (El Yacoubi et al., 2000; Griebel et al., 1991; Prediger, da Silva, Batista, Bittencourt, & Takahashi, 2006) in the EPM, light/dark box, or open field tests. In summary, A_1 receptor agonists mostly seem to have anxiolytic properties, and A_1 receptor antagonists tend to be anxiogenic.

In contrast, previous data about the effects of pharmacological modulations of A_{2A} receptors on anxiety-like behaviors appear to be ambiguous. An A_{2A} receptor agonist, DPMA (N6-[2-(3,5-Dimethoxyphenyl)-2-(2-methylphenyl)-ethyl]adenosine), had no effect on anxiety-like behaviors in the EPM test (Jain et al., 1995; Thorsell, Johnson, & Heilig, 2007).

Another agonist, CGS21680 (2-[*p*-(2-carboxyethyl)-phenethylamino]-5′-*N*-ethylcarboxamido adenosine) had anxiolytic effects in the EPM test, although it was susceptible to differing interpretations because of the robust decrease in locomotor activity at the same doses (El Yacoubi et al., 2000). In addition, it has been shown that DMPX (3,7-dimethyl-1-propargylxanthine), ZM241385 (4-(2-[7-amino-2-(2-furyl) [1,2,4]-triazolo[2,3-a][1,3,5]triazin-5-yl amino]ethyl) phenol), and SCH58261 (5-amino-7-(2-phenylethyl)-2-(2-furyl)-pyrazolo(4,3-e)-1,2,4-triazolo(1,5-c)pyrimidine), all of which are A_{2A} antagonists, did not affect anxiety-like parameters in the EPM test (El Yacoubi et al., 2000; Jain et al., 1995; Prediger et al., 2004).

However, acute administration of istradefylline ((*E*)-1,3-diethyl-8-(3,4-dimethoxystyryl)-7-methyl-3,7-dihydo-1*H*-purine-2,6-dione, formerly known as KW-6002) at 3 mg/kg or higher was recently reported to significantly increase the time spent in the open arms and the number of entries into the open arms in the EPM test. Istradefylline increased locomotor activity in the open field at a dose of 10 mg/kg, but not at a dose of 3 mg/kg or lower. Moreover, following daily treatment of istradefylline for 3 weeks, the anxiolytic-like behaviors tended to increase in the open arm performance of the EPM test and risk assessment behavior, another index which increased as the anxiety level, significantly decreased in the test (Kobayashi et al., 2014). These results suggest that istradefylline possesses anxiolytic properties. The anxiolytic-like action of istradefylline was determined to depend on the blockade of A_{2A} receptors, because this compound is a highly selective A_{2A} receptor antagonist without significant affinities for other receptors, channels, transporters, and enzymes relating to neurotransmission (Saki, Yamada, Koshimura, Sasaki, & Kanda, 2013). These results seem to be contradictory to previous data, which showed that some antagonists of this receptor did not affect the anxiety-like behaviors in the EPM test (El Yacoubi et al., 2000; Jain et al., 1995; Prediger et al., 2004). This discrepancy is not clearly explained at the moment, but it may be due to differences in potency, efficacy, and/or selectivity for modulating the A_{2A} receptor function, or differences in the effects on motor activity. In addition, although A_{2A} receptor KO mice have also been reported to be "more anxious" than their wild-type littermates in some animal models including the EPM test (El Yacoubi et al., 2000; Ledent et al., 1997), this "more anxious" phenotype may be explained by adaptive or compensatory mechanisms associated with the complete deficits in A_{2A} receptor-mediated signal throughout development. Incidentally, there are close similarities between the blockade of A_{2A} receptors and the blockade of serotonin transporters; some selective serotonin reuptake inhibitors (SSRIs) have beneficial effects on several types of anxiety in addition to depression, whereas the genetical deletion of serotonin transporters in mice showed a less-depressive, but more anxious-like phenotype (Haenisch & Bonisch, 2011).

3. ADENOSINE RECEPTORS AND DEPRESSION

3.1. Adenosine receptors and depression in humans

3.1.1 Effect of caffeine

From the epidemiological point of view, it remains unclear whether the habit of drinking coffee or caffeine-intake affects depression-like conditions. An analysis of medical histories of 3600 adult twins failed to show significant associations between lifelong consumption and lifetime histories of various psychiatric disorders including major depression (Kendler, Myers, & Gardner, 2006). In hospitalized adult psychiatric patients, high caffeine consumers tended to have higher scores on some scales of depression than moderate and low consumers. However, the influence of other factors, such as weight and cigarette consumption, also should be considered, because these factors were also closely correlated with caffeine intake (Greden, Fontaine, Lubetsky, & Chamberlin, 1978; Rihs, Muller, & Baumann, 1996). On the other hand, in an analysis of a large epidemiological database ($N=3223$), the consumption of caffeine, even at low doses, was associated with a reduced risk of depression, when compared to the incidence of depression in those with no caffeine intake (Smith, 2009).

It is well known that depressive symptoms often lead to suicide attempts, and some reports have considered the relationship between suicide and caffeine intake. Specifically, the consumption of coffee and/or tea was related to a lower risk of suicide, with the risk becoming progressively lower at higher levels of coffee intake (Klatsky, Armstrong, & Friedman, 1993). In addition, a 10-year follow-up study in 86,626 female registered nurses in United States also suggested a strong inverse association between moderate coffee drinking (2–6 cups/day) and risk of suicide (Kawachi, Willett, Colditz, Stampfer, & Speizer, 1996). However, another study, in which 43,166 subjects were followed-up for the mean 14.6 years, suggests that the effect of caffeine on suicide was dose-dependent (Tanskanen et al., 2000): although they could find a trend connecting lower suicide rates to low and moderate doses of caffeine, suicide was significantly increased in heavy coffee drinkers (≥ 8 cups/day). Nonetheless, an appropriate amount of coffee drinking seems to have preferable, or at least not unfavorable, effects on the risk for and condition of depression, although the relationship between the effects of caffeine and adenosine A_{2A} receptor antagonism remains unclear.

3.2. Adenosine receptors and depression-like behaviors in animals

As is the case with anxiety, the features of the human depression cannot be always recreated in experimental animals. However, there are valuable experimental paradigms using relatively short-term exposure to inescapable and/or uncontrollable stress. These paradigms include the forced swim test (FST), the tail suspension test (TST), and the learned helplessness (LH) test, and they have been widely used in order to evaluate the function of A_{2A} receptors, because these tests were considered to be able to reliably detect depressant/antidepressant drug response (Duman, 2010).

3.2.1 *Effect of nonselective pharmacological modulation of adenosine receptors*

The nonselective activation of adenosine receptors in the brain seems to induce depressive-like symptoms in experimental animals, because adenosine, its analogues, and inhibitors of adenosine degradation, all caused depressant-like behavioral effects in some models of depression. Specifically, both adenosine and 2-chloroadenosine, a nonselective adenosine receptor agonist, prolonged the immobilization period, which was regarded as "behavioral despair" and represented the index of the depression state in the mouse FST. This adenosine-induced immobility was reversed by classical antidepressants such as imipramine and desipramine. Dipyridamole, which was known to inhibit adenosine uptake, potentiated the depressant-like effects of adenosine (Kulkarni & Mehta, 1985). In addition, the inhibition of adenosine deaminase by EHNA (erythro-9-(2-hydroxy-3-nonyl)adenine) and the elevation of adenosine levels in the brain also mimic and coordinate with the effect of inescapable shock (IES) on shuttle-escape performance in the rat LH test (Woodson, Minor, & Job, 1998). However, the pharmacological effects of the nonselective activation of adenosine receptors on depression are somewhat controversial. According to a series of studies by Kaster and his colleagues, administrations of adenosine, either peripherally or intracerebroventricularly, decreased the immobilization period in the mouse FST and TST. These antidepressant-like effects of adenosine were inhibited by both the A_1 receptor-selective antagonist DPCPX and the A_{2A} receptor-selective antagonist ZM241385.

Moreover, these effects appeared to be mediated at least in part through an interaction with the NO–cGMP pathway, serotonin 1_A receptor, or the opioid system (Kaster, Budni, Santos, & Rodrigues, 2007; Kaster et al., 2004; Kaster, Rosa, Santos, & Rodrigues, 2005; Kaster, Santos, & Rodrigues, 2005)

Caffeine effectively reduced immobility time for mice in the FST (El Yacoubi, Costentin, & Vaugeois, 2003). Caffeine also partially reversed the behavioral depression induced by reserpine, which mediated depletion of monoamine neurotransmitters in rats during the FST (Huang & Minor, 2000). In addition, both caffeine and theophylline completely reversed escape deficits in inescapably shocked rats in the LH test (Minor, Chang, & Winslow, 1994; Minor, Winslow, & Chang, 1994; Yamada et al., 2014). These findings suggest that the nonselective pharmacological inhibition of adenosine receptors ameliorates the depressant-like behaviors.

3.2.2 Effects of genetical manipulations of adenosine A_{2A} receptors

The genetic inactivation of the A_{2A} receptor in mice reduced the immobilization period both in the FST and the TST (El Yacoubi et al., 2001, 2003), although these mice were found to display reduced locomotor activities in an open field when compared to control mice (Chen et al., 1999; El Yacoubi et al., 2000; Ledent et al., 1997). Given that many existing antidepressants, such as tricyclic antidepressants (TCA), which can reduce immobility time, also tended to decrease motor activity, it was suggested that the A_{2A} receptor KO mice were in a "lower depressive" state than the control mice.

3.2.3 Effects of pharmacological inhibition of adenosine A_{2A} receptors

The single administration of A_{2A} receptor antagonists SCH58261, ZM241385 and istradefylline into mice reduced the immobilization period or suppressed the "behavioral despair" in the FST and the TST (El Yacoubi et al., 2001, 2003; Yamada, Kobayashi, Mori, Jenner, & Kanda, 2013). Subchronic oral administration of istradefylline into rats, following a schedule based on the original method of Porsolt, Le Pichon, and Jalfre (1977), also significantly reduced the immobilization period in the FST in a dose dependent manner, with the minimum effective dose (MED) being 0.16 mg/kg. This implies that istradefylline, which has already been prescribed for patients with Parkinson's disease, exhibits antidepressant-like effects in a dose range that is similar or lower than the range required for anti-PD effects: In nonclinical studies of the antiparkinsonian activity of orally administered istradefylline, its ED_{50} value for the reversal of reserpine-induced catalepsy in mice was found to be 0.26 mg/kg (Shiozaki et al., 1999), and the MED for the potentiation of rotational behavior induced by L-DOPA in 6-OHDA-induced hemiparkinsonian rats was 3 mg/kg (Koga, Kurokawa, Ochi, Nakamura, & Kuwana, 2000). The reduction of the immobilization period by istradefylline, as well as TCAs, in the rat FST was suppressed through a combined administration with corticosterone. By contrast, the reduction produced by the dopamine D_2 receptor agonist quinpirole in the FST was not affected by corticosterone (Yamada et al., 2013). These results suggest that the decrease in immobility time produced by istradefylline, which was reported to increase the spontaneous motor activity in an open field at 10 mg/kg or higher (Kobayashi et al., 2014), was not due to the motor activity, but probably on the antidepressant activity, whereas the effects of quinpirole in this test were a false-positive due to its psychostimulant activity (D'Aquila, Collu, Pani, Gessa, & Serra, 1994).

The antidepressant-like effect of istradefylline was also demonstrated in the rat LH test (Yamada et al., 2014), another animal model of depression. This model has better face validity and higher predictive validity than the FST and the TST (Deussing, 2006; Duman, 2010; Sherman, Sacquitne, & Petty, 1982). When orally administered acutely or chronically for 2 weeks, istradefylline significantly ameliorated the IES-induced escape deficit. The efficacy was comparable to that of chronic treatment with the TCA desipramine or the SSRI fluoxetine. Additionally, the A_1 selective antagonist DPCPX did not affect the IES-induced escape deficit in the same experiment. These findings indicate that an A_{2A} receptor antagonists may provide a rapid onset-type antidepressant drug in contrast to the delayed onset-type existing drugs, such as SSRIs, because in the LH test, the latter was effective only after chronic administration for several weeks (Yamada et al., 2014).

The effort-related choice test has been considered to be a model for the evaluation of symptoms such as psychomotor slowing, fatigue, or anergia in depression (Salamone, Correa, Nunes, Randall, & Pardo, 2012). In this test, effort-related choice behaviors are evaluated by a concurrent choice task: lever pressings for preferred food, which is a high reinforcement/cost option versus less preferred chow, which is a low reinforcement/cost option. The dopamine D_2 receptor antagonist haloperidol is known to shift this choice behavior, reducing lever-pressing for food and increasing chow intake. Systemic administrations of a selective A_{2A}

antagonists, MSX-3 (3,7-dihydro-8-[(1*E*)-2-(3-methoxyphenyl)ethenyl]-7-methyl-3-[3-(phosphonooxy)propyl]-1-(2-propynyl)-1*H*-purine-2,6-dione) or istradefylline canceled the haloperidol-induced shift of the choice behavior on this test (Farrar et al., 2007; Salamone et al., 2009), but those of an A_1 antagonists such as DPCPX did not.

Taken together, A_{2A} antagonists appear to exhibit a broad spectrum of beneficial effects on depression-like symptoms in model animals.

4. MECHANISMS OF ACTIONS AFTER THE BLOCKADE OF ADENOSINE A_{2A} RECEPTORS

4.1. Sites of the actions

Adenosine A_{2A} receptors are abundantly expressed in striatum of the basal ganglia (Rosin, Robeva, Woodard, Guyenet, & Linden, 1998; Svenningsson et al., 1999). By blocking receptors in this brain region, A_{2A} receptor antagonists are expected to improve the motor symptoms in patients with Parkinson's disease (Mori & Shindou, 2003). Depression model rats, which were subjected to unpredictable chronic mild stress for 40 days, presented an increase in A_{2A} receptor binding in the striatum (Crema et al., 2013). Therefore, the antidepressant-like effects of A_{2A} receptor antagonists may be associated with the normalization of the striatal A_{2A} receptor-upregulations.

Regarding other regions in the brain, A_{2A} receptors are also detected to some extent in nucleus accumbens (NAc) and olfactory tubercule. Moreover, mild but distinct levels of A_{2A} receptors are found in the cortex, amygdala, hippocampus, hypothalamus, thalamus, and cerebellum (Dixon, Gubitz, Sirinathsinghji, Richardson, & Freeman, 1996; Moreau & Huber, 1999; Rosin et al., 1998; Svenningsson et al., 1999). Some of these regions are thought to be important for the regulation of the anxiety and depression states. Among these regions, the NAc is suggested to be involved in the effects of A_{2A} receptors on depression in several reports. When the specific A_{2A} receptor agonist CGS21680 was infused into the NAc of normal rats, the animals acquired an escape deficit similar to the deficit induced by IES-stress in the rat LH test (Yamada et al., 2014). Additionally, effort-related choice behaviors are considered to be regulated by dopaminergic "reward system" activity in the NAc, and local infusions of the A_{2A} receptor agonist CGS 21680 into the NAc reallocate the animal's operant behavior from high reinforcement/cost tasks to low reinforcement/cost tasks (Font et al., 2008; Mingote et al., 2008). Therefore, there seems to be little doubt that the NAc is involved in the effect of A_{2A} receptors on depression, and possibly on anxiety. However, since there was no confirmation or denial regarding the involvement of other sites, further research is needed.

4.2. Interaction between dopamine D_2 receptors and adenosine A_{2A} receptors

In some experiments, the dopamine D_2 receptor antagonist haloperidol and an A_{2A} receptor blockade mutually canceled their effects on depression-like behaviors (El Yacoubi et al., 2001; Farrar et al., 2007). Considering the effectiveness of some drugs in activating dopaminergic signaling for the management of depression (Barone, 2011) and the coexpression of A_{2A} receptors with dopamine D_2 receptors in some neuronal cells (Ongini & Fredholm, 1996), the results of D_2 antagonists may support the hypothesis that the antidepressant-like effects of A_{2A} receptor antagonists might be attributed to the stimulation of D_2 receptor-mediated neurotransmission by a direct negative interaction between the A_{2A} and D_2 receptors.

However, the amelioration of behavioral despair by A_{2A} antagonists, but not the D_2 receptor agonist quinpirole, in the FST was reversed by corticosterone. Furthermore, quinpirole also could not improve the IES stress-induced escape deficit in the LH test without a robust increase in motor activity (Yamada et al., 2014). These results suggest that the antidepressant-like effects of A_{2A} antagonists are qualitatively different from those of dopaminergic agents and, moreover, are independent of the interaction of D_2/A_{2A} receptors.

4.3. Monoaminergic neurotransmitters

Considering the close relationship between therapeutic agents for mental illness and monoaminergic receptors and transporters (Gillman, 2007), it may be hypothesized that the activation of the monoaminergic system, especially serotonergic and noradrenergic components, by A_{2A} receptor blockade would contribute to this relationship. However, the situation is somewhat complicated. For example, PD125944 (N6-[2-(3,5-dimethoxyphenyl)-2-(methylphenyl)ethyl]adenosine), an A_{2A} receptor agonist, was shown to increase basal serotonin release in the hippocampus using *in vivo* microdialysis in freely moving rats (Okada et al., 2001). In addition, restraint stress to rats caused a marked increase in serotonin (and dopamine) levels in the hippocampus, and caffeine, probably via blockade of the A_{2A} receptor, normalized the upregulation of the serotonin levels (Yamato et al., 2002). On the contrary, the serotonin $_{2A/2C}$ receptor antagonist methysergide, the alpha2 adrenergic receptor antagonist yohimbine, and the beta adrenergic receptor antagonist propranolol could not affect the antidepressant-like effects of istradefylline on IES-induced escape deficits in the rat LH model (Yamada et al., 2014). The latter results suggest that, based on the results of the antagonists used so far, neither serotonergic nor adrenergic neurotransmission is involved in the antidepressant-like effect of A_{2A} antagonists.

4.4. The HPA axis

The HPA axis is closely associated with stress-induced disorders, and thus, the indirect modulation of this endocrine system plays an important role in the antidepressant action of TCAs and SSRIs (Pariante & Miller, 2001; Reul, Labeur, Grigoriadis, De Souza, & Holsboer, 1994; Reul, Stec, Soder, & Holsboer, 1993). In addition, the more direct control of the HPA axis by the glucocorticoid receptor antagonist mifepristone or the glucocorticoid-synthesis inhibitor metyrapone relieved the symptoms of depression in clinical studies (Flores, Kenna, Keller, Solvason, & Schatzberg, 2006; Jahn et al., 2004). Interestingly, adenosine controls the release of corticotrophin and cortisol/corticosterone, the latter being major adrenal cortex hormones in humans/rodents (Chau, Rose, & Koos, 1999; Jegou et al., 2003; Scaccianoce, Navarra, Di Sciullo, Angelucci, & Endroczi, 1989). Given that corticosterone suppressed the antidepressant-like effects of A_{2A} antagonists in the rat FST (Yamada et al., 2013), A_{2A} antagonists may affect stress-induced disorders as a consequence of direct and/or indirect modulation of the HPA axis. The neurohormonal analysis parameters of the A_{2A} receptor KO mice may also support this hypothesis about the relationship between the A_{2A} receptor and the HPA axis. Specifically, A_{2A} receptor gene disruption induced a significant increase in alpha-melanocyte-stimulating hormone content in the amygdala and cerebral cortex. Moreover, in the anterior lobe of the pituitary, a substantial increase in ACTH was observed, and plasma corticosterone levels were also significantly higher in the A_{2A} receptor KO mice (Jegou et al., 2003) than in the wild-type littermates.

The hippocampus has already been recognized as a key region in the control of the HPA axis (Pariante & Lightman, 2008), and according to electrophysiologic experiments using slice preparations, A_{2A} receptors appear to modulate hippocampal excitability (Duster, Prickaerts, & Blokland, 2014). Moreover, although subchronic restraint stress caused an increase in

A_{2A} receptor density and induced loss of synaptophysin, a synaptic marker, in rat hippocampal nerve terminals, the blockade of A_{2A} receptors by a the selective antagonist SCH 58261 (7-(2-phenylethyl)-5-amino-2-(2-furyl)-pyrazolo-[4,3-e]-1,2,4-triazolo[1,5-c]pyrimidine) relieves the stress-induced loss of synaptophysin (Cunha, Canas, Oliveira, & Cunha, 2006). Therefore, the effects of A_{2A} antagonists may depend, at least in part, on the direct or indirect control of the HPA axis.

4.5. Brain-derived neurotrophic factor

A polymorphism in the brain-derived neurotrophic factor (BDNF) gene appears to alter susceptibility to neuropsychiatric disorders including depression (Sen et al., 2003). Moreover, in postmortem human brains, a decrease in BDNF in the hippocampus was associated with morbidity of depression (Chen, Dowlatshahi, MacQueen, Wang, & Young, 2001). Based on these observations, the "neurotrophin hypothesis of depression" recently became a focus of attention (Martinowich, Manji, & Lu, 2007). Therefore, it may be interesting to investigate the correlation between A_{2A} receptor function and BDNF, but the situation is somewhat complicated. The activation of A_{2A} receptors has been shown to facilitate the excitatory action of BDNF on synaptic transmission in rat hippocampal CA1 slices (Diogenes, Fernandes, Sebastiao, & Ribeiro, 2004). Conversely, the genetic or pharmacological blockade of A_{2A} receptors suppressed the BDNF-induced potentiation of synaptic transmission and reduced the normal BDNF level in the mouse hippocampus (Tebano et al., 2008). These results suggest that the A_{2A} receptor appears to negatively control BDNF and its transactivation of the downstream pathway, although the A_{2A} receptor blockade generally exhibited the antidepressant activities. Further research is needed to elucidate the relationship between A_{2A} receptors and BDNF.

5. CONCLUSION

Recently, istradefylline became the first therapeutic agent targeting adenosine A_{2A} receptors, and other compounds with a similar mechanism will be available in near future. Therefore, it is important to investigate the effects of the A_{2A} receptor's modulation on anxiety and depression, given their high complication rate with Parkinson's disease (Reijnders, Ehrt, Weber, Aarsland, & Leentjens, 2008). Further studies using animal models would clarify the potential risks and benefits of A_{2A} receptor antagonists for these mental symptoms in patients with Parkinson's disease.

Conversely, the introduction of these agents into clinics as anti-Parkinson's disease drugs can potentially broaden our understanding of the role of A_{2A} receptors in these mental disorders.

REFERENCES

Alsene, K., Deckert, J., Sand, P., & de Wit, H. (2003). Association between A2a receptor gene polymorphisms and caffeine-Induced anxiety. *Neuropsychopharmacology, 28*(9), 1694–1702.

Baldwin, H. A., & File, S. E. (1989). Caffeine-Induced anxiogenesis: The role of adenosine, benzodiazepine and noradrenergic receptors. *Pharmacology, Biochemistry, and Behavior, 32*(1), 181–186.

Barone, P. (2011). Treatment of depressive symptoms in Parkinson's disease. *European Journal of Neurology, 18*(Suppl 1), 11–15.

Bhattacharya, S. K., Satyan, K. S., & Chakrabarti, A. (1997). Anxiogenic action of caffeine: An experimental study in rats. *Journal of Psychopharmacology, 11*(3), 219–224.

Boulenger, J. P., Uhde, T. W., Wolff, E. A., 3rd, & Post, R. M. (1984). Increased sensitivity to caffeine in patients with panic disorders. Preliminary evidence. *Archives of General Psychiatry, 41*(11), 1067–1071.

Bruce, M. S. (1990). The anxiogenic effects of caffeine. *Postgraduate Medical Journal, 66*(Suppl 2), S18–S24.

Charney, D. S., Heninger, G. R., & Jatlow, P. I. (1985). Increased anxiogenic effects of caffeine in panic disorders. *Archives of General Psychiatry, 42*(3), 233–243.

Chau, A., Rose, J. C., & Koos, B. J. (1999). Adenosine modulates corticotropin and cortisol release during hypoxia in fetal sheep. *American Journal of Obstetrics & Gynecology, 180*(5), 1272–1277.

Chen, B., Dowlatshahi, D., MacQueen, G. M., Wang, J. F., & Young, L. T. (2001). Increased hippocampal BDNF immunoreactivity in subjects treated with antidepressant medication. *Biological Psychiatry, 50*(4), 260–265.

Chen, J. F., Huang, Z., Ma, J., Zhu, J., Moratalla, R., Standaert, D., et al. (1999). A_{2A} adenosine receptor deficiency attenuates brain injury induced by transient focal ischemia in mice. *The Journal of Neuroscience, 19*(21), 9192–9200.

Chen, J., Rinaldo, L., Lim, S. J., Young, H., Messing, R. O., & Choi, D. S. (2007). The type 1 equilibrative nucleoside transporter regulates anxiety-like behavior in mice. *Genes, Brain, and Behavior, 6*(8), 776–783.

Costenla, A. R., Diogenes, M. J., Canas, P. M., Rodrigues, R. J., Nogueira, C., Maroco, J., et al. (2011). Enhanced role of adenosine A_{2A} receptors in the modulation of LTP in the rat hippocampus upon ageing. *European Journal of Neuroscience, 34*(1), 12–21.

Crema, L. M., Pettenuzzo, L. F., Schlabitz, M., Diehl, L., Hoppe, J., Mestriner, R., et al. (2013). The effect of unpredictable chronic mild stress on depressive-like behavior and on hippocampal A_1 and striatal A_{2A} adenosine receptors. *Physiology & Behavior, 109*, 1–7.

Cunha, G. M., Canas, P. M., Oliveira, C. R., & Cunha, R. A. (2006). Increased density and synapto-protective effect of adenosine A_{2A} receptors upon sub-chronic restraint stress. *Neuroscience, 141*(4), 1775–1781.

d'Alcantara, P., Ledent, C., Swillens, S., & Schiffmann, S. N. (2001). Inactivation of adenosine A_{2A} receptor impairs long term potentiation in the accumbens nucleus without altering basal synaptic transmission. *Neuroscience, 107*(3), 455–464.

D'Aquila, P. S., Collu, M., Pani, L., Gessa, G. L., & Serra, G. (1994). Antidepressant-like effect of selective dopamine D1 receptor agonists in the behavioural despair animal model of depression. *European Journal of Pharmacology, 262*, 107–111.

Deckert, J., Nothen, M. M., Franke, P., Delmo, C., Fritze, J., Knapp, M., et al. (1998). Systematic mutation screening and association study of the A1 and A2a adenosine receptor genes in panic disorder suggest a contribution of the A2a gene to the development of disease. *Molecular Psychiatry*, *3*(1), 81–85.

Deussing, J. M. (2006). Animal models of depression. *Drug Discovery Today: Disease Models*, *3*(4), 375–383.

Diogenes, M. J., Fernandes, C. C., Sebastiao, A. M., & Ribeiro, J. A. (2004). Activation of adenosine A_{2A} receptor facilitates brain-derived neurotrophic factor modulation of synaptic transmission in hippocampal slices. *The Journal of Neuroscience*, *24*(12), 2905–2913.

Dixon, A. K., Gubitz, A. K., Sirinathsinghji, D. J., Richardson, P. J., & Freeman, T. C. (1996). Tissue distribution of adenosine receptor mRNAs in the rat. *British Journal of Pharmacology*, *118*(6), 1461–1468.

Duman, C. H. (2010). Models of depression. *Vitamins & Hormones*, *82*, 1–21.

Duster, R., Prickaerts, J., & Blokland, A. (2014). Purinergic signaling and hippocampal long-term potentiation. *Current Neuropharmacology*, *12*(1), 37–43.

El Yacoubi, M., Costentin, J., & Vaugeois, J. M. (2003). Adenosine A_{2A} receptors and depression. *Neurology*, *61*(11 Suppl 6), S82–S87.

El Yacoubi, M., Ledent, C., Parmentier, M., Bertorelli, R., Ongini, E., Costentin, J., et al. (2001). Adenosine A_{2A} receptor antagonists are potential antidepressants: Evidence based on pharmacology and A_{2A} receptor knockout mice. *British Journal of Pharmacology*, *134*(1), 68–77.

El Yacoubi, M., Ledent, C., Parmentier, M., Costentin, J., & Vaugeois, J. M. (2000). The anxiogenic-like effect of caffeine in two experimental procedures measuring anxiety in the mouse is not shared by selective A_{2A} adenosine receptor antagonists. *Psychopharmacology*, *148*(2), 153–163.

Farrar, A. M., Pereira, M., Velasco, F., Hockemeyer, J., Muller, C. E., & Salamone, J. D. (2007). Adenosine A_{2A} receptor antagonism reverses the effects of dopamine receptor antagonism on instrumental output and effort-related choice in the rat: Implications for studies of psychomotor slowing. *Psychopharmacology*, *191*(3), 579–586.

Ferré, S. (2008). An update on the mechanisms of the psychostimulant effects of caffeine. *Journal of Neurochemistry*, *105*(4), 1067–1079.

Flores, B. H., Kenna, H., Keller, J., Solvason, H. B., & Schatzberg, A. F. (2006). Clinical and biological effects of mifepristone treatment for psychotic depression. *Neuropsychopharmacology*, *31*(3), 628–636.

Florio, C., Prezioso, A., Papaioannou, A., & Vertua, R. (1998). Adenosine A_1 receptors modulate anxiety in CD1 mice. *Psychopharmacology*, *136*(4), 311–319.

Font, L., Mingote, S., Farrar, A. M., Pereira, M., Worden, L., Stopper, C., et al. (2008). Intra-Accumbens injections of the adenosine A_{2A} agonist CGS 21680 affect effort-Related choice behavior in rats. *Psychopharmacology*, *199*(4), 515–526.

Gillman, P. K. (2007). Tricyclic antidepressant pharmacology and therapeutic drug interactions updated. *British Journal of Pharmacology*, *151*(6), 737–748.

Gimenez-Llort, L., Fernandez-Teruel, A., Escorihuela, R. M., Fredholm, B. B., Tobena, A., Pekny, M., et al. (2002). Mice lacking the adenosine A1 receptor are anxious and aggressive, but are normal learners with reduced muscle strength and survival rate. *European Journal of Neuroscience*, *16*(3), 547–550.

Greden, J. F., Fontaine, P., Lubetsky, M., & Chamberlin, K. (1978). Anxiety and depression associated with caffeinism among psychiatric inpatients. *The American Journal of Psychiatry*, *135*(8), 963–966.

Green, P. J., & Suls, J. (1996). The effects of caffeine on ambulatory blood pressure, heart rate, and mood in coffee drinkers. *Journal of Behavioral Medicine*, *19*(2), 111–128.

Griebel, G., Saffroy-Spittler, M., Misslin, R., Remmy, D., Vogel, E., & Bourguignon, J. J. (1991). Comparison of the behavioural effects of an adenosine A_1/A_2-receptor

antagonist, CGS 15943A, and an A1-selective antagonist, DPCPX. *Psychopharmacology, 103*(4), 541–544.
Haenisch, B., & Bonisch, H. (2011). Depression and antidepressants: Insights from knockout of dopamine, serotonin or noradrenaline re-uptake transporters. *Pharmacology & Therapeutics, 129*, 352–368.
Hamilton, S. P., Slager, S. L., De Leon, A. B., Heiman, G. A., Klein, D. F., Hodge, S. E., et al. (2004). Evidence for genetic linkage between a polymorphism in the adenosine 2A receptor and panic disorder. *Neuropsychopharmacology, 29*(3), 558–565.
Haskell, C. F., Kennedy, D. O., Wesnes, K. A., & Scholey, A. B. (2005). Cognitive and mood improvements of caffeine in habitual consumers and habitual non-consumers of caffeine. *Psychopharmacology, 179*(4), 813–825.
Huang, Q. J., & Minor, T. R. (2000). Adenosine mediates reserpine-induced behavioral depression in rats. *Brain, Behavior, and Immunity, 14*, 101.
Imaizumi, M., Miyazaki, S., & Onodera, K. (1994). Effects of xanthine derivatives in a light/dark test in mice and the contribution of adenosine receptors. *Methods and Findings in Experimental and Clinical Pharmacology, 16*(9), 639–644.
Jahn, H., Schick, M., Kiefer, F., Kellner, M., Yassouridis, A., & Wiedemann, K. (2004). Metyrapone as additive treatment in major depression: A double-blind and placebo-controlled trial. *Archives of General Psychiatry, 61*(12), 1235–1244.
Jain, N., Kemp, N., Adeyemo, O., Buchanan, P., & Stone, T. W. (1995). Anxiolytic activity of adenosine receptor activation in mice. *British Journal of Pharmacology, 116*(3), 2127–2133.
Jegou, S., El Yacoubi, M., Mounien, L., Ledent, C., Parmentier, M., Costentin, J., et al. (2003). Adenosine A_{2A} receptor gene disruption provokes marked changes in melanocortin content and pro-opiomelanocortin gene expression. *Journal of Neuroendocrinology, 15*(12), 1171–1177.
Johansson, B., Halldner, L., Dunwiddie, T. V., Masino, S. A., Poelchen, W., Gimenez-Llort, L., et al. (2001). Hyperalgesia, anxiety, and decreased hypoxic neuroprotection in mice lacking the adenosine A_1 receptor. *Proceedings of the National Academy of Sciences, 98*(16), 9407–9412.
Kaster, M. P., Budni, J., Santos, A. R., & Rodrigues, A. L. (2007). Pharmacological evidence for the involvement of the opioid system in the antidepressant-like effect of adenosine in the mouse forced swimming test. *European Journal of Pharmacology, 576*(1–3), 91–98.
Kaster, M. P., Rosa, A. O., Rosso, M. M., Goulart, E. C., Santos, A. R., & Rodrigues, A. L. (2004). Adenosine administration produces an antidepressant-like effect in mice: Evidence for the involvement of A_1 and A_{2A} receptors. *Neuroscience Letters, 355*(1–2), 21–24.
Kaster, M. P., Rosa, A. O., Santos, A. R., & Rodrigues, A. L. (2005). Involvement of nitric oxide-cGMP pathway in the antidepressant-like effects of adenosine in the forced swimming test. *The International Journal of Neuropsychopharmacology, 8*(4), 601–606.
Kaster, M. P., Santos, A. R., & Rodrigues, A. L. (2005). Involvement of 5-HT_{1A} receptors in the antidepressant-like effect of adenosine in the mouse forced swimming test. *Brain Research Bulletin, 67*(1–2), 53–61.
Kawachi, I., Willett, W. C., Colditz, G. A., Stampfer, M. J., & Speizer, F. E. (1996). A prospective study of coffee drinking and suicide in women. *Archives of Internal Medicine, 156*(5), 521–525.
Kendler, K. S., Myers, J. O., & Gardner, C. (2006). Caffeine intake, toxicity and dependence and lifetime risk for psychiatric and substance use disorders: An epidemiologic and co-twin control analysis. *Psychological Medicine, 36*(12), 1717–1725.
Klatsky, A. L., Armstrong, M. A., & Friedman, G. D. (1993). Coffee, tea, and mortality. *Annals of Epidemiology, 3*(4), 375–381.
Kobayashi, M., Kase, J., Yamada, K., Mori, A., Jenner, P., & Kanda, T. (2014). Anxiolytic-like activity of the adenosine A2A receptor antagonist, istradefylline (KW-6002) in the rat elevated plus maze. Submitted.

Koga, K., Kurokawa, M., Ochi, M., Nakamura, J., & Kuwana, Y. (2000). Adenosine A_{2A} receptor antagonists KF17837 and KW-6002 potentiate rotation induced by dopaminergic drugs in hemi-Parkinsonian rats. *European Journal of Pharmacology*, *408*, 249–255.

Koran, L. M., Aboujaoude, E., & Gamel, N. N. (2009). Double-blind study of dextroamphetamine versus caffeine augmentation for treatment-resistant obsessive-compulsive disorder. *Journal of Clinical Psychiatry*, *70*(11), 1530–1535.

Kulkarni, S. K., & Mehta, A. K. (1985). Purine nucleoside-mediated immobility in mice: Reversal by antidepressants. *Psychopharmacology*, *85*(4), 460–463.

Kulkarni, S. K., Singh, K., & Bishnoi, M. (2007). Involvement of adenosinergic receptors in anxiety related behaviours. *Indian Journal of Experimental Biology*, *45*(5), 439–443.

Lam, P., Hong, C. J., & Tsai, S. J. (2005). Association study of A_{2a} adenosine receptor genetic polymorphism in panic disorder. *Neuroscience Letters*, *378*(2), 98–101.

Lang, U. E., Lang, F., Richter, K., Vallon, V., Lipp, H. P., Schnermann, J., et al. (2003). Emotional instability but intact spatial cognition in adenosine receptor 1 knock out mice. *Behavioural Brain Research*, *145*(1–2), 179–188.

Ledent, C., Vaugeois, J. M., Schiffmann, S. N., Pedrazzini, T., El Yacoubi, M., Vanderhaeghen, J. J., et al. (1997). Aggressiveness, hypoalgesia and high blood pressure in mice lacking the adenosine A2a receptor. *Nature*, *388*(6643), 674–678.

Lieberman, H. R., Tharion, W. J., Shukitt-Hale, B., Speckman, K. L., & Tulley, R. (2002). Effects of caffeine, sleep loss, and stress on cognitive performance and mood during U.S. Navy SEAL training. Sea-Air-Land. *Psychopharmacology*, *164*(3), 250–261.

Lieberman, H. R., Wurtman, R. J., Emde, G. G., & Coviella, I. L. (1987). The effects of caffeine and aspirin on mood and performance. *Journal of Clinical Psychopharmacology*, 7(5), 315–320.

Loke, W. H. (1988). Effects of caffeine on mood and memory. *Physiology & Behavior*, *44*(3), 367–372.

Lopes, L. V., Sebastiao, A. M., & Ribeiro, J. A. (2011). Adenosine and related drugs in brain diseases: Present and future in clinical trials. *Current Topics in Medicinal Chemistry*, *11*(8), 1087–1101.

Maemoto, T., Tada, M., Mihara, T., Ueyama, N., Matsuoka, H., Harada, K., et al. (2004). Pharmacological characterization of FR194921, a new potent, selective, and orally active antagonist for central adenosine A_1 receptors. *Journal of Pharmacological Sciences*, *96*(1), 42–52.

Martinowich, K., Manji, H., & Lu, B. (2007). New insights into BDNF function in depression and anxiety. *Nature Neuroscience*, *10*(9), 1089–1093.

Mingote, S., Font, L., Farrar, A. M., Vontell, R., Worden, L. T., Stopper, C. M., et al. (2008). Nucleus accumbens adenosine A_{2A} receptors regulate exertion of effort by acting on the ventral striatopallidal pathway. *The Journal of Neuroscience*, *28*(36), 9037–9046.

Minor, T. R., Chang, W. C., & Winslow, J. L. (1994). Stress and adenosine: I Effect of methylxanthine and amphetamine stimulants on learned helplessness in rats. *Behavioral Neuroscience*, *108*(2), 254–264.

Minor, T. R., Winslow, J. L., & Chang, W. C. (1994). Stress and adenosine: II. Adenosine analogs mimic the effect of inescapable shock on shuttle-escape performance in rats. *Behavioral Neuroscience*, *108*(2), 265–276.

Moreau, J. L., & Huber, G. (1999). Central adenosine A_{2A} receptors: An overview. *Brain Research. Brain Research Reviews*, *31*(1), 65–82.

Mori, A., & Shindou, T. (2003). Modulation of GABAergic transmission in the striatopallidal system by adenosine A_{2A} receptors: A potential mechanism for the antiparkinsonian effects of A_{2A} antagonists. *Neurology*, *61*(11 Suppl 6), S44–S48.

Okada, M., Nutt, D. J., Murakami, T., Zhu, G., Kamata, A., Kawata, Y., et al. (2001). Adenosine receptor subtypes modulate two major functional pathways for hippocampal serotonin release. *The Journal of Neuroscience*, *21*(2), 628–640.

Olesen, J., Gustavsson, A., Svensson, M., Wittchen, H. U., Jonsson, B., CDBE2010 study group, et al. (2012). The economic cost of brain disorders in Europe. *European Journal of Neurology, 19*(1), 155–162.

Ongini, E., & Fredholm, B. B. (1996). Pharmacology of adenosine A_{2A} receptors. *Trends in Pharmacological Sciences, 17*(10), 364–372.

Pariante, C. M., & Lightman, S. L. (2008). The HPA axis in major depression: Classical theories and new developments. *Trends in Neurosciences, 31*(9), 464–468.

Pariante, C. M., & Miller, A. H. (2001). Glucocorticoid receptors in major depression: Relevance to pathophysiology and treatment. *Biological Psychiatry, 49*(5), 391–404.

Porsolt, R. D., Le Pichon, M., & Jalfre, M. (1977). Depression: A new animal model sensitive to antidepressant treatments. *Nature, 266*, 730–732.

Prediger, R. D., Batista, L. C., & Takahashi, R. N. (2004). Adenosine A_1 receptors modulate the anxiolytic-like effect of ethanol in the elevated plus-maze in mice. *European Journal of Pharmacology, 499*(1–2), 147–154.

Prediger, R. D., da Silva, G. E., Batista, L. C., Bittencourt, A. L., & Takahashi, R. N. (2006). Activation of adenosine A1 receptors reduces anxiety-like behavior during acute ethanol withdrawal (hangover) in mice. *Neuropsychopharmacology, 31*(10), 2210–2220.

Rebola, N., Lujan, R., Cunha, R. A., & Mulle, C. (2008). Adenosine A_{2A} receptors are essential for long-term potentiation of NMDA-EPSCs at hippocampal mossy fiber synapses. *Neuron, 57*(1), 121–134.

Reijnders, J. S., Ehrt, U., Weber, W. E., Aarsland, D., & Leentjens, A. F. (2008). A systematic review of prevalence studies of depression in Parkinson's disease. *Movement Disorders, 23*(2), 183–189.

Reul, J. M., Labeur, M. S., Grigoriadis, D. E., De Souza, E. B., & Holsboer, F. (1994). Hypothalamic-pituitary-adrenocortical axis changes in the rat after long-term treatment with the reversible monoamine oxidase-A inhibitor moclobemide. *Neuroendocrinology, 60*(5), 509–519.

Reul, J. M., Stec, I., Soder, M., & Holsboer, F. (1993). Chronic treatment of rats with the antidepressant amitriptyline attenuates the activity of the hypothalamic-pituitary-adrenocortical system. *Endocrinology, 133*(1), 312–320.

Rihs, M., Muller, C., & Baumann, P. (1996). Caffeine consumption in hospitalized psychiatric patients. *European Archives of Psychiatry and Clinical Neuroscience, 246*(2), 83–92.

Rosin, D. L., Robeva, A., Woodard, R. L., Guyenet, P. G., & Linden, J. (1998). Immunohistochemical localization of adenosine A_{2A} receptors in the rat central nervous system. *Journal of Comparative Neurology, 401*(2), 163–186.

Saki, M., Yamada, K., Koshimura, E., Sasaki, K., & Kanda, T. (2013). In vitro pharmacological profile of the A_{2A} receptor antagonist istradefylline Naunyn-Schmiedeberg's. *Archives of Pharmacology, 386*, 963–972.

Salamone, J. D., Correa, M., Nunes, E. J., Randall, P. A., & Pardo, M. (2012). The behavioral pharmacology of effort-related choice behavior: Dopamine, adenosine and beyond. *Journal of the Experimental Analysis of Behavior, 97*(1), 125–146.

Salamone, J. D., Farrar, A. M., Font, L., Patel, V., Schlar, D. E., Nunes, E. J., et al. (2009). Differential actions of adenosine A_1 and A_{2A} antagonists on the effort-related effects of dopamine D_2 antagonism. *Behavioural Brain Research, 201*(1), 216–222.

Scaccianoce, S., Navarra, D., Di Sciullo, A., Angelucci, L., & Endroczi, E. (1989). Adenosine and pituitary-adrenocortical axis activity in the rat. *Neuroendocrinology, 50*(4), 464–468.

Schiffmann, S. N., Fisone, G., Moresco, R., Cunha, R. A., & Ferré, S. (2007). Adenosine A_{2A} receptors and basal ganglia physiology. *Progress in Neurobiology, 83*(5), 277–292.

Sen, S., Nesse, R. M., Stoltenberg, S. F., Li, S., Gleiberman, L., Chakravarti, A., et al. (2003). A BDNF coding variant is associated with the NEO personality inventory domain neuroticism, a risk factor for depression. *Neuropsychopharmacology, 28*(2), 397–401.

Sherman, A. D., Sacquitne, J. L., & Petty, F. (1982). Specificity of the learned helplessness model of depression. *Pharmacology, Biochemistry, and Behavior, 16*(3), 449–454.
Shiozaki, S., Ichikawa, S., Nakamura, J., Kitamura, S., Yamada, K., & Kuwana, Y. (1999). Actions of adenosine A2A receptor antagonist KW-6002 on drug-induced catalepsy and hypokinesia caused by reserpine or MPTP. *Psychopharmacology, 147*, 90–95.
Smith, A. P. (2009). Caffeine, cognitive failures and health in a non-working community sample. *Human Psychopharmacology, 24*(1), 29–34.
Stern, K. N., Chait, L. D., & Johanson, C. E. (1989). Reinforcing and subjective effects of caffeine in normal human volunteers. *Psychopharmacology, 98*(1), 81–88.
Svenningsson, P., Le Moine, C., Fisone, G., & Fredholm, B. B. (1999). Distribution, biochemistry and function of striatal adenosine A_{2A} receptors. *Progress in Neurobiology, 59*(4), 355–396.
Tanskanen, A., Tuomilehto, J., Viinamaki, H., Vartiainen, E., Lehtonen, J., & Puska, P. (2000). Heavy coffee drinking and the risk of suicide. *European Journal of Epidemiology, 16*(9), 789–791.
Tebano, M. T., Martire, A., Potenza, R. L., Gro, C., Pepponi, R., Armida, M., et al. (2008). Adenosine A_{2A} receptors are required for normal BDNF levels and BDNF-induced potentiation of synaptic transmission in the mouse hippocampus. *Journal of Neurochemistry, 104*(1), 279–286.
Thorsell, A1., Johnson, J., & Heilig, M. (2007). Effect of the adenosine A_{2a} receptor antagonist 3,7-dimethyl-propargylxanthine on anxiety-like and depression-like behavior and alcohol consumption in Wistar Rats. *Alcoholism, Clinical and Experimental Research, 31*(8), 1302–1307.
Wei, C. J., Augusto, E., Gomes, C. A., Singer, P., Wang, Y., Boison, D., et al. (2014). Regulation of fear responses by striatal and extrastriatal adenosine A_{2A} receptors in forebrain. *Biological Psychiatry, 75*, 855–863.
Wittchen, H. U., Jacobi, F., Rehm, J., Gustavsson, A., Svensson, M., Jonsson, B., et al. (2011). The size and burden of mental disorders and other disorders of the brain in Europe 2010. *European Neuropsychopharmacology, 21*(9), 655–679.
Woodson, J. C., Minor, T. R., & Job, R. F. (1998). Inhibition of adenosine deaminase by erythro-9-(2-hydroxy-3-nonyl)adenine (EHNA) mimics the effect of inescapable shock on escape learning in rats. *Behavioral Neuroscience, 112*(2), 399–409.
Yamada, K., Hattori, E., Shimizu, M., Sugaya, A., Shibuya, H., & Yoshikawa, T. (2001). Association studies of the cholecystokinin B receptor and A2a adenosine receptor genes in panic disorder. *Journal of Neural Transmission, 108*(7), 837–848.
Yamada, K., Kobayashi, M., Mori, A., Jenner, P., & Kanda, T. (2013). Antidepressant-like activity of the adenosine A_{2A} receptor antagonist, istradefylline (KW-6002), in the forced swim test and the tail suspension test in rodents. *Pharmacology, Biochemistry, and Behavior, 114–115*, 23–30.
Yamada, K., Kobayashi, M., Shiozaki, S., Ohta, T., Mori, A., Jenner, P., et al. (2014). Antidepressant activity of the adenosine A_{2A} receptor antagonist, istradefylline (KW-6002) on learned helplessness in rats. *Psychopharmacology, 231*, 2839–2849.
Yamato, T., Yamasaki, S., Misumi, Y., Kino, M., Obata, T., & Aomine, M. (2002). Modulation of the stress response by coffee: An in vivo microdialysis study of hippocampal serotonin and dopamine levels in rat. *Neuroscience Letters, 332*(2), 87–90.
Zangrossi, H., Jr., Leite, J. R., & Graeff, F. G. (1992). Anxiolytic effect of carbamazepine in the elevated plus-maze: Possible role of adenosine. *Psychopharmacology, 106*(1), 85–89.

CHAPTER SIXTEEN

The Adenosine Neuromodulation System in Schizophrenia

Daniel Rial*, Diogo R. Lara†, Rodrigo A. Cunha*,‡,1

*CNC-Center for Neuroscience and Cell Biology, University of Coimbra, Coimbra, Portugal

†Laboratório de Neuroquímica e Psicofarmacologia, Departamento de Biologia Celular e Molecular, Faculdade de Biociências, Pontifícia Universidade Católica do Rio Grande do Sul, Porto Alegre, Brazil

‡Faculty of Medicine, University of Coimbra, Coimbra, Portugal

[1]Corresponding author: e-mail address: cunharod@gmail.com

Contents

Abstract

The management of schizophrenia endophenotypes, namely positive, negative, and cognitive symptoms is still an open goal, justifying the search of novel therapeutic avenues. We now review the evidence supporting the interest in targeting the adenosine modulation system to counteract the core features of schizophrenia. This interest is forwarded by the combined ability of strategies aimed at bolstering adenosine levels

International Review of Neurobiology, Volume 119
ISSN 0074-7742
http://dx.doi.org/10.1016/B978-0-12-801022-8.00016-7

together with the increasingly recognized impact of adenosine A_{2A} receptors to control dopaminergic signaling, working memory, and behavioral sensitization; this is further heralded by the suggested clinical effectiveness of therapies increasing extracellular adenosine such as dipyridamole and allopurinol and the emergent recognition of a role for adenosine in neurodevelopment. Finally, the combined role of A_1 and A_{2A} receptors in assisting the implementation of adaptive changes and encoding of information salience in neuronal circuits together with the adaptive alterations of A_1 and A_{2A} receptor density upon brain dysfunction prompts the novel working hypothesis that the parallel imbalance of adenosine formation and of A_1 and A_{2A} receptors blurs the adequate encoding of information salience in neuronal circuits, which we propose to be a core pathogenic feature in the development of schizophrenia endophenotypes. This proposal should also provide a rationale to assist the design of future therapeutic intervention targeting the adenosine modulation system to manage schizophrenia endophenotypes: these should not be based only on an attempt to target adenosine kinase-A_1 receptors or only A_{2A} receptors, but should instead simultaneously target these two arms of the adenosine modulation system.

1. CLINICAL FEATURES OF SCHIZOPHRENIA

Schizophrenia is a chronic and often disabling mental illness with a prevalence of around 0.7% of the population (Mcgrath, Saha, Chant, & Welham, 2008). The clinical presentation of this syndrome involves symptoms divided in "positive" (presence of abnormalities), "negative" (absence of normal features), and cognitive (Insel, 2010). Positive symptoms refer to delusions (thought disturbance) and hallucinations (perceptual disturbances), which are the so-called psychotic manifestations. Negative symptoms include affect flattening, apathy, poor speech, lack of pleasure, and social withdrawal, whereas cognitive symptoms are expressed as deficits in attention, executive functioning, memory, and psychomotor speed of processing. These clusters of symptoms are not necessarily present in all patients and differ in course, responses to treatment, and impact on daily functioning (Foussias, Agid, Fervaha, & Remington, 2013; Rajji, Ismail, & Mulsant, 2009). These characteristics illustrate the clinical complexity and heterogeneity of this disorder.

The clinical presentation is commonly accompanied by psychiatric and clinical comorbidities. Substance use disorders, particularly heavy cigarette smoking, co-occur in at least 50% of patients (Thoma & Daum, 2013). More recently, all anxiety disorders, such as obsessive–compulsive disorder (OCD), social phobia, and posttraumatic stress disorders (PTSD) were found to be 2–4 × more prevalent in schizophrenic subjects (10–15%) than in the general population (2–5%) (Achim et al., 2011). Clinical comorbidities also show an interesting pattern in this patient population. There is a clear excess of type-2 diabetes mellitus, constipation, and Parkinson's disease, possibly influenced by antipsychotic treatment, but also epilepsy, chronic obstructive pulmonary disorders (COPD), and liver diseases (Schoepf, Uppal, Potluri, & Heun, 2014; Smith, Langan, Mclean, Guthrie, & Mercer, 2013). In contrast, cardiovascular disease, hypertension, hyperlipidemia, cataract, and cancer are less common than in the general population (Schoepf et al., 2014; Smith et al., 2013), which is particularly striking given their massive smoking habits and low level of physical activity. It is still unclear to what degree these lower rates are due to under-diagnosis, as other common conditions (e.g., type-2 DM) could be identified, but their leading mortality causes are cardiovascular disease and cancer (Crump, Winkleby, Sundquist, & Sundquist, 2013).

Schizophrenia usually manifests itself in adolescence or young adulthood through the emergence of a psychotic episode. However, a neurodevelopmental basis from schizophrenia has been highlighted in the past three decades (Lewis & Levitt, 2002; Weinberger, 1987). Patients often have neurodevelopment delays and cognitive impairment since early childhood (Sørensen et al., 2010). Longitudinal population studies have also evidenced maturation deficits in the first year of life and persistently reduced IQ in children that eventually develop the illness (Reichenberg et al., 2010; Woodberry, Giuliano, & Seidman, 2008). Childhood deficits in verbal memory, gross motor skills, and attention were also predictive of later development of the disorder in the offspring of schizophrenic subjects (Erlenmeyer-Kimling et al., 2000).

1.1. Neurodevelopmental basis of schizophrenia

The concept of schizophrenia as a disorder of neurodevelopment posits that a transient insult or impairment in development early in life may only manifest itself when later compensatory mechanisms may fail to control circuit and neurotransmitter dysfunctions (Insel, 2010; Thompson & Levitt, 2010). This pathophysiological process has also been coined the "two-hit hypothesis" and the latter phase could result from "developmental allostasis" (Thompson & Levitt, 2010). Under this perspective, the disorder begins in prenatal or perinatal life, with disruption of normal brain development (e.g., neuronal proliferation, migration) by genetic and/or environmental factors, such as viral infections, malnutrion, vitamin D deficiency, and hypoxia

(Mcgrath, Burne, Féron, MacKay-Sim, & Eyles, 2010; Tandon, Keshavan, & Nasrallah, 2008). Later, social adversities in childhood and adolescence, such as being an ethnic minority or a victim of bullying, may also promote the emergence of psychotic symptoms (Varese et al., 2012). These factors may act synergistically with cannabis use, possibly impairing pruning of the neuronal arbor, myelin deposition, and the normal balance between inhibitory and excitatory pathways. Such deregulations may translate into a prodromal phase characterized by social and cognitive deficits (Morgan et al., 2014) and ultimately into the eruption of psychotic symptoms, which is typically followed by a relapsing–remitting course that leads to chronic disability (Insel, 2010).

1.2. Therapeutic management of schizophrenia

For more than 50 years, the main treatment strategy has been the regular use of antipsychotics, which share dopamine D2 receptor (D2R) antagonism as the main mechanism of action (Ginovart & Kapur, 2012). Clinical response to antipsychotics depends on the level of D2R occupancy and start rapidly when this occupation takes place effectively (Ginovart & Kapur, 2012; Narendran et al., 2009). Except for a higher efficacy of clozapine and a marginal superiority of olanzapine and amisulpride, all antipsychotics show similar efficacy, but distinct adverse event profiles. Notably, the currently used antipsychotics are effective mostly for positive symptoms (Ginovart & Kapur, 2012).

This inability of the currently used antipsychotics to effectively manage negative and cognitive symptoms has contributed to the emergence of several alternative mechanisms to explain the etiology of schizophrenia. Glutamate enhancing agents such as D-serine, *N*-acetylcysteine, and sarcosine may produce beneficial effects mostly for negative symptoms when added to antipsychotics, but not to clozapine (Singh & Singh, 2011). Sodium benzoate, which may increase D-serine by inhibiting D-amino acid oxidase, may also afford similar benefits (Lane et al., 2013). A recent study also found a rapid, robust, and sustained effect of a single administration of sodium nitroprusside in schizophrenic patients, aiming at increasing nitric oxide (Hallak et al., 2013). Intriguingly, this effect does not seem to involve the soluble guanylate cyclase/cGMP pathway (Issy, Pedrazzi, Yoneyama, & Del-Bel, 2014). Other interesting treatment alternatives are the α-2 adrenergic blockade by mianserin and mirtazapine for negative symptoms and a beneficial addition of folate and vitamin B12 in a subgroup of patients (Roffman et al., 2013).

There has been a strong emphasis on preventing conversion from the prodromal phase to full-blown psychosis (Stafford, Jackson, Mayo-Wilson, Morrison, & Kendall, 2013) with both pharmacological and psychological approaches. The most impressive result so far was achieved by the administration of long-chain omega-3 polyunsaturated fatty acids, which was associated with a 12-month conversion to psychosis in 2 of 41 (4.9%) individuals, compared to 11 of 40 (27.5%) individuals in the placebo group (Amminger et al., 2010). However, this treatment is not effective in established schizophrenia (Fusar-Poli & Berger, 2012).

2. MORPHOLOGICAL AND NEUROCHEMICAL FEATURES OF SCHIZOPHRENIA

Anatomical brain abnormalities have been well studied in schizophrenia (Honea, Crow, Passingham, & MacKay, 2005; Shenton, Whitford, & Kubicki, 2010). The most common findings comparing patients with schizophrenia and healthy controls are decreased gray matter volume in the left superior temporal gyrus and the left medial temporal lobe (Honea et al., 2005). However, many factors may contribute to these differences, such as genetics, obstetric complications, birth injuries, early brain insults, substance abuse, antipsychotic treatment, and several environmental factors (Bromet & Fennig, 1999; MacDonald & Schulz, 2009). Unaffected relatives of schizophrenia patients show similar but less pronounced gray matter abnormalities, suggesting a significant genetic susceptibility to these brain alterations (Oertel-Knöchel et al., 2012; Tian et al., 2011).

Abnormalities of both cortical and deeper brain structures, including the white matter, are present in the first episode and become more pronounced as the illness progresses (Delisi, 2008; Ellison-Wright et al., 2008; Fornito et al., 2009; Gogtay, 2008; Gogtay & Rapoport, 2008; Hulshoff Pol & Kahn, 2008; Olabi et al., 2011; Pantelis et al., 2005). This brain tissue loss has been estimated to be −0.5% a year for typical patients, which is twice the rate of healthy individuals (Hulshoff Pol & Kahn, 2008). This decline is even steeper in childhood-onset cases, which represent a more severe phenotype (Gogtay & Rapoport, 2008).

2.1. Synaptic changes in schizophrenia

At the cellular level, one of the most notable alterations found in schizophrenia, which has only been seldom highlighted (Blennow, Davidsson, Gottfries, Ekman, & Heilig, 1996; Field, Walker, & Conn, 2011; Glantz,

Gilmore, Lieberman, & Jarskog, 2006; Seshadri, Zeledon, & Sawa, 2013; Yin, Chen, Sathyamurthy, Xiong, & Mei, 2012) is an alteration of synaptic connectivity. One of the initial observations supporting this maladaptive alteration of synaptic contacts was the reported 30% reduction of the number of synaptic spines in the striatum of schizophrenics when compared to control subjects (Roberts, Conley, Kung, Peretti, & Chute, 1996). The same group further detailed this loss as being most evident for symmetric synapses, suggesting an imbalance between excitatory and inhibitory transmission (Kung, Conley, Chute, Smialek, & Roberts, 1998). Notably, accumulating evidence reinforces the causality between schizophrenia and the decreased density of presynaptic proteins in different brain structures. Thus, the *postmortem* analysis of the hippocampus of schizophrenics shows reduced levels of synapsin-1 (Browning, Dudek, Rapier, Leonard, & Freedman, 1993) and rab3 density is reduced in the thalamus and other cortical areas (Davidsson et al., 1999). Accordingly, the staining of SNAP-25 and synaptophysin are also reduced in the prefrontal cortex of schizophrenic patients (Karson et al., 1999) and synaptophysin immunolabeling is reduced in different layers of the dentate gyrus (Chambers, Thomas, Saland, Neve, & Perrone-Bizzozero, 2005). The prefrontal cortex of schizophrenics also displays changes in synaptic proteins (Fung, Sivagnanasundaram, & Weickert, 2011), namely a reduced density of phosphorylated syntaxin 1, and reduced binding of syntaxin 1 to SNAP-25 and MUNC18, leading to a decrease in SNARE complex formation (Castillo, Ghose, Tamminga, & Ulery-Reynolds, 2010). Additionally, genomic convergence analysis of the cerebellum identified 23 genes with altered expression involved with the presynaptic compartment, especially with vesicular transport (Mudge et al., 2008). Interestingly, the density of several presynaptic proteins in the anterior cingulate cortex of schizophrenics was correlated with the treatment and treatment response of different antipsychotics (Barksdale, Lahti, & Roberts, 2014).

Beside these presynaptic alterations in schizophrenia, several studies also report changes in postsynaptic components (De Bartolomeis, Latte, Tomasetti, & Iasevoli, 2014). Thus, single nucleotide polymorphisms of the regulatory scaffold protein synapse-associated protein 97 gene (SAP97) displayed a significant association with schizophrenia (Uezato et al., 2012) and mutations of the DLG4 gene (encoding the postsynaptic density protein 95, PSD95) confers susceptibility to schizophrenia (Cheng et al., 2010). Also, there is a decreased the immunolabeling of PSD95, PSD93, and SAP102 (Hahn et al., 2006; Meador-Woodruff, Clinton, Beneyto, &

Mccullumsmith, 2003; Ohnuma et al., 2000) and a deregulated synaptonuclear shuttling of the postsynaptic density protein proSAP2/Shank3 (Grabrucker et al., 2014). Indeed, functional gene group analysis identifies synaptic gene groups as risk factor for schizophrenia (Lips et al., 2012). Accordingly, the analysis of recurrent genomic copy number variants implicates specific abnormalities of postsynaptic signaling complexes in the pathogenesis of schizophrenia (Fromer et al., 2014; Guilmatre et al., 2009; Kenny et al., 2013; Kirov et al., 2012). Furthermore, there are also deficits of proteins involved in the structural maintenance of synapses, as typified by the association of schizophrenia with the neurexin 1 gene (Rujescu et al., 2009; Todarello et al., 2014) as well as with the CYFIP1 gene, critical for the maintenance of dendritic complexity and the stabilization of mature spines (Pathania et al., 2014). Additionally, proteins involved in the vectorial transport of cargoes to synapses, such as dysbindin, are recognized as susceptibility genes for schizophrenia (Larimore et al., 2011), leading to hampered synaptic responses (Carlson et al., 2011), which are isoform specific (Talbot et al., 2011). Overall, these changes are all indicative of an abnormal synaptic connectivity corresponding to a synaptic desynchronization, leading to an accelerated synaptic pruning, which would be part of the ethiology of schizophrenia (Boksa, 2012; Mirnics, Middleton, Marquez, Lewis, & Levitt, 2000).

2.2. Glia changes

The functioning of neuronal networks is not only dependent on the direct connectivity between neurons, but also on the support of this connectivity ensured by glia cells. In particular, the functional adaptability of synapses is critically dependent on the coincidence of information arrival, as typified by the importance of spike time dependent forms of plasticity (STDP) for information encoding (reviewed in Caporale & Dan, 2008; Feldman, 2012). STDP is critically dependent on the speed of information flow through axons, arguably one of the main functions of oligodendrocytes (Fields, 2005). Accordingly oligodendrocyte dysfunction is tightly associated with the emergence of mood and memory impairments and has been argued to be at the core of schizophrenia-like symptoms (Karoutzou, Emrich, & Dietrich, 2008; Mitterauer & Kofler-Westergren, 2011; Roussos & Haroutunian, 2014; Stewart & Davis, 2004; Tkachev et al., 2003).

Apart from the role of oligodendrocytes, there is an additional role of astrocytes in the control of information flow in brain circuits. This is best heralded by the emergence and rapid consolidation of the tripartite synapse concept, whereby the astrocytic processes enwrapping synapses have a decisive role in the flow of information and in the adaptability of synapses (reviewed in Halassa, Fellin, & Haydon, 2007; Halassa & Haydon, 2010; Perea, Navarrete, & Araque, 2009). The initial studies focused on the quantification of putative changes of the density of glia cells in schizophrenia patients and they essentially reported an increased density of glial cells in schizophrenic patients (Benes, Davidson, & Bird, 1986; Cotter, Pariante, & Everall, 2001; Rothermundt et al., 2007; Stark, Uylings, Sanz-Arigita, & Pakkenberg, 2004; Steiner et al., 2008; Weis & Llenos, 2004). However, other groups failed to show the same increase of the density of glial cells in schizophrenic patients and others instead observed a decrease in the glial markers (Niizato, Iritani, Ikeda, & Arai, 2001; Rajkowska et al., 2002; Rajkowska, Selemon, & Goldman-Rakic, 1998; Steffek, Mccullumsmith, Haroutunian, & Meador-Woodruff, 2008; Williams et al., 2013). This divergence of conclusions probably stems from the analysis of different astrocytic markers in different brain regions, a contention supported by systematic studies showing the differential alteration of different astrocytic markers in different brain regions of schizophrenics (Katsel et al., 2011). Notably, genomic analysis has repeatedly identified glia constellations of genes in association with schizophrenia (Duncan et al., 2014; Goudriaan et al., 2013) and the chronic consumption of antipsychotics can substantially alter gliosis (Konopaske et al., 2008); these observations have redirected attention to the study of the role of particular astrocytic functions, rather than astrocytic density, as possible contributors for the development of schizophrenia (reviewed in Schnieder & Dwork, 2011). Some studies documented changes in astrocytes, such as a decrease in myo-inositol (a glial marker), of glutamine synthetase (Chang et al., 2007; Steffek et al., 2008) and of glycogen metabolism (Lavoie, Allaman, Petit, Do, & Magistretti, 2011) that could contribute to metabolic deregulation associated with schizophrenia (Kondziella et al., 2006; Pennington et al., 2008; Rao, Kim, Harry, Rapoport, & Reese, 2013). Other studies have documented changes in the production of gliotransmitters, such as D-serine (Sacchi et al., 2008), that could hamper the correct functioning of the tripartite synapse (Mitterauer, 2005).

Probably one of the most robust changes related to maladaptive changes of astrocytes in schizophrenia is the alteration of the expression, density, and activity of EAATs (excitatory amino acid transporters), namely of the glutamate transporter 1 (GLT-1 or EAAT2), which are located in astrocytes and are responsible for almost 90% of the reuptake of extracellular glutamate in the brain (Haugeto et al., 1996). Thus, there is a susceptibility locus for schizophrenia within or near the glutamate transporter 1 (GLT-1) gene (Deng et al., 2004), which is deregulated in schizophrenic patients (Poletti et al., 2013; Shao & Vawter, 2008; Spangaro et al., 2012). This is in notable agreement with the observed upregulation of glutamate transporter 1 (GLT-I) mRNA, protein, and function that has been consistently shown in the cortex of schizophrenic patients (Huerta, Mccullumsmith, Haroutunian, Giménez-Amaya, & Meador-Woodruff, 2006; Matute, Melone, Vallejo-Illarramendi, & Conti, 2005; Ohnuma et al., 2000; Rao, Kellom, Reese, Rapoport, & Kim, 2012; Shan et al., 2013; Simpson, Slater, & Deakin, 1998; Smith, Haroutunian, Davis, & Meador-Woodruff, 2001). Furthermore, GLT-1 upregulation impairs adaptation of the startle reflex in adult rats, which was prevented by the glutamate transport inhibitor DHK (Bellesi, Melone, Gubbini, Battistacci, & Conti, 2009). Finally, psychotomimetics, such as phencyclidine, increase GLT-1 levels and activity (Fattorini et al., 2008), and the antipsychotic clozapine specifically downregulates GLT-1 expression and function (Melone, Bragina, & Conti, 2003; Vallejo-Illarramendi, Torres-Ramos, Melone, Conti, & Matute, 2005). Apart from this robust association between GLT-1 and schizophrenia, other studies have also reported alterations of other glutamate transporters, namely EAAT1, in different brain regions in schizophrenia, albeit less consistently: there is a decreased EAAT1 expression and EAAT1 glycosylation in the dorsolateral prefrontal cortex (Bauer, Haroutunian, Meador-Woodruff, & Mccullumsmith, 2010), but there are increased mRNA levels of EAAT1 in the anterior cingulate cortex and thalamus (Smith et al., 2001). As will be further discussed in this review, the emphasis in the importance of this single astrocytic alteration (upregulation of GLT-1) can only be overemphasized in view of the importance of GLT-1, which is underscored by the impact of modifying GLT-I activity on synaptic plasticity as well as on neurodegeneration (reviewed in Sattler & Rothstein, 2006; Tzingounis & Wadiche, 2007).

2.3. Main neurotransmitter systems affected in schizophrenia

Based on imaging, neurochemical and therapeutic findings in patients with schizophrenia and in animal models of this disease, there are currently two major neurotransmitter systems that have been implicated in schizophrenia, namely the dopamine and the glutamate systems.

2.3.1 Dopamine

The deregulation of the dopamine system plays a critical role in psychosis, diminished cognition, abnormal reward function, and movement disorders, all of which are manifested in schizophrenia. The recent advent of neurochemical imaging studies has been paramount to confirm the dopamine hypothesis of schizophrenia: indeed, such studies allowed showing that acute psychotic episodes are associated with an increase of dopamine synthesis, dopamine release, and resting-state synaptic dopamine concentrations (Laruelle, 1998). Of great importance, the dopamine release correlates positively with the severity of positive symptoms and with the subsequent response to therapeutic intervention based on the blockade of dopamine actions. In particular, schizophrenic patients present a greater occupation of dopamine D2R (Laruelle, Gelernter, & Innis, 1998) and a greater fraction of D2R occupied by endogenous dopamine than normal control subjects (Abi-Dargham et al., 2000). The currently most used antipsychotic agents also validate the dopaminergic hypothesis of schizophrenia: in fact, the first drugs with therapeutic effectiveness to manage positive symptoms, were serendipitously discovered in the mid-1950s and were characterized as being D2R antagonists (Snyder, 1981). Conversely, dopaminomimetic drugs (receptor agonists or amphetamine-like drugs) induce positive symptoms similar to those observed in acute paranoid crisis (Sayed & Garrison, 1983). The chronic use of dopaminomimetic drugs leads to severe states of hallucinations (Srisurapanont et al., 2011) and cognitive deficits (Nordahl, Salo, & Leamon, 2003; Rund et al., 2004). Likewise, the administration of amphetamine to rodents induces responses such as stereotyped behavior (Kurokawa, Mizuno, Shibasaki, & Ohkuma, 2010), and decreases the prepulse inhibition (PPI) of the acoustic startle response (Arai et al., 2008), a sensory gating response that is characteristically blunted in schizophrenics (Powell, Weber, & Geyer, 2012). Additionally, dopaminomimetic drugs also trigger different types of cognitive deficits ranging from working memory to reversal learning in the attention set shifting (Izquierdo et al., 2010; Nagai et al., 2007), which are characteristic cognitive endophenotypes of schizophrenia (Castner, Goldman-Rakic, & Williams, 2004; Green et al., 2004; Keefe & Harvey, 2012; Lett, Voineskos, Kennedy, Levine, & Daskalakis, 2014). However, in spite of this solid support for the involvement of dopamine in schizophrenia, the dopaminergic hypothesis might not be sufficient to account for all symptoms of schizophrenia since: (1) antipsychotics have no effect on negative and cognitive symptoms; (2) some patients are refractory to therapeutic strategies based on the use of dopamine receptor antagonists.

2.3.2 Glutamate

The involvement of the glutamate system in the etiology of schizophrenia rose with the observed psychotic effects of phencyclidine and ketamine (two antagonists of *N*-methyl-D-aspartate receptors, NMDA-R) that are remarkably similar to those found in schizophrenia (Itil, Keskiner, Kiremitci, & Holden, 1967; Luby, Cohen, Rosenbaum, Gottlieb, & Kelley, 1959). Notably, the hypothesis of glutamatergic hypofunction in schizophrenia cannot only explain the positive symptoms of schizophrenia, but also provides a mechanistic explanation for the negative and cognitive symptoms characteristic of the premorbid stage of schizophrenia (Neill et al., 2010; Ranganath, Minzenberg, & Ragland, 2008). In fact, synaptic NMDA-R play a key role to trigger different forms of synaptic plasticity that are considered to be the neurophysiological basis of learning and memory (Lynch, 2004; Martin, Grimwood, & Morris, 2000). Additionally, a reduced NMDA-R function also impairs the interactions between multiple brain regions (Fitzgerald, 2012; Greene, 2001; Gunduz-Bruce, 2009); this is a hallmark of the schizophrenic brain, where there is an unbalanced connectivity between different brain regions rather than a locus of dysfunction present in a defined brain region (Field et al., 2011; Lisman et al., 2008). Indeed, the glutamatergic hypofunction in schizophrenia has been particularly explored as a mechanistic basis of alterations in the thalamocortical loop resulting in an exaggerated sensory flooding and psychotic symptoms and the well-known dopaminergic dysfunction (Fitzgerald, 2012; Greene, 2001). Notably, the glutamatergic hypothesis of schizophrenia can actually account for the dopaminergic alterations (Javitt, 2007; Jentsch, Taylor, Elsworth, Redmond, & Roth, 1999). In fact, phencyclidine, ketamine, and MK-801 can all trigger the release of dopamine, thus driving dopaminergic hyperactivation (De Bartolomeis, Fiore, & Iasevoli, 2005; Krystal et al., 1994; Svensson, 2000). Additionally, the chronic administration of NMDA-R antagonists disturbs transmission in brain circuits, leading to changes in dopaminergic transmission similar to these found in schizophrenia (Javitt, 2007; Jentsch et al., 1999). From the mechanistic point of view, it is worth stressing that glutamatergic hypofunction is in line with the observed upregulation of astrocytic glutamate transporters, namely GLT-1, which constitutes one of the most robust neurochemical traits of schizophrenia (see above). Also, evidence from both animal models and human studies implicates a dysfunction of NMDA-R both in disease progression and symptoms of schizophrenia (Snyder & Gao, 2013). Therapeutic interventions also provide additional support to the role of the glutamatergic system in schizophrenia, as gauged from the improvements of symptoms observed when using D-cycloserine and D-serine (coagonists binding at the glycine modulatory domain of NMDA-R) (Goff et al., 1999; Tsai, Yang, Chung, Lange, & Coyle, 1998). The use of bitopertin, an inhibitor of glycine transporter, also had promising effects on the negative symptoms of schizophrenia (Hashimoto, Malchow, Falkai, & Schmitt, 2013). Additionally, a meta-analysis concluded that both amantadine and memantine (low affinity NMDA-R antagonist) were effective to manage cognitive dysfunction in schizophrenic patients (Kishi & Iwata, 2013). Overall, this hypothesis of glutamatergic hypofunction in schizophrenia seems rather compelling to explain the diversity of symptoms as well as their evolution; however, it is important to stress that there are still more unanswered rather than resolved questions, to make a harmonious link between the synaptic effects of NMDA-R, the impact of NMDA-R on global circuit functioning and the behavioral effects of NMDA-R, namely because there are several populations of NMDA-R often with opposite roles and there are different cell types endowed with NMDA-R (Hardingham & Bading, 2010; Köhr, 2006), namely astrocytes (Dzamba, Honsa, & Anderova, 2013; Halassa & Haydon, 2010; Verkhratsky & Kirchhoff, 2007).

2.3.3 GABA

The GABAergic system is the main inhibitory neurotransmitter system in brain circuits, but has not been under tight scrutiny as a possible key player in schizophrenia (Heresco-Levy, 2003; Lewis, 2013). Notably, the parvalbumin-positive, fast-firing GABAergic interneurons that provide recurrent inhibition to cortical-limbic pyramidal neurons, seem to be most sensitive to NMDA-R hypofunction (Coyle, Basu, Benneyworth, Balu, & Konopaske, 2012; Gonzalez-Burgos & Lewis, 2012; Nakazawa et al., 2012). The cortical GABAergic system is also the most affected during brain maturation, for instance upon exposure to psychoactive drugs (Berghuis et al., 2007; Grantyn, Henneberger, Jüttner, Meler, & Kirischuk, 2011; Miller, 1986; Ramamoorthi & Lin, 2011; Silva et al., 2013), and it has been proposed as a key mediator between the first and second hit argued to be required trigger schizophrenia (Volk & Lewis, 2013). Thus, as now occurs in several neurodegenerative disorders ranging from Alzheimer's disease to depression (e.g., Brambilla, Perez, Barale, Schettini, & Soares, 2003; Lodge & Grace, 2011; Verret et al., 2012), there is a renewed interest in the plastic alteration of the GABAergic system (reviewed by Caroni, Donato, & Muller, 2012), which is expected to be extended to schizophrenia in the near future.

3. THE ADENOSINE NEUROMODULATION SYSTEM

The biased presentation of some features pertinent to the emergence of schizophrenia was mostly aimed to introduce the potential interest to target the adenosine modulation system as a promising novel strategy to alleviate the burden of schizophrenia. This will require an initial presentation of the adenosine modulation, which will be followed by a more directed description of the evidence linking this system with some of the main neurochemical, neurophysiological, and behavioral features of schizophrenia. Since our attempt will mostly focus on the role of the receptors operated by adenosine, it is important to call the readers attention to the important work developed by Detlev Boison over the years, who has championed the idea that the manipulation of the metabolism of adenosine might be a promising avenue to manage schizophrenia, as elaborated in an elegant and recent review (Boison, Singer, Shen, Feldon, & Yee, 2012).

Adenosine is probably best known as a metabolic product of ATP catabolism. Accordingly, the production of adenosine is tightly coupled to the use of energy, with an increased production of adenosine upon stressful situations or increased workload (Cunha, 2001; Dunwiddie & Masino, 2001; Fredholm, Chen, Cunha, Svenningsson, & Vaugeois, 2005). Thus, adenosine fulfills two parallel roles in brain circuits, acting as a homeostatic paracrine adaptor to coordinate activity in different cellular elements in response to stressful situations, as well as an important activity-dependent signal to adapt cellular function to changes of workload (Cunha, 2001, 2005, 2008; Dunwiddie & Masino, 2001; Fredholm et al., 2005). In other words, adenosine plays a dual role: (1) as a neuromodulator controlling information flow in neuronal circuits as a function of its activity (Dunwiddie & Masino, 2001; Fredholm et al., 2005; Tomé, Silva, & Cunha, 2010), with a prominent role in the definition of information salience (Cunha, 2008); (2) as a controller of the neuropathological consequences of brain insults, with a documented impact on the neurochemical, neurophysiological, and behavioral features of different brain disorders such as Parkinson's or Alzheimer's disease, epilepsy, or ischemia as detailed in different reviews (Chen et al., 2007; Cunha, 2005; Fredholm et al., 2005; Gomes, Kaster, Tomé, Agostinho, & Cunha, 2011).

Adenosine acts through metabotropic receptors that are mainly inhibitory A_1 and facilitatory A_{2A} receptors ($A_{2A}R$) in the brain (Fredholm et al., 2005). Both A_1 and $A_{2A}R$ are mostly located in synapses in cortical circuits (Rebola,

Canas, Oliveira, & Cunha, 2005; Rebola, Pinheiro, Oliveira, Malva, & Cunha, 2003). Inhibitory A_1R decrease the release of neurotransmitters, mainly of glutamate (e.g., Ambrósio, Malva, Carvalho, & Carvalho, 1997; Dunwiddie & Haas, 1985; Thompson, Haas, & Gähwiler, 1992), and they also decrease excitability by controlling AMPA and NMDA receptor recruitment and trafficking (de Mendonça, Sebastião, & Ribeiro, 1995) as well as potassium currents, namely after-hyperpolarizing currents (Greene & Haas, 1985). These receptors are tonically activated by endogenous extracellular adenosine: this on-going break actually constitutes a hurdle for the engagement of plastic changes of efficiency of brain circuits, thus acting as a low-pass filter (Cunha, 2008). Disruptions of this break system are expected to trigger increased excitation and increased noise, as well as greater susceptibility to neuronal damage, as elegantly demonstrated by the genetic manipulation of the key enzymatic system responsible for controlling the extracellular levels of adenosine–adenosine kinase (reviewed in Boison, 2013). The acute activation of A_1R is also recognized to afford a robust neuroprotection (de Mendonça, Sebastião, & Ribeiro, 2000; Dunwiddie & Masino, 2001; Fredholm, 1997; Rudolphi, Schubert, Parkinson, & Fredholm, 1992). Although the therapeutic potential of A_1R is somehow hampered by their desensitization over time (Abbracchio, Fogliatto, Paoletti, Rovati, & Cattabeni, 1992; Coelho et al., 2006; Hettinger, Leid, & Murray, 1998; Ruiz et al., 1996), its importance to control the spreading of neurodegenerative disorders has been solidly documented using pharmacological strategies (Fedele, Li, Lan, Fredholm, & Boison, 2006; Gouder, Fritschy, & Boison, 2003; Pignataro, Simon, & Boison, 2007), whereas the genetic manipulation of A_1R is somehow confounded by adaptive processes (Olsson et al., 2004).

The role of $A_{2A}R$ is rather different, since they are not recruited under most conditions (Cunha, 2008). In fact, $A_{2A}R$ are selectively recruited by a particular source of adenosine, originated from the extracellular catabolism by ecto-nucleotidases of ATP (Augusto et al., 2013; Cunha, Correia-de-Sá, Sebastião, & Ribeiro, 1996; Rebola, Lujan, Cunha, & Mulle, 2008) that can be released together with neurotransmitters (Cunha, Almeida, & Ribeiro, 2001; Pankratov, Lalo, & Verkhratsky, 2006; White & MacDonald, 1990). Since ATP is released in disproportionally larger amounts from synaptic compartments in conditions triggering synaptic plasticity (Cunha, Vizi, Ribeiro, & Sebastião, 1996; Wieraszko, Goldsmith, & Seyfried, 1989), it is only under conditions of synaptic plasticity that $A_{2A}R$ are recruited to enhance the release of glutamate (Lopes, Cunha, Kull, Fredholm, & Ribeiro, 2002; Rodrigues,

Alfaro, Rebola, Oliveira, & Cunha, 2005), bolster the activation of NMDA receptors (Azdad et al., 2009; Rebola et al., 2008) and inhibit the function of presynaptic inhibitory systems such as A_1R (Ciruela et al., 2006; Lopes, Cunha, & Ribeiro, 1999) or cannabinoid CB1 receptors (Ferré et al., 2010; Lerner, Horne, Stella, & Kreitzer, 2010; Martire et al., 2011). Thus, $A_{2A}R$ play a particular role in the control of synaptic plasticity processes involving the recruitment of NMDA-R (Costenla et al., 2011; d'Alcantara, Ledent, Swillens, & Schiffmann, 2001; Rebola et al., 2008). Also in accordance with their ability to bolster glutamate release and NMDA-R activation, the overactivation of $A_{2A}R$ triggers neurodegeneration (Cunha, 2005; Chen et al., 2007; Chen, Eltzschig, & Fredholm, 2013; Gomes et al., 2011; Stone et al., 2011), namely when involving synaptotoxicity (Canas, Porciúncula, et al., 2009; Cognato et al., 2010; Duarte, Agostinho, Carvalho, & Cunha, 2012; Gonçalves, Simões, Cunha, & Pereira De Almeida, 2013; Silva, Porciúncula, Canas, Oliveira, & Cunha, 2007). The control of neurodegeneration might also involve the control microglia, the main cell type involved in brain neuroinflammatory responses that contributes to the progression of neurodegeneration (Kettenmann, Hanisch, Noda, & Verkhratsky, 2011; Kettenmann, Kirchhoff, & Verkhratsky, 2013; Ladeby et al., 2005). Thus, $A_{2A}R$ are localized in microglia cells (Dai et al., 2010; Rebola et al., 2011; Saura et al., 2005) and control the recruitment (Orr, Orr, Li, Gross, & Traynelis, 2009), the proliferation (Gomes et al., 2013), and the impact of microglia and inflammatory mediators on neuronal function (Rebola et al., 2011; Simões et al., 2012; Stone & Behan, 2007). $A_{2A}R$ are also located in astrocytes (Daré, Schulte, Karovic, Hammarberg, & Fredholm, 2007; Matos, Augusto, Santos-Rodrigues, et al., 2012; Nishizaki et al., 2002) where they control the uptake of glutamate (Matos, Augusto, Santos-Rodrigues, et al., 2012) and likely several other functions thanks to their ability to control Na^+/K^+-ATPase (Matos, Augusto, Agostinho, Cunha, & Chen, 2013), the main energizing systems supporting astrocytic functions (Benarroch, 2011; Kirischuk, Parpura, & Verkhratsky, 2012). Finally, $A_{2A}R$ also play a role in defining hemodynamic responses thanks to their involvement in the vasodilation of brain vessels (Kusano et al., 2010; Ngai, Coyne, Meno, West, & Winn, 2001; Shin, Park, & Hong, 2000).

3.1. Adenosine and dopamine

The interest in the adenosine neuromodulation system in mood-related disorders was prompted by the particularly high density of $A_{2A}R$ in dopamine-rich brain regions, namely in the basal ganglia (Schiffmann, Fisone, Moresco, Cunha, & Ferré, 2007; Svenningsson, Le Moine, Fisone, & Fredholm, 1999). Albeit some studies have focused on the ability of adenosine receptors to control the release of dopamine (Borycz et al., 2007; Jin, Johansson, & Fredholm, 1993; Quarta et al., 2004), most of the studies exploring a putative control by the adenosine neuromodulation system of the dopaminergic system have centered on receptor–receptor interactions (Fuxe, Ferre, Zoli, & Agnati, 1998; Fuxe et al., 2001). Direct receptor–receptor interactions were first documented in the 1980s with the proposal that receptors could interact at the membrane level, leading to more intricate intracellular biochemical pathways interactions (Agnati, Fuxe, Zini, Lenzi, & Hokfelt, 1980). Taking into account all the evidence, the current prevalent view is that basically there are two different types of receptor–receptor interaction: (1) one relies at second messenger where the receptors target the same intracellular signaling cascade (Stoof & Kebabian, 1984); (2) the second type of interaction takes place at the membrane level with a molecular cross talk, related to the ability of receptors to form receptor heteromers (Ciruela et al., 2004; Hillion et al., 2002). The two types of receptor–receptor interaction have been revealed in behavioral studies showing an antagonistic interaction between adenosine and dopamine present in the brain, which mainly involve D2R and $A_{2A}R$ (Fuxe et al., 1998, 2001). The first indication that the mechanism might involve antagonistic intramembranous $A_{2A}R$–D2R interactions was obtained in studies using striatal membrane preparations showing that the $A_{2A}R$ agonist CGS21680 reduced the binding affinity of D2R agonists (Ferré, Von Euler, Johansson, Fredholm, & Fuxe, 1991). The continuous exploration of the features of these $A_{2A}R$–D2R interaction revealed several additional important features such as the coaggregation, -internalization, and -desensitization of the D2R–$A_{2A}R$ complex (Hillion et al., 2002), which occur in the same time frame (Torvinen et al., 2005). It was also shown that in the absence of $A_{2A}R$ (using $A_{2A}R$ knockout mice), D2R desensitization is increased (Al-Hasani et al., 2011). Taken all together, it is clear that the modulatory actions of $A_{2A}R$ and D2R are bidirectional and the interplay between these two systems is intensive. Notably, there are several different types of neuronal responses controlled by D2R–$A_{2A}R$ complexes, which are expected to have a robust impact on the functioning of striatal circuits, such as: (1) the regulation of the synaptic Ca^{2+} influx (Higley & Sabatini, 2010); (2) the NMDA-R-mediated excitation in the nucleus accumbens (Azdad et al., 2009); (3) the internalization of receptors via interactions with β-arrestin2 (Borroto-Escuela et al., 2011).

In accordance with the purported role of dopamine in schizophrenia, D2R–A_{2A}R complexes are expected to be of particular relevance for the therapeutic of schizophrenia: A_{2A}R agonists would reduce D2R receptor signaling through A_{2A}R–D2R heteromers thus counteracting the dopaminergic hyperfunction in schizophrenia (Ferré, 1997; Ferré, O'Connor, Snaprud, Ungerstedt, & Fuxe, 1994). An indirect support to this scenario was provided by an open clinical trial showing that dypiridamole (an inhibitor of adenosine uptake, increasing the availability of adenosine in the synaptic cleft) used as an adjuvant in the haloperidol (a typical antipsychotic, an antagonist of D2 receptors) significantly decreased the positive symptoms of schizophrenia and also a trend of reduction of the negative symptoms was observed (Akhondzadeh, Shasavand, Jamilian, Shabestari, & Kamalipour, 2000).

3.2. Adenosine and glutamatergic synaptic functions

Apart from this tight coupling of A_{2A}R and D2R, the adenosine modulation system also has a particular relevance in the control of glutamatergic systems. In fact, the release of glutamate in striatal and cortical regions is under dual control of inhibitory A_1R and facilitatory A_{2A}R (Ciruela et al., 2006; Lopes et al., 2002). Additionally, both A_1R and A_{2A}R can control the recruitment of AMPA and NMDA-R (de Mendonça et al., 1995; Dias, Ribeiro, & Sebastião, 2012; Rebola et al., 2008; Wirkner et al., 2004). Most importantly, the combined action of A_1R and A_{2A}R is organized to assist encoding information salience in cortical networks (reviewed in Cunha, 2008), although this has not been tested in different brain circuits, namely in thalamic circuits (Bekar et al., 2008; Ramlackhansingh et al., 2011; Rosin, Robeva, Woodard, Guyenet, & Linden, 1998; Ulrich & Huguenard, 1995). This makes the adenosine modulation system particularly well positioned to rebalance the glutamate hypofunction that is characteristic of schizophrenia.

Apart from this functional ability of the adenosine modulation system to potentially correct aberrant glutamate transmission, adenosine receptors have an additional potential to prevent the synaptotoxicity that seems to be a promident hallmark of schizophrenia (Blennow et al., 1996; Field et al., 2011; Glantz et al., 2006; Seshadri et al., 2013; Yin et al., 2012). Indeed, both A_1R and A_{2A}R are most abundantly located in synapses (Rebola, Canas, et al., 2005; Rebola, Pinheiro, et al., 2003), mainly in glutamatergic synapses (Costenla et al., 2011; Rebola, Rodrigues, et al., 2005).

Furthermore, $A_{2A}R$ have a tight association with synaptic damage: thus $A_{2A}R$ are upregulated in synapses upon brain insults (Cognato et al., 2010; Duarte et al., 2012; Rebola, Porciúncula, et al., 2005) and the blockade of $A_{2A}R$ can prevent synaptic damage (Canas, Porciúncula, et al., 2009; Silva et al., 2007). This makes $A_{2A}R$ antagonists a potentially relevant neuroprotective strategy with relevance for different neurodegenerative and neuropsychiatric disorders where synaptic damage is an early pathogenic event (reviewed in Cunha et al., 2008; Gomes et al., 2011), such as in schizophrenia.

3.3. Adenosine and glial cells

Another potential role of the adenosine neuromodulation system that is directly relevant to appreciating its potential therapeutic interest in schizophrenia is the ability of $A_{2A}R$ to control astrocytic glutamate transport. Indeed, several studies have now documented the presence of $A_{2A}R$ in astrocytes (Matos, Augusto, Santos-Rodrigues, et al., 2012; Nishizaki et al., 2002). Although several possible roles have been proposed (reviewed in Boison, Chen, & Fredholm, 2010; Daré et al., 2007), we will focus here on the ability of astrocytic $A_{2A}R$ to control GLT-1 expression and activity (Matos et al., 2013; Matos, Augusto, Machado, et al., 2012; Matos, Augusto, Santos-Rodrigues, et al., 2012), since the upregulation of GLT-1 mRNA, protein, and function has been consistently shown in the cortex of schizophrenic patients (Huerta et al., 2006; Matute et al., 2005; Ohnuma et al., 2000; Rao et al., 2012; Shan et al., 2013; Simpson et al., 1998; Smith et al., 2001). Notably, we found that the selective elimination of astrocytic $A_{2A}R$ (using GFAP-driven $A_{2A}R$ knockout mice), lead to an upregulation and increased activity of GLT-1 (Matos, Augusto, Machado, et al., 2012; Matos, Augusto, Santos-Rodrigues, et al., 2012), as observed in the brain of schizophrenic patients. Further arguing for a potentially important role of astrocytic $A_{2A}R$ in schizophrenia, we found that GFAP-driven $A_{2A}R$ knockout mice displayed an enhanced behavioral sensitization upon exposure to psychoactive drugs and a decreased working memory, two endophenotypes of schizophrenia (Matos et al., submitted). Interestingly, the mechanism of this astrocytic based alteration (selective astrocytic $A_{2A}R$ genetic elimination) involved an astrocytic-to-neuron wave of communication, with an increased $A_{2A}R$ enhancement of the evoked release of glutamate, increased plasma membrane density of NR2B-containing NMDA-R and internalyzation of AMPA receptors (see Fig. 16.1), in tight accordance

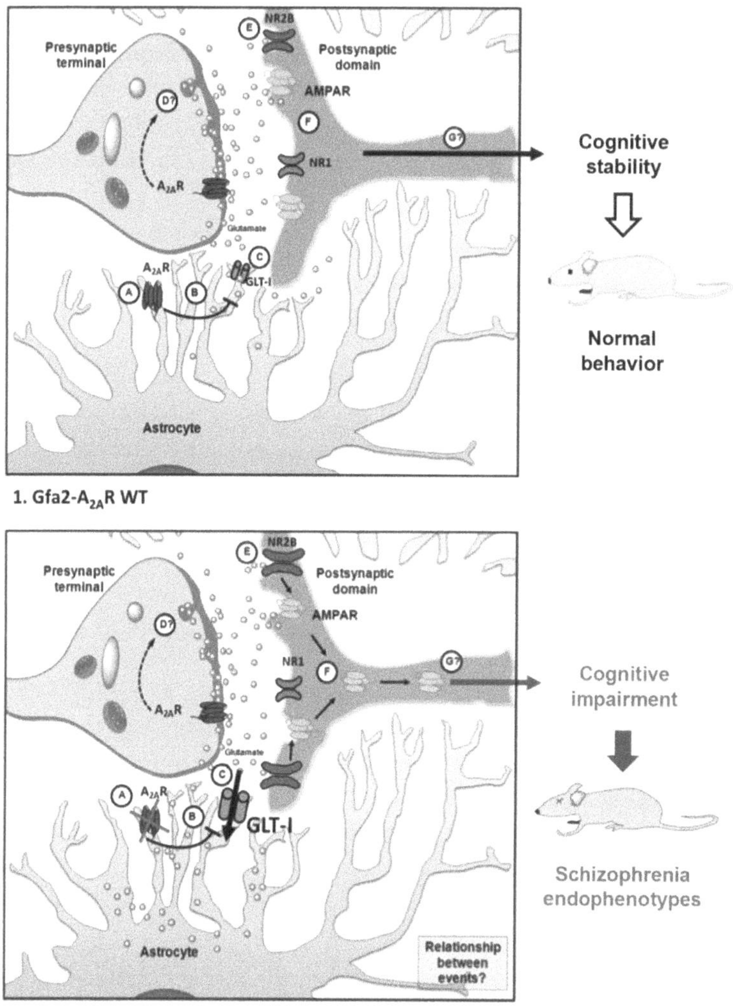

2. Gfa2-$A_{2A}R$ KO

Figure 16.1 Proposed model of glutamatergic dysfunction and subsequent working memory impairment and emergence of schizophrenia-like endophenotypes in mice with a selective deletion of adenosine A_{2A} receptors ($A_{2A}R$) in astrocytes (Gfa2-$A_{2A}R$ knockout mice). (1) In WT mice, astrocytic $A_{2A}R$ play a crucial role inhibiting the astrocytic uptake of glutamate through GLT-1 transporters (A–B). This important step may allow preventing an abnormal accumulation of perisynaptic glutamate levels (C) and triggering an adaption of the neuronal glutamatergic system, with a lack of aberrant $A_{2A}R$-induced bolstering of synaptic glutamate levels neuronal (D) possibly triggering changes in the density of NR2B-containing NMDA-R (E) and internalization of AMPA-R (F). The normal trafficking of AMPA-R may account for the normal cognitive profile of Gfa2-$A_{2A}R$ WT mice (G?). (2) The lack of $A_{2A}R$ in astrocytes of Gfa2-$A_{2A}R$

(Continued)

with the previously reported malfunction in the astrocyte-neuron communication upon repeated exposure to a glutamatergic antagonist (MK-801, an animal model of schizophrenia) (Kondziella, Brenner, Eyjolfsson, & Sonnewald, 2007). These observations herald a key role for astrocytic $A_{2A}R$ in the genesis of adaptive changes (which also involve synaptic $A_{2A}R$) that seem sufficient to trigger schizophrenia-like endophenotypes. Certainly, additional work will be required to identify the signaling systems responsible for the astrocytic-to-neuron wave of adaptive changes, but purines emerge as potential candidates based on the innovative findings of Phillip Haydon's group (Hines, Schmitt, Hines, Moss, & Haydon, 2013; Pascual et al., 2005; Schmitt, Sims, Dale, & Haydon, 2012). This hypothesis is further supported by the impact of the manipulation of astrocytic adenosine kinase on the emergence of schizophrenia-like endophenotypes (Shen et al., 2012; Yee, Singer, Chen, Feldon, & Boison, 2007). Overall, albeit the relation between all these concurrent findings still remains to be adequately integrated, it seems that the astrocytic purinergic system can be responsible by the emergence and amplification of adaptive changes associated with schizophrenia. This is particularly interesting since it prompts a shift from neurocentric to astrocentric mechanisms as possible ethiological basis of schizophrenia. Such a refocus on astrocytes as the central cellular elements initially affected in schizophrenia is well in line with the integrative role of astrocytes and its impact of higher brain functions (Halassa & Haydon, 2010; Li et al., 2012; Suzuki et al., 2011), two features that are also characteristic of schizophrenia.

3.4. Adenosine and brain maturation

A final link between the adenosine modulation system and the main abnormal features associated with schizophrenia, is the impact of adenosine in

Figure 16.1—Cont'd knockout mice (A) removes the inhibition of GLT-1 transporters (B), leading to enhanced levels of GLT-I transporters. These astrocytic events trigger a constellation of neuronal adaptive changes in glutamatergic synapses (C). Thus, the function of presynaptic $A_{2A}R$ seems bolstered (D) likely contributing to enhanced levels of synaptic glutamate; this is accompanied by a reorganization of the postsynaptic glutamate receptors typified by an enhanced density of NR2B-containing NMDA-R (E) leading to the internalization of AMPA-R (F); altogether, these changes are associated with cognitive impairment and the emergence of endophenotypes characteristic of schizophrenia in Gfa2-$A_{2A}R$ knockout mice (G?) such as enhanced response to psychotomimetics and decreased working memory performance. *This figure was prepared by Marco Matos together with Jiang-Fan Chen and summarizes finding currently submitted for publication (Matos et al., submitted).*

brain development. The two-hit hypothesis of schizophrenia stipulates the requirement for an initial modification of brain development that would predispose for a greater susceptibility to secondary alterations affecting mature brain circuits to allow the emergence of symptoms (Shenton, Dickey, Frumin, & Mccarley, 2001). Thus, a putative impact of adenosine on brain development might be an additional argument to support the interest of targeting this system in the realm of schizophrenia.

The first description of a robust impact of adenosine in brain development was provided by the group of Scott Rivkees. Based on the rationale that both inflammation and hypoxia trigger an increase of the extracellular levels of adenosine (e.g., Park, Van Wylen, Rubio, & Berne, 1987; Rodríguez-Núñez et al., 2001), and considering that either hypoxia or inflammation triggers have been explored experimentally as first hit triggers in animal models of schizophrenia (Boksa, 2004; Fendt et al., 2008; Meyer, Feldon, & Yee, 2009), Rivkees' group launched a series of elegant studies mingling genetic and pharmacological manipulations in a postnatal *in vivo* setting, complemented by studies in cell cultures, to highlight the impact of adenosine on brain development (reviewed in Rivkees & Wendler, 2011; Rivkees, Zhao, Porter, & Turner, 2001). They showed that the administration of an A_1R agonist during the early postnatal period triggered a leukomalacia together with diffuse modifications of the organization of the gray matter as well as of white matter tracts (Turner, Yan, Schwartz, Othman, & Rivkees, 2002). Accordingly, adenosine acting through A_1R was the causal factor responsible for ventricular dilation and the periventricular white matter injury upon hypoglycemic or hypoxic–ischemic encephalopathy (Back et al., 2006; Kim, Yu, Fredholm, & Rivkees, 2005; Turner et al., 2003). Thus, there is a perinatal period when A_1R activation is deleterious, since they become neuroprotective from 7 days onward (Adén, Leverin, Hagberg, & Fredholm, 2001). Indeed, there is a tight control of adenosine formation in the embryo and in early life, as gauged by the high density of adenosine deaminase in the placenta (Nagy, Geiger, & Staines, 1990) and the marked early life ontogenic changes of adenosine kinase expression in the brain (Studer et al., 2006). In contrast to the deleterious role of A_1R activation during the early postnatal period, $A_{2A}R$ activation seems to be required for an adequate brain development. In fact, $A_{2A}R$ blockade leads to a delayed insertion of interneurons in hippocampal circuits, which entrains deficits in memory performance until aldulthood (Silva et al., 2013). However, the overactivation of $A_{2A}R$ is deleterious in the neonate, in conditions such as brain hypoxia/ischemia (Adén et al., 2003; Bona, Adén, Gilland, Fredholm, & Hagberg, 1997; Yang et al., 2013).

The analysis of the cellular effects operated by adenosine receptors during development has also revealed opposite effects of A_1R and $A_{2A}R$. Thus, A_1R are present in axons during early brain development (Swanson, Drazba, & Rivkees, 1995; Swanson, Krahl, Liu, Drazba, & Rivkees, 1998) and their activation inhibits neurite outgrowth in cultured neurons (Shaban, Smith, & Stone, 1998; Thevananther, Rivera, & Rivkees, 2001) and can trigger the apoptosis of embryonic neurons (Wakade, Palmer, McCauley, Przywara, & Wakade, 1995); in contrast, $A_{2A}R$ activation bolsters neuritogenesis (Cheng, Shih, & Chern, 2002; O'Driscoll & Gorman, 2005; Sun et al., 2010). In parallel, in accordance with the impact of A_1 receptors on periventricular white matter injury in neonates, A_1R are located in oligodendrocytes and their activation stimulates migration of olygodendrocytes and promotes myelinization (Asghari, Azarnia, Mirnajafi-Zadeh, & Javan, 2013; Casadó, Mallol, Lluis, Franco, & Canela, 1991; Othman, Yan, & Rivkees, 2003; Stevens, Porta, Haak, Gallo, & Fields, 2002); in contrast, $A_{2A}R$ activation inhibits the proliferation and activation of immature olygodendrocytes (Coppi, Cellai, Maraula, Pugliese, & Pedata, 2013; Genovese et al., 2009; Melani et al., 2009).

Overall, these observations provide initial clues that support a fine-tuning role for adenosine to control different aspects of brain development. However, the different roles of A_1R and $A_{2A}R$ prompt the need of additional studies to build an integrated scenario to comprehend the role of the adenosine modulation system during brain development.

4. IMPACT OF MANIPULATING THE ADENOSINE SYSTEM IN ANIMAL MODELS OF SCHIZOPHRENIA

The description of the adenosine modulation system makes it evident that there is a potential for this system to control schizophrenia. Remarkably, this is confirmed by the wealth of studies that actually explored the impact of manipulating the adenosine neuromodulation system to modify the behavioral responses that are recorded in animal models as characteristic endophenotypes pertinent to schizophrenia. Based on the clinical findings, the most frequently measured endpoints are behavioral sensitization, startle responses, and working memory alteration (Amann et al., 2010; Featherstone, Kapur, & Fletcher, 2007; Moghaddam & Javitt, 2012; Turetsky et al., 2007; van Snellenberg, 2009). We will subsequently provide a short review to document the ability of the adenosine neuromodulation system to affect this triad of endophenotypes characteristic of schizophrenia.

4.1. Behavioral sensitization

Behavioral sensitization corresponds to a progressive augmentation of responses to repeated drug administration that results from neuroadaptive plastic processes of brain circuits (Kalivas, Pierce, Cornish, & Sorg, 1998; Robinson & Berridge, 1993). It is most often measured as an augmented motor-stimulant response upon repeated, intermittent exposure to a particular psychotomimetic drug, which can last up to 1 year after the exposure to the drug (Paulson, Camp, & Robinson, 1991). The influence of dopamine and glutamate in this sensitization process is crucial and it involves different circuits from different brain regions, namely of the medial prefrontal cortex, ventral tegmental area, nucleus accumbens, hippocampus, and basolateral amygdala (Steketee & Kalivas, 2011). There is now robust evidence from different studies supporting an important role of $A_{2A}R$ in the long-term adaptive responses to repeated dopaminergic stimulation both in rodents (Bastia et al., 2005) and also in nonhuman primates (Bibbiani et al., 2003). The use of $A_{2A}R$ knockout ($A_{2A}R$-KO) confirmed this involvement of $A_{2A}R$ in the processes of behavioral sensitization to repeated treatment either with amphetamine, cocaine, MK-801, opioids, or L-DOPA mice, which are attenuated/eliminated in the absence of the $A_{2A}R$ (Augusto et al., 2013; Brown, Short, Cowen, Ledent, & Lawrence, 2009; Chen et al., 2003; Fredduzzi et al., 2002; Ruiz-Medina, Ledent, Carretón, & Valverde, 2011; Shen et al., 2008; Soria et al., 2006; Xiao et al., 2006). Interestingly, the $A_{2A}R$ agonist CGS21680 can also attenuate the development of behavioral sensitization induced by methamphetamine (Shimazoe, Yoshimatsu, Kawashimo, & Watanabe, 2000) or D2R agonists (Hobson, Merritt, & Bachtell, 2012). Likewise, CGS21680 reduces the psychomotor stimulating effects of dopaminomimetic agents such as amphetamine, including stereotypies (Poleszak & Malec, 2000; Rimondini, Ferre, Ogren, & Fuxe, 1997), blocks conditioned avoidance response in rats and is equieffective with haloperidol and raclopride in the apomorphine-induced climbing behavior in mice (Kafka & Corbett, 1996). In addition, systemic, intrastriatal, or intraaccumbal administration of CGS21680 to rodents reduce spontaneous locomotor activity, and at higher doses, induce catalepsy (Barraco, Martens, Parizon, & Normile, 1993; Hauber & Koch,

1997; Kafka & Corbett, 1996; Rimondini et al., 1997). These apparent contradictory findings probably result from the opposite role on behavioral sensitization of presynaptic and postsynaptic $A_{2A}R$ in corticostriatal synapses (Shen et al., 2008, 2013), whereby endogenous adenosine predominantly activates presynaptic $A_{2A}R$ that promote behavioral sensitization (Bastia et al., 2005; Shen et al., 2008, 2013; Xiao et al., 2006) (hence justifying the beneficial effect of the genetic or pharmacological blockade of $A_{2A}R$), whereas the exogenous activation of $A_{2A}R$ with CGS2160 recruits a pool of post synaptic $A_{2A}R$ that is not tonically activated by endogenous adenosine and decreases D2R-mediated development of behavioral sensitization (hence justifying the beneficial effect of $A_{2A}R$ agonists).

4.2. Startle

The startle reflex is a defensive immobilization response to a sensory stimulus (Turpin, 1986). One of the advantages of the startle reflex is its translational sensibility from animals to humans (Braff & Geyer, 1990; Geyer & Braff, 1987). The measurement of PPI (prepulse inhibition) is especially important for the schizophrenia spectrum of diseases, because it relies on the fact that schizophrenic patients loose the ability to filter irrelevant from relevant stimuli (Braff et al., 1978; Mcghie & Chapman, 1961; Venables, 1960) thus being unable to regulate and separate environmental relevant inputs allocating the attentional resources (Swerdlow, Braff, Hartston, Perry, & Geyer, 1996). The role of $A_{2A}R$ in startle and PPI has been explored in a handful of studies. It was observed that the $A_{2A}R$ agonist CGS21680 has no effects *per se* (Hauber & Koch, 1997; Sills, Azampanah, & Fletcher, 2001) but can attenuate the impact of some psychotomimetic drugs on PPI (Andersen, Fuxe, Werge, & Gerlach, 2002; Sills et al., 2001; Wardas, Konieczny, & Pietraszek, 2003). There were also no effects *per se* of the nonselective antagonist of adenosine receptors caffeine on PPI in animals (Bakshi, Geyer, Taaid, & Swerdlow, 1995), whereas high caffeine intake blunte PPI in humans (Swerdlow et al., 2000) and abolishe the effect of controlled attention on PPI in human volunteers at a dose of 4 mg/kg (Flaten & Elden, 1999); the variability of these responses may be related to gender and to polymorphisms of $A_{2A}R$ (Domschke et al., 2012; Gajewska et al., 2013). The selective pharmacological blockade of $A_{2A}R$ produced no effects of PPI (Bleickardt, Lashomb, Merkel, & Hodgson, 2012), albeit the injection of the $A_{2A}R$ antagonist MSX-3 directly into the nucleus accumbens reduce

PPI (Nagel et al., 2003), as is also observed in global A_{2A}R-KO mice (Wang, Short, Ledent, Lawrence, & van den Buuse, 2003). However, there were minor changes of the startle reflex (Wei et al., 2013) or PPI (Singer, Wei, Chen, Boison, & Yee, 2013) in striatum-selective A_{2A}R-KO mice. By contrast, increasing adenosine availability by blocking adenosine kinase restore the apomorphine-induced PPI disturbance (Shen et al., 2012). This is in agreement with the ability of A_1R agonists to prevent the impact of psychotomimetics on PPI (e.g., Gotoh et al., 2002; Sills, Azampanah, & Fletcher, 1999). Thus, there seems to be an ability of A_1R to control startling and PPI, which might be therapeutically exploited by increasing the extracellular levels of adenosine through the inhibition of adenosine kinase, whereas the role of A_{2A}R is still far from clear.

4.3. Memory

Schizophrenics demonstrate incapacitating cognitive dysfunction (Elvevag & Goldberg, 2000) not related with intellectual disability (Gray, Mcmahon, & Gold, 2013). The general trend of these deficits are more marked in later life and are an important treatment target, because such deficits are the best predictor of a patient's level of interpersonal skills, occupational functioning, and self care (Bowie & Harvey, 2005). The cognitive impairments in schizophrenics are diverse (since deficits in attention, working memory, verbal fluency, processing speed, executive functions, and verbal memory), but the most notable modification is the inability to effectively acquire and properly handle information (Cirillo & Seidman, 2003) consistent with a predominant and prominent deterioration of working memory (Amann et al., 2010; van Snellenberg, 2009). The involvement of adenosine receptors in the control of cognitive processes is compelling, as testified by the impact of the nonselective antagonist of adenosine receptors caffeine on memory performance (Borota et al., 2014; Cunha & Agostinho, 2010), which is reenforced by the ability of A_{2A}R antagonists to prevent memory deterioration in aging and upon diverse neurodegenerative and neuropsychiatric conditions (reviewed in Cunha & Agostinho, 2010). This role of A_{2A}R in the control of memory processes is heralded by the ability of A_{2A}R antagonists to control synaptic plasticity processes in hippocampal circuits (Costenla et al., 2011; Dias, Rombo, Ribeiro, Henley, & Sebastiao, 2013; Rebola et al., 2008), the purported neurophysiologic basis of learning and memory (Lynch, 2004; Martin et al., 2000). The detrimental role of an overactivation of $A_{2A}R$ on memory performance was directly demonstrated by the observation that the infusion of CGS21680 into the posterior cingulate cortex of rodents is sufficient to cause an amnesic effect at all doses tested, an effect blocked by an $A_{2A}R$ antagonist (Pereira et al., 2005). Similar observations were made in the specific context of working memory: thus transgenic rats overexpressing the human $A_{2A}R$ display a deterioration of working memory performance (Gimenez-Llort et al., 2007). Conversely, the genetic elimination of $A_{2A}R$ in $A_{2A}R$-KO mice leads to an improvement of working memory performance (Augusto et al., 2013; Shen et al., 2012; Wang, Ma, & van den Buuse, 2006; Wei, Singer, et al., 2011). Accordingly, the pharmacological blockade of $A_{2A}R$ also alleviates working memory deficits (Canas, Porciúncula, et al., 2009; Kadowaki Horita, Kobayashi, Mori, Jenner, & Kanda, 2013; Moeller et al., 2012; Ning et al., 2013), as does the nonselective adenosine receptor antagonist caffeine in humans (Haller et al., 2013; Koppelstaetter et al., 2008); this is in agreement with the impact of caffeine and selective $A_{2A}R$ antagonists to control attention (Beste, Stock, Ness, Epplen, & Arning, 2012) with particular interest for conditions such as attention deficit and hyperactivity disorders (Molero et al., 2013; Pandolfo, Machado, Köfalvi, Takahashi, & Cunha, 2013; Pires et al., 2009; Prediger, Fernandes, & Takahashi, 2005). Finally, it was also reported that the overexpression of adenosine kinase (the enzyme responsible for adenosine clearance) induce working memory deficits and attention disturbances (Shen et al., 2012). Overall, these observations indicate a robust impact of $A_{2A}R$ on working memory performance.

5. IMPACT OF CAFFEINE AND OTHER DRUGS ACTING ON THE ADENOSINE MODULATION SYSTEM IN SCHIZOPHRENIC PATIENTS

In the absence of clinical trials directly testing adenosine receptor ligands in schizophrenic patients, an interesting alternative to gauge the possible relevance of the adenosine modulation system in schizophrenia is to explore the impact of the consumption of drugs that affect the adenosine modulation system. One such drug is caffeine, the most widely consumed psychoactive drug, which only known molecular mechanisms of action in nontoxic doses is the antagonism of adenosine receptors (Fredholm et al., 1999, 2005).

Caffeine has long been shown to worsen psychosis and cognitive deficits in schizophrenia patients (De Freitas & Schwartz, 1979; Lucas et al., 1990) despite ongoing treatment with antipsychotics. However, two similar studies failed to find robust differences in symptoms and behaviors of patients who were switched to decaffeinated coffee (Koczapski, Paredes, Kogan, Ledwidge, & Higenbottam, 1989; Mayo, Falkowski, & Jones, 1993). High caffeine intake has been reported to produce psychotic symptoms that were resolved after lowering the dose (Hedges, Woon, & Hoopes, 2009). There are also case reports supporting that coffee or caffeinated energy drinks exacerbated psychotic symptoms in individual patients (Cerimele, Stern, & Jutras-Aswad, 2010; Tibrewal & Dhillon, 2011). In a similar vein, reduction of caffeine intake was important to reverse the lack of response to clozapine treatment (Dratcu, Grandison, McKay, Bamidele, & Vasudevan, 2007). Regarding the regular intake of caffeine, Gurpegui, Aguilar, Martínez-Ortega, Diaz, and De Leon (2004) observed fewer but heavier caffeine consumers in schizophrenia, but much of this association was explained by heavy smoking in schizophrenia. Interestingly, the expected deficiency of sensorimotor gating, evaluated as a disturbed prepulse inhibition or P50 evoked potential, which is characteristic of schizophrenic individuals (Potter, Summerfelt, Gold, & Buchanan, 2006), is mimicked by theophylline in healthy volunteers (Ghisolfi et al., 2002). Together, these data suggest that heavy caffeine consumption may exacerbate or even produce psychotic symptoms in susceptible individuals, and may negatively influence the treatment with clozapine.

Additional information on the possible therapeutic interest of the adenosine modulation system in schizophrenia can be inferred from open clinical trials carried out with drugs known to affect the adenosine system, namely with: (1) allopurinol, a xanthine oxidase inhibitor used in gout that causes a back flux increase of adenosine levels (Pacher, Nivorozhkin, & Szabó, 2006); (2) dipyridamole, an inhibitor of nonconcentrative nucleoside transporters, used as an antithrombotic to manage pulmonary hypertension that also increases adenosine levels (Chakrabarti & Freedman, 2008); (3) propentofylline, a mixed inhibitor of nucleoside transporters and of A2R, that has been developed for dementia (Kittner, Rössner, & Rother, 1997), but is currently not approved for clinical use. Albeit a meta-analysis of most trials suggeste that adenosine modulators afforded a benefit compared to placebo in the positive and negative syndrome scale (Hirota & Kishi, 2013), the small sample size of most studies still warns caution to translate this trend into clinical practice. The first study showed that dipyridamole was beneficial for positive symptoms of schizophrenia patient on haloperidol therapy (Akhondzadeh et al., 2000). This was confirmed in a double-blind trial comparing olanzapine (20 mg/day) to dipyridamole monotherapy (200 mg/day), which did not support a significant antipsychotic effect of dipyridamole monotherapy, the results provide some evidence for examining dipyridamole (200 mg/day) as adjunct to symptomatic antipsychotic-treated schizophrenia patients (Wonodi et al., 2011). Additionally, a 8-weeks double-blind and placebo-controlled trial indicate that the combination of risperidone (6 mg/day) and propentofylline (3 × 300 mg/day) show a significant superiority over risperidone alone in the treatment of positive symptoms, general psychopathology symptoms as well as in the total scores in the positive and negative syndrome scale (Salimi et al., 2008). As for allopurinol, the first trial tested the effect of allopurinol as odd-on therapy in patients refractory to typical antipsychotics showing an improvement in half of the patients (Lara et al., 2001), confirmed in a subsequent trial revealing the ability of add-on allopurinol to alleviate mostly positive symptoms and decreasing illness duration in patients medicated with chlorpromazine (550 mg/day) or equivalent (Brunstein, Ghisolfi, Ramos, & Lara, 2005). Another trial testing add-on allopurinol also reported a benefit for positive and general symptoms, but not negative symptoms (Akhondzadeh, Safarcherati, & Amini, 2005). Finally, a more recent double-blind trial of adjunctive allopurinol for schizophrenia also concluded that add-on allopurinol reduce total scores in the positive and negative syndrome scale in a subset of schizophrenia patients (Dickerson et al., 2009).

Overall, the trend for an exacerbation by caffeine and an attenuation by dipyridamole/propentofylline and allopurinol of schizophrenia symptoms support a putative role for deficient levels of adenosine in the brain of schizophrenic patients and are supportive of the adenosine hypofunction hypothesis of schizophrenia.

5.1. Modification of the adenosine neuromodulation system in schizophrenia

The planning of a therapeutic intervention based on the manipulation of the adenosine modulation system to manage schizophrenia-related endophenotypes, should not only be based on the impact of the adenosine receptors in control animals (as described so far), but should mainly take into account the alterations of the adenosine receptors that may occur in schizophrenia. This is particularly relevant given that the adenosine system is known to undergo major changes upon brain insults, although it is still unclear if this is associated with the etiology of brain dysfunction or if it is instead an adaptive process attempting to counteract brain dysfunction (reviewed in Cunha, 2005; Fredholm et al., 2005; Gomes et al., 2011). Thus, upon brain damage, there is an upregulation of $A_{2A}R$ accompanied by a downregulation of A_1R, as observed in animal models of aging (Canas, Duarte, Rodrigues, Köfalvi, & Cunha, 2009; Cunha et al., 2005; Sebastião, Cunha, de Mendonça, & Ribeiro, 2000; Rebola, Sebastião, et al., 2003), in Alzheimer's disease (Arendash et al., 2006; Espinosa et al., 2013), in epilepsy (Cognato et al., 2010; Rebola, Porciúncula, et al., 2005), in attention deficit and hyperactivity disorder (Pandolfo et al., 2013), in diabetic encephalopathy (Duarte et al., 2012; Duarte, Carvalho, Cunha, & Gruetter, 2009), or upon exposure to repeated stress (Cunha, Canas, Oliveira, & Cunha, 2006). A similar increase of the density of $A_{2A}R$ seems to occur in the diseased human brain (Albasanz, Rodríguez, Ferrer, & Martín, 2006; Albasanz, Perez, Barrachina, Ferrer, & Martín, 2008; Angulo et al., 2003; Calon et al., 2004; Deckert et al., 2003; Kurumaji & Toru, 1998; Ramlackhansingh et al., 2011; Rissanen et al., 2013; but see Villar-Menéndez et al., 2014), whereas the changes of A_1R density are not so consistent (Albasanz et al., 2008; Albasanz, Rodríguez, Ferrer, & Martín, 2007; Angulo et al., 2003; Boy et al., 2008; Deckert et al., 1996; Fukumitsu et al., 2008; Jaarsma, Sebens, & Korf, 1991; Kalaria, Sromek, Wilcox, & Unnerstall, 1990; Ułas et al., 1993).

The information about changes in the adenosine neuromodulation system in the brain of schizophrenic patients is surprisingly scarce. Thus, there are two postmortem studies that used different approaches to converge in the conclusion that there is an upregulation of striatal $A_{2A}R$ in the brain of schizophrenic patients compared to controls (Deckert et al., 2003; Kurumaji & Toru, 1998). This is in agreement with the observed positive association between $A_{2A}R$ upregulation and baseline P50 amplitude in schizophrenic patients (Zhang et al., 2012), whereas a recent study presented data to support that a reduction of striatal $A_{2A}R$ could define a particular group of schizophrenic patients (Villar-Menéndez et al., 2014). This tentative association between an upregulation of striatal $A_{2A}R$ with schizophrenia is further strengthened by the evidence indicating that the $A_{2A}R$ gene, located in the 22q12–13 region, is a candidate gene for susceptibility to schizophrenia (Deckert et al., 1996; Hong, Liu, Liu, Liao, & Tsai, 2005; Luu et al., 2008; Ottoni et al., 2005). It has also been reported that A_1R polymorphisms were also enriched in a Japanese population of schizophrenic patients. It still remains to be investigated if there are adaptive changes of

A_1R or of $A_{2A}R$ in different brain regions associated with schizophrenia endophenotypes, as well as changes in the different enzymes regulating the adenosine systems, such as adenosine kinase or ectonucleotidases (see Aliagas et al., 2013).

6. PROPOSED ADENOSINE HYPOTHESIS OF SCHIZOPHRENIA

The possible role of the adenosine modulation system in schizophrenia has evolved over the years. The first proposal was focused on the therapeutic interest of $A_{2A}R$ agonists, based on their antidopaminergic action through $A_{2A}R$–D2R heteromers (Ferré et al., 1994). A second proposal, integrating the impact of adenosine on development, on the control of glutamatergic synapses and on the clinical experience with dipyridamole, was forwarded by Diogo Lara (Lara, Dall'igna, Ghisolfi, & Brunstein, 2006; Lara & Souza, 2000) and essentially proposed an 'adenosine hypofunction'. A recent series of state-of-the-art studies led Detlev Boison developed this 'adenosine hypofunction' hypothesis of schizophrenia, narrowing the defect to the overactivation of adenosine kinase and bringing the astrocyte for the first time to the center stage of schizophrenia research (Boison et al., 2012). In parallel, two groups of researchers proposed a potential therapeutic interest of $A_{2A}R$ antagonists in the management of schizophrenia endophenotypes (Cunha et al., 1998; Wardas, 2008).

The present review compiled the available information to propose a novel adenosine hypothesis of schizophrenia: this corresponds to a hypothesis of imbalance of the adenosine modulation system, that actually corresponds to the merging of the previous two hypothesis. Thus, we propose that there is a parallel decrease of A_1R function, caused by the decreased density of A_1R and by the increased activity of adenosine kinase that reduces the levels of adenosine selective engaged in the activation of A_1R (see Cunha, 2005 for detailed review); this is accompanied by a deregulation of $A_{2A}R$, with an decreased density of astrocytic $A_{2A}R$ and an upregulation of neuronal $A_{2A}R$. Based on our previous general hypothesis on the integrated role of adenosine A_1R and $A_{2A}R$ in assisting encoding information salience in neuronal circuits, together with our new data showing a key role of astrocytic $A_{2A}R$ to set in motion an astrocyte-to-neuron process of readaptation of glutamatergic synapses and $A_{2A}R$ therein (see Fig. 16.1), this new proposal essentially argues that the parallel imbalance of A_1R and $A_{2A}R$ blurs the adequate encoding of information salience in neuronal circuits, which we propose to be a core pathogenic feature in the development of schizophrenia endophenotypes. This proposal is not only of scientific interest but should also provide a rationale to assist the design of future therapeutic intervention targeting the adenosine modulation system to manage schizophrenia endophenotypes: these should not be based only on an attempt to target adenosine kinase-A_1R or only $A_{2A}R$, but should instead simultaneously target these two arms of the adenosine modulation system.

ACKNOWLEDGMENTS

This work was supported by Fundação para a Ciência e a Tecnologia (PTDC/SAU-NSC/122254/2010), DARPA (09-68-ESR-FP-010 AND W911NF-10-1-0059), and a joint Portuguese-Brazilian grant (CAPES-FCT) and CNPq (Ciência sem Fronteiras, Brazil).

REFERENCES

Abbracchio, M. P., Fogliatto, G., Paoletti, A. M., Rovati, G. E., & Cattabeni, F. (1992). Prolonged in vitro exposure of rat brain slices to adenosine analogues: Selective desensitization of adenosine A1 but not A2 receptors. *European Journal of Pharmacology*, *227*(3), 317–324.

Abi-Dargham, A., Rodenhiser, J., Printz, D., Zea-Ponce, Y., Gil, R., Kegeles, L. S., et al. (2000). Increased baseline occupancy of D2 receptors by dopamine in schizophrenia. *Proceedings of the National Academy of Sciences of the United States of America*, *97*, 8104–8109.

Achim, A. M., Maziade, M., Raymond, E., Olivier, D., Mérette, C., & Roy, M. A. (2011). How prevalent are anxiety disorders in schizophrenia? A meta-analysis and critical review on asignificant association. *Schizophrenia Bulletin*, *37*, 811–821.

Adén, U., Halldner, L., Lagercrantz, H., Dalmau, I., Ledent, C., & Fredholm, B. B. (2003). Aggravated brain damage after hypoxic ischemia in immature adenosine A2A knockout mice. *Stroke*, *34*(3), 739–744.

Adén, U., Leverin, A. L., Hagberg, H., & Fredholm, B. B. (2001). Adenosine A_1 receptor agonism in the immature rat brain and heart. *European Journal of Pharmacology*, *426*(3), 185–192.

Agnati, L. F., Fuxe, K., Zini, I., Lenzi, P., & Hokfelt, T. (1980). Aspects on receptor regulation and isoreceptor identification. *Medical Biology*, *58*, 182–187.

Akhondzadeh, S., Safarcherati, A., & Amini, H. (2005). Beneficial antipsychotic effects of allopurinol as add-on therapy for schizophrenia: A double blind, randomized and placebo controlled trial. *Progress in Neuro-Psychopharmacology & Biological Psychiatry*, *29*(2), 253–259.

Akhondzadeh, S., Shasavand, E., Jamilian, H., Shabestari, O., & Kamalipour, A. (2000). Dipyridamole in the treatment of schizophrenia: Adenosine-dopamine receptor interactions. *Journal of Clinical Pharmacy and Therapeutics*, *25*, 131–137.

Albasanz, J. L., Perez, S., Barrachina, M., Ferrer, I., & Martín, M. (2008). Up-regulation of adenosine receptors in the frontal cortex in Alzheimer's disease. *Brain Pathology*, *18*(2), 211–219.

Albasanz, J. L., Rodríguez, A., Ferrer, I., & Martín, M. (2006). Adenosine A2A receptors are up-regulated in Pick's disease frontal cortex. *Brain Pathology*, *16*(4), 249–255.

Albasanz, J. L., Rodríguez, A., Ferrer, I., & Martín, M. (2007). Up-regulation of adenosine A1 receptors in frontal cortex from Pick's disease cases. *European Journal of Neuroscience*, *26*(12), 3501–3508.

Al-Hasani, R., Foster, J. D., Metaxas, A., Ledent, C., Hourani, S. M., Kitchen, I., et al. (2011). Increased desensitization of dopamine D2 receptor-mediated response in the ventral tegmental area in the absence of adenosine A_{2A} receptors. *Neuroscience, 190*, 103–111.

Aliagas, E., Villar-Menéndez, I., Sévigny, J., Roca, M., Romeu, M., Ferrer, I., et al. (2013). Reduced striatal ecto-nucleotidase activity in schizophrenia patients supports the "adenosine hypothesis" *Purinergic Signal, 9*(4), 599–608.

Amann, L. C., Gandal, M. J., Halene, T. B., Ehrlichman, R. S., White, S. L., McCarren, H. S., et al. (2010). Mouse behavioral endophenotypes for schizophrenia. *Brain Research Bulletin, 83*, 147–161.

Ambrósio, A. F., Malva, J. O., Carvalho, A. P., & Carvalho, C. M. (1997). Inhibition of N-, P/Q- and other types of Ca2+ channels in rat hippocampal nerve terminals by the adenosine A1 receptor. *European Journal of Pharmacology, 340*(2–3), 301–310.

Amminger, G. P., Schäfer, M. R., Papageorgiou, K., Klier, C. M., Cotton, S. M., Harrigan, S. M., et al. (2010). Long-chain omega-3 fatty acids for indicated prevention of psychotic disorders: A randomized, placebo-controlled trial. *Archives of General Psychiatry, 67*, 146–154.

Andersen, M. B., Fuxe, K., Werge, T., & Gerlach, J. (2002). The adenosine A2A receptor agonist CGS 21680 exhibits antipsychotic-like activity in Cebus apella monkeys. *Behavioral Pharmacology, 13*(8), 639–644.

Angulo, E., Casadó, V., Mallol, J., Canela, E. I., Viñals, F., Ferrer, I., et al. (2003). A1 adenosine receptors accumulate in neurodegenerative structures in Alzheimer disease and mediate both amyloid precursor protein processing and tau phosphorylation and translocation. *Brain Pathology, 13*(4), 440–451.

Arai, S., Takuma, K., Mizoguchi, H., Ibi, D., Nagai, T., Takahashi, K., et al. (2008). Involvement of pallidotegmental neurons in methamphetamine- and MK-801-induced impairment of prepulse inhibition of the acoustic startle reflex in mice: Reversal by $GABA_B$ receptor agonist baclofen. *Neuropsychopharmacology, 33*, 3164–3175.

Asghari, A. A., Azarnia, M., Mirnajafi-Zadeh, J., & Javan, M. (2013). Adenosine A1 receptor agonist, N6-cyclohexyladenosine, protects myelin and induces remyelination in an experimental model of rat optic chiasm demyelination; electrophysiological and histopathological studies. *Journal of the Neurological Sciences, 325*(1–2), 22–28.

Augusto, E., Matos, M., Sévigny, J., El-Tayeb, A., Bynoe, M. S., Muller, C. E., et al. (2013). Ecto-5'-nucleotidase (CD73)-mediated formation of adenosine is critical for the striatal adenosine A2A receptor functions. *Journal of Neuroscience, 33*, 11390–11399.

Azdad, K., Gall, D., Woods, A. S., Ledent, C., Ferré, S., & Schiffmann, S. N. (2009). Dopamine D2 and adenosine A_{2A} receptors regulate NMDA-mediated excitation in accumbens neurons through A_{2A}-D2 receptor heteromerization. *Neuropsychopharmacology, 34*, 972–986.

Back, S. A., Craig, A., Luo, N. L., Ren, J., Akundi, R. S., Ribeiro, I., et al. (2006). Protective effects of caffeine on chronic hypoxia-induced perinatal white matter injury. *Annals of Neurology, 60*(6), 696–705.

Bakshi, V. P., Geyer, M. A., Taaid, N., & Swerdlow, N. R. (1995). A comparison of the effects of amphetamine, strychnine and caffeine on prepulse inhibition and latent inhibition. *Behavioral Pharmacology, 6*(8), 801–809.

Barksdale, K. A., Lahti, A. C., & Roberts, R. C. (2014). Synaptic proteins in the postmortem anterior cingulate cortex in schizophrenia: Relationship to treatment and treatment response. *Neuropsychopharmacology*.

Barraco, R. A., Martens, K. A., Parizon, M., & Normile, H. J. (1993). Adenosine A_{2a} receptors in the nucleus accumbens mediate locomotor depression. *Brain Research Bulletin, 31*, 397–404.

Bastia, E., Xu, Y. H., Scibelli, A. C., Day, Y. J., Linden, J., Chen, J. F., et al. (2005). A crucial role for forebrain adenosine A_{2A} receptors in amphetamine sensitization. *Neuropsychopharmacology, 30*, 891–900.

Bauer, D., Haroutunian, V., Meador-Woodruff, J. H., & McCullumsmith, R. E. (2010). Abnormal glycosylation of EAAT1 and EAAT2 in prefrontal cortex of elderly patients with schizophrenia. *Schizophrenia Research, 117*, 92–98.
Bekar, L., Libionka, W., Tian, G. F., Xu, Q., Torres, A., Wang, X., et al. (2008). Adenosine is crucial for deep brain stimulation-mediated attenuation of tremor. *Nature Medicine, 14*(1), 75–80.
Bellesi, M., Melone, M., Gubbini, A., Battistacci, S., & Conti, F. (2009). GLT-1 upregulation impairs prepulse inhibition of the startle reflex in adult rats. *Glia, 57*, 703–713.
Benarroch, E. E. (2011). Na+, K+-ATPase: Functions in the nervous system and involvement in neurologic disease. *Neurology, 76*, 287–293.
Benes, F. M., Davidson, J., & Bird, E. D. (1986). Quantitative cytoarchitectural studies of the cerebral cortex of schizophrenics. *Archives of General Psychiatry, 43*, 31–35.
Berghuis, P., Rajnicek, A. M., Morozov, Y. M., Ross, R. A., Mulder, J., Urbán, G. M., et al. (2007). Hardwiring the brain: Endocannabinoids shape neuronal connectivity. *Science, 316*, 1212–1216.
Beste, C., Stock, A. K., Ness, V., Epplen, J. T., & Arning, L. (2012). Differential effects of ADORA2A gene variations in pre-attentive visual sensory memory subprocesses. *European Neuropsychopharmacology, 22*(8), 555–561.
Bibbiani, F., Oh, J. D., Petzer, J. P., Castagnoli, N., Jr., Chen, J. F., Schwarzschild, M. A., et al. (2003). A2A antagonist prevents dopamine agonist-induced motor complications in animal models of Parkinson's disease. *Experimental Neurology, 184*, 285–294.
Bleickardt, C. J., Lashomb, A. L., Merkel, C. E., & Hodgson, R. A. (2012). Adenosine A_{2A} Receptor Antagonists Do Not Disrupt Rodent Prepulse Inhibition: An Improved Side Effect Profile in the Treatment of Parkinson's Disease. *Parkinson's Disease, 2012*, 591094.
Blennow, K., Davidsson, P., Gottfries, C. G., Ekman, R., & Heilig, M. (1996). Synaptic degeneration in thalamus in schizophrenia. *Lancet, 348*, 692–693.
Boison, D. (2013). Adenosine kinase: Exploitation for therapeutic gain. *Pharmacological Reviews, 65*(3), 906–943.
Boison, D., Chen, J. F., & Fredholm, B. B. (2010). Adenosine signaling and function in glial cells. *Cell Death and Differentiation, 17*(7), 1071–1082.
Boison, D., Singer, P., Shen, H. Y., Feldon, J., & Yee, B. K. (2012). Adenosine hypothesis of schizophrenia-opportunities for pharmacotherapy. *Neuropharmacology, 62*(3), 1527–1543.
Boksa, P. (2004). Animal models of obstetric complications in relation to schizophrenia. *Brain Research Reviews, 45*(1), 1–17.
Boksa, P. (2012). Abnormal synaptic pruning in schizophrenia: Urban myth or reality? *Journal of Psychiatry & Neuroscience, 37*, 75–77.
Bona, E., Adén, U., Gilland, E., Fredholm, B. B., & Hagberg, H. (1997). Neonatal cerebral hypoxia-ischemia: The effect of adenosine receptor antagonists. *Neuropharmacology, 36*(9), 1327–1338.
Borota, D., Murray, E., Keceli, G., Chang, A., Watabe, J. M., Ly, M., et al. (2014). Post-study caffeine administration enhances memory consolidation in humans. *Nature Neuroscience, 17*, 201–203.
Borroto-Escuela, D. O., Romero-Fernandez, W., Tarakanov, A. O., Ciruela, F., Agnati, L. F., & Fuxe, K. (2011). On the existence of a possible A_{2A}-D2-beta-Arrestin2 complex: A_{2A} agonist modulation of D2 agonist-induced β-arrestin2 recruitment. *Journal of Molecular Biology, 406*, 687–699.
Borycz, J., Pereira, M. F., Melani, A., Rodrigues, R. J., Köfalvi, A., Panlilio, L., et al. (2007). Differential glutamate-dependent and glutamate-independent adenosine A1 receptor-

mediated modulation of dopamine release in different striatal compartments. *Journal of Neurochemistry, 101*, 355–363.
Bowie, C. R., & Harvey, P. D. (2005). Cognition in schizophrenia: Impairments, determinants, and functional importance. *The Psychiatric Clinics of North America, 28*(613–633), 626.
Boy, C., Meyer, P. T., Kircheis, G., Holschbach, M. H., Herzog, H., Elmenhorst, D., et al. (2008). Cerebral A1 adenosine receptors (A1AR) in liver cirrhosis. *European Journal of Nuclear Medicine and Molecular Imaging, 35*(3), 589–597.
Braff, D. L., & Geyer, M. A. (1990). Sensorimotor gating and schizophrenia. Human and animal model studies. *Archives of General Psychiatry, 47*, 181–188.
Braff, D., Stone, C., Callaway, E., Geyer, M., Glick, I., & Bali, L. (1978). Prestimulus effects on human startle reflex in normals and schizophrenics. *Psychophysiology, 15*, 339–343.
Brambilla, P., Perez, J., Barale, F., Schettini, G., & Soares, J. C. (2003). GABAergic dysfunction in mood disorders. *Molecular Psychiatry, 8*(8), 721–737, 715.
Bromet, E. J., & Fennig, S. (1999). Epidemiology and natural history of schizophrenia. *Biological Psychiatry, 46*, 871–881.
Brown, R. M., Short, J. L., Cowen, M. S., Ledent, C., & Lawrence, A. J. (2009). A differential role for the adenosine A2A receptor in opiate reinforcement vs opiate-seeking behavior. *Neuropsychopharmacology, 34*(4), 844–856.
Browning, M. D., Dudek, E. M., Rapier, J. L., Leonard, S., & Freedman, R. (1993). Significant reductions in synapsin but not synaptophysin specific activity in the brains of some schizophrenics. *Biological Psychiatry, 34*, 529–535.
Brunstein, M. G., Ghisolfi, E. S., Ramos, F. L., & Lara, D. R. (2005). A clinical trial of adjuvant allopurinol therapy for moderately refractory schizophrenia. *Journal of Clinical Psychiatry, 66*(2), 213–219.
Calon, F., Dridi, M., Hornykiewicz, O., Bédard, P. J., Rajput, A. H., & Di Paolo, T. (2004). Increased adenosine A2A receptors in the brain of Parkinson's disease patients with dyskinesias. *Brain, 127*(Pt 5), 1075–1084.
Canas, P. M., Duarte, J. M. N., Rodrigues, R. J., Köfalvi, A., & Cunha, R. A. (2009). Modification upon aging of the density of presynaptic modulation systems in the hippocampus. *Neurobiology of Aging, 30*, 1877–1884.
Canas, P. M., Porciúncula, L. O., Cunha, G. M. A., Silva, C. G., Machado, N. J., Oliveira, J. M. A., et al. (2009). Adenosine A2A receptor blockade prevents synaptotoxicity and memory dysfunction caused by β-amyloid peptides via p38 mitogen-activated protein kinase pathway. *Journal of Neuroscience, 29*, 14741–14751.
Caporale, N., & Dan, Y. (2008). Spike timing-dependent plasticity: A Hebbian learning rule. *Annual Review of Neuroscience, 31*, 25–46.
Carlson, G. C., Talbot, K., Halene, T. B., Gandal, M. J., Kazi, H. A., Schlosser, L., et al. (2011). Dysbindin-1 mutant mice implicate reduced fast-phasic inhibition as a final common disease mechanism in schizophrenia. *Proceedings of the National Academy of Sciences of the United States of America, 108*, 962–970.
Caroni, P., Donato, F., & Muller, D. (2012). Structural plasticity upon learning: Regulation and functions. *Nature Reviews. Neuroscience, 13*(7), 478–490.
Casadó, V., Mallol, J., Lluis, C., Franco, R., & Canela, E. I. (1991). Adenosine receptors in myelin fractions and subfractions: The effect of the agonist (R)-phenylisopropyladenosine on myelin membrane microviscosity. *Journal of Neurochemistry, 57*(5), 1623–1629.
Castillo, M. A., Ghose, S., Tamminga, C. A., & Ulery-Reynolds, P. G. (2010). Deficits in syntaxin 1 phosphorylation in schizophrenia prefrontal cortex. *Biological Psychiatry, 67*, 208–216.
Castner, S. A., Goldman-Rakic, P. S., & Williams, G. V. (2004). Animal models of working memory: Insights for targeting cognitive dysfunction in schizophrenia. *Psychopharmacology, 174*, 111–125.

Cerimele, J. M., Stern, A. P., & Jutras-Aswad, D. (2010). Psychosis following excessive ingestion of energy drinks in a patient with schizophrenia. *The American Journal of Psychiatry*, *167*, 353.
Chakrabarti, S., & Freedman, J. E. (2008). Dipyridamole, cerebrovascular disease, and the vasculature. *Vascular Pharmacology*, *48*(4–6), 143–149.
Chambers, J. S., Thomas, D., Saland, L., Neve, R. L., & Perrone-Bizzozero, N. I. (2005). Growth-associated protein 43 (GAP-43) and synaptophysin alterations in the dentate gyrus of patients with schizophrenia. *Progress in Neuro-Psychopharmacology & Biological Psychiatry*, *29*, 283–290.
Chang, L., Friedman, J., Ernst, T., Zhong, K., Tsopelas, N. D., & Davis, K. (2007). Brain metabolite abnormalities in the white matter of elderly schizophrenic subjects: Implication for glial dysfunction. *Biological Psychiatry*, *62*, 1396–1404.
Chen, J. F., Eltzschig, H. K., & Fredholm, B. B. (2013). Adenosine receptors as drug targets—What are the challenges? *Nature Reviews. Drug Discovery*, *12*(4), 265–286.
Chen, J. F., Fredduzzi, S., Bastia, E., Yu, L., Moratalla, R., Ongini, E., et al. (2003). Adenosine A_{2A} receptors in neuroadaptation to repeated dopaminergic stimulation: Implications for the treatment of dyskinesias in Parkinson's disease. *Neurology*, *61*(Suppl 6), S74–S81.
Chen, J. F., Sonsalla, P. K., Pedata, F., Melani, A., Domenici, M. R., Popoli, P., et al. (2007). Adenosine A2A receptors and brain injury: Broad spectrum of neuroprotection, multifaceted actions and "fine tuning" modulation. *Progress in Neurobiology*, *83*(5), 310–331.
Cheng, M. C., Lu, C. L., Luu, S. U., Tsai, H. M., Hsu, S. H., Chen, T. T., et al. (2010). Genetic and functional analysis of the DLG4 gene encoding the post-synaptic density protein 95 in schizophrenia. *PLoS One*, *5*, e15107.
Cheng, H. C., Shih, H. M., & Chern, Y. (2002). Essential role of cAMP-response element-binding protein activation by A2A adenosine receptors in rescuing the nerve growth factor-induced neurite outgrowth impaired by blockage of the MAPK cascade. *The Journal of Biological Chemistry*, *277*(37), 33930–33942.
Cirillo, M. A., & Seidman, L. J. (2003). Verbal declarative memory dysfunction in schizophrenia: From clinical assessment to genetics and brain mechanisms. *Neuropsychology Review*, *13*, 43–77.
Ciruela, F., Burgueno, J., Casado, V., Canals, M., Marcellino, D., Goldberg, S. R., et al. (2004). Combining mass spectrometry and pull-down techniques for the study of receptor heteromerization. Direct epitope-epitope electrostatic interactions between adenosine A2A and dopamine D2 receptors. *Analytical Chemistry*, *76*, 5354–5363.
Ciruela, F., Casadó, V., Rodrigues, R. J., Lujan, R., Burgueño, J., Canals, M., et al. (2006). Presynaptic control of striatal glutamatergic neurotransmission by adenosine A1-A2A receptor heteromers. *Journal of Neuroscience*, *26*, 2080–2087.
Cognato, G. P., Agostinho, P. M., Hockemeyer, J., Müller, C. E., Souza, D. O., & Cunha, R. A. (2010). Caffeine and adenosine A2A receptor antagonists prevent memory impairment and synaptotoxicity in adult rats triggered by a convulsive episode in early life. *Journal of Neurochemistry*, *112*, 453–462.
Coppi, E., Cellai, L., Maraula, G., Pugliese, A. M., & Pedata, F. (2013). Adenosine A_{2A} receptors inhibit delayed rectifier potassium currents and cell differentiation in primary purified oligodendrocyte cultures. *Neuropharmacology*, *73*, 301–310.
Costenla, A. R., Diógenes, M. J., Canas, P. M., Rodrigues, R. J., Nogueira, C., Maroco, J., et al. (2011). Enhanced role of adenosine A2A receptors in the modulation of LTP in the rat hippocampus upon aging. *European Journal of Neuroscience*, *34*, 12–21.
Cotter, D. R., Pariante, C. M., & Everall, I. P. (2001). Glial cell abnormalities in major psychiatric disorders: The evidence and implications. *Brain Research Bulletin*, *55*, 585–595.

Coyle, J. T., Basu, A., Benneyworth, M., Balu, D., & Konopaske, G. (2012). Glutamatergic synaptic dysregulation in schizophrenia: Therapeutic implications. *Handbook of Experimental Pharmacology, 213*, 267–295.

Crump, C., Winkleby, M. A., Sundquist, K., & Sundquist, J. (2013). Comorbidities and mortality in persons with schizophrenia: A Swedish national cohort study. *The American Journal of Psychiatry, 170*, 324–333.

Cunha, R. A. (2001). Adenosine as a neuromodulator and as a homeostatic regulator in the nervous system: Different roles, different sources and different receptors. *Neurochemistry International, 38*, 107–125.

Cunha, R. A. (2005). Neuroprotection by adenosine in the brain: From A1 receptor activation to A2A receptor blockade. *Purinergic Signalling, 1*, 111–134.

Cunha, R. A., & Agostinho, P. M. (2010). Chronic caffeine consumption prevents memory disturbance in different animal models of memory decline. *Journal of Alzheimer Disease, 20*(Supplement 1), 95–116.

Cunha, R. A., Almeida, T., & Ribeiro, J. A. (2001). Parallel modification of adenosine extracellular metabolism and modulatory action in the hippocampus of aged rats. *Journal of Neurochemistry, 76*, 372–382.

Cunha, G. M. A., Canas, P. M., Oliveira, C. R., & Cunha, R. A. (2006). Increased density and synapto-protective effect of adenosine A2A receptors upon sub-chronic restraint stress. *Neuroscience, 141*, 1775–1781.

Cunha, R. A., Correia-de-Sá, P., Sebastião, A. M., & Ribeiro, J. A. (1996). Preferential activation of excitatory adenosine receptors at rat hippocampal and neuromuscular synapses by adenosine formed from released adenine nucleotides. *British Journal of Pharmacology, 119*, 253–260.

Cunha, R. A., Vizi, E. S., Ribeiro, J. A., & Sebastião, A. M. (1996). Preferential release of ATP and its extracellular catabolism as a source of adenosine upon high- but not low-frequency stimulation of rat hippocampal slices. *Journal of Neurochemistry, 67*, 2180–2187.

d'Alcantara, P., Ledent, C., Swillens, S., & Schiffmann, S. N. (2001). Inactivation of adenosine A2A receptor impairs long term potentiation in the accumbens nucleus without altering basal synaptic transmission. *Neuroscience, 107*(3), 455–464.

Dai, S. S., Zhou, Y. G., Li, W., An, J. H., Li, P., Yang, N., et al. (2010). Local glutamate level dictates adenosine A_{2A} receptor regulation of neuroinflammation and traumatic brain injury. *Journal of Neuroscience, 30*(16), 5802–5810.

Daré, E., Schulte, G., Karovic, O., Hammarberg, C., & Fredholm, B. B. (2007). Modulation of glial cell functions by adenosine receptors. *Physiology & Behavior, 92*(1–2), 15–20.

Davidsson, P., Gottfries, J., Bogdanovic, N., Ekman, R., Karlsson, I., Gottfries, C. G., et al. (1999). The synaptic-vesicle-specific proteins rab3a and synaptophysin are reduced in thalamus and related cortical brain regions in schizophrenic brains. *Schizophrenia Research, 40*, 23–29.

de Bartolomeis, A., Fiore, G., & Iasevoli, F. (2005). Dopamine-glutamate interaction and antipsychotics mechanism of action: Implication for new pharmacological strategies in psychosis. *Current Pharmaceutical Design, 11*(27), 3561–3594.

de Bartolomeis, A., Latte, G., Tomasetti, C., & Iasevoli, F. (2014). Glutamatergic postsynaptic density protein dysfunctions in synaptic plasticity and dendritic spines morphology: Relevance to schizophrenia and other behavioral disorders pathophysiology, and implications for novel therapeutic approaches. *Molecular Neurobiology, 49*, 484–511.

Deckert, J., Brenner, M., Durany, N., Zöchling, R., Paulus, W., Ransmayr, G., et al. (2003). Up-regulation of striatal adenosine A_{2A} receptors in schizophrenia. *Neuroreport, 14*(3), 313–316.

Deckert, J., Nöthen, M. M., Rietschel, M., Wildenauer, D., Bondy, B., Ertl, M. A., et al. (1996). Human adenosine A2a receptor (A2aAR) gene: System- atic mutation screening in patients with schizophrenia. *Journal of Neural Transmission, 103*, 1447–1455.

De Freitas, B., & Schwartz, G. (1979). Effects of caffeine in chronic psychiatric patients. *The American Journal of Psychiatry*, *136*, 1337–1338.

DeLisi, L. E. (2008). The concept of progressive brain change in schizophrenia: Implications for understanding schizophrenia. *Schizophrenia Bulletin*, *34*, 312–321.

de Mendonça, A., Sebastião, A. M., & Ribeiro, J. A. (1995). Inhibition of NMDA receptor-mediated currents in isolated rat hippocampal neurones by adenosine A1 receptor activation. *Neuroreport*, *6*(8), 1097–1100.

de Mendonça, A., Sebastião, A. M., & Ribeiro, J. A. (2000). Adenosine: Does it have a neuroprotective role after all? *Brain Research Reviews*, *33*(2–3), 258–274.

Deng, X., Shibata, H., Ninomiya, H., Tashiro, N., Iwata, N., Ozaki, N., et al. (2004). Association study of polymorphisms in the excitatory amino acid transporter 2 gene (SLC1A2) with schizophrenia. *BMC Psychiatry*, *4*, 21.

Dias, R. B., Ribeiro, J. A., & Sebastião, A. M. (2012). Enhancement of AMPA currents and GluR1 membrane expression through PKA-coupled adenosine A_{2A} receptors. *Hippocampus*, *22*(2), 276–291.

Dias, R. B., Rombo, D. M., Ribeiro, J. A., Henley, J. M., & Sebastiao, A. M. (2013). Adenosine: Setting the stage for plasticity. *Trends in Neurosciences*, *36*, 248–257.

Dickerson, F. B., Stallings, C. R., Origoni, A. E., Sullens, A., Khushalani, S., Sandson, N., et al. (2009). A double-blind trial of adjunctive allopurinol for schizophrenia. *Schizophrenia Research*, *109*(1–3), 66–69.

Domschke, K., Gajewska, A., Winter, B., Herrmann, M. J., Warrings, B., Mühlberger, A., et al. (2012). ADORA2A Gene variation, caffeine, and emotional processing: A multi-level interaction on startle reflex. *Neuropsychopharmacology*, *37*(3), 759–769.

Dratcu, L., Grandison, A., McKay, G., Bamidele, A., & Vasudevan, V. (2007). Clozapine-resistant psychosis, smoking, and caffeine: Managing the neglected effects of substances that our patients consume every day. *American Journal of Therapeutics*, *14*, 314–318.

Duarte, J. M. N., Agostinho, P. M., Carvalho, R. A., & Cunha, R. A. (2012). Caffeine consumption prevents diabetes-induced memory impairment and synaptotoxicity in the hippocampus of NONcNZO10/LtJ diabetic mice. *PLoS One*, *7*, e21899.

Duarte, J. M. N., Carvalho, R. A., Cunha, R. A., & Gruetter, R. (2009). Caffeine consumption attenuates neurochemical modifications in the hippocampus of streptozotocin-induced diabetic rats. *Journal of Neurochemistry*, *111*, 368–379.

Duncan, L. E., Holmans, P. A., Lee, P. H., O'Dushlaine, C. T., Kirby, A. W., Smoller, J. W., et al. (2014). Pathway analyses implicate glial cells in schizophrenia. *PLoS One*, *9*, e89441.

Dunwiddie, T. V., & Haas, H. L. (1985). Adenosine increases synaptic facilitation in the in vitro rat hippocampus: Evidence for a presynaptic site of action. *Journal Physiology*, *369*, 365–377.

Dunwiddie, T. V., & Masino, S. A. (2001). The role and regulation of adenosine in the central nervous system. *Annual Review of Neuroscience*, *24*, 31–55.

Dzamba, D., Honsa, P., & Anderova, M. (2013). NMDA Receptors in glial cells: Pending questions. *Current Neuropharmacology*, *11*(3), 250–262.

Elvevag, B., & Goldberg, T. E. (2000). Cognitive impairment in schizophrenia is the core of the disorder. *Critical Reviews in Neurobiology*, *14*, 1–21.

Erlenmeyer-Kimling, L., Rock, D., Roberts, S. A., Janal, M., Kestenbaum, C., Cornblatt, B., et al. (2000). Attention, memory, and motor skills as childhood predictors of schizophrenia-related psychoses: The New York High-Risk Project. *The American Journal of Psychiatry*, *157*, 1416–1422.

Espinosa, J., Rocha, A., Nunes, F., Costa, M. S., Schein, V., Kazlauckas, V., et al. (2013). Caffeine consumption prevents memory impairment, neuronal damage and adenosine A2a receptors up-regulation in the hippocampus of a rat model of sporadic dementia. *Journal of Alzheimer Disease*, *34*, 509–518.

Fattorini, G., Melone, M., Bragina, L., Candiracci, C., Cozzi, A., Pellegrini-Giampietro, D. E., et al. (2008). GLT-1 expression and Glu uptake in rat cerebral cortex are increased by phencyclidine. *Glia*, *56*, 1320–1327.

Featherstone, R. E., Kapur, S., & Fletcher, P. J. (2007). The amphetamine-induced sensitized state as a model of schizophrenia. *Progress in Neuro-Psychopharmacology & Biological Psychiatry*, *31*(8), 1556–1571.

Fedele, D. E., Li, T., Lan, J. Q., Fredholm, B. B., & Boison, D. (2006). Adenosine A1 receptors are crucial in keeping an epileptic focus localized. *Experimental Neurology*, *200*(1), 184–190.

Feldman, D. E. (2012). The spike-timing dependence of plasticity. *Neuron*, *75*, 556–571.

Ferré, S. (1997). Adenosine-dopamine interactions in the ventral striatum. Implications for the treatment of schizophrenia. *Psychopharmacology*, *133*(2), 107–120.

Ferré, S., Lluís, C., Justinova, Z., Quiroz, C., Orru, M., Navarro, G., et al. (2010). Adenosine-cannabinoid receptor interactions. Implications for striatal function. *British Journal of Pharmacology*, *160*(3), 443–453.

Ferré, S., O'Connor, W. T., Snaprud, P., Ungerstedt, U., & Fuxe, K. (1994). Antagonistic interaction between adenosine A2A receptors and dopamine D2 receptors in the ventral striopallidal system. Implications for the treatment of schizophrenia. *Neuroscience*, *63*(3), 765–773.

Ferré, S., von Euler, G., Johansson, B., Fredholm, B. B., & Fuxe, K. (1991). Stimulation of high-affinity adenosine A_2 receptors decreases the affinity of dopamine D2 receptors in rat striatal membranes. *Proceedings of the National Academy of Sciences of the United States of America*, *88*, 7238–7241.

Field, J. R., Walker, A. G., & Conn, P. J. (2011). Targeting glutamate synapses in schizophrenia. *Trends in Molecular Medicine*, *17*, 689–698.

Fields, R. D. (2005). Myelination: An overlooked mechanism of synaptic plasticity? *The Neuroscientist*, *11*, 528–531.

Fitzgerald, P. J. (2012). The NMDA receptor may participate in widespread suppression of circuit level neural activity, in addition to a similarly prominent role in circuit level activation. *Behavioural Brain Research*, *230*, 291–298.

Flaten, M. A., & Elden, A. (1999). Caffeine and prepulse inhibition of the acoustic startle reflex. *Psychopharmacology*, *147*(3), 322–330.

Foussias, G., Agid, O., Fervaha, G., & Remington, G. (2013). Negative symptoms of schizophrenia: Clinical features, relevance to real world functioning and specificity versus other CNS disorders. *European Neuropsychopharmacology*, *24*(5), 693–709.

Fredduzzi, S., Moratalla, R., Monopoli, A., Cuellar, B., Xu, K., Ongini, E., et al. (2002). Persistent behavioral sensitization to chronic L-DOPA requires A_{2A} adenosine receptors. *Journal of Neuroscience*, *22*, 1054–1062.

Fredholm, B. B., Chen, J. F., Cunha, R. A., Svenningsson, P., & Vaugeois, J. M. (2005). Adenosine and brain function. *International Review of Neurobiology*, *63*, 191–270.

Fromer, M., Pocklington, A. J., Kavanagh, D. H., Williams, H. J., Dwyer, S., Gormley, P., et al. (2014). De novo mutations in schizophrenia implicate synaptic networks. *Nature*, *506*, 179–184.

Fukumitsu, N., Ishii, K., Kimura, Y., Oda, K., Hashimoto, M., Suzuki, M., et al. (2008). Adenosine A_1 receptors using 8-dicyclopropylmethyl-1-[(11)C]methyl-3-propylxanthine PET in Alzheimer's disease. *Annals of Nuclear Medicine*, *22*(10), 841–847.

Fung, S. J., Sivagnanasundaram, S., & Weickert, C. S. (2011). Lack of change in markers of presynaptic terminal abundance alongside subtle reductions in markers of presynaptic terminal plasticity in prefrontal cortex of schizophrenia patients. *Biological Psychiatry*, *69*, 71–79.

Fusar-Poli, P., & Berger, G. (2012). Eicosapentaenoic acid interventions in schizophrenia: Meta-analysis of randomized, placebo-controlled studies. *Journal of Clinical Psychopharmacology*, *32*, 179–185.

Fuxe, K., Ferre, S., Zoli, M., & Agnati, L. F. (1998). Integrated events in central dopamine transmission as analyzed at multiple levels. Evidence for intramembrane adenosine A_{2A}/dopamine D2 and adenosine A_1/dopamine D1 receptor interactions in the basal ganglia. *Brain Research Reviews*, *26*, 258–273.

Fuxe, K., Stromberg, I., Popoli, P., Rimondini-Giorgini, R., Torvinen, M., Ogren, S. O., et al. (2001). Adenosine receptors and Parkinson's disease. Relevance of antagonistic adenosine and dopamine receptor interactions in the striatum. *Advances in Neurology*, *86*, 345–353.

Gajewska, A., Blumenthal, T. D., Winter, B., Herrmann, M. J., Conzelmann, A., Mühlberger, A., et al. (2013). Effects of ADORA2A gene variation and caffeine on prepulse inhibition: A multi-level risk model of anxiety. *Progress in Neuro-Psychopharmacology & Biological Psychiatry*, *40*, 115–121.

Genovese, T., Melani, A., Esposito, E., Mazzon, E., Di Paola, R., Bramanti, P., et al. (2009). The selective adenosine A2A receptor agonist CGS 21680 reduces JNK MAPK activation in oligodendrocytes in injured spinal cord. *Shock*, *32*(6), 578–585.

Geyer, M. A., & Braff, D. L. (1987). Startle habituation and sensorimotor gating in schizophrenia and related animal models. *Schizophrenia Bulletin*, *13*, 643–668.

Ghisolfi, E. S., Prokopiuk, A. S., Becker, J., Ehlers, J. A., Belmonte-de-Abreu, P., Souza, D. O., et al. (2002). The adenosine antagonist theophylline impairs p50 auditory sensory gating in normal subjects. *Neuropsychopharmacology*, *27*(4), 629–637.

Gimenez-Llort, L., Schiffmann, S. N., Shmidt, T., Canela, L., Camon, L., Wassholm, M., et al. (2007). Working memory deficits in transgenic rats overexpressing human adenosine A2A receptors in the brain. *Neurobiology of Learning and Memory*, *87*, 42–56.

Ginovart, N., & Kapur, S. (2012). Role of dopamine D_2 receptors for antipsychotic activity. *Handbook of Experimental Pharmacology*, *212*, 27–52.

Glantz, L. A., Gilmore, J. H., Lieberman, J. A., & Jarskog, L. F. (2006). Apoptotic mechanisms and the synaptic pathology of schizophrenia. *Schizophrenia Research*, *81*, 47–63.

Goff, D. C., Tsai, G., Levitt, J., Amico, E., Manoach, D., Schoenfeld, D. A., et al. (1999). A placebo-controlled trial of D-cycloserine added to conventional neuroleptics in patients with schizophrenia. *Archives of General Psychiatry*, *56*, 21–27.

Gogtay, N., & Rapoport, J. L. (2008). Childhood-onset schizophrenia: Insights from neuroimaging studies. *Journal of the American Academy of Child and Adolescent Psychiatry*, *47*, 1120–1124.

Gomes, C. A., Ferreira, R., George, J., Sanches, R., Rodrigues, D. I., Gonçalves, N., et al. (2013). Activation of microglial cells triggers a release of brain-derived neurotrophic factor (BDNF) inducing their proliferation in an adenosine A2A receptor-dependent manner. *Journal of Neuroinflammation*, *10*, 16.

Gomes, C. V., Kaster, M. P., Tomé, A. R., Agostinho, P. M., & Cunha, R. A. (2011). Adenosine receptors and brain diseases: Neuroprotection and neurodegeneration. *Biochimica et Biophysica Acta - Biomembranes*, *1808*, 1380–1399.

Gonçalves, N., Simões, A. T., Cunha, R. A., & Pereira de Almeida, L. (2013). Caffeine and adenosine A2A receptor inactivation decrease striatal neuropathology in a lentiviral-based model of Machado-Joseph disease. *Annals of Neurology*, *73*, 655–666.

Gonzalez-Burgos, G., & Lewis, D. A. (2012). NMDA receptor hypofunction, parvalbumin-positive neurons, and cortical gamma oscillations in schizophrenia. *Schizophrenia Bulletin*, *38*(5), 950–957.

Gotoh, L., Kawanami, N., Nakahara, T., Hondo, H., Motomura, K., Ohta, E., et al. (2002). Effects of the adenosine A_1 receptor agonist N^6-cyclopentyladenosine on phencyclidine-induced behavior and expression of the immediate-early genes in the discrete brain regions of rats. *Brain Research. Molecular Brain Research*, *100*(1–2), 1–12.

Gouder, N., Fritschy, J. M., & Boison, D. (2003). Seizure suppression by adenosine A1 receptor activation in a mouse model of pharmacoresistant epilepsy. *Epilepsia*, *44*(7), 877–885.

Goudriaan, A., De Leeuw, C., Ripke, S., Hultman, C. M., Sklar, P., Sullivan, P. F., et al. (2013). Specific Glial Functions Contribute to Schizophrenia Susceptibility. *Schizophrenia Bulletin*, *40*(4), 925–935.

Grabrucker, S., Proepper, C., Mangus, K., Eckert, M., Chhabra, R., Schmeisser, M. J., et al. (2014). The PSD protein ProSAP2/Shank3 displays synapto-nuclear shuttling which is deregulated in a schizophrenia-associated mutation. *Experimental Neurology*, *253*, 126–137.

Grantyn, R., Henneberger, C., Jüttner, R., Meler, J. C., & Kirischuk, S. (2011). Functional hallmarks of GABAergic synapse maturation and the diverse roles of neurotrophins. *Frontiers in Cellular Neuroscience*, *5*, 1–12.

Gray, B. E., McMahon, R. P., & Gold, J. M. (2013). General intellectual ability does not explain the general deficit in schizophrenia. *Schizophrenia Research*, *147*, 315–319.

Green, M. F., Nuechterlein, K. H., Gold, J. M., Barch, D. M., Cohen, J., Essock, S., et al. (2004). Approaching a consensus cognitive battery for clinical trials in schizophrenia: The NIMH-MATRICS conference to select cognitive domains and test criteria. *Biological Psychiatry*, *56*, 301–307.

Greene, R. (2001). Circuit analysis of NMDAR hypofunction in the hippocampus, in vitro, and psychosis of schizophrenia. *Hippocampus*, *11*, 569–577.

Greene, R. W., & Haas, H. L. (1985). Adenosine actions on CA1 pyramidal neurones in rat hippocampal slices. *Journal Physiology*, *366*, 119–127.

Guilmatre, A., Dubourg, C., Mosca, A. L., Legallic, S., Goldenberg, A., Drouin-Garraud, V., et al. (2009). Recurrent rearrangements in synaptic and neurodevelopmental genes and shared biologic pathways in schizophrenia, autism, and mental retardation. *Archives of General Psychiatry*, *66*, 947–956.

Gunduz-Bruce, H. (2009). The acute effects of NMDA antagonism: From the rodent to the human brain. *Brain Research Reviews*, *60*, 279–286.

Gurpegui, M., Aguilar, M. C., Martínez-Ortega, J. M., Diaz, F. J., & de Leon, J. (2004). Caffeine intake in outpatients with schizophrenia. *Schizophrenia Bulletin*, *30*, 935–945.

Hahn, C. G., Wang, H. Y., Cho, D. S., Talbot, K., Gur, R. E., Berrettini, W. H., et al. (2006). Altered neuregulin 1-erbB4 signaling contributes to NMDA receptor hypofunction in schizophrenia. *Nature Medicine*, *12*, 824–828.

Halassa, M. M., Fellin, T., & Haydon, P. G. (2007). The tripartite synapse: Roles for gliotransmission in health and disease. *Trends in Molecular Medicine*, *13*, 54–63.

Halassa, M. M., & Haydon, P. G. (2010). Integrated brain circuits: Astrocytic networks modulate neuronal activity and behavior. *Annual Review of Physiology*, *72*, 335–355.

Hallak, J. E., Maia-de-Oliveira, J. P., Abrao, J., Evora, P. R., Zuardi, A. W., Crippa, J. A., et al. (2013). Rapid improvement of acute schizophrenia symptoms after intravenous sodium nitroprusside: A randomized, double-blind, placebo-controlled trial. *JAMA Psychiatry*, *70*, 668–676.

Haller, S., Rodriguez, C., Moser, D., Toma, S., Hofmeister, J., Sinanaj, I., et al. (2013). Acute caffeine administration impact on working memory-related brain activation and functional connectivity in the elderly: A BOLD and perfusion MRI study. *Neuroscience*, *250*, 364–371.

Hardingham, G. E., & Bading, H. (2010). Synaptic versus extrasynaptic NMDA receptor signalling: Implications for neurodegenerative disorders. *Nature Reviews. Neuroscience*, *11*(10), 682–696.

Hashimoto, K., Malchow, B., Falkai, P., & Schmitt, A. (2013). Glutamate modulators as potential therapeutic drugs in schizophrenia and affective disorders. *European Archives of Psychiatry and Clinical Neuroscience*, *263*, 367–377.

Hauber, W., & Koch, M. (1997). Adenosine A_{2a} receptors in the nucleus accumbens modulate prepulse inhibition of the startle response. *Neuroreport*, *8*, 1515–1518.

Haugeto, O., Ullensvang, K., Levy, L. M., Chaudhry, F. A., Honore, T., Nielsen, M., et al. (1996). Brain glutamate transporter proteins form homomultimers. *The Journal of Biological Chemistry*, *271*, 27715–27722.

Hedges, D. W., Woon, F. L., & Hoopes, S. P. (2009). Caffeine-induced psychosis. *CNS Spectrums*, *14*, 127–129.

Heresco-Levy, U. (2003). Glutamatergic neurotransmission modulation and the mechanisms of antipsychotic atypicality. *Progress in Neuro-Psychopharmacology & Biological Psychiatry*, *27*, 1113–1123.

Hettinger, B. D., Leid, M., & Murray, T. F. (1998). Cyclopentyladenosine-induced homologous down-regulation of A1 adenosine receptors (A1AR) in intact neurons is accompanied by receptor sequestration but not a reduction in A1AR mRNA expression or G protein alpha-subunit content. *Journal of Neurochemistry*, *71*(1), 221–230.

Higley, M. J., & Sabatini, B. L. (2010). Competitive regulation of synaptic Ca^{2+} influx by D2 dopamine and A_{2A} adenosine receptors. *Nature Neuroscience*, *13*, 958–966.

Hillion, J., Canals, M., Torvinen, M., Casado, V., Scott, R., Terasmaa, A., et al. (2002). Coaggregation, cointernalization, and codesensitization of adenosine A_{2A} receptors and dopamine D2 receptors. *Journal of Biological Chemistry*, *277*, 18091–18097.

Hines, D. J., Schmitt, L. I., Hines, R. M., Moss, S. J., & Haydon, P. G. (2013). Antidepressant effects of sleep deprivation require astrocyte-dependent adenosine mediated signaling. *Translational Psychiatry*, *3*, e212.

Hirota, T., & Kishi, T. (2013). Adenosine hypothesis in schizophrenia and bipolar disorder: A systematic review and meta-analysis of randomized controlled trial of adjuvant purinergic modulators. *Schizophrenia Research*, *149*(1–3), 88–95.

Hobson, B. D., Merritt, K. E., & Bachtell, R. K. (2012). Stimulation of adenosine receptors in the nucleus accumbens reverses the expression of cocaine sensitization and cross-sensitization to dopamine D2 receptors in rats. *Neuropharmacology*, *63*(6), 1172–1181.

Honea, R., Crow, T. J., Passingham, D., & Mackay, C. E. (2005). Regional deficits in brain volume in schizophrenia: A meta-analysis of voxel-based morphometry studies. *The American Journal of Psychiatry*, *162*, 2233–2235.

Hong, C. J., Liu, H. C., Liu, T. Y., Liao, D. L., & Tsai, S. J. (2005). Association studies of the adenosine A2a receptor (1976T>C) genetic polymorphism in Parkinson's disease and schizophrenia. *Journal of Neural Transmission*, *112*, 1503–1510.

Huerta, I., McCullumsmith, R. E., Haroutunian, V., Giménez-Amaya, J. M., & Meador-Woodruff, J. H. (2006). Expression of excitatory amino acid transporter interacting protein transcripts in the thalamus in schizophrenia. *Synapse*, *59*, 394–402.

Insel, T. R. (2010). Rethinking schizophrenia. *Nature*, *468*, 188–193.

Issy, A. C., Pedrazzi, J. F., Yoneyama, B. H., & Del-Bel, E. A. (2014). Critical role of nitric oxide in the modulation of prepulse inhibition in Swiss mice. *Psychopharmacology*, *231*, 663–672.

Itil, T., Keskiner, A., Kiremitci, N., & Holden, J. M. (1967). Effect of phencyclidine in chronic schizophrenics. *Canadian Psychiatric Association Journal*, *12*(2), 209–212.

Izquierdo, A., Belcher, A. M., Scott, L., Cazares, V. A., Chen, J., O'Dell, S. J., et al. (2010). Reversal-specific learning impairments after a binge regimen of methamphetamine in rats: Possible involvement of striatal dopamine. *Neuropsychopharmacology*, *35*, 505–514.

Jaarsma, D., Sebens, J. B., & Korf, J. (1991). Reduction of adenosine A1-receptors in the perforant pathway terminal zone in Alzheimer hippocampus. *Neuroscience Letters*, *121*(1–2), 111–114.

Javitt, D. C. (2007). Glutamate and schizophrenia: Phencyclidine, N-methyl-D-aspartate receptors, and dopamine-glutamate interactions. *International Review of Neurobiology*, *78*, 69–108.

Jentsch, J. D., Taylor, J. R., Elsworth, J. D., Redmond, D. E., Jr., & Roth, R. H. (1999). Altered frontal cortical dopaminergic transmission in monkeys after subchronic phencyclidine exposure: Involvement in frontostriatal cognitive deficits. *Neuroscience, 90*, 823–832.

Jin, S., Johansson, B., & Fredholm, B. B. (1993). Effects of adenosine A1 and A2 receptor activation on electrically evoked dopamine and acetylcholine release from rat striatal slices. *The Journal of Pharmacology and Experimental Therapeutics, 267*(2), 801–808.

Kadowaki Horita, T., Kobayashi, M., Mori, A., Jenner, P., & Kanda, T. (2013). Effects of the adenosine A2A antagonist istradefylline on cognitive performance in rats with a 6-OHDA lesion in prefrontal cortex. *Psychopharmacology, 230*(3), 345–352.

Kafka, S. H., & Corbett, R. (1996). Selective adenosine A_{2A} receptor/dopamine D2 receptor interactions in animal models of schizophrenia. *European Journal of Pharmacology, 295*, 147–154.

Kalaria, R. N., Sromek, S., Wilcox, B. J., & Unnerstall, J. R. (1990). Hippocampal adenosine A1 receptors are decreased in Alzheimer's disease. *Neuroscience Letters, 118*(2), 257–260.

Kalivas, P. W., Pierce, R. C., Cornish, J., & Sorg, B. A. (1998). A role for sensitization in craving and relapse in cocaine addiction. *Journal of Psychopharmacology, 12*, 49–53.

Karoutzou, G., Emrich, H. M., & Dietrich, D. E. (2008). The myelin-pathogenesis puzzle in schizophrenia: A literature review. *Molecular Psychiatry, 13*, 245–260.

Karson, C. N., Mrak, R. E., Schluterman, K. O., Sturner, W. Q., Sheng, J. G., & Griffin, W. S. (1999). Alterations in synaptic proteins and their encoding mRNAs in prefrontal cortex in schizophrenia: A possible neurochemical basis for 'hypofrontality'. *Molecular Psychiatry, 4*, 39–45.

Katsel, P., Byne, W., Roussos, P., Tan, W., Siever, L., & Haroutunian, V. (2011). Astrocyte and glutamate markers in the superficial, deep, and white matter layers of the anterior cingulate gyrus in schizophrenia. *Neuropsychopharmacology, 36*, 1171–1177.

Keefe, R. S., & Harvey, P. D. (2012). Cognitive impairment in schizophrenia. *Handbook of Experimental Pharmacology, 213*, 11–37.

Kenny, E. M., Cormican, P., Furlong, S., Heron, E., Kenny, G., Fahey, C., et al. (2013). Excess of rare novel loss-of-function variants in synaptic genes in schizophrenia and autism spectrum disorders. *Molecular Psychiatry*, in press.

Kettenmann, H., Hanisch, U. K., Noda, M., & Verkhratsky, A. (2011). Physiology of microglia. *Physiological Reviews, 91*(2), 461.

Kettenmann, H., Kirchhoff, F., & Verkhratsky, A. (2013). Microglia: New roles for the synaptic stripper. *Neuron*, 77(1), 10–18.

Kim, M., Yu, Z. X., Fredholm, B. B., & Rivkees, S. A. (2005). Susceptibility of the developing brain to acute hypoglycemia involving A1 adenosine receptor activation. *American Journal of Physiology. Endocrinology and Metabolism, 289*(4), E562–E569.

Kirischuk, S., Parpura, V., & Verkhratsky, A. (2012). Sodium dynamics: Another key to astroglial excitability? *Trends in Neurosciences, 35*, 497–506.

Kirov, G., Pocklington, A. J., Holmans, P., Ivanov, D., Ikeda, M., Ruderfer, D., et al. (2012). De novo CNV analysis implicates specific abnormalities of postsynaptic signalling complexes in the pathogenesis of schizophrenia. *Molecular Psychiatry, 17*, 142–153.

Kishi, T., & Iwata, N. (2013). NMDA receptor antagonists interventions in schizophrenia: Meta-analysis of randomized, placebo-controlled trials. *Journal of Psychiatric Research, 47*(9), 1143–1149.

Kittner, B., Rössner, M., & Rother, M. (1997). Clinical trials in dementia with propentofylline. *Annals of the New York Academy of Sciences, 826*, 307–316.

Koczapski, A., Paredes, J., Kogan, C., Ledwidge, B., & Higenbottam, J. (1989). Effects of caffeine on behavior of schizophrenic inpatients. *Schizophrenia Bulletin, 15*, 339–344.

Köhr, G. (2006). NMDA receptor function: Subunit composition versus spatial distribution. *Cell and Tissue Research, 326*(2), 439–446.

Kondziella, D., Brenner, E., Eyjolfsson, E. M., Markinhuhta, K. R., Carlsson, M. L., & Sonnewald, U. (2006). Glial-neuronal interactions are impaired in the schizophrenia model of repeated MK801 exposure. *Neuropsychopharmacology*, *31*, 1880–1887.

Kondziella, D., Brenner, E., Eyjolfsson, E. M., & Sonnewald, U. (2007). How do glial-neuronal interactions fit into current neurotransmitter hypotheses of schizophrenia? *Neurochemistry International*, *50*, 291–301.

Konopaske, G. T., Dorph-Petersen, K. A., Sweet, R. A., Pierri, J. N., Zhang, W., Sampson, A. R., et al. (2008). Effect of chronic antipsychotic exposure on astrocyte and oligodendrocyte numbers in macaque monkeys. *Biological Psychiatry*, *63*, 759–765.

Koppelstaetter, F., Poeppel, T. D., Siedentopf, C. M., Ischebeck, A., Verius, M., Haala, I., et al. (2008). Does caffeine modulate verbal working memory processes? An fMRI study. *NeuroImage*, *39*(1), 492–499.

Krystal, J. H., Karper, L. P., Seibyl, J. P., Freeman, G. K., Delaney, R., Bremner, J. D., et al. (1994). Subanesthetic effects of the noncompetitive NMDA antagonist, ketamine, in humans. Psychotomimetic, perceptual, cognitive, and neuroendocrine responses. *Archives of General Psychiatry*, *51*, 199–214.

Kung, L., Conley, R., Chute, D. J., Smialek, J., & Roberts, R. C. (1998). Synaptic changes in the striatum of schizophrenic cases: A controlled postmortem ultrastructural study. *Synapse*, *28*, 125–139.

Kurokawa, K., Mizuno, K., Shibasaki, M., & Ohkuma, S. (2010). Regulation of ryanodine receptors by dopamine D1 receptors during methamphetamine-induced place conditioning. *Journal of Neurochemistry*, *115*, 1206–1214.

Kurumaji, A., & Toru, M. (1998). An increase in [^{3}H] CGS21680 binding in the striatum of postmortem brains of chronic schizophrenics. *Brain Research*, *808*, 320–323.

Kusano, Y., Echeverry, G., Miekisiak, G., Kulik, T. B., Aronhime, S. N., Chen, J. F., et al. (2010). Role of adenosine A2 receptors in regulation of cerebral blood flow during induced hypotension. *Journal of Cerebral Blood Flow and Metabolism*, *30*(4), 808–815.

Ladeby, R., Wirenfeldt, M., Garcia-Ovejero, D., Fenger, C., Dissing-Olesen, L., Dalmau, I., et al. (2005). Microglial cell population dynamics in the injured adult central nervous system. *Brain Research Reviews*, *48*(2), 196–206.

Lane, H. Y., Lin, C. H., Green, M. F., Hellemann, G., Huang, C. C., Chen, P. W., et al. (2013). Add-on treatment of benzoate for schizophrenia: A randomized, double-blind, placebo-controlled trial of D-amino acid oxidase inhibitor. *JAMA Psychiatry*, *70*, 1267–1275.

Lara, D. R., Brunstein, M. G., Ghisolfi, E. S., Lobato, M. I., Belmonte-de-Abreu, P., & Souza, D. O. (2001). Allopurinol augmentation for poorly responsive schizophrenia. *International Clinical Psychopharmacology*, *16*(4), 235–237.

Lara, D. R., Dall'Igna, O. P., Ghisolfi, E. S., & Brunstein, M. G. (2006). Involvement of adenosine in the neurobiology of schizophrenia and its therapeutic implications. *Progress in Neuro-Psychopharmacology & Biological Psychiatry*, *30*, 617–629.

Lara, D. R., & Souza, D. O. (2000). Schizophrenia: A purinergic hypothesis. *Medical Hypotheses*, *54*, 157–166.

Larimore, J., Tornieri, K., Ryder, P. V., Gokhale, A., Zlatic, S. A., Craige, B., et al. (2011). The schizophrenia susceptibility factor dysbindin and its associated complex sort cargoes from cell bodies to the synapse. *Molecular Biology of the Cell*, *22*, 4854–4867.

Laruelle, M. (1998). Imaging dopamine transmission in schizophrenia. A review and meta-analysis. *The Quarterly Journal of Nuclear Medicine*, *42*, 211–221.

Laruelle, M., Gelernter, J., & Innis, R. B. (1998). D2 receptors binding potential is not affected by Taq1 polymorphism at the D2 receptor gene. *Molecular Psychiatry*, *3*, 261–265.

Lavoie, S., Allaman, I., Petit, J. M., Do, K. Q., & Magistretti, P. J. (2011). Altered glycogen metabolism in cultured astrocytes from mice with chronic glutathione deficit; relevance for neuroenergetics in schizophrenia. *PLoS One*, *6*, e22875.

Lerner, T. N., Horne, E. A., Stella, N., & Kreitzer, A. C. (2010). Endocannabinoid signaling mediates psychomotor activation by adenosine A2A antagonists. *Journal of Neuroscience*, *30*(6), 2160–2164.

Lett, T. A., Voineskos, A. N., Kennedy, J. L., Levine, B., & Daskalakis, Z. J. (2014). Treating working memory deficits in schizophrenia: A review of the neurobiology. *Biological Psychiatry*, *75*, 361–370.

Lewis, D. A. (2013). Inhibitory neurons in human cortical circuits: Substrate for cognitive dysfunction in schizophrenia. *Current Opinion in Neurobiology*, *26C*, 22–26.

Lewis, D. A., & Levitt, P. (2002). Schizophrenia as a disorder of neurodevelopment. *Annual Review of Neuroscience*, *25*, 409–432.

Li, Y. K., Wang, F., Wang, W., Luo, Y., Wu, P. F., Xiao, J. L., et al. (2012). Aquaporin-4 deficiency impairs synaptic plasticity and associative fear memory in the lateral amygdala: Involvement of downregulation of glutamate transporter-1 expression. *Neuropsychopharmacology*, *37*, 1867–1878.

Lips, E. S., Cornelisse, L. N., Toonen, R. F., Min, J. L., Hultman, C. M., International Schizophrenia Consortium, et al. (2012). Functional gene group analysis identifies synaptic gene groups as risk factor for schizophrenia. *Molecular Psychiatry*, *17*, 996–1006.

Lisman, J. E., Coyle, J. T., Green, R. W., Javitt, D. C., Benes, F. M., Heckers, S., et al. (2008). Circuit-based framework for understanding neurotransmitter and risk gene interactions in schizophrenia. *Trends in Neurosciences*, *31*, 234–242.

Lodge, D. J., & Grace, A. A. (2011). Hippocampal dysregulation of dopamine system function and the pathophysiology of schizophrenia. *Trends in Pharmacological Sciences*, *32*(9), 507–513.

Lopes, L. V., Cunha, R. A., Kull, B., Fredholm, B. B., & Ribeiro, J. A. (2002). Adenosine A2A receptor facilitation of hippocampal synaptic transmission is dependent on tonic A1 receptor inhibition. *Neuroscience*, *112*, 319–329.

Lopes, L. V., Cunha, R. A., & Ribeiro, J. A. (1999). Crosstalk between A1 and A2A adenosine receptors in the hippocampus and cortex of young adult and old rats. *Journal of Neurophysiology*, *82*, 3196–3203.

Luby, E. D., Cohen, B. D., Rosenbaum, G., Gottlieb, J. S., & Kelley, R. (1959). Study of a new schizophrenomimetic drug; sernyl. *Archives of Neurology and Psychiatry*, *81*, 363–369.

Lucas, P. B., Pickar, D., Kelsoe, J., Rapaport, M., Pato, C., & Hommer, D. (1990). Effects of the acute administration of caffeine in patients with schizophrenia. *Biological Psychiatry*, *28*, 35–40.

Luu, S. U., Liao, H. M., Hung, T. W., Liu, B. Y., Cheng, M. C., Liao, D. L., et al. (2008). Mutation analysis of adenosine A2a receptor gene and interaction study with dopamine D2 receptor gene in schizophrenia. *Psychiatric Genetics*, *18*(1), 43.

MacDonald, A. W., & Schulz, S. C. (2009). What we know: Findings that every theory of schizophrenia should explain. *Schizophrenia Bulletin*, *35*, 93–108.

Martin, S. J., Grimwood, P. D., & Morris, R. G. (2000). Synaptic plasticity and memory: An evaluation of the hypothesis. *Annual Review of Neuroscience*, *23*, 649–711.

Martire, A., Tebano, M. T., Chiodi, V., Ferreira, S. G., Cunha, R. A., Köfalvi, A., et al. (2011). Presynaptic adenosine A2A receptors control cannabinoid CB1 receptor-mediated inhibition of striatal glutamatergic neurotransmission. *Journal of Neurochemistry*, *116*, 273–280.

Matos, M., Augusto, E., Agostinho, P., Cunha, R. A., & Chen, J. F. (2013). Antagonistic interaction between adenosine A2A receptors and $Na+/K+-$ATPase-$\alpha 2$ controlling glutamate uptake in astrocytes. *Journal of Neuroscience*, *33*, 18492–18502.

Matos, M., Augusto, E., Machado, N. J., Santos-Rodrigues, A., Cunha, R. A., & Agostinho, P. (2012). Astrocytic adenosine A2A receptors control the β-amyloid peptide-induced decrease of glutamate uptake. *Journal of Alzheimer Disease*, *31*, 555–567.

Matos, M., Augusto, E., Santos-Rodrigues, A., Schwarzschild, M. A., Chen, J. F., Cunha, R. A., et al. (2012). Adenosine A2A receptors modulate glutamate uptake in cultured astrocytes and gliosomes. *Glia*, *60*, 702–716.

Matute, C., Melone, M., Vallejo-Illarramendi, A., & Conti, F. (2005). Increased expression of the astrocytic glutamate transporter GLT-1 in the prefrontal cortex of schizophrenics. *Glia*, *49*, 451–500.

Mayo, K. M., Falkowski, W., & Jones, C. A. (1993). Caffeine: Use and effects in long-stay psychiatric patients. *British Journal of Psychiatry*, *162*, 543–545.

McGhie, A., & Chapman, J. (1961). Disorders of attention and perception in early schizophrenia. *The British Journal of Medical Psychology*, *34*, 103–116.

McGrath, J. J., Burne, T. H., Féron, F., Mackay-Sim, A., & Eyles, D. W. (2010). Developmental vitamin D deficiency and risk of schizophrenia: A 10-year update. *Schizophrenia Bulletin*, *36*, 1073–1078.

McGrath, J., Saha, S., Chant, D., & Welham, J. (2008). Schizophrenia: A concise overview of incidence, prevalence, and mortality. *Epidemiologic Reviews*, *30*, 67–76.

Meador-Woodruff, J. H., Clinton, S. M., Beneyto, M., & McCullumsmith, R. E. (2003). Molecular abnormalities of the glutamate synapse in the thalamus in schizophrenia. *Annals of the New York Academy of Sciences*, *1003*, 75–93.

Melani, A., Cipriani, S., Vannucchi, M. G., Nosi, D., Donati, C., Bruni, P., et al. (2009). Selective adenosine A2a receptor antagonism reduces JNK activation in oligodendrocytes after cerebral ischaemia. *Brain*, *132*(Pt 6), 1480–1495.

Melone, M., Bragina, L., & Conti, F. (2003). Clozapine-induced reduction of glutamate transport in the frontal cortex is not mediated by GLAST and EAAC1. *Molecular Psychiatry*, *8*, 12–13.

Meyer, U., Feldon, J., & Yee, B. K. (2009). A review of the fetal brain cytokine imbalance hypothesis of schizophrenia. *Schizophrenia Bulletin*, *35*(5), 959–972.

Miller, M. W. (1986). Effects of alcohol on the generation and migration of cerebral cortical neurons. *Science*, *233*(4770), 1308–1311.

Mirnics, K., Middleton, F. A., Marquez, A., Lewis, D. A., & Levitt, P. (2000). Molecular characterization of schizophrenia viewed by microarray analysis of gene expression in prefrontal cortex. *Neuron*, *28*, 53–67.

Mitterauer, B. (2005). Nonfunctional glial proteins in tripartite synapses: A pathophysiological model of schizophrenia. *The Neuroscientist*, *11*, 192–198.

Mitterauer, B. J., & Kofler-Westergren, B. (2011). Possible effects of synaptic imbalances on oligodendrocyte-axonic interactions in schizophrenia: A hypothetical model. *Frontiers in Psychiatry*, *2*, 15.

Moeller, F. G., Steinberg, J. L., Lane, S. D., Kjome, K. L., Ma, L., Ferre, S., et al. (2012). Increased Orbitofrontal Brain Activation after Administration of a Selective Adenosine A (2A) Antagonist in Cocaine Dependent Subjects. *Frontiers in Psychiatry*, *3*, 44.

Moghaddam, B., & Javitt, D. (2012). From revolution to evolution: The glutamate hypothesis of schizophrenia and its implication for treatment. *Neuropsychopharmacology*, *37*, 4–15.

Molero, Y., Gumpert, C., Serlachius, E., Lichtenstein, P., Walum, H., Johansson, D., et al. (2013). A study of the possible association between adenosine A2A receptor gene polymorphisms and attention-deficit hyperactivity disorder traits. *Genes, Brain, and Behavior*, *12*(3), 305–310.

Morgan, C., Reininghaus, U., Reichenberg, A., Frissa, S., SELCoH Study Team, Hotopf, M., et al. (2014). Adversity, cannabis use and psychotic experiences: Evidence of cumulative and synergistic effects. *The British Journal of Psychiatry*, *204*, 346–353.

Mudge, J., Miller, N. A., Khrebtukova, I., Lindquist, I. E., May, G. D., Huntley, J. J., et al. (2008). Genomic convergence analysis of schizophrenia: mRNA sequencing reveals altered synaptic vesicular transport in post-mortem cerebellum. *PLoS One*, *3*, e3625.

Nagai, T., Takuma, K., Dohniwa, M., Ibi, D., Mizoguchi, H., Kamei, H., et al. (2007). Repeated methamphetamine treatment impairs spatial working memory in rats: Reversal by clozapine but not haloperidol. *Psychopharmacology*, *194*, 21–32.

Nagel, J., Schladebach, H., Koch, M., Schwienbacher, I., Muller, C. E., & Hauber, W. (2003). Effects of an adenosine A2A receptor blockade in the nucleus accumbens on locomotion, feeding, and prepulse inhibition in rats. *Synapse*, *49*, 279–286.

Nagy, J. I., Geiger, J. D., & Staines, W. A. (1990). Adenosine deaminase and purinergic neuroregulation. *Neurochemistry International*, *16*(3), 211–221.

Nakazawa, K., Zsiros, V., Jiang, Z., Nakao, K., Kolata, S., Zhang, S., et al. (2012). GABAergic interneuron origin of schizophrenia pathophysiology. *Neuropharmacology*, *62*(3), 1574–1583.

Narendran, R., Frankle, W. G., Mason, N. S., Rabiner, E. A., Gunn, R. N., Searle, G. E., et al. (2009). Positron emission tomography imaging of amphetamine-induced dopamine release in the human cortex: A comparative evaluation of the high affinity dopamine D2/3 radiotracers [^{11}C]FLB 457 and [^{11}C]fallypride. *Synapse*, *63*, 447–461.

Neill, J. C., Barnes, S., Cook, S., Grayson, B., Idris, N. F., McLean, S. L., et al. (2010). Animal models of cognitive dysfunction and negative symptoms of schizophrenia: Focus on NMDA receptor antagonism. *Pharmacology & Therapeutics*, *128*, 419–432.

Ngai, A. C., Coyne, E. F., Meno, J. R., West, G. A., & Winn, H. R. (2001). Receptor subtypes mediating adenosine-induced dilation of cerebral arterioles. *American Journal of Physiology. Heart and Circulatory Physiology*, *280*(5), H2329–H2335.

Niizato, K., Iritani, S., Ikeda, K., & Arai, H. (2001). Astroglial function of schizophrenic brain: A study using lobotomized brain. *Neuroreport*, *12*, 1457–1460.

Ning, Y. L., Yang, N., Chen, X., Xiong, R. P., Zhang, X. Z., Li, P., et al. (2013). Adenosine A2A receptor deficiency alleviates blast-induced cognitive dysfunction. *Journal of Cerebral Blood Flow and Metabolism*, *33*(11), 1789–1798.

Nishizaki, T., Nagai, K., Nomura, T., Tada, H., Kanno, T., Tozaki, H., et al. (2002). A new neuromodulatory pathway with a glial contribution mediated via A_{2a} adenosine receptors. *Glia*, *39*(2), 133–147.

Nordahl, T. E., Salo, R., & Leamon, M. (2003). Neuropsychological effects of chronic methamphetamine use on neurotransmitters and cognition: A review. *The Journal of Neuropsychiatry and Clinical Neurosciences*, *15*, 317–325.

O'Driscoll, C. M., & Gorman, A. M. (2005). Hypoxia induces neurite outgrowth in PC12 cells that is mediated through adenosine A2A receptors. *Neuroscience*, *131*(2), 321–329.

Oertel-Knöchel, V., Knöchel, C., Matura, S., Rotarska-Jagiela, A., Magerkurth, J., Prvulovic, D., et al. (2012). Cortical-basal ganglia imbalance in schizophrenia patients and unaffected first-degree relatives. *Schizophrenia Research*, *138*, 120–127.

Ohnuma, T., Kato, H., Arai, H., Faull, R. L., McKenna, P. J., & Emson, P. C. (2000). Gene expression of PSD95 in prefrontal cortex and hippocampus in schizophrenia. *Neuroreport*, *11*, 3133–3137.

Olabi, B., Ellison-Wright, I., McIntosh, A. M., Wood, S. J., Bullmore, E., & Lawrie, S. M. (2011). Are there progressive brain changes in schizophrenia? A meta-analysis of structural magnetic resonance imaging studies. *Biological Psychiatry*, *70*, 88–96.

Olsson, T., Cronberg, T., Rytter, A., Asztély, F., Fredholm, B. B., Smith, M. L., et al. (2004). Deletion of the adenosine A1 receptor gene does not alter neuronal damage following ischaemia in vivo or in vitro. *European Journal of Neuroscience*, *20*(5), 1197–1204.

Orr, A. G., Orr, A. L., Li, X. J., Gross, R. E., & Traynelis, S. F. (2009). Adenosine A_{2A} receptor mediates microglial process retraction. *Nature Neuroscience*, *12*(7), 872–878.

Othman, T., Yan, H., & Rivkees, S. A. (2003). Oligodendrocytes express functional A1 adenosine receptors that stimulate cellular migration. *Glia*, *44*(2), 166–172.
Ottoni, G. L., Lucchese, I. C., Martins, F., Grillo, R. W., Bogo, M. R., & Lara, D. R. (2005). Association between 2592C>TINS polymorphism of adeno- sine A2A receptor gene and schizophrenia. *Schizophrenia Bulletin*, *31*, 274.
Pacher, P., Nivorozhkin, A., & Szabó, C. (2006). Therapeutic effects of xanthine oxidase inhibitors: Renaissance half a century after the discovery of allopurinol. *Pharmacological Reviews*, *58*(1), 87–114.
Pandolfo, P., Machado, N. J., Köfalvi, A., Takahashi, R. N., & Cunha, R. A. (2013). Caffeine regulates frontocorticostriatal dopamine transporter density and improves attention and cognitive deficits in an animal model of attention deficit hyperactivity disorder. *European Neuropsychopharmacology*, *23*, 317–328.
Pankratov, Y., Lalo, U., & Verkhratsky, A. (2006). North RA Vesicular release of ATP at central synapses. *Pflügers Archiv*, *452*(5), 589–597.
Pantelis, C., Yucel, M., Wood, S. J., Velakoulis, D., Sun, D., Berger, G., et al. (2005). Structural brain imaging evidence for multiple pathological processes at different stages of brain development in schizophrenia. *Schizophrenia Bulletin*, *31*, 672–696.
Park, T. S., Van Wylen, D. G., Rubio, R., & Berne, R. M. (1987). Increased brain interstitial fluid adenosine concentration during hypoxia in newborn piglet. *Journal of Cerebral Blood Flow and Metabolism*, 7(2), 178–183.
Pascual, O., Casper, K. B., Kubera, C., Zhang, J., Revilla-Sanchez, R., Sul, J. Y., et al. (2005). Astrocytic purinergic signaling coordinates synaptic networks. *Science*, *310*(5745), 113–116.
Pathania, M., Davenport, E. C., Muir, J., Sheehan, D. F., López-Doménech, G., & Kittler, J. T. (2014). The autism and schizophrenia associated gene CYFIP1 is critical for the maintenance of dendritic complexity and the stabilization of mature spines. *Translational Psychiatry*, *4*, e374.
Paulson, P. E., Camp, D. M., & Robinson, T. E. (1991). Time course of transient behavioral depression and persistent behavioral sensitization in relation to regional brain monoamine concentrations during amphetamine withdrawal in rats. *Psychopharmacology*, *103*(4), 480–492.
Pennington, K., Beasley, C. L., Dicker, P., Fagan, A., English, J., Pariante, C. M., et al. (2008). Prominent synaptic and metabolic abnormalities revealed by proteomic analysis of the dorsolateral prefrontal cortex in schizophrenia and bipolar disorder. *Molecular Psychiatry*, *13*, 1102–1117.
Perea, G., Navarrete, M., & Araque, A. (2009). Tripartite synapses: Astrocytes process and control synaptic information. *Trends in Neurosciences*, *32*, 421–431.
Pereira, G. S., Rossato, J. I., Sarkis, J. J., Cammarota, M., Bonan, C. D., & Izquierdo, I. (2005). Activation of adenosine receptors in the posterior cingulate cortex impairs memory retrieval in the rat. *Neurobiology of Learning and Memory*, *83*, 217–223.
Pignataro, G., Simon, R. P., & Boison, D. (2007). Transgenic overexpression of adenosine kinase aggravates cell death in ischemia. *Journal of Cerebral Blood Flow and Metabolism*, *27*(1), 1–5.
Pires, V. A., Pamplona, F. A., Pandolfo, P., Fernandes, D., Prediger, R. D., & Takahashi, R. N. (2009). Adenosine receptor antagonists improve short-term object-recognition ability of spontaneously hypertensive rats: A rodent model of attention-deficit hyperactivity disorder. *Behavioral Pharmacology*, *20*(2), 134–145.
Poleszak, E., & Malec, D. (2000). Influence of adenosine receptor agonists and antagonists on amphetamine-induced stereotypy in rats. *Polish Journal of Pharmacology*, *52*, 423–429.
Poletti, S., Radaelli, D., Bosia, M., Buonocore, M., Pirovano, A., Lorenzi, C., et al. (2013). Effect of glutamate transporter EAAT2 gene variants and gray matter deficits on working memory in schizophrenia. *Europan Psychiatry*, *29*(4), 219–225.

Potter, D., Summerfelt, A., Gold, J., & Buchanan, R. W. (2006). Review of clinical correlates of P50 sensory gating abnormalities in patients with schizophrenia. *Schizophrenia Bulletin*, *32*, 692–700.

Powell, S. B., Weber, M., & Geyer, M. A. (2012). Genetic models of sensorimotor gating: Relevance to neuropsychiatric disorders. *Current Topics in Behavioral Neurosciences*, *12*, 251–318.

Prediger, R. D., Fernandes, D., & Takahashi, R. N. (2005). Blockade of adenosine A2A receptors reverses short-term social memory impairments in spontaneously hypertensive rats. *Behavioural Brain Research*, *159*(2), 197–205.

Quarta, D., Borycz, J., Solinas, M., Patkar, K., Hockemeyer, J., Ciruela, F., et al. (2004). Adenosine receptor-mediated modulation of dopamine release in the nucleus accumbens depends on glutamate neurotransmission and N-methyl-D-aspartate receptor stimulation. *Journal of Neurochemistry*, *91*(4), 873–880.

Rajji, T. K., Ismail, Z., & Mulsant, B. H. (2009). Age at onset and cognition in schizophrenia: Meta-analysis. *The British Journal of Psychiatry*, *195*, 286–293.

Rajkowska, G., Miguel-Hidalgo, J. J., Makkos, Z., Meltzer, H., Overholser, J., & Stockmeier, C. (2002). Layer-specific reductions in GFAP-reactive astroglia in the dorsolateral prefrontal cortex in schizophrenia. *Schizophrenia Research*, *57*, 127–138.

Rajkowska, G., Selemon, L. D., & Goldman-Rakic, P. S. (1998). Neuronal and glial somal size in the prefrontal cortex: A postmortem morphometric study of schizophrenia and Huntington disease. *Archives of General Psychiatry*, *55*, 215–224.

Ramamoorthi, K., & Lin, Y. (2011). The contribution of GABAergic dysfunction to neurodevelopmental disorders. *Trends in Molecular Medicine*, *17*, 452–462.

Ramlackhansingh, A. F., Bose, S. K., Ahmed, I., Turkheimer, F. E., Pavese, N., & Brooks, D. J. (2011). Adenosine 2A receptor availability in dyskinetic and nondyskinetic patients with Parkinson disease. *Neurology*, *76*(21), 1811–1816.

Ranganath, C., Minzenberg, M. J., & Ragland, J. D. (2008). The cognitive neuroscience of memory function and dysfunction in schizophrenia. *Biological Psychiatry*, *64*, 18–25.

Rao, J. S., Kellom, M., Reese, E. A., Rapoport, S. I., & Kim, H. W. (2012). Dysregulated glutamate and dopamine transporters in postmortem frontal cortex from bipolar and schizophrenic patients. *Journal of Affective Disorders*, *136*, 63–71.

Rao, J. S., Kim, H. W., Harry, G. J., Rapoport, S. I., & Reese, E. A. (2013). Increased neuroinflammatory and arachidonic acid cascade markers, and reduced synaptic proteins, in the postmortem frontal cortex from schizophrenia patients. *Schizophrenia Research*, *147*, 24–31.

Rebola, N., Canas, P. M., Oliveira, C. R., & Cunha, R. A. (2005). Different synaptic and subsynaptic localization of adenosine A2A receptors in the hippocampus and striatum of the rat. *Neuroscience*, *132*, 893–903.

Rebola, N., Lujan, R., Cunha, R. A., & Mulle, C. (2008). Long-term potentiation of NMDA-EPSCs at hippocampal mossy fiber synapses: An essential role for adenosine A2A receptors. *Neuron*, *57*, 121–134.

Rebola, N., Pinheiro, P. C., Oliveira, C. R., Malva, J. O., & Cunha, R. A. (2003). Subcellular localization of adenosine A1 receptors in nerve terminals and synapses of the rat hippocampus. *Brain Research*, *987*, 49–58.

Rebola, N., Porciúncula, L. O., Lopes, L. V., Oliveira, C. R., Soares-da-Silva, P., & Cunha, R. A. (2005). Long-term effect of convulsive behavior on the density of adenosine A1 and A2A receptors in the rat cerebral cortex. *Epilepsia*, *46*(Suppl.5), 159–165.

Rebola, N., Rodrigues, R. J., Lopes, L. V., Richardson, P. J., Oliveira, C. R., & Cunha, R. A. (2005). Adenosine A1 and A2A receptors are co-expressed in pyramidal neurons and co-localized in glutamatergic nerve terminals of the rat hippocampus. *Neuroscience*, *133*, 79–83.

Rebola, N., Sebastião, A. M., de Mendonça, A., Oliveira, C. R., Ribeiro, J. A., & Cunha, R. A. (2003). Enhanced adenosine A2A receptor facilitation of synaptic transmission in the hippocampus of aged rats. *Journal of Neurophysiology, 90*, 1295–1303.

Rebola, N., Simões, A. P., Canas, P. M., Tomé, A. R., Andrade, G. M., Barry, C. E., et al. (2011). Adenosine A2A receptors control neuroinflammation and consequent hippocampal neuronal dysfunction. *Journal of Neurochemistry, 117*, 100–111.

Reichenberg, A., Caspi, A., Harrington, H., Houts, R., Keefe, R. S. E., Murray, R. M., et al. (2010). Static and dynamic cognitive deficits in childhood preceding adult schizophrenia: A 30-year study. *The American Journal of Psychiatry, 167*, 160–169.

Rimondini, R., Ferre, S., Ogren, S. O., & Fuxe, K. (1997). Adenosine A_{2A} agonists: A potential new type of atypical antipsychotic. *Neuropsychopharmacology, 17*, 82–91.

Rissanen, E., Virta, J. R., Paavilainen, T., Tuisku, J., Helin, S., Luoto, P., et al. (2013). Adenosine A_{2A} receptors in secondary progressive multiple sclerosis: A $[^{11}C]$TMSX brain PET study. *Journal of Cerebral Blood Flow and Metabolism, 33*(9), 1394–1401.

Rivkees, S. A., & Wendler, C. C. (2011). Adverse and protective influences of adenosine on the newborn and embryo: Implications for preterm white matter injury and embryo protection. *Pediatric Research, 69*(4), 271–278.

Rivkees, S. A., Zhao, Z., Porter, G., & Turner, C. (2001). Influences of adenosine on the fetus and newborn. *Molecular Genetics and Metabolism, 74*(1–2), 160–171.

Roberts, R. C., Conley, R., Kung, L., Peretti, F. J., & Chute, D. J. (1996). Reduced striatal spine size in schizophrenia: A postmortem ultrastructural study. *Neuroreport, 7*, 1214–1218.

Robinson, T. E., & Berridge, K. C. (1993). The neural basis of drug craving: An incentive-sensitization theory of addiction. *Brain Research Reviews, 18*, 247–291.

Rodrigues, R. J., Alfaro, T. M., Rebola, N., Oliveira, C. R., & Cunha, R. A. (2005). Co-localization and functional interaction between adenosine A2A and metabotropic group 5 receptors in glutamatergic nerve terminals of the rat striatum. *Journal of Neurochemistry, 92*, 433–441.

Rodríguez-Núñez, A., Cid, E., Rodríguez-García, J., Camiña, F., Rodríguez-Segade, S., & Castro-Gago, M. (2001). Concentrations of nucleotides, nucleosides, purine bases, oxypurines, uric acid, and neuron-specific enolase in the cerebrospinal fluid of children with sepsis. *Journal of Child Neurology, 16*(9), 704–706.

Roffman, J. L., Lamberti, J. S., Achtyes, E., Macklin, E. A., Galendez, G. C., Raeke, L. H., et al. (2013). Randomized multicenter investigation of folate plus vitamin B12 supplementation in schizophrenia. *JAMA Psychiatry, 70*, 481–489.

Rosin, D. L., Robeva, A., Woodard, R. L., Guyenet, P. G., & Linden, J. (1998). Immunohistochemical localization of adenosine A2A receptors in the rat central nervous system. *The Journal of Comparative Neurology, 401*(2), 163–186.

Rothermundt, M., Ohrmann, P., Abel, S., Siegmund, A., Pedersen, A., Ponath, G., et al. (2007). Glial cell activation in a subgroup of patients with schizophrenia indicated by increased S100B serum concentrations and elevated myo-inositol. *Progress in Neuro-Psychopharmacology & Biological Psychiatry, 31*, 361–364.

Roussos, P., & Haroutunian, V. (2014). Schizophrenia: Susceptibility genes and oligodendroglial and myelin related abnormalities. *Frontiers in Cellular Neuroscience, 8*, 5.

Rudolphi, K. A., Schubert, P., Parkinson, F. E., & Fredholm, B. B. (1992). Adenosine and brain ischemia. *Cerebrovascular and Brain Metabolism Reviews, 4*(4), 346–369.

Ruiz, A., Sanz, J. M., Gonzalez-Calero, G., Fernández, M., Andrés, A., Cubero, A., et al. (1996). Desensitization and internalization of adenosine A1 receptors in rat brain by in vivo treatment with R-PIA: Involvement of coated vesicles. *Biochimica et Biophysica Acta, 1310*(1), 168–174.

Ruiz-Medina, J., Ledent, C., Carretón, O., & Valverde, O. (2011). The A2a adenosine receptor modulates the reinforcement efficacy and neurotoxicity of MDMA. *Journal of Psychopharmacology*, *25*(4), 550–564.
Rujescu, D., Ingason, A., Cichon, S., Pietiläinen, O. P., Barnes, M. R., Toulopoulou, T., et al. (2009). Disruption of the neurexin 1 gene is associated with schizophrenia. *Human Molecular Genetics*, *18*, 988–996.
Rund, B. R., Melle, I., Friis, S., Larsen, T. K., Midboe, L. J., Opjordsmoen, S., et al. (2004). Neurocognitive dysfunction in first-episode psychosis: Correlates with symptoms, premorbid adjustment, and duration of untreated psychosis. *The American Journal of Psychiatry*, *161*, 466–472.
Sacchi, S., Bernasconi, M., Martineau, M., Mothet, J. P., Ruzzene, M., Pilone, M. S., et al. (2008). pLG72 modulates intracellular D-serine levels through its interaction with D-amino acid oxidase: Effect on schizophrenia susceptibility. *The Journal of Biological Chemistry*, *283*, 22244–22256.
Salimi, S., Fotouhi, A., Ghoreishi, A., Derakhshan, M. K., Khodaie-Ardakani, M. R., Mohammadi, M. R., et al. (2008). A placebo controlled study of the propentofylline added to risperidone in chronic schizophrenia. *Progress in Neuro-Psychopharmacology & Biological Psychiatry*, *32*(3), 726–732.
Sattler, R., & Rothstein, J. D. (2006). Regulation and dysregulation of glutamate transporters. *Handbook of Experimental Pharmacology*, *175*, 277–303.
Saura, J., Angulo, E., Ejarque, A., Casadó, V., Tusell, J. M., Moratalla, R., et al. (2005). Adenosine A2A receptor stimulation potentiates nitric oxide release by activated microglia. *Journal of Neurochemistry*, *95*(4), 919–929.
Sayed, Y., & Garrison, J. M. (1983). The dopamine hypothesis of schizophrenia and the antagonistic action of neuroleptic drugs—A review. *Psychopharmacology Bulletin*, *19*, 283–288.
Schiffmann, S. N., Fisone, G., Moresco, R., Cunha, R. A., & Ferré, S. (2007). Adenosine A2A receptors and basal ganglia physiology. *Progress in Neurobiology*, *83*, 277–292.
Schmitt, L. I., Sims, R. E., Dale, N., & Haydon, P. G. (2012). Wakefulness affects synaptic and network activity by increasing extracellular astrocyte-derived adenosine. *Journal of Neuroscience*, *32*(13), 4417–4425.
Schnieder, T. P., & Dwork, A. J. (2011). Searching for neuropathology: Gliosis in schizophrenia. *Biological Psychiatry*, *69*, 134–139.
Schoepf, D., Uppal, H., Potluri, R., & Heun, R. (2014). Physical comorbidity and its relevance on mortality in schizophrenia: A naturalistic 12-year follow-up in general hospital admissions. *European Archives of Psychiatry and Clinical Neuroscience*, *264*, 3–28.
Sebastião, A. M., Cunha, R. A., de Mendonça, A., & Ribeiro, J. A. (2000). Modification of adenosine modulation of synaptic transmission in the hippocampus of aged rats. *British Journal of Pharmacology*, *131*, 1629–1634.
Seshadri, S., Zeledon, M., & Sawa, A. (2013). Synapse-specific contributions in the cortical pathology of schizophrenia. *Neurobiology of Disease*, *53*, 26–35.
Shaban, M., Smith, R. A., & Stone, T. W. (1998). Adenosine receptor-mediated inhibition of neurite outgrowth from cultured sensory neurons is via an A1 receptor and is reduced by nerve growth factor. *Developmental Brain Research*, *105*(2), 167–173.
Shan, D., Lucas, E. K., Drummond, J. B., Haroutunian, V., Meador-Woodruff, J. H., & McCullumsmith, R. E. (2013). Abnormal expression of glutamate transporters in temporal lobe areas in elderly patients with schizophrenia. *Schizophrenia Research*, *144*, 1–8.
Shao, L., & Vawter, M. P. (2008). Shared gene expression alterations in schizophrenia and bipolar disorder. *Biological Psychiatry*, *64*, 89–97.
Shen, H. Y., Canas, P. M., Garcia-Sanz, P., Lan, J. Q., Boison, D., Moratalla, R., et al. (2013). Adenosine A2A receptors in striatal glutamatergic terminals and GABAergic

neurons oppositely modulate psychostimulant action and DARPP-32 phosphorylation. *PLoS One*, *8*, e80902.

Shen, H. Y., Coelho, J. E., Ohtsuka, N., Canas, P. M., Day, Y. J., Huang, Q. Y., et al. (2008). A critical role of the adenosine A2A receptor in extra-striatal neurons in modulating psychomotor activity as revealed by opposite phenotypes of striatum- and forebrain-A2A receptor knockouts. *Journal of Neuroscience*, *28*, 2970–2975.

Shen, H. Y., Singer, P., Lytle, N., Wei, C. J., Lan, J. Q., Williams-Karnesky, R. L., et al. (2012). Adenosine augmentation ameliorates psychotic and cognitive endophenotypes of schizophrenia. *The Journal of Clinical Investigation*, *122*, 2567–2577.

Shenton, M. E., Dickey, C. C., Frumin, M., & McCarley, R. W. (2001). A review of MRI findings in schizophrenia. *Schizophrenia Research*, *49*(1–2), 1–52.

Shenton, M. E., Whitford, T. J., & Kubicki, M. (2010). Structural neuroimaging in schizophrenia from methods to insights to treatments. *Dialogues in Clinical Neuroscience*, *12*, 317–332.

Shimazoe, T., Yoshimatsu, A., Kawashimo, A., & Watanabe, S. (2000). Roles of adenosine A_1 and A_{2A} receptors in the expression and development of methamphetamine-induced sensitization. *European Journal of Pharmacology*, *388*, 249–254.

Shin, H. K., Park, S. N., & Hong, K. W. (2000). Implication of adenosine A2A receptors in hypotension-induced vasodilation and cerebral blood flow autoregulation in rat pial arteries. *Life Sciences*, *67*(12), 1435–1445.

Sills, T. L., Azampanah, A., & Fletcher, P. J. (1999). The adenosine A1 receptor agonist N6-cyclopentyladenosine blocks the disruptive effect of phencyclidine on prepulse inhibition of the acoustic startle response in the rat. *European Journal of Pharmacology*, *369*(3), 325–329.

Sills, T. L., Azampanah, A., & Fletcher, P. J. (2001). The adenosine A2A agonist CGS 21680 reverses the reduction in prepulse inhibition of the acoustic startle response induced by phencyclidine, but not by apomorphine and amphetamine. *Psychopharmacology*, *156*(2–3), 187–193.

Silva, C. G., Métin, C., Fazeli, W., Machado, N. J., Darmopil, S., Launay, P. S., et al. (2013). Adenosine receptor antagonists including caffeine alter fetal brain development in mice. *Science Translational Medicine*, *5*, 59–70.

Silva, C. G., Porciúncula, L. O., Canas, P. M., Oliveira, C. R., & Cunha, R. A. (2007). Blockade of adenosine A2A receptors prevents staurosporine-induced apoptosis of rat hippocampal neurons. *Neurobiology of Disease*, *27*, 182–189.

Simões, A. P., Duarte, J. A., Agasse, F., Canas, P. M., Tomé, A. R., Agostinho, P., et al. (2012). Blockade of adenosine A2A receptors prevents interleukin-1β-induced exacerbation of neuronal toxicity through a p38 MAPK pathway. *Journal of Neuroinflammation*, *9*, 204.

Simpson, M. D., Slater, P., & Deakin, J. F. (1998). Comparison of glutamate and gamma-aminobutyric acid uptake binding sites in frontal and temporal lobes in schizophrenia. *Biological Psychiatry*, *15*, 423–427.

Singer, P., Wei, C. J., Chen, J. F., Boison, D., & Yee, B. K. (2013). Deletion of striatal adenosine A_{2A} receptor spares latent inhibition and prepulse inhibition but impairs active avoidance learning. *Behavioural Brain Research*, *242*, 54–61.

Singh, S. P., & Singh, V. (2011). Meta-analysis of the efficacy of adjunctive NMDA receptor modulators in chronic schizophrenia. *CNS Drugs*, *25*, 859–885.

Smith, R. E., Haroutunian, V., Davis, K. L., & Meador-Woodruff, J. H. (2001). Expression of excitatory amino acid transporter transcripts in the thalamus of subjects with schizophrenia. *The American Journal of Psychiatry*, *158*, 1393–1399.

Smith, D. J., Langan, J., McLean, G., Guthrie, B., & Mercer, S. W. (2013). Schizophrenia is associated with excess multiple physical-health comorbidities but low levels of recorded cardiovascular disease in primary care: Cross-sectional study. *BMJ Open*.

Snyder, S. H. (1981). Dopamine receptors, neuroleptics, and schizophrenia. *The American Journal of Psychiatry*, *138*, 460–464.

Snyder, M. A., & Gao, W. J. (2013). NMDA hypofunction as a convergence point for progression and symptoms of schizophrenia. *Frontiers in Cellular Neuroscience*, *7*, 31.

Sørensen, H. J., Mortensen, E. L., Schiffman, J., Reinisch, J. M., Maeda, J., & Mednick, S. A. (2010). Early developmental milestones and risk of schizophrenia: A 45-year follow-up of the Copenhagen Perinatal Cohort. *Schizophrenia Research*, *118*, 41–47.

Soria, G., Castañé, A., Ledent, C., Parmentier, M., Maldonado, R., & Valverde, O. (2006). The lack of A2A adenosine receptors diminishes the reinforcing efficacy of cocaine. *Neuropsychopharmacology*, *31*(5), 978–987.

Spangaro, M., Bosia, M., Zanoletti, A., Bechi, M., Cocchi, F., Pirovano, A., et al. (2012). Cognitive dysfunction and glutamate reuptake- effect of EAAT2 polymorphism in schizophrenia. *Neuroscience Letters*, *522*, 151–155.

Srisurapanont, M., Arunpongpaisal, S., Wada, K., Marsden, J., Ali, R., & Kongsakon, R. (2011). Comparisons of methamphetamine psychotic and schizophrenic symptoms: A differential item functioning analysis. *Progress in Neuro-Psychopharmacology & Biological Psychiatry*, *35*, 959–964.

Stafford, M. R., Jackson, H., Mayo-Wilson, E., Morrison, A. P., & Kendall, T. (2013). Early interventions to prevent psychosis: Systematic review and meta-analysis. *BMJ*, *346*, f185.

Stark, A. K., Uylings, H. B., Sanz-Arigita, E., & Pakkenberg, B. (2004). Glial cell loss in the anterior cingulate cortex, a subregion of the prefrontal cortex, in subjects with schizophrenia. *The American Journal of Psychiatry*, *161*, 882–888.

Steffek, A. E., McCullumsmith, R. E., Haroutunian, V., & Meador-Woodruff, J. H. (2008). Cortical expression of glial fibrillary acidic protein and glutamine synthetase is decreased in schizophrenia. *Schizophrenia Research*, *103*, 71–82.

Steiner, J., Bernstein, H. G., Bielau, H., Farkas, N., Winter, J., Dobrowolny, H., et al. (2008). S100B-immunopositive glia is elevated in paranoid as compared to residual schizophrenia: A morphometric study. *Journal of Psychiatric Research*, *42*, 868–876.

Steketee, J. D., & Kalivas, P. W. (2011). Drug wanting: Behavioral sensitization and relapse to drug-seeking behavior. *Pharmacological Reviews*, *63*, 348–365.

Stevens, B., Porta, S., Haak, L. L., Gallo, V., & Fields, R. D. (2002). Adenosine: A neuron-glial transmitter promoting myelination in the CNS in response to action potentials. *Neuron*, *36*(5), 855–868.

Stewart, D. G., & Davis, K. L. (2004). Possible contributions of myelin and oligodendrocyte dysfunction to schizophrenia. *International Review of Neurobiology*, *59*, 381–424.

Stone, T. W., & Behan, W. M. (2007). Interleukin-1beta but not tumor necrosis factor-alpha potentiates neuronal damage by quinolinic acid: Protection by an adenosine A2A receptor antagonist. *Journal of Neuroscience Research*, *85*(5), 1077–1085.

Stoof, J. C., & Kebabian, J. W. (1984). Two dopamine receptors: Biochemistry, physiology and pharmacology. *Life Sciences*, *35*, 2281–2296.

Studer, F. E., Fedele, D. E., Marowsky, A., Schwerdel, C., Wernli, K., Vogt, K., et al. (2006). Shift of adenosine kinase expression from neurons to astrocytes during postnatal development suggests dual functionality of the enzyme. *Neuroscience*, *142*(1), 125–137.

Sun, C. N., Chuang, H. C., Wang, J. Y., Chen, S. Y., Cheng, Y. Y., Lee, C. F., et al. (2010). The A2A adenosine receptor rescues neuritogenesis impaired by p53 blockage via KIF2A, a kinesin family member. *Developmental Neurobiology*, *70*(8), 604–621.

Suzuki, A., Stern, S. A., Bozdagi, O., Huntley, G. W., Walker, R. H., Magistretti, P. J., et al. (2011). Astrocyte-neuron lactate transport is required for long-term memory formation. *Cell*, *144*, 810–823.

Svenningsson, P., Le Moine, C., Fisone, G., & Fredholm, B. B. (1999). Distribution, biochemistry and function of striatal adenosine A2A receptors. *Progress in Neurobiology, 59*(4), 355–396.

Svensson, T. H. (2000). Dysfunctional brain dopamine systems induced by psychotomimetic NMDA-receptor antagonists and the effects of antipsychotic drugs. *Brain Research Reviews, 31*(2–3), 320–329.

Swanson, T. H., Drazba, J. A., & Rivkees, S. A. (1995). Adenosine A1 receptors are located predominantly on axons in the rat hippocampal formation. *The Journal of Comparative Neurology, 363*(4), 517–531.

Swanson, T. H., Krahl, S. E., Liu, Y. Z., Drazba, J. A., & Rivkees, S. A. (1998). Evidence for physiologically active axonal adenosine receptors in the rat corpus callosum. *Brain Research, 784*(1–2), 188–198.

Swerdlow, N. R., Braff, D. L., Hartston, H., Perry, W., & Geyer, M. A. (1996). Latent inhibition in schizophrenia. *Schizophrenia Research, 20*, 91–103.

Swerdlow, N. R., Eastvold, A., Gerbranda, T., Uyan, K. M., Hartman, P., Doan, Q., et al. (2000). Effects of caffeine on sensorimotor gating of the startle reflex in normal control subjects: Impact of caffeine intake and withdrawal. *Psychopharmacology, 151*(4), 368–378.

Talbot, K., Louneva, N., Cohen, J. W., Kazi, H., Blake, D. J., & Arnold, S. E. (2011). Synaptic dysbindin-1 reductions in schizophrenia occur in an isoform-specific manner indicating their subsynaptic location. *PLoS One, 6*, e16886.

Tandon, R., Keshavan, M. S., & Nasrallah, H. A. (2008). Schizophrenia, "just the facts" what we know in 2008. 2. Epidemiology and etiology. *Schizophrenia Research, 102*, 1–18.

Thevananther, S., Rivera, A., & Rivkees, S. A. (2001). A1 adenosine receptor activation inhibits neurite process formation by Rho kinase-mediated pathways. *Neuroreport, 12*(14), 3057–3063.

Thoma, P., & Daum, I. (2013). Comorbid substance use disorder in schizophrenia: A selective overview of neurobiological and cognitive underpinnings. *Psychiatry and Clinical Neurosciences, 67*, 367–383.

Thompson, S. M., Haas, H. L., & Gähwiler, B. H. (1992). Comparison of the actions of adenosine at pre- and postsynaptic receptors in the rat hippocampus in vitro. *Journal Physiology, 451*, 347–363.

Thompson, B. L., & Levitt, P. (2010). Now you see it, now you don't-closing in on allostasis and developmental basis of psychiatric disorders. *Neuron, 65*, 437–439.

Tian, L., Meng, C., Yan, H., Zhao, Q., Liu, Q., Yan, J., et al. (2011). Convergent evidence from multimodal imaging reveals amygdala abnormalities in schizophrenic patients and their first-degree relatives. *PLoS One, 6*, e28794.

Tibrewal, P., & Dhillon, R. (2011). Caffeine induced psychotic exacerbation. *The Australian and New Zealand Journal of Psychiatry, 45*, 179–180.

Tkachev, D., Mimmack, M. L., Ryan, M. M., Wayland, M., Freeman, T., Jones, P. B., et al. (2003). Oligodendrocyte dysfunction in schizophrenia and bipolar disorder. *Lancet, 362*, 798–805.

Todarello, G., Feng, N., Kolachana, B. S., Li, C., Vakkalanka, R., Bertolino, A., et al. (2014). Incomplete penetrance of NRXN1 deletions in families with schizophrenia. *Schizophrenia Research.*

Tomé, A. R., Silva, H., & Cunha, R. A. (2010). Role of the purinergic neuromodulation system in epilepsy. *Open Neuroscience Journal, 4*, 64–83.

Torvinen, M., Torri, C., Tombesi, A., Marcellino, D., Watson, S., Lluis, C., et al. (2005). Trafficking of adenosine A_{2A} and dopamine D2 receptors. *Journal of Molecular Neuroscience, 25*, 191–200.

Tsai, G., Yang, P., Chung, L. C., Lange, N., & Coyle, J. T. (1998). D-serine added to antipsychotics for the treatment of schizophrenia. *Biological Psychiatry, 44*, 1081–1089.

Turetsky, B. I., Calkins, M. E., Light, G. A., Olincy, A., Radant, A. D., & Swerdlow, N. R. (2007). Neurophysiological endophenotypes of schizophrenia: The viability of selected candidate measures. *Schizophrenia Bulletin*, *33*(1), 69–94.

Turner, C. P., Seli, M., Ment, L., Stewart, W., Yan, H., Johansson, B., et al. (2003). A1 adenosine receptors mediate hypoxia-induced ventriculomegaly. *Proceedings of the National Academy of Sciences of the United States of America*, *100*(20), 11718–11722.

Turner, C. P., Yan, H., Schwartz, M., Othman, T., & Rivkees, S. A. (2002). A1 adenosine receptor activation induces ventriculomegaly and white matter loss. *Neuroreport*, *13*(9), 1199–1204.

Turpin, G. (1986). Effects of stimulus intensity on autonomic responding: The problem of differentiating orienting and defense reflexes. *Psychophysiology*, *23*, 1–14.

Tzingounis, A. V., & Wadiche, J. I. (2007). Glutamate transporters: Confining runaway excitation by shaping synaptic transmission. *Nature Reviews. Neuroscience*, *8*, 935–947.

Uezato, A., Kimura-Sato, J., Yamamoto, N., Iijima, Y., Kunugi, H., & Nishikawa, T. (2012). Further evidence for a male-selective genetic association of synapse-associated protein 97 (SAP97) gene with schizophrenia. *Behavioral and Brain Functions*, *8*, 2.

Ułas, J., Brunner, L. C., Nguyen, L., & Cotman, C. W. (1993). Reduced density of adenosine A1 receptors and preserved coupling of adenosine A1 receptors to G proteins in Alzheimer hippocampus: A quantitative autoradiographic study. *Neuroscience*, *52*(4), 843–854.

Ulrich, D., & Huguenard, J. R. (1995). Purinergic inhibition of GABA and glutamate release in the thalamus: Implications for thalamic network activity. *Neuron*, *15*(4), 909–918.

Vallejo-Illarramendi, A., Torres-Ramos, M., Melone, M., Conti, F., & Matute, C. (2005). Clozapine reduces GLT-1 expression and glutamate uptake in astrocyte cultures. *Glia*, *50*, 276–279.

van Snellenberg, J. X. (2009). Working memory and long-term memory deficits in schizophrenia: Is there a common substrate? *Psychiatry Research*, *174*, 89–96.

Varese, F., Smeets, F., Drukker, M., Lieverse, R., Lataster, T., Viechtbauer, W., et al. (2012). Childhood adversities increase the risk of psychosis: A meta-analysis of patient-control, prospective- and cross-sectional cohort studies. *Schizophrenia Bulletin*, *38*, 661–671.

Venables, P. H. (1960). The effect of auditory and visual stimulation on the skin potential response of schizophrenics. *Brain*, *83*, 77–92.

Verkhratsky, A., & Kirchhoff, F. (2007). NMDA Receptors in glia. *The Neuroscientist*, *13*(1), 28–37.

Verret, L., Mann, E. O., Hang, G. B., Barth, A. M., Cobos, I., Ho, K., et al. (2012). Inhibitory interneuron deficit links altered network activity and cognitive dysfunction in Alzheimer model. *Cell*, *149*(3), 708–721.

Villar-Menéndez, I., Díaz-Sánchez, S., Blanch, M., Albasanz, J. L., Pereira-Veiga, T., Monje, A., et al. (2014). Reduced striatal adenosine A2A receptor levels define a molecular subgroup in schizophrenia. *Journal of Psychiatric Research*, *51*, 49–59.

Volk, D. W., & Lewis, D. A. (2013). Prenatal ontogeny as a susceptibility period for cortical GABA neuron disturbances in schizophrenia. *Neuroscience*, *248C*, 154–164.

Wakade, T. D., Palmer, K. C., McCauley, R., Przywara, D. A., & Wakade, A. R. (1995). Adenosine-induced apoptosis in chick embryonic sympathetic neurons: A new physiological role for adenosine. *Journal Physiology*, *488*(Pt 1), 123–138.

Wang, J. H., Ma, Y. Y., & van den Buuse, M. (2006). Improved spatial recognition memory in mice lacking adenosine A_{2A} receptors. *Experimental Neurology*, *199*, 438–445.

Wang, J. H., Short, J., Ledent, C., Lawrence, A. J., & van den Buuse, M. (2003). Reduced startle habituation and prepulse inhibition in mice lacking the adenosine A_{2A} receptor. *Behavioural Brain Research*, *143*, 201–207.

Wardas, J. (2008). Potential role of adenosine A2A receptors in the treatment of schizophrenia. *Frontiers in Bioscience*, *13*, 4071–4096.

Wardas, J., Konieczny, J., & Pietraszek, M. (2003). Influence of CGS 21680, a selective adenosine A_{2A} agonist, on the phencyclidine-induced sensorimotor gating deficit and motor behaviour in rats. *Psychopharmacology*, *168*(3), 299–306.

Wei, C. J., Augusto, E., Gomes, C. A., Singer, P., Wang, Y., Boison, D., et al. (2013). Regulation of fear responses by striatal and extra-striatal adenosine A2A receptors in forebrain. *Biological Psychiatry*.

Wei, C. J., Singer, P., Coelho, J., Boison, D., Feldon, J., Yee, B. K., et al. (2011). Selective inactivation of adenosine A_{2A} receptors in striatal neurons enhances working memory and reversal learning. *Learning & Memory*, *18*(7), 459–474.

Weinberger, D. R. (1987). Implications of normal brain development for the pathogenesis of schizophrenia. *Archives of General Psychiatry*, *44*, 660–669.

Weis, S., & Llenos, I. C. (2004). GFAP-immunopositive astrocytes in schizophrenia. *Schizophrenia Research*, *67*, 293–295.

White, T. D., & MacDonald, W. F. (1990). Neural release of ATP and adenosine. *Annals of the New York Academy of Sciences*, *603*, 287–298.

Wieraszko, A., Goldsmith, G., & Seyfried, T. N. (1989). Stimulation-dependent release of adenosine triphosphate from hippocampal slices. *Brain Research*, *485*(2), 244–250.

Williams, M. R., Hampton, T., Pearce, R. K., Hirsch, S. R., Ansorge, O., Thom, M., et al. (2013). Astrocyte decrease in the subgenual cingulate and callosal genu in schizophrenia. *European Archives of Psychiatry and Clinical Neuroscience*, *263*, 41–52.

Wirkner, K., Gerevich, Z., Krause, T., Günther, A., Köles, L., Schneider, D., et al. (2004). Adenosine A2A receptor-induced inhibition of NMDA and GABAA receptor-mediated synaptic currents in a subpopulation of rat striatal neurons. *Neuropharmacology*, *46*(7), 994–1007.

Wonodi, I., Gopinath, H. V., Liu, J., Adami, H., Hong, L. E., Allen-Emerson, R., et al. (2011). Dipyridamole monotherapy in schizophrenia: Pilot of a novel treatment approach by modulation of purinergic signaling. *Psychopharmacology*, *218*(2), 341–345.

Woodberry, K. A., Giuliano, A. J., & Seidman, L. J. (2008). Premorbid IQ in schizophrenia: A meta-analytic review. *The American Journal of Psychiatry*, *165*, 579–587.

Xiao, D., Bastia, E., Xu, Y. H., Benn, C. L., Cha, J. H., Peterson, T. S., et al. (2006). Forebrain adenosine A2A receptors contribute to L-3,4-dihydroxyphenylalanine-induced dyskinesia in hemiparkinsonian mice. *Journal of Neuroscience*, *26*(52), 13548–13555.

Yang, Z. J., Wang, B., Kwansa, H., Heitmiller, K. D., Hong, G., Carter, E. L., et al. (2013). Adenosine A2A receptor contributes to ischemic brain damage in newborn piglet. *Journal of Cerebral Blood Flow and Metabolism*, *33*(10), 1612–1620.

Yee, B. K., Singer, P., Chen, J. F., Feldon, J., & Boison, D. (2007). Transgenic overexpression of adenosine kinase in brain leads to multiple learning impairments and altered sensitivity to psychomimetic drugs. *European Journal of Neuroscience*, *26*(11), 3237–3252.

Yin, D. M., Chen, Y. J., Sathyamurthy, A., Xiong, W. C., & Mei, L. (2012). Synaptic dysfunction in schizophrenia. *Advances in Experimental Medicine and Biology*, *970*, 493–516.

Zhang, J., Abdallah, C. G., Wang, J., Wan, X., Liang, C., Jiang, L., et al. (2012). Upregulation of adenosine A2A receptors induced by atypical antipsychotics and its correlation with sensory gating in schizophrenia patients. *Psychiatry Research*, *200*(2–3), 126–132.

INDEX

Note: Page numbers followed by "*f*" indicate figures and "*t*" indicate tables.

A

O

P

T

W

CONTENTS OF RECENT VOLUMES

Volume 39

Volume 40

Volume 41

Volume 42

Volume 43

Volume 44

Volume 45

Volume 46

Volume 47

Volume 48

Volume 49

Volume 50

Volume 51

Volume 52

Volume 53

Volume 54

Volume 55

Volume 56

Volume 57

Volume 58

Volume 59

Volume 60

Volume 61

Volume 62

Volume 63

Volume 64

Volume 65

Volume 66

Volume 67

Volume 68

Volume 69

Volume 70

Volume 71

Volume 72

Volume 73

Volume 74

Volume 75

Volume 76

Volume 77

Volume 78

Volume 79

Volume 80

Volume 81

Volume 82

Volume 83

Volume 84

Volume 85

Volume 86

Volume 87

Volume 88

Volume 89

Volume 90

Volume 91

Volume 92

Volume 93

Volume 94

Volume 95

Volume 96

Volume 97

Volume 98

Volume 99

Volume 100

Volume 101

Volume 105

Volume 106

Volume 107

Volume 108

Volume 109

Volume 110

Volume 111

Volume 112

Volume 113

Volume 114

Volume 115

Volume 116

Volume 117

Volume 118

FSC
www.fsc.org
MIX
Papier | Fördert gute Waldnutzung
FSC® C083411

Kolibri 360 GmbH
Ferdinand-Jühlke-Straße 7
99095 Erfurt, Deutschland
produktsicherheit@kolibri360.de